T0122629

Advances in Intelligent Systems and Computing

Volume 730

Series editor

Janusz Kacprzyk, Polish Academy of Sciences, Warsaw, Poland
e-mail: kacprzyk@ibspan.waw.pl

About this Series

The series "Advances in Intelligent Systems and Computing" contains publications on theory, applications, and design methods of Intelligent Systems and Intelligent Computing. Virtually all disciplines such as engineering, natural sciences, computer and information science, ICT, economics, business, e-commerce, environment, healthcare, life science are covered. The list of topics spans all the areas of modern intelligent systems and computing.

The publications within "Advances in Intelligent Systems and Computing" are primarily textbooks and proceedings of important conferences, symposia and congresses. They cover significant recent developments in the field, both of a foundational and applicable character. An important characteristic feature of the series is the short publication time and world-wide distribution. This permits a rapid and broad dissemination of research results.

Advisory Board

Chairman

Nikhil R. Pal, Indian Statistical Institute, Kolkata, India
e-mail: nikhil@isical.ac.in

Members

Rafael Bello Perez, Universidad Central "Marta Abreu" de Las Villas, Santa Clara, Cuba
e-mail: rbellop@uclv.edu.cu

Emilio S. Corchado, University of Salamanca, Salamanca, Spain
e-mail: escorchado@usal.es

Hani Hagras, University of Essex, Colchester, UK
e-mail: hani@essex.ac.uk

László T. Kóczy, Széchenyi István University, Győr, Hungary
e-mail: koczy@sze.hu

Vladik Kreinovich, University of Texas at El Paso, El Paso, USA
e-mail: vladik@utep.edu

Chin-Teng Lin, National Chiao Tung University, Hsinchu, Taiwan
e-mail: ctlin@mail.nctu.edu.tw

Jie Lu, University of Technology, Sydney, Australia
e-mail: Jie.Lu@uts.edu.au

Patricia Melin, Tijuana Institute of Technology, Tijuana, Mexico
e-mail: epmelin@hafsamx.org

Nadia Nedjah, State University of Rio de Janeiro, Rio de Janeiro, Brazil
e-mail: nadia@eng.uerj.br

Ngoc Thanh Nguyen, Wroclaw University of Technology, Wroclaw, Poland
e-mail: Ngoc-Thanh.Nguyen@pwr.edu.pl

Jun Wang, The Chinese University of Hong Kong, Shatin, Hong Kong
e-mail: jwang@mae.cuhk.edu.hk

More information about this series at http://www.springer.com/series/11156

Anna M. Gil-Lafuente · José M. Merigó
Bal Kishan Dass · Rajkumar Verma
Editors

Applied Mathematics and Computational Intelligence

Springer

Editors
Anna M. Gil-Lafuente
Department of Business
University of Barcelona
Barcelona
Spain

José M. Merigó
Department of Management Control
  and Information Systems
University of Chile
Santiago
Chile

Bal Kishan Dass
Department of Mathematics
University of Delhi
Delhi
India

Rajkumar Verma
Department of Applied Sciences
Delhi Technical Campus
Greater Noida, Uttar Pradesh
India

ISSN 2194-5357   ISSN 2194-5365   (electronic)
Advances in Intelligent Systems and Computing
ISBN 978-3-319-75791-9   ISBN 978-3-319-75792-6   (eBook)
https://doi.org/10.1007/978-3-319-75792-6

Library of Congress Control Number: 2018934343

© Springer International Publishing AG, part of Springer Nature 2018
This work is subject to copyright. All rights are reserved by the Publisher, whether the whole or part of the material is concerned, specifically the rights of translation, reprinting, reuse of illustrations, recitation, broadcasting, reproduction on microfilms or in any other physical way, and transmission or information storage and retrieval, electronic adaptation, computer software, or by similar or dissimilar methodology now known or hereafter developed.
The use of general descriptive names, registered names, trademarks, service marks, etc. in this publication does not imply, even in the absence of a specific statement, that such names are exempt from the relevant protective laws and regulations and therefore free for general use.
The publisher, the authors and the editors are safe to assume that the advice and information in this book are believed to be true and accurate at the date of publication. Neither the publisher nor the authors or the editors give a warranty, express or implied, with respect to the material contained herein or for any errors or omissions that may have been made. The publisher remains neutral with regard to jurisdictional claims in published maps and institutional affiliations.

Printed on acid-free paper

This Springer imprint is published by the registered company Springer International Publishing AG
part of Springer Nature
The registered company address is: Gewerbestrasse 11, 6330 Cham, Switzerland

Preface

The 24th International Conference of the 'Forum for Interdisciplinary Mathematics (FIM)' entitled *Applied Mathematics and Computational Intelligence* took place in Barcelona, Spain, November 18–20, 2015, and was co-organized by the *University of Barcelona* (Spain), the *Spanish Royal Academy of Economic and Financial Sciences* (Spain), and the *Forum for Interdisciplinary Mathematics* (India).

The Forum is a registered trust in India. It is, in effect, an India-based international society of scholars working in mathematical sciences and its partner areas (a partner area is defined as one where some knowledge of mathematical sciences is desirable to carry out research and development). The society was incepted in 1975 by a group of University of Delhi intellectuals led by Professor Bhu Dev Sharma. In 2015, the FIM is running into 42th year of active standing. Right from the beginning, FIM had the support and association of India's great mathematicians and also users of mathematics from different disciplines in the country and abroad.

The Forum began holding conferences right from the beginning. It started at the national level. The first conference was held in 1975 at 'Calcutta University, Calcutta (India).' The second conference was held at Rajasthan University, Jaipur (India), in 1976. With the General Secretary, Professor Bhu Dev Sharma taking-up a chair abroad, the holding of conferences at the national had a period of interruption. Later, it was decided to hold international conference every year alternating between India and outside.

The process of holding international conferences began in 1995 and is continuing unabated. In such a way, this 24th International Conference entitled *Applied Mathematics and Computational Intelligence* continues and extends the series of international conferences organized by FIM. Previous international conferences were held at Calcutta University, Calcutta, India (July 1995); Rajasthan University, Jaipur, India (June 1996); University of Southern Maine, USA (July 1997); Banaras Hindu University, India (December 1997); University of Mysore, India (December 1998); University of South Alabama, USA (December 1999); Indian Institute of Technology, Mumbai, India (December 2000); University of Wollongong, Australia (December 2001); University of Allahabad, Allahabad, India (December 2002); University of Southern Maine, USA (October 2003); Institute of

Engineering & Technology, Lucknow, India (December 2004); Auburn University, Auburn, AL, USA (December 2005); Tomar Polytechnic Institute, Tomar, Portugal (September 2006); IIT, Madras, Chennai, India (January 2007); University of Science & Technology of China, Shanghai (May 2007); Memphis University, USA (May 2008); University of West Bohemia, Czech Republic (May 2009); Jaypee University of Information Technology, Waknaghat, HP, India (August 2009); Patna University, Patna, Bihar, India (December 2010); Alcorn University, Montreal, Canada (June 2011); Panjab University, Chandigarh, India (December 2012); Waseda University, City of Kitakyushu, Japan (November 2013); NITK, Surathkal, Karnataka, India (December 2014).

Starting with the 8th International Conference at the University of Wollongong, Australia, the Forum has started organizing and funding a symposium solely for the purpose of encouraging and awarding young researchers consisting of new Ph.D. awardees and aspirants, also known as '*Professor R. S. Varma Memorial Student Competition*' (RSVMSC). These awards are well structured, critiqued, and judged by the leading scholars of various fields, and at the conclusion of which a certificate and cash award (presently Rs. 25,000.00) are provided to the winners. In a very short time, RSVMSC has become popular among young investigators in India as FIM has appreciably realized their participation at its conferences. Among other activities, FIM publishes the following scientific publications:

1. Journal of Combinatorics, Information and System Sciences
2. Research Monographs and Lecture Notes with Springer

This international conference aims to bring together the foremost experts from different disciplines, young researchers, academics, and students to discuss new research ideas and present recent advances in *interdisciplinary mathematics, statistics, computational intelligence, economics, and computer science.*

FIM-AMCI-2015 received a large number of papers from all over the world. They were carefully reviewed by experts, and only high-quality papers were selected for oral or poster presentation during conference days. This book comprises a selection of papers presented at the conference. We believe it is a good example of the excellent work of the associates and the significant progress about this line of research in recent times.

This book is organized according to four general tracks of the conference: Mathematical Foundations, Computational Intelligence and Optimization Techniques, Modeling and Simulation Techniques, Applications in Business and Engineering.

Finally, we would like to express our sincere thanks to all the plenary speakers, authors, reviewers, and participants at the conference, organizations, and institutional sponsors for their help, support, and contributions to the success of the event.

The AMCI 2015-FIM XXIV Conference is supported by:

November 2017

Anna M. Gil-Lafuente
José M. Merigó
Bal Kishan Dass
Rajkumar Verma

Organization

Honorary Committee

Special thanks to the members of the Honorary Committee for their support in the organization of the AMCI 2015-FIM XXIV.

BhuDev Sharma	Former Prof. of Mathematics, Clark Atlanta University, Atlanta, GA, USA
Jaume Gil Aluja	President Royal Academy of Economic and Financial Sciences, Spain
B. K. Dass	Former Professor and Head, Department of Mathematics, Delhi University, Delhi, India
S. C. Malik	Professor of Statistics, M.D. University, Rohtak, India

Scientific Committee

Thanks to all the members of the Scientific Committee for their kind support in the organization of the AMCI 2015-FIM XXIV, Barcelona, Spain.

Mario Aguer Hortal, Spain
Luis Amiguet Molina, Spain
Xavier Bertran Roura, Spain
Claudio Bonilla, Chile
Sefa Boria Reverter, Spain
Jose Manuel Brotons Martínez, Spain
Huayou Chen, China
Bernard De Baets, Belgium
José Antonio Redondo López, Spain
Maria Àngels Farreras Noguer, Spain
Aurelio Fernández Bariviera, Spain
Joao Ferreira, Portugal
Joan Carles Ferrer Comalat, Spain

Beatriz Flores Romero, Mexico
Irene García Rondón, Cuba
Vasile Georgescu, Romania
Jaume Gil Aluja, Spain
Anna M. Gil-Lafuente, Spain
Jaime Gil Lafuente, Spain
Federico Gonzalez Santoyo, Mexico
Francesc Granell Trías, Spain
Salvatore Greco, Italy
Montserrat Guillen, Spain
Korkmaz Imanov, Azerbaijan
Angel Juan, Spain
Janusz Kacprzyk, Poland

Tomonori Kawano, Japan
Yuriy P. Kondratenko, Ukraine
Sigifredo Laengle, Chile
Huchang Liao, China
Vicente Liern Carrión, Spain
Salvador Linares Mustarós, Spain
Peide Liu, China
Gino Loyola, Chile
Sebastia Massanet, Spain
Gaspar Mayor, Spain
José M. Merigó, Chile
Onofre Martorell Cunill, Spain
Radko Mesiar, Slovakia
Jaime Miranda, Chile
Daniel Palacios-Marques, Spain
Witold Pedrycz, Canada
Ding-Hong Peng, China
Marta Peris-Ortiz, Spain
Ali Emrouznejad, UK
Kurt J. Engemann, USA
Hiroshi Sakai, Japan
Jonas Saparaukas, Lithuania
Byeong Seok Ahn, South Korea

Shun-Feng Su, China
Baiqin Sun, China
Vicenç Torra, Spain
Shusaku Tsumoto, Japan
David Urbano, Spain
Oscar Valero, Spain
Rajkumar Verma, India
Rashmi Verma, India
Emilio Vizuete Luciano, Spain
Junzo Watada, Japan
Guiwu Wei, China
Yejun Xu, China
Zeshui Xu, China
Ronald R. Yager, USA
Dejian Yu, China
Shouzhen Zeng, China
Ligang Zhou, China
Giuseppe Zollo, Italy
Gustavo Zurita, Chile
Luis Martínez, Spain
Javier Martin, Spain
Manoj Shani, India

Organizing Committee

Special thanks to all the members of the Organizing Committee for their support during the preparation of the AMCI 2015-FIM XXIV International Conference.

Co-chair of the Organizing Committee

Anna M. Gil-Lafuente, Spain
José M. Merigó Lindahl, Chile
B. K. Dass, India
Rajkumar Verma, India

Organizing Committee

Francisco J. Arroyo, Spain
Fabio Blanco, Colombia
Sefa Boria, Spain
Elena Rondós Casas, Spain
Kusum Deep, India
Jaime Gil Lafuente, Spain

Suresh Hegde, India
Aras Keropyan, Turkey
Salvador Linares, Spain
Alexander López Guauque, Colombia
Carolina Luis Bassa, Spain
Suresh C. Malik, India
Ramon Poch, Spain
Lourdes Souto, Cuba
Emili Vizuete, Spain
Binyamin Yusoff, Malaysia

Contents

Modeling and Simulation Techniques

Applications in Business and Engineering

Mathematical Foundations

Best Proximity Point Theorems
for Generalized Contractive Mappings

S. Arul Ravi$^{(\boxtimes)}$ and A. Anthony Eldred

Research Department of Mathematics, St. Joseph's College (Autonomous),
Tiruchirappalli, India
ammaarulravi@gmail.com, anthonyeldred@yahoo.co.in

Abstract. Recently, J. Calallero (Fixed Point Theory and Applications 2012, 2012:231) observed best proximity results for Geraghty-contractions by using the P-property. In this paper we introduce the notion of Boyd and wong result and Generalized weakly contractive mapping and show the existence and uniqueness of the best proximity point of such contractions in the setting of a metric space.

Keywords: Best proximity point · P-property
Boyd and Wong contraction · Generalized weakly contractive

1 Introduction

In nonlinear functional analysis, fixed point theory and best proximity point theory play an important role in the establishment of the existence of a certain differential and integral equations. As a consequence fixed point theory is very much useful for various quantitative sciences that involve such equations. The most remarkable paper in this field was reported by Banach in 1922 [3]. In his paper Banach proved that every contraction in a complete metric space has a unique fixed point. Following this paper many have extended and generalized this remarkable fixed point theorem of Banach by changing either the conditions of the mappings or the construction of the space. In particular, one of the notable generalizations of Banach fixed point theorem was introduced by Geraghty [7].

Theorem 1 [7]. Let (X, d) be a complete metric space and $f : X \to X$ be an operator. Suppose that there exists $\beta : [0, \infty) \to (0, 1)$ satisfying if f satisfies the following inequality:

$$d(f(x), f(y)) \leq \beta(d(x, y))d(x, y) \text{ for any } x, y \in X,$$

then f has unique fixed point.

It is natural that some mapping, especially non-self mappings defined on a complete metric space (X, d), do not necessarily possess a fixed point, that is $d(x, f(x)) > 0$ for all $x \in X$. In such situations it is reasonable to search for

© Springer International Publishing AG, part of Springer Nature 2018
A. M. Gil-Lafuente et al. (Eds.): FIM 2015, AISC 730, pp. 3–13, 2018.
https://doi.org/10.1007/978-3-319-75792-6_1

the existence and uniqueness of a point $x^* \in X$ such that $d(x^*, f(x^*))$ is an approximation of an $x \in X$ such that $d(x, f(x)) = 0$.

In other words one speculates to determine an approximate solution x^* that is optimal in the sense that the distance between x^* and $f(x^*)$ is minimum. Here the point x^* is called the best proximity point. In this paper we generalize and improve certain results of Caballero et al. [6].

2   Prelimineries

Let (X, d) be a metric space and A and B be nonempty subsets of a metric space X. A mapping $f : A \to B$ is called a k-contraction if there exists $k \in (0, 1)$ such that

$$d(f(x), f(y)) \le kd(x, y) \text{ for any } x, y \in A.$$

It is clear that a k-contraction coincides with the celebrated Banach fixed point theorem if one takes $A = B$ where A is a complete subset of X.

Let A and B be nonempty subsets of a metric space (X, d). we denote by A_0 and B_0 the following sets:

$$A_0 = \{x \in A : d(x, y) = d(A, B), \quad for some \quad y \in B\}$$
$$B_0 = \{y \in B : d(x, y) = d(A, B), \quad for some \quad x \in A\} where$$
$$d(A, B) = \inf\{d(x, y) : x \in A, y \in B\}.$$

Definition 1 [8]. Let (A, B) be a pair of nonempty subsets of a metric space (X, d) with $A_0 \ne \emptyset$. Then the pair (A, B) is said to have the P-property if and only if for any $x_1, x_2 \in A_0$ and $y_1, y_2 \in B_0$ $d(x_1, y_1) = d(A, B)$ and $d(x_2, y_2) = d(A, B)$ implies that $d(x_1, x_2) = d(y_1, y_2)$.

It can be easily seen that for any nonempty subset A of (X, d), the pair (A, A) has the P-property. In [11] Sankarraj has proved that any pair (A, B) of nonempty closed convex subsets of a real Hilbert space H satisfies P-property.

Now we introduce the class of those functions $\beta : [0, \infty) \to [0, 1)$ satisfying the following condition: $\beta(t_n) \to 1 \Rightarrow t_n \to 0$.

Definition 2 [6]. Let A and B be nonempty subsets of a metric space (X, d). A mapping $f : A \to B$ is said to be a Geraghty contraction if there exists $\beta \in F$ such that

$$d(f(x), f(y)) \le \beta(d(x, y))d(x, y) \text{ for any } x, y \in A.$$

Remark 1. Notice that since $\beta : [0, \infty) \to (0, 1)$, we have

$$d(f(x), f(y)) \le \beta(d(x, y))d(x, y) < d(x, y) \text{ for any } x, y \in A \text{ with } x \ne y.$$

Theorem 2 [6]. Let (A, B) be a pair of nonempty closed subsets of a complete metric space (X, d) such that A_0 is nonempty. Let $f : A \to B$ be a continuous Geraghty contraction satisfying $f(A_0) \subseteq B_0$. Suppose that the pair (A, B) has the P-property. Then there exists a unique $x^* \in A$ such that $d(x^*, f(x^*)) = d(A, B)$.

We would like to extend the result of Caballero and explore the best proximity point based on the well known result of Boyd and Wong [5].

Theorem 3 [1]. Let X be a complete metric space and let $f : X \to X$ satisfy

$$d(f(x), f(y)) \leq \psi(d(x, y))$$

where $\psi : R^+ \to R^+$ is upper semi-continuous from the right and satisfies $0 \leq \psi(t) < t$. Then f has a unique fixed point. Further if $x_0 \in X$ and $x_{n+1} = f(x_n)$, then $\{x_n\}$ converges to the fixed point.

A mapping $f : X \to X$ is said to be contractive if

$$d(f(x), f(y)) < d(x, y) \quad for each \quad x, y \in X \quad with \quad x \neq y. \tag{1}$$

3 Main Results

Theorem 4 Let (A, B) be a pair of nonempty closed subsets of a complete metric space (X, d) such that $A_0 \neq \emptyset$. Let $f : A \to B$ be such that $f(A_0) \subseteq B_0$. Suppose

$$d(f(x), f(y)) \leq \psi(d(x, y)) \text{ for each } x, y \in A,$$

where $\psi : R^+ \to [0, \infty)$ is upper semi-continuous from the right satisfies $0 \leq \psi(t) < t$ for $t > 0$. Furthermore the pair (A, B) has the P-property. Then there exists a unique $x^* \in A$ such that $d(x^*, f(x^*)) = d(A, B)$.

Proof. Regarding that A_0 is nonempty, we take $x_0 \in A_0$.

Since $f(x_0) \in f(A_0) \subseteq B_0$, we can find $x_1 \in A_0$ such that $d(x_1, f(x_0)) = d(A, B)$. Analogously regarding the assumption $f(x_1) \in f(A_0) \subseteq B_0$, we determine $x_2 \in A_0$ such that $d(x_2, f(x_1)) = d(A, B)$.

Recursively we obtain a sequence $\{x_n\}$ in A_0 satisfying

$$d(x_{n+1}, f(x_n)) = d(A, B) \quad for any \quad n \in N \tag{2}$$

Since (A, B) has the P-property we derive that

$$d(x_n, x_{n+1}) = d(f(x_{n-1}), f(x_n)) \quad for any \quad n \in N \tag{3}$$

If there exists $n_0 \in N$ such that $d(x_{n_0}, x_{n_0+1}) = 0$, then the proof is completed. Indeed

$$0 = d(x_{n_0}, x_{n_0+1}) = d(f(x_{n_0-1}), f(x_{n_0})) \tag{4}$$

and consequently $f(x_{n_0-1}) = f(x_{n_0})$.

On the other hand due to 2 we have $d(x_{n_0}, f(x_{n_0-1})) = d(A, B)$.

Therefore we conclude that

$$d(A, B) = d(x_{n_0}, f(x_{n_0-1})) = d(x_{n_0}, f(x_{n_0})) \tag{5}$$

For the rest of the proof we suppose that $d(x_n, x_{n+1}) > 0$ for any $n \in N$.

Since f is contractive, for any $n \in N$, we have that

$$d(x_{n+1}, x_{n+2}) = d(f(x_n), f(x_{n+1})) \leq \psi(d(x_n, x_{n+1})) < d(x_n, x_{n+1}) \quad (6)$$

consequently $\{d(x_n, x_{n+1})\}$ is monotonically decreasing sequence and bounded below and so we have $\lim_{n \to \infty} d(x_n, x_{n+1}) = r$ exists.

Let $\lim_{n \to \infty} d(x_n, x_{n+1}) = r \geq 0$.

Assume that $r > 0$. Then from 1 we have $d(x_{n+1}, x_{n+2}) \leq \psi(d(x_n, x_{n+1}))$ which implies that $r \leq \psi(r) \Rightarrow r = 0$.

That is

$$\lim_{n \to \infty} d(x_n, x_{n+1}) = 0 \quad (7)$$

Notice that since $d(x_{n+1}, f(x_n)) = d(A, B)$ for any $n \in N$, for fixed $p, q \in N$, we have $d(x_p, f(x_{p-1})) = d(x_q, f(x_{q-1})) = d(A, B)$ and since (A, B) satisfies the P-property, $d(x_p, x_q) = d(f(x_{p-1}), f(x_{q-1}))$.

In what follows, we prove that $\{x_n\}$ is cauchy sequence.

On the contrary, assume that we have

$$\epsilon = \limsup_{m,n \to \infty} d(x_n, x_m) > 0 \quad (8)$$

Then there exists $\epsilon > 0$, such that for any $k \in N$, there exists $m_k > n_k \geq k$, such that

$$d(x_{m_k}, x_{n_k}) \geq \epsilon \quad (9)$$

Furthermore assume that for each k, m_k is the smallest number greater than n_k for which 9 holds. In view of 6, there exists k_0 such that $k \geq k_0$ implies that $d(x_k, x_{k+1}) \geq \epsilon$.

For such k, we have

$$\begin{aligned}
\epsilon &\leq d(x_{m_k}, x_{n_k}) \\
&\leq d(x_{m_k}, x_{m_{k-1}}) + d(x_{m_{k-1}}, x_{n_k}) \\
&\leq d(x_{m_k}, x_{m_{k-1}}) + \epsilon \\
&\leq d(x_k, x_{k-1}) + \epsilon.
\end{aligned}$$

This proves $\lim_{k \to \infty} d(x_{m_k}, x_{n_k}) = \epsilon$.

On the other hand

$$\begin{aligned}
d(x_{m_k}, x_{n_k}) &\leq d(x_{m_k}, x_{m_{k+1}}) + d(x_{m_{k+1}}, x_{n_{k+1}}) + d(x_{n_{k+1}}, x_{n_k}) \\
&\leq 2d(x_k, x_{k+1}) + \psi(d(x_{m_k}, x_{n_k})).
\end{aligned}$$

Since $\lim_{k \to \infty} d(x_k, x_{k+1}) = 0$, the above inequality yields

$$\epsilon \leq \limsup_{m,n \to \infty} d(x_{m_k}, x_{n_k}) \leq \limsup_{m,n \to \infty} \psi(d(x_{m_k}, x_{n_k})) \leq \psi(c).$$

It follows that $\epsilon \leq \psi(\epsilon)$, a contradiction.

Therefore $\{x_n\}$ is a cauchy sequence.

Since $\{x_n\} \subset A$ and A is closed subset of the complete metric space (X, d), we can find $x^* \in A$ such that $x_n \to x^*$.

Since the mapping is contractive and continuous, we have $f(x_n) \to f(x^*)$.

This implies that $d(x_n, x_{n+1}) \to d(x^*, f(x^*))$.

Taking into consideration that the sequence $\{d(x_{n+1}, f(x_n))\}$ is a constant sequence with the value $d(A, B)$, we deduce that $d(x^*, f(x^*)) = d(A, B)$.

This means that x^* is a best proximity point of f.

This proves the existence of our result.

For the uniqueness, suppose that x_1 and x_2 are two best proximity points of f with $x_1 \neq x_2$. This means that $d(x_i, f(x_i)) = d(A, B)$ for $i = 1, 2$.

Using the P-property, we have $d(x_1, x_2) = d(f(x_1), f(x_2))$

Using the fact that f is contractive and continuous, we have

$d(x_1, x_2)$
$= d(f(x_1), f(x_2))$
$\leq \psi(d(x_1, x_2))$
$< d(x_1, x_2)$ which is a contradiction.

Therefore $x_1 = x_2$.

This completes the proof.

In the following result we introduce the concept of generalized weakly contractive mapping and find best proximity point based on the work of Choudhury [4].

Definition 3 [10]. A mapping $f : X \to X$, where (X, d) is a metric space, is said to be weakly contractive if for any $x, y \in X$, then

$$d(f(x), f(y) \leq d(x, y) - \phi(d(x, y)) \tag{10}$$

where $\phi : [0, \infty) \to [0, \infty)$ is continuous and nondecreasing function such that $\phi(t) = 0$ if and only if $t = 0$. If one takes $\phi(t) = (1 - k)t$, where $0 < k < t$, a weak contraction reduces to a Banach contraction.

In [2] Alber and Guerre proved that if $f : \Omega \to \Omega$ is a weakly contractive self-map, where Ω is a closed convex subset of a Hilbert space, then f has a unique fixed point in Ω. Later, in [10] Rhodes proved that the existence of a unique fixed point for a weakly contractive self-map could be achieved even in a complete metric space setting.

Definition 4 [12]. Let A, B be nonempty subsets of a metric space X.

A map $f : A \to B$ is said to be weakly contractive mapping if

$$d(f(x), f(y)) \leq d(x, y) - \psi(d(x, y)), \text{ for all } x, y \in A,$$

where $\psi : [0, \infty) \to [0, \infty)$ is a continuous and nondecreasing function such that ψ is positive on $(0, \infty), \psi(0) = 0$ and $\lim_{n \to \infty} \psi(t) = \infty$. If A is bounded, then the infinity condition can be omitted.

Note that

$$d(f(x), f(y)) \leq d(x, y) - \psi(d(x, y)) < d(x, y) \text{ if } x, y \in A \text{ with } x \neq y.$$

That is f is a contractive map. The notion called the P-property was introduced in [11] and was used to prove a extended version of Banach's contraction principle.

Theorem 5 [12]. Let (A, B) be a pair of nonempty closed subsets of a complete metric space (X, d) such that A_0 is nonempty. Let $f : A \to B$ be a weakly contractive mapping satisfying $f(A_0) \subseteq B_0$. Assume that the pair (A, B) has the p-property. Then there exists a unique $x^* \in A$ such that $d(x^*, f(x^*)) = d(A, B)$.

Definition 5 [9]. A function $\psi : [0, \infty) \to [0, \infty)$ is called an altering function if the following properties are satisfied:

(a) ψ is monotone increasing and continuous
(b) $\psi(t) = 0$ if and only if $t = 0$.

Definition 6 [4]. Let (X, d) be a metric space, f a self-mapping of X. We shall call f a generalized weakly contractive mapping if for all $x, y \in X$, then

$$\psi(d(f(x), f(y))) \leq \psi(m(x, y)) - \phi(\max\{d(x, y), d(y, f(y))\})$$

where

$$m(x, y) = \max\{d(x, y), d(x, f(x)), d(y, f(y)), \frac{1}{2}[d(x, f(y)) + d(y, f(x))]\}$$

and ψ is an altering distance function also $\phi : [0, \infty) \to [0, \infty)$ is a continuous function with $\phi(t) = 0$ if and only if $t = 0$. A generalized weakly contractive mapping is more general than that satisfying $d(f(x), f(y)) \leq km(x, y)$ for some constant $0 \leq k < 1$ and is included in those mappings which satisfy

$$d(f(x), f(y)) < m(x, y).$$

Definition 7. Let A, B be nonempty subsets of a metric space X. A map $f : A \to B$ is said to be a generalized weakly contractive mapping if for all $x, y \in A$, then

$$\psi(d(f(x), f(y))) \leq \psi(m(x, y)) - \phi(\max\{d(x, y), d(y, f(y)) - d(A, B))\})$$

where

$$m(x, y) = \max\{d(x, y), d(x, f(x)) - d(A, B), d(y, f(y)) - d(A, B),$$
$$\frac{1}{2}[d(x, f(y)) + d(y, f(x))] - d(A, B)\}.$$

A generalized weakly contractive mapping is more general than that satisfying $d(f(x), f(y)) \leq km(x, y)$ for some constant $0 \leq k < 1$ and is included in those mappings which satisfy
$$d(f(x), f(y)) < m(x, y).$$

Theorem 6. Let (A, B) be a pair of nonempty closed subsets of a complete metric space (X, d) such that A_0 is nonempty. Let $f : A \to B$ be such that $f(A_0) \subseteq B_0$. Suppose

$$\psi(d(f(x), f(y)) \leq \psi(m(x, y)) - \phi(\max\{d(x, y), d(y, f(y)) - d(A, B))\}) \quad (11)$$

Furthermore the pair (A, B) has the p-property. Then there exists a unique x^* in A such that $d(x^*, f(x^*)) = d(A, B)$.

Proof. Choose $x_0 \in A$.

Since $f(x_0) \in f(A_0) \subseteq B_0$, there exists $x_1 \in A_0$ such that $d(x_1, f(x_0)) = d(A, B)$.

Analogously regarding the assumption, $f(x_1) \in f(A_0) \subseteq B_0$, we determine $x_2 \in A_0$ such that $d(x_2, f(x_1)) = d(A, B)$.

Recursively we obtain a sequence $\{x_n\}$ in A_0 satisfying

$$d(x_{n+1}, f(x_n)) = d(A, B) \quad for any \quad n \in N \quad (12)$$

Claim: $d(x_n, x_{n+1}) \to 0$.

If $x_N = x_{N+1}$, then x_N is a best proximity point.

By the P-property, we have

$$d(x_{n+1}, x_{n+2}) = d(f(x_n), f(x_{n+1})).$$

Hence we assume that $x_n \neq x_{n+1}$ for all $n \in N$.

Since $d(x_{n+1}, f(x_n)) = d(A, B)$, from (11) we have for all $n \in N$

$$\begin{aligned}
\psi(d(x_{n+1}, x_{n+2})) &= \psi(d(f(x_n), f(x_{n+1}))) \\
&\leq \psi(max\{d(x_n, x_{n+1}), d(x_n, f(x_n)) - d(A, B), d(x_{n+1}, f(x_{n+1})) - d(A, B), \\
&\quad \frac{1}{2}[d(x_n, f(x_{n+1})) + d(x_{n+1}, f(x_n))] - d(A, B)\}) \\
&\quad -\phi(\max\{d(x_n, x_{n+1}), d(x_{n+1}, f(x_{n+1})) - d(A, B)\}) \\
&\leq \psi(max\{d(x_n, x_{n+1}), d(x_n, f(x_n)) - d(A, B), d(x_{n+1}, f(x_{n+1})) - d(A, B), \\
&\quad \frac{1}{2}(d(x_n, f(x_{n+1}))) - d(A, B)\}) \\
&\quad -\phi(\max\{d(x_n, x_{n+1}), d(x_{n+1}, f(x_{n+1})) - d(A, B)\})
\end{aligned}$$

Since

$$\begin{aligned}
\frac{1}{2}(d(x_n, f(x_{n+1}))) - d(A, B) &\leq \frac{1}{2}(d(x_n, x_{n+1}) + d(x_{n+1}, f(x_{n+1}))) - d(A, B) \\
&\leq \max\{d(x_n, x_{n+1}), d(x_{n+1}, f(x_{n+1})) - d(A, B)\} \\
d(x_n, f(x_n)) - d(A, B) &\leq d(x_n, x_{n+1}) + d(x_{n+1}, f(x_n)) - d(A, B) \\
&= d(x_n, x_{n+1})
\end{aligned}$$

It follows that

$$\begin{aligned}
\psi(d(f(x_n), f(x_{n+1}))) &\leq \psi(\max\{d(x_n, x_{n+1}), d(x_{n+1}, f(x_{n+1})) - d(A, B)\}) \\
&\quad -\phi(\max\{d(x_n, x_{n+1}), d(x_{n+1}, f(x_{n+1})) - d(A, B)\})
\end{aligned}$$

$$\psi(d(x_{n+1}, x_{n+2})) \leq \psi(\max\{d(x_n, x_{n+1}), d(x_{n+1}, x_{n+2})\})$$
$$-\phi(\max\{d(x_n, x_{n+1}), d(x_{n+1}, x_{n+2})\}) \qquad (13)$$

Suppose that $d(x_n, x_{n+1}) \leq d(x_{n+1}, x_{n+2})$, for some positive integer n. Then from 13 we have

$$\psi(d(x_{n+1}, x_{n+2}) \leq \psi(d(x_{n+1}, x_{n+2})) - \phi(d(x_{n+1}, x_{n+2})),$$

that is

$$\phi(d(x_{n+1}, x_{n+2})) \leq 0,$$

which implies that $d(x_{n+1}, x_{n+2}) = 0$, contradicting our assumption.

Therefore $d(x_{n+1}, x_{n+2}) < d(x_n, x_{n+1})$ for any $n \in N$ and hence $\{d(x_n, x_{n+1})\}$ is monotone decreasing sequence of non-negative real numbers, hence there exists $r \geq 0$ such that $\lim_{n \to \infty} d(x_n, x_{n+1}) = r$. In view of the facts from 13 for any $n \in N$, we have

$$\psi(d(x_{n+1}, x_{n+2})) \leq \psi(d(x_n, x_{n+1})) - \phi(d(x_n, x_{n+1})),$$

Taking the limit as $n \to \infty$ in the above inequality and using the continuities of ψ and ϕ we have $\psi(r) \leq \psi(r) - \phi(r)$ which implies $\phi(r) = 0$.

Hence

$$\lim_{n \to \infty} d(x_n, x_{n+1}) = 0 \qquad (14)$$

Next we show that $\{x_n\}$ is a cauchy sequence.

If otherwise there exists an $\epsilon > 0$ for which we can find two sequences of positive integers $\{m_k\}$ and $\{n_k\}$ such that for all positive integers $k, n_k > m_k > k$,

$$d(x_{m_k}, x_{n_k}) \geq \epsilon \text{ and } d(x_{m_k}, x_{n_k-1}) < \epsilon.$$

Now

$$\epsilon \leq d(x_{m_k}, x_{n_k}) \leq d(x_{m_k}, x_{n_k-1}) + d(x_{n_k-1}, x_{n_k}).$$

that is

$$\epsilon \leq d(x_{m_k}, x_{n_k}) < \epsilon + d(x_{n_k-1}, x_{n_k}),$$

Taking the limit as $k \to \infty$ in the above inequality and using 14 we have

$$\lim_{k \to \infty} d(x_{m_k}, x_{n_k}) = \epsilon \qquad (15)$$

Again

$$d(x_{m_k}, x_{n_k}) \leq d(x_{m_k}, x_{m_k+1}) + d(x_{m_k+1}, x_{n_k+1}) + d(x_{n_k+1}, x_{n_k})$$

Taking the limit as $k \to \infty$ in the above inequalities and using 14 and 15 we have

$$\lim_{k \to \infty} d(x_{m_k+1}, x_{n_k+1}) = \epsilon \qquad (16)$$

Again

$$d(x_{m_k}, x_{n_k}) \leq d(x_{m_k}, x_{n_k+1}) + d(x_{n_k+1}, x_{n_k}) \text{ and}$$
$$d(x_{m_k}, x_{n_k+1}) \leq d(x_{m_k}, x_{n_k}) + d(x_{n_k}, x_{n_k+1})$$

Letting $k \to \infty$ in the above inequalities and using 14 and 15 we have

$$\lim_{k \to \infty} d(x_{m_k}, x_{n_k+1}) = \epsilon \tag{17}$$

Similarly

$$\lim_{k \to \infty} d(x_{n_k}, x_{m_k+1}) = \epsilon \tag{18}$$

For $x = x_{m_k}, y = y_{m_k}$, we have

$$d(x_{m_k}, f(x_{m_k})) - d(A, B) \leq d(x_{m_k}, x_{m_k+1}) + d(x_{m_k+1}, f(x_{m_k})) - d(A, B)$$
$$= d(x_{m_k}, x_{m_k+1})$$

similarly

$$d(x_{n_k}, f(x_{n_k})) - d(A, B) = d(x_{n_k}, x_{n_k+1}).$$

Also

$$d(x_{m_k}, f(x_{n_k})) - d(A, B) = d(x_{m_k}, x_{n_k+1}) \text{ and}$$
$$d(x_{n_k}, f(x_{m_k})) - d(A, B) = d(x_{n_k}, x_{m_k+1}).$$

From 11 we have

$$\psi(d(x_{m_k+1}, x_{n_k+1})) = \psi(d(f(x_{m_k}), f(x_{n_k})))$$
$$\leq \psi(\max\{d(x_{m_k}, x_{n_k}), d(x_{m_k}, f(x_{m_k})) - d(A, B), d(x_{n_k}, f(x_{n_k})) - d(A, B),$$
$$\frac{1}{2}[d(x_{m_k}, f(x_{n_k})) + d(x_{n_k}, f(x_{m_k})) - d(A, B)]\})$$
$$-\phi(\max\{d(x_{m_k}, x_{n_k}), d(x_{n_k}, f(x_{n_k})) - d(A, B)\})$$
$$\leq \psi(\max\{d(x_{m_k}, x_{n_k}), d(x_{m_k}, x_{m_k+1}), d(x_{n_k}, x_{n_k+1}),$$
$$\frac{1}{2}[d(x_{m_k}, x_{n_k+1}) + d(x_{n_k}, x_{m_k+1})]\})$$
$$-\phi(\max\{d(x_{m_k}, x_{n_k}), d(x_{n_k}, x_{n_k+1})\})$$

It follows that

$$\psi(d(f(x_{m_k}), f(x_{n_k}))) \leq \psi(\max\{d(x_{m_k}, x_{n_k}), d(x_{n_k}, f(x_{n_k+1})),$$
$$\frac{1}{2}[d(x_{m_k}, x_{n_k+1}) + d(x_{n_k}, x_{m_k+1})]\})$$
$$-\phi(\max\{d(x_{m_k}, x_{n_k}), d(x_{n_k}, f(x_{n_k+1}))\})$$
$$\psi(d(x_{m_k+1}, x_{n_k+1})) \leq \psi(\max\{d(x_{m_k}, x_{n_k}), d(x_{n_k}, x_{n_k+1})\})$$
$$-\phi(\max\{d(x_{m_k}, x_{n_k}), d(x_{n_k}, x_{n_k+1})\})$$

From 14, 15, 17, 18 and Letting $k \to \infty$ in the above inequalities and using the continuities of ψ and ϕ,
we have $\psi(\epsilon) \leq \psi(\epsilon) - \phi(\epsilon)$
which is contradiction by virtue of property of ϕ.

Hence $\{x_n\}$ is a cauchy sequence.

Since $\{x_n\} \subset A$ and A is a closed subset of the complete metric space (X, d), there exists x^* in A such that $x_n \to x^*$.

Putting $x = x_n$ and $y = x^*$ in 11 and since

$$d(x_n, f(x^*)) \leq d(x_n, x^*) + d(x^*, f(x_n)) \text{ and}$$
$$d(x^*, f(x_n)) \leq d(x^*, f(x^*)) + d(f(x^*), f(x_n))$$

we have

$$\psi(d(x_{n+1}, f(x^*))) - d(A, B) \leq \psi(d(f(x_n), f(x^*)))$$
$$\leq \psi(\max\{d(x_n, x^*), d(x_n, x_{n+1}), d(x^*, f(x^*)) - d(A, B),$$
$$\frac{1}{2}[d(x_n, f(x^*)) + d(x^*, f(x_n))] - d(A, B)\})$$
$$-\phi(\max\{d(x_n, x^*), d(x^*, f(x^*)) - d(A, B)\})$$

Taking the limit as $n \to \infty$ in the above inequality and using the continuities of ψ and ϕ, we have $\psi(d(x^*, f(x^*)) - d(A, B)) \leq \psi(d(x^*, f(x^*)) - d(A, B)) - \phi(d(x^*, f(x^*)) - d(A, B))$. Which implies that $d(x^*, f(x^*)) - d(A, B)$.

Hence x^* is a best proximity point of f.

For the uniqueness

Let p and q be two best proximity points of f and suppose that $p \neq q$.

Then putting $x = p$ and $y = q$ in (11) we obtain

$$\psi(d(f(p), f(q))) \leq \psi(\max\{d(p, q), d(p, f(p)) - d(A, B), d(q, f(q)) - d(A, B),$$
$$\frac{1}{2}[d(p, f(q)) + d(q, f(p))] - d(A, B)\})$$
$$-\phi(\max\{d(p, q), d(q, f(q)) - d(A, B)\})$$

That is

$$\psi(d(p, q)) \leq \psi(d(p, q)) - \phi(d(p, q))$$

which is contradiction by virtue of a property of ϕ.

Therefore $p = q$.

This completes the proof.

References

1. Eldred, A.A.: Existence and Convergence of Best Proximity Theorems. Ph.D. thesis. IIT Madras (2007)
2. Alber, Y.I., Guerre Delabriere, S.: Principle of weakly contractive maps in Hilbert spaces. In: New Results in Operator Theory and its Applications, vol. 98, pp. 7–22. Birkhäuser, Basel (1997)

3. Banach, S.: Sur les operations dans les ensembles abstraits er leur applications aux equations integrals. Fundam. Math. **3**, 133–181 (1922)
4. Choudhury, B.S., Konar, P., Rhoades, B.E., Metiya, N.: Fixed point theorems for genealized weakly contractive mapping. Nonlinear Anal. **74**(6), 2116–2126 (2011)
5. Boyd, W.D., Wong, J.S.W.: On nonlinear contractions. Proc. Am. Math. Soc. **20**, 458–464 (1969)
6. Caballero, J., Harjani, J., Sadarangani, K.: A best proximity point theorem for Geraghty-contractions. Fixed Point Theory Appl. **231** (2012)
7. Geraghty, M.: On contractive mappings. Proc. Am. Math. Soc. **40**, 604–608 (1973)
8. Karapinar, E.: On best proximity point of ψ-Geraghty contractions. Fixed Point Theory Appl. **200** (2013)
9. Khan, M.S., Sessa, S., Sweleh, M.S.: Fixed point theorems by altering distance between the points. Bull. Aust. Math. Soc. **30**, 1–9 (1984)
10. Rhodes, B.E.: Some theorems on weakly contractive maps. Nonlinear Anal. Theory Methods Appl. **47**(4), 2683–2693 (2001)
11. Sanksr, R.V.: Banach contraction principle for non-self mappings, Preprint
12. Sankar, R.V.: Best proximity point theorem for weakly contractive non-self mappings. Nonlinear Anal. **74**, 4804–4808 (2011)

The Method of Optimal Nonlinear Extrapolation of Vector Random Sequences on the Basis of Polynomial Degree Canonical Expansion

Vyacheslav S. Shebanin[1], Yuriy P. Kondratenko[2(✉)], and Igor P. Atamanyuk[1]

[1] Mykolaiv National Agrarian University,
Georgiy Gongadze Street 9, Mykolaiv 54000, Ukraine
{rector, atamanyuk}@mnau.edu.ua
[2] Petro Mohyla Black Sea National University,
68th Desantnykiv Street 10, Mykolaiv 54003, Ukraine
y_kondrat2002@yahoo.com

Abstract. The given work is dedicated to the solving of important scientific and technical problem of forming of the method of the optimal (in mean-square sense) extrapolation of the realizations of vector random sequences for the accidental quantity of the known values used for prognosis and for various order of nonlinear stochastic relations. Prognostic model is synthesized on the basis of polynomial degree canonical expansion of vector random sequence. The formula for the determination of the mean-square error of the extrapolation which allows us to estimate the accuracy of the solving of the prognostication problem with the help of the introduced method is obtained. The block diagrams of the algorithms of the determination of the parameters of the introduced method are also presented in the work. Taking into account the recurrent character of the processes of the estimation of the future values of the investigated sequence the method is quite simple in calculating respect. The introduced method of extrapolation as well as the vector canonical expansion assumed as its basis doesn't put any essential limitations on the class of prognosticated random sequences (linearity, Markovian property, stationarity, scalarity, monotony etc.).

Keywords: Optimal nonlinear · Extrapolation · Vector random sequences
Polynomial canonical expansion

1 Introduction

The peculiarity of the wide range of applied problems in different spheres of science and techniques is the probabilistic nature of the investigated phenomenon or the presence of the influence of random factors on the investigated object as a result of what the process of changing of its state also takes probabilistic character. The objects of such a class which relate to the objects with randomly variable conditions of functioning (RVCF) are investigated, for example, during the solving of the problems

© Springer International Publishing AG, part of Springer Nature 2018
A. M. Gil-Lafuente et al. (Eds.): FIM 2015, AISC 730, pp. 14–25, 2018.
https://doi.org/10.1007/978-3-319-75792-6_2

of technical diagnostics [4], radiolocation, medical diagnostics [5], robotics and automation [13, 20], forecasting control of reliability [15], weather forecasting [17], information security, synthesis of the models of chemical kinetics, management of technological processes, motion control [14], etc. The characteristic peculiarity of these problems is the presence of the preliminary stage of gathering of the information about the object of investigation. Random character of external influence and coordinates (input and output) of the objects with RVCF under the conditions of sufficient statistic data volume determines the necessity and reasonability of the usage of deductive [10] methods of random sequences prognosis for their solving.

It is known that the most general extrapolation form for the solving of the problems of the prognosis is the mathematical model in the form of Kolmogorov-Gabor polynomial [9]. Such a model allows taking into account the accidental number of random sequence measurements and the order of degree nonlinearity. But its practical application is limited with significant difficulties connected with the forming of the large quantity of equations for the determination of the extrapolator parameters. Existing optimal methods which are used during the solving of applied problems are obtained for the definite classes of random sequences, in particular, the methods of Kolmogorov [12] and Wiener [21] are for stationary processes, Kalman's filter-extrapolator [11, 18] is for markovian random sequences, methods of Pugachev [19], Kudritsky [16] are for non-stationary gaussian sequences etc. It should be mentioned that their application allows to obtain optimal results only for the sequences with definite a priori known characteristics.

Thus the theoretically substantiated solutions of the problem of the prognosis of random sequences exist but the known methods and models are based on the usage of appropriate limitations which don't permit to obtain maximal accuracy of extrapolation and can't be used in practice for the objects with RVCF under the most general assumptions concerning the degree of nonlinear stochastic relations and the quantity of measurements used for the prognosis.

2   Statement of the Problem

Vector random sequence $\left\{\vec{X}\right\} = X_h(i)$, $h = \overline{1,H}$ describing the time change of H interconnected parameters of a certain object with randomly changeable conditions of functioning is completely designated in the discrete series of points t_i, $i = \overline{1,I}$ by moment functions $M\left[X_l^{\nu}(i)\right]$, $M\left[X_l^{\nu}(i)X_h^{\mu}(j)\right]$, $i,j = \overline{1,I}$; $l,h = \overline{1,H}$; $\nu,\mu = \overline{1,N}$. It is necessary to get the optimal estimations $x_h^*(i)$, $i = \overline{k+1,I}$, $h = \overline{1,H}$ of the future values of the investigated random sequence for each its constituent $X_h(i)$ provided that the values $x_h^{\mu}(j)$, $j = \overline{1,k}$, $\mu = \overline{1,N}$, $h = \overline{1,H}$ in the first k points of observation are known.

3 Solution

The most universal approach to the solving of a stated problem from the point of view of the limitations put on a random process is the usage of the apparatus of canonical expansions [16, 19]. For the vector case such an expansion with full account of correlated relations between the constituents is of the form [1]:

$$
X_h(i) = M[X_h(i)] + \sum_{v=1}^{i} \sum_{\lambda=1}^{H} V_v^{(\lambda)} \varphi_{hv}^{(\lambda)}(i), \quad i = \overline{1, I}, \tag{1}
$$

where

$$
V_v^{(\lambda)} = X_\lambda(v) - M[X_\lambda(v)] - \sum_{\mu=1}^{v-1} \sum_{j=1}^{H} V_\mu^{(j)} \varphi_{\lambda\mu}^{(j)}(v)
$$
$$
- \sum_{j=1}^{\lambda-1} V_v^{(j)} \varphi_{\lambda v}^{(j)}(v), \quad v = \overline{1, I}; \tag{2}
$$

$$
\varphi_{hv}^{(\lambda)}(i) = \frac{M\left[V_v^{(\lambda)}(X_h(i) - M[X_h(i)])\right]}{M\left[\left\{V_v^{(\lambda)}\right\}^2\right]} = \frac{1}{D_\lambda(v)} \left(M[X_\lambda(v)X_h(i)]\right.
$$
$$
- M[X_\lambda(v)]M[X_h(i)] - \sum_{\mu=1}^{v-1} \sum_{j=1}^{H} D_j(\mu) \varphi_{\lambda\mu}^{(j)}(v) \varphi_{h\mu}^{(j)}(i)
$$
$$
- \sum_{j=1}^{\lambda-1} D_j(v) \varphi_{\lambda v}^{(j)}(v) \varphi_{hv}^{(j)}(i), \quad \lambda = \overline{1, h}, v = \overline{1, i}. \tag{3}
$$

$$
D_\lambda(v) = M\left[\left\{V_v^{(\lambda)}\right\}^2\right] = M\left[\{X_\lambda(v)\}^2\right] - M^2[X_\lambda(v)]
$$
$$
- \sum_{\mu=1}^{v-1} \sum_{j=1}^{H} D_j(\mu) \left\{\varphi_{\lambda\mu}^{(j)}(v)\right\}^2 - \sum_{j=1}^{\lambda-1} D_j(v) \left\{\varphi_{\lambda v}^{(j)}(v)\right\}^2, \quad v = \overline{1, I}; \tag{4}
$$

Coordinate functions $\varphi_{hv}^{(\lambda)}(i)$, $h, \lambda = \overline{1, H}; v, i = \overline{1, I}$ have the following characteristics:

$$
\varphi_{hv}^{(\lambda)}(i) = \begin{cases} 1, & \text{for } (h = \lambda) \wedge (v = i); \\ 0, & \text{for } (i < v) \vee ((h < \lambda) \wedge (v = i)). \end{cases} \tag{5}
$$

The algorithm of the prognosis on the basis of canonical expansion (1) is of the form [6]:

$$m_{x;h}^{(\mu,l)}(i) = \begin{cases} M[X_h(i)], & \text{if } \mu = 0; \\ m_{x;h}^{(\mu,l-1)}(i) + \left[x_l(\mu) - m_{x;l}^{(\mu,l-1)}(\mu)\right]\varphi_{h\mu}^{(l)}(i), & \text{if } l \neq 1; \\ m_{x;h}^{(\mu,H)}(i) + \left[x_1(\mu) - m_{x;1}^{(\mu-1,H)}(\mu)\right]\varphi_{h\mu}^{(1)}(i), & \text{for } l = 1; \end{cases} \quad (6)$$

$m_{x;h}^{(\mu,l)}(i) = M\left[X_h(i)/x_\lambda(v), \lambda = \overline{1,H}, v = \overline{1,\mu-1}; x_j(\mu), \quad j = 1,l\right], h = \overline{1,H}, i = \overline{k,I}$ - is linear optimal by the criterion of the minimum of the mean-square error of prognosis the estimation of the future values of the investigated sequence provided that the values $x_\lambda(v), \lambda = \overline{1,H}, v = \overline{1,\mu-1}; x_j(\mu), j = \overline{1,l}$ are known.

The only shortcoming of the algorithm (6) within the framework of problem statement is that the given solution as well as the canonical expansion assumed as its basis uses for prognosis only correlated functions.

The increase of the volume of a priori information about the investigated process is possible in the algorithm of the prognosis by means of the usage of the appropriate nonlinear expansion [2]:

$$X_h(i) = M[X_h(i)] + \sum_{v=1}^{i-1}\sum_{l=1}^{H}\sum_{\lambda=1}^{N} W_{vl}^{(\lambda)}\beta_{l\lambda}^{(h,1)}(v,i) + \sum_{l=1}^{h-1}\sum_{\lambda=1}^{N} W_{il}^{(\lambda)}\beta_{l\lambda}^{(h,1)}(i,i) + W_{ih}^{(1)}, \quad (7)$$

$$W_{vl}^{(\lambda)} = X_l^\lambda(v) - M\left[X_l^\lambda(v)\right] - \sum_{\mu=1}^{v-1}\sum_{m=1}^{H}\sum_{j=1}^{N} W_{\mu m}^{(j)}\beta_{mj}^{(l,\lambda)}(\mu,v)$$

$$- \sum_{m=1}^{l-1}\sum_{j=1}^{N} W_{vm}^{(j)}\beta_{mj}^{(l,\lambda)}(v,v) - \sum_{j=1}^{\lambda-1} W_{vl}^{(j)}\beta_{lj}^{(l,\lambda)}(v,v), v = \overline{1,I}; \quad (8)$$

$$D_{l,\lambda}(v) = M\left[\left\{W_{vl}^{(\lambda)}\right\}^2\right] = M\left[X_l^{2\lambda}(v)\right] - M^2\left[X_l^\lambda(v)\right]$$

$$- \sum_{\mu=1}^{v-1}\sum_{m=1}^{H}\sum_{j=1}^{N} D_{mj}(\mu)\left\{\beta_{mj}^{(l,\lambda)}(\mu,v)\right\}^2 - \sum_{m=1}^{l-1}\sum_{j=1}^{N} D_{mj}(v)\left\{\beta_{mj}^{(l,\lambda)}(v,v)\right\}^2 \quad (9)$$

$$- \sum_{j=1}^{\lambda-1} D_{lj}(v)\left\{\beta_{lj}^{(l,\lambda)}(v,v)\right\}^2, v = \overline{1,I};$$

$$\beta_{l\lambda}^{(h,s)}(v,i) = \frac{M\left[W_{vl}^{(\lambda)}\left(X_h^s(i) - M[X_h^s(i)]\right)\right]}{M\left[\left\{W_{vl}^{(\lambda)}\right\}^2\right]}$$

$$= \frac{1}{D_{l\lambda}(v)}\left(M\left[X_l^\lambda(v)X_h^s(i)\right] - M\left[X_l^\lambda(v)\right]M\left[X_h^s(i)\right]\right.$$

$$- \sum_{\mu=1}^{v-1}\sum_{m=1}^{H}\sum_{j=1}^{N} D_{mj}(\mu)\beta_{mj}^{(l,\lambda)}(\mu,v)\beta_{mj}^{(h,s)}(\mu,i) \tag{10}$$

$$- \sum_{m=1}^{l-1}\sum_{j=1}^{N} D_{mj}(v)\beta_{mj}^{(l,\lambda)}(v,v)\beta_{mj}^{(h,s)}(v,i)$$

$$- \sum_{j=1}^{\lambda-1} D_{lj}(v)\beta_{lj}^{(l,\lambda)}(v,v)\beta_{lj}^{(h,s)}(v,i),\ \lambda = \overline{1,h},\ v = \overline{1,i}.$$

Random sequence $X_h(i), i = \overline{1,I}; h = \overline{1,H}$ is presented with the help of $H \times N$ arrays $\{W_l^{(\lambda)}\}, \lambda = \overline{1,N}; l = \overline{1,H}$ of uncorrelated centered random coefficients $W_{vl}^{(\lambda)}, v = \overline{1,I}; \lambda = \overline{1,N}; l = \overline{1,H}$. Each of these coefficients contains information about appropriate value of $X_l^\lambda(v)$ and coordinate functions $\beta_{l\lambda}^{(h,s)}(v,i)$ describe probabilistic relations of $\lambda + s$ degree between the constituents $X_l(i)$ and $X_h(i)$ in the sections t_v and t_i.

Let's assume that as a result of measurement the first value $x_1(1)$ of the constituent $X_1(1)$ of the sequence $\{\vec{X}\}$ in the point t_1 is known. Consequently, the values of the coefficients $W_{11}^{(\lambda)}, \lambda = \overline{1,N}$ are known:

$$w_{11}^{(\lambda)} = x_1^\lambda(1) - M\left[X_1^\lambda(1)\right] - \sum_{j=1}^{\lambda-1} w_{11}^{(j)}\beta_{1j}^{(1,\lambda)}(1,1),\ \lambda = \overline{1,N}. \tag{11}$$

The substitution of $w_{11}^{(1)}$ into (7) allows to get polynomial canonical expansion of the a posteriori random sequence $\left\{\vec{X}^{(1,1)}\right\} = X_h(i/x_1(1))$:

$$X_h^{(1,1)}(i) = X_h(i/x_1(1)) = M[X_h(i)] + (x_1(1) - M[X_1(1)])\beta_{11}^{(h,1)}(1,i)$$

$$+ \sum_{\lambda=2}^{N} W_{11}^{(\lambda)}\beta_{1\lambda}^{(h,1)}(1,i) + \sum_{l=2}^{H}\sum_{\lambda=1}^{N} W_{1l}^{(\lambda)}\beta_{l\lambda}^{(h,1)}(1,i)$$

$$+ \sum_{v=2}^{i-1}\sum_{l=1}^{H}\sum_{\lambda=1}^{N} W_{vl}^{(\lambda)}\beta_{l\lambda}^{(h,1)}(v,i) + \sum_{l=1}^{h-1}\sum_{\lambda=1}^{N} W_{il}^{(\lambda)}\beta_{l\lambda}^{(h,1)}(i,i) + W_{ih}^{(1)},\ i = \overline{1,I}.$$

$$\tag{12}$$

The application of the operation of mathematical expectation to (12) gives the optimal (by the criterion of the minimum of the mean-square error of extrapolation) estimation of the future values of the sequence $\left\{\bar{X}\right\}$ provided that for the determination of the given estimation one value of $x(1)$ is used:

$$m_{x;1,h}^{(1,1)}(1,i) = M[X_h(i/x_1(1))] = M[X_h(i)] + (x_1(1) - M[X_1(1)])\beta_{11}^{(h,1)}(1,i). \quad (13)$$

Taking into account that the coordinate functions $\beta_{l\lambda}^{(h,s)}(v,i)$; $l,h = \overline{1,H}$; $\lambda, s = \overline{1,N}$; $v, i = \overline{1,I}$ are determined from the condition of the minimum of the mean-square error of approximation in the spaces between the random values $X_l^{\lambda}(v)$ and $X_h^s(i)$, the expression (13) can be generalized in case of prognostication $x_h^s(i)$, $s = \overline{1,N}$ $h = \overline{1,H}$, $i = \overline{2,I}$:

$$m_{x;1,h}^{(1,1)}(s,i) = M\left[X_h^s(i/x_1(1))\right] = M\left[X_h^s(i)\right] + (x_1(1) - M[X_1(1)])\beta_{11}^{(h,s)}(1,i). \quad (14)$$

where $m_{x;1,h}^{(1,1)}(s,i)$ is the optimal estimation of the future value $x_h^s(i)$ provided that for the prognosis the value $x(1)$ is used.

Fixation in (12) of the second value $w_1^{(2)}$ gives canonical expansion to the a posteriori sequence $\left\{\bar{X}^{(1,2)}\right\} = X_h\left(i/x_1(1), x_1^2(1)\right)$:

$$X_h^{(1,2)}(i) = X_h\left(i/x_1(1), x_1^2(1)\right) = M[X_h(i)] + (x_1(1) - M[X_1(1)])$$
$$\times \beta_{11}^{(h,1)}(1,i) + \left[x_1^2(1) - (x_1(1) - M[X_1(1)])\beta_{11}^{(1,2)}(1,1)\right]\beta_{12}^{(h,1)}(1,i)$$
$$+ \sum_{\lambda=3}^{N} W_{11}^{(\lambda)}\beta_{1\lambda}^{(h,1)}(1,i) + \sum_{l=2}^{H}\sum_{\lambda=1}^{N} W_{1l}^{(\lambda)}\beta_{l\lambda}^{(h,1)}(1,i) \quad (15)$$
$$+ \sum_{v=2}^{i-1}\sum_{l=1}^{H}\sum_{\lambda=1}^{N} W_{vl}^{(\lambda)}\beta_{l\lambda}^{(h,1)}(v,i) + \sum_{l=1}^{h-1}\sum_{\lambda=1}^{N} W_{il}^{(\lambda)}\beta_{l\lambda}^{(h,1)}(i,i) + W_{ih}^{(1)}, \ i = \overline{1,I}.$$

The application of the operation of mathematical expectation to (15) allows to get the algorithm of extrapolation by two values $x_1(1)$, $x_1^2(1)$ with the usage of the expression (14):

$$m_{x;1,h}^{(1,2)}(s,i) = M\left[X_h^s(i/x_1(1), x_1^2(1))\right] = m_{x;1,h}^{(1,1)}(s,i)$$
$$+ \left(x_1^2(1) - m_{x;1,1}^{(1,1)}(2,1)\right)\beta_{12}^{(h,s)}(1,i), \ i = \overline{1,I}. \quad (16)$$

For the first section t_1 at the last recurrent cycle as a posteriori information about the constituent $X_1(i)$ the value $x_1^N(1)$ is used

$$m_{x;1,h}^{(1,N)}(s,i) = M\left[X_h^s\big(i/x_1(1), x_1^2(1)\ldots, x_1^N(1)\big)\right] = m_{x;1,h}^{(1,N-1)}(s,i)$$
$$+ \left(x_1^N(1) - m_{x,1,1}^{(1,N-1)}(N,1)\right)\beta_{1N}^{(h,s)}(1,i),\ i = \overline{1,I}. \tag{17}$$

After what the prognosis is specified be means of the application of the value $x_2(1)$

$$m_{x;2,h}^{(1,1)}(s,i) = M\left[X_h^s\big(i/x_1(1), x_1^2(1)\ldots, x_1^N(1), x_2(1)\big)\right] = m_{x;1,h}^{(1,N)}(s,i)$$
$$+ \left(x_2(1) - m_{x;1,2}^{(1,N)}(1,1)\right)\beta_{21}^{(h,s)}(1,i),\ i = \overline{1,I}. \tag{18}$$

For the section t_1 the last recurrent cycle is the operation

$$m_{x;H,h}^{(1,N)}(s,i) = M\left[X_h^s\big(i/x_1(1), x_1^2(1), \ldots, x_1^N(1), \ldots, x_H^2(1), \ldots, x_H^{N-1}(1)\big)\right]$$
$$= m_{x;H,h}^{(1,N-1)}(s,i) + \left(x_H^N(1) - m_{x;H,H}^{(1,N-1)}(N,1)\right)\beta_{HN}^{(h,s)}(1,i),\ i = \overline{1,I}.$$

Then the transfer to a new section of t_2 is performed and the specification of prognostic values is carried out by means of the usage of the value $x_1(2)$.

The generalization of the obtained regularity allows to write down the algorithm of the prognosis (of extrapolation) for random number of the known values:

$$m_{x;j,h}^{(\mu,l)}(s,i) = \begin{cases} M[X_h(i)],\ \text{if}\ \mu = 0; \\ m_{x;j,h}^{(\mu,l-1)}(s,i) + \left(x_j^l(\mu) - m_{x;j,j}^{(\mu,l-1)}(l,\mu)\right)\beta_{j,l}^{(h,s)}(\mu,i),\ \text{if}\ l > 1, j < H; \\ m_{x;j,h}^{(\mu,N)}(s,i) + \left(x_{j+1}(\mu) - m_{x;j,j+1}^{(\mu,1)}(N,\mu)\right)\beta_{j+1,1}^{(h,s)}(\mu,i), \\ \text{if}\ l = 1, j < H; \\ m_{x;H,h}^{(\mu,N)}(s,i) + \left(x_1(\mu+1) - m_{x;H,1}^{(\mu,N)}(N,\mu+1)\right) \\ \times \beta_{1,1}^{(h,s)}(\mu+1,i),\ \text{for}\ l = 1, j = H; \end{cases} \tag{19}$$

$$m_{x;j,h}^{(\mu,l)}(1,i) = M\left[X_h(i)/x_\lambda^n(v),\ \lambda = \overline{1,H}, n = \overline{1,N}, v = \overline{1,\mu-1};\ x_\lambda^n(\mu), \lambda = \overline{1,j}, n = \overline{1,l}\right]$$

- is the optimal (in a mean-square sense) estimation of the future values of the investigated random sequence provided that for the prognosis the a posteriori information is applied $x_\lambda^n(v)$, $\lambda = \overline{1,H}$, $n = \overline{1,N}$, $v = \overline{1,\mu-1}$; $x_\lambda^n(\mu)$, $\lambda = \overline{1,j}$, $n = \overline{1,l}$.

The diagram in Fig. 1 reflects the peculiarities of the calculating process during the usage of the prognostic model (19).

The expression for the mean-square error of the extrapolation with the help of the algorithm (19) by the known values $x_j^n(\mu)$, $\mu = \overline{1,k}$; $j = \overline{1,H}$; $n = \overline{1,N}$ is of the form

$$E_h^{(k,N)}(i) = M\left[X_h^2(i)\right] - M^2\left[X_h(i)\right] - \sum_{\mu=1}^{k}\sum_{j=1}^{H}\sum_{n=1}^{N}D_{jn}(\mu)\left\{\beta_{jn}^{(h,1)}(\mu,i)\right\}^2 \tag{20}$$

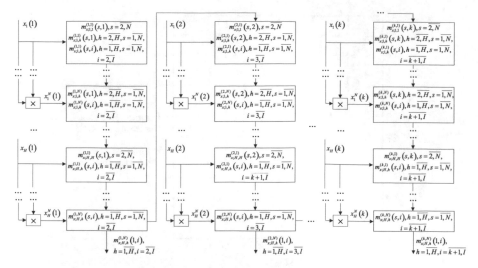

Fig. 1. Diagram of the procedure of the forming of the future values of a random sequence with the help of the algorithm (19)

The mean-square error of the extrapolation $E_h^{(k,N)}(i)$ is equal to the dispersion of the a posteriori random sequence

$$X_h^{(k,N)}(i) = X\left(i/x_l^v(j),\, v=\overline{1,N}, j=\overline{1,k}, l=\overline{1,H}\right) = m_{H,h}^{(k,N)}(1,i) +$$
$$\sum_{v=k+1}^{i-1} \sum_{l=1}^{H} \sum_{\lambda=1}^{N} W_{vl}^{(\lambda)} \beta_{l\lambda}^{(h,1)}(v,i) + \sum_{l=1}^{h-1} \sum_{\lambda=1}^{N} W_{il}^{(\lambda)} \beta_{l\lambda}^{(h,1)}(i,i) + W_{ih}^{(1)},\, i=\overline{k+1,I}.$$

The application of the developed method of extrapolation of the random sequences on the basis of the prognostic model (19) presupposes the realization of the following stages:

Phase 1. Gathering of statistic data about the investigated random sequence;
Phase 2. Estimation of the moment functions $M[X_l^v(i)]$, $M[X_l^v(i)X_h^\mu(j)]$, $i,j=\overline{1,I}$; $l,h=\overline{1,H}$; $v,\mu=\overline{1,N}$ on the basis of the cumulated realizations of the random sequence;
Phase 3. Forming of the canonical expansion (7) for the investigated vector random sequence $\{\vec{X}\} = X_h(i), i,j=\overline{1,I}$; $h=\overline{1,H}$;
Phase 4. Calculation of the estimations of the future values of the extrapolated realization on the basis of the algorithm of prognosis (19);
Phase 5. Estimation of the quality of the prognosis problem solving for the investigated sequence with the help of the expression (20).

In case of the absence of stochastic relations between the constituents the prognostic model (19) is simplified to H expressions [3, 7, 8] for the extrapolation of scalar sequences.

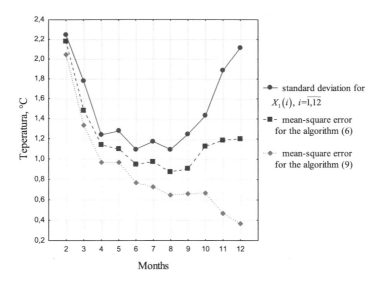

Fig. 2. Mean-square error of the extrapolation of the realizations of the sequence $X_1(i)$, $i = \overline{1, 12}$ with the help of the algorithms (6), (19)

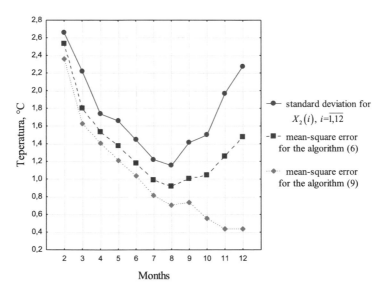

Fig. 3. Mean-square error of the extrapolation of the realizations of the sequence $X_2(i)$, $i = \overline{1, 12}$ with the help of the algorithms (6), (19)

The introduced method is approbated for the prognostication of the random sequences describing the change of the average monthly temperature of the air in the cities of Odessa and Kiev (Ukraine). The values of the average monthly (from January to December) temperature for one hundred years (1910–2009 years) were used as statistic data [22].

Numerical experiment was organized in the following way. On the basis of 99 realizations of the random sequences $X_1(i), X_2(i) \, i = \overline{1,12}$ the parameters of the algorithms were determined (6), (18); for the one remaining realization (from one hundred of those that were available in the base of statistic data) the estimation of future values was calculated and the error of prognosis was determined. The procedure contained one hundred iterations for $N = 4$, at the same time the forecast realization was withdrawn from the training sample and investigated in previous experiment realization was placed on its spot.

In Figs. 2 and 3 mean-square errors of extrapolation of the realizations of the random sequences $X_1(i), X_2(i) \, i = \overline{1,12}$ obtained as a result of numerical experiment with the help of the algorithms (6), (19) are presented.

The results of the numerical experiment show the considerable gain in the accuracy of the prognostication with the help of the method (19) in comparison with (6) at the expense of the usage of nonlinear relations.

4 Conclusions

Thereby the discrete algorithm of the nonlinear extrapolation of the vector random sequence that doesn't put any significant limitations on the class of the investigated sequences: stationarity, Markovian property, linearity, monotony etc. is synthesized by the authors. The universality of the obtained solution is determined by that a canonical expansion exists and describes precisely in the points of discrecity any random process with a final dispersion. The algorithm allows to use the stochastic relations of the random order of nonlinearity and random quantity of measuring results. The given discrete algorithm is optimal in the sense of mean-square criterion.

Taking into account the recurrent character of the calculations of the extrapolator parameters, its implementation with a computer is quite simple. The results of the numerical experiment confirm high accuracy of the developed method of prognostication.

So long as the majority of the investigated physical, technical, economic or other real processes are stochastic, the introduced method has the widest possibilities of the application during the solving of the management problems in different spheres of science and techniques: forecasting control of engineering devices reliability, medical diagnostics, radiolocation, management of technological objects etc.

References

1. Atamanyuk, I.P.: Canonical expansion of vector random process with full account of the mutually correlated relations for each constituent. J. Tech. Sci. **8**, 119–121 (1998)
2. Atamanyuk, I.P.: Vector polynomial canonical expansion of a random process. Tech. Diagn. Nondestr. Control **1**, 36–40 (2003)
3. Atamanyuk, I.P.: The algorithm to determine the optimal parameters of a wiener filter-extrapolator for non-stationary stochastic processes observed with errors. J. Cybern. Syst. Anal. **4**, 154–159 (2011)
4. Atamanyuk, I., Kondratenko, Y.: Computer's analysis method and reliability assessment of fault-tolerance operation of information systems. In: Batsakis, S., et al. (eds.) Proceedings of the 11th International Conference ICT in Education, Research and Industrial Applications: Integration, Harmonization and Knowledge Transfer, ICTERI-2015, CEUR-WS, Lviv, Ukraine, May 14–16, vol. 1356, pp. 507–522 (2015)
5. Atamanyuk, I., Kondratenko, Y.: Calculation method for a computer's diagnostics of cardiovascular diseases based on canonical decompositions of random sequences. In: Batsakis, S., et al. (eds.) Proceedings of the 11th International Conference ICT in Education, Research and Industrial Applications: Integration, Harmonization and Knowledge Transfer, ICTERI-2015, CEUR-WS, Lviv, Ukraine, May 14–16, vol. 1356, pp. 108–120 (2015)
6. Atamanyuk, I.P., Kondratenko, Y.P.: The synthesis of optimal linear stochastic systems of control on the basis of the apparatus of canonical decompositions of random sequences. Control. Syst. Mach. **1**, 8–12 (2012)
7. Atamanyuk, I.P., Kondratenko, Y.P.: The algorithm of optimal nonlinear extrapolation of the realizations of random process with the filtration of errors changes. Electron. Model. **4**, 23–40 (2012)
8. Atamanyuk, I.P., Kondratenko, V.Y., Kozlov, O.V., Kondratenko, Y.P.: The algorithm ofoptimal polynomial extrapolation of random processes. In: Engemann, K.J., et al. (eds.) Proceedings of International Conference Modeling and Simulation in Engineering, Economics and Management, MS 2012. Lecture Notes in Business Information Processing, vol. 115, pp. 78–87. Springer, Heidelberg (2012). https://doi.org/10.1007/978-3-642-30433-0_9
9. Box, G.E.P., Jenkins, G.M.: Time Series Analysis, Forecasting and Control. Holden-Day, San Francisco (1970)
10. Ivakhnenko, A.G., Ivakhnenko, G.A.: The review of problems solvable by algorithms of the group method of data handling. Int. J. Patt. Recog. Image Anal. Adv. Mathem. Theory Appl. **5**(4), 527–535 (1995)
11. Kalman, R.E.: A new approach to linear filtering and prediction problems. Trans. ASME–J. Basic Eng. **82**, 35–45 (1960)
12. Kolmogorov, A.N.: Interpolation and extrapolation of stationary random sequences. J. Proc. Acad. Sci. USSR **5**, 3–14 (1941)
13. Kondratenko, Y.P.: Robotics, automation and information systems: future perspectives and correlation with culture, sport and life science. In: Gil-Lafuente, A.M., Zopounidis, C. (eds.) Decision Making and Knowledge Decision Support Systems. Lecture Notes in Economics and Mathematical Systems, vol. 675, pp. 43–56. Springer, Cham (2015). https://doi.org/10.1007/978-3-319-03907-7_6
14. Kondratenko, Y.P., Timchenko, V.L.: Increase in navigation safety by developing distributed man-machine control systems. In: Proceedings of the Third International Offshore and Polar Engineering Conference, Singapore, vol. 2, pp. 512–519 (1993)

15. Kudritsky, V.D.: Predictive Control of Radioelectronic Devices. Technics, Kyiv (1982)
16. Kudritsky, V.D.: Filtering, Extrapolation and Recognition Realizations of Random Functions. FADA Ltd., Kyiv (2001)
17. Mandel, J., Beezley, J.D., Kochanski, A.K., Kondratenko, V.Y., Kim, M.: Assimilation of perimeter data and coupling with fuel moisture in a wildland fire–atmosphere DDDAS. Proced. Comput. Sci. **9**, 1100–1109 (2012)
18. Simon, D.: Training fuzzy systems with the extended Kalman filter. Fuzzy Sets Syst. **132**, 189–199 (2002)
19. Pugachev, V.S.: The Theory of Random Functions and its Application. Fitmatgiz, Moscow (1962)
20. Tkachenko, A.N., Brovinskaya, N.M., Kondratenko, Y.P.: Evolutionary adaptation of control processes in robots operating in non-stationary environments. Mech. Mach. Theory **18**(4), 275–278 (1983). Printed in Great Britain, https://doi.org/10.1016/0094-114X(83)90118-0
21. Wiener, N.: Extrapolation, Interpolation, and Smoothing of Stationary Time Series: With Engineering Applications. MIT Press, New York (1949)
22. datamarket.com/data/set/1loo/average-monthly-temperatures-across-the-world#q=

Check for
updates

Elastic-Plastic Analysis for a Functionally Graded Rotating Cylinder Under Variation in Young's Modulus

Manoj Sahni[1]($\boxtimes$) and Ritu Sahni[2]

[1] Department of Mathematics, School of Technology,
PDPU, Gandhinagar 382007, Gujarat, India
manoj_sahni17@rediffmail.com
[2] Centre for Engineering and Enterprise, UIAR,
Gandhinagar 382007, Gujarat, India
ritusrivastava1981@gmail.com

Abstract. In engineering applications, pure metals are rarely used because the application may require a material with different properties that is hard as well as ductile. The functionally graded materials are the materials obtained from the composition of two or more different materials, different in properties from the constituent material, to enhance the strength of the resultant material. The concept was introduced in Japan during a space plane project in 1984. Since then, a lot of research work has done in this area under various profiles and under various conditions.

In this paper, the study of the behaviour of variation of Young's modulus is studied against radii. The axisymmetric case is considered in which the Young's modulus is a function of radial co-ordinate only. The radial and circumferential stresses are calculated for different radii ratio and with the parametric change in Young's modulus. An analytical solution for stresses is developed and the results are compared with those available in literature.

Keywords: Rotating cylinder · Young's modulus · Internal pressure
FGM

1 Introduction

In the development of our civilization, materials have played an important role and the society even associate ages with them. Materials have been classified in groups based on the structure or properties. With the development of the industries, much new class of materials is developed called as composite materials. The research on composite materials has started in the past 50 to 60 years. The composite materials are homogeneous mixture of two or more materials with significantly different physical and chemical properties. The development for new materials has been discussed a lot in the scientific community. All engineering and science disciplines need to know about the behavior of materials under external responses. Research has been tremendously increased to study the behavior of materials under various profiles like variation of thickness, density, Poisson's ratio, Young's modulus, etc.

© Springer International Publishing AG, part of Springer Nature 2018
A. M. Gil-Lafuente et al. (Eds.): FIM 2015, AISC 730, pp. 26–39, 2018.
https://doi.org/10.1007/978-3-319-75792-6_3

With the technology development and research in scientific community, the innovative composite materials are developed called as functionally graded materials (FGMs). The term FGM was coined in 1986 by a group of Japanese scientist. Since then FGMs have played very important role in many industrial and defense applications such as compact disc drives, automotive, aircrafts, turbine rotors, flywheels, missiles, aerospace and stainless steel cladding of nuclear pressure vessels [1, 2]. Therefore composites made of FGMs have been attracting considerable attention in recent years.

A new beam finite element for the analysis of functionally graded materials is given by Chakrabarty et al. [3]. Steady-state creep of thick-walled cylindrical vessels made of functionally graded materials subjected to internal pressure is investigated by You et al. [4]. Bayat et al. [5] solved the problem of a functionally graded rotating disc with axisymmetric bending and steady-state thermal loading. The material properties of the disc were assumed to be graded in the direction of radial by a power law distribution of volume fraction of the constituent materials. Nejad et al. [6] used the infinitesimal theory of elasticity for solving the problem of a one-dimensional steady-state thermal stresses in a rotating functionally graded pressurized thick-walled hollow cylinder and obtained the closed form solution under generalized plain strain and plane stress assumptions, respectively. The study by Callioglu et al. [7] deals with stress analysis of functionally graded rotating annular discs subjected to internal pressure and various temperature distributions, such as uniform T0(reference temperature), linearly increasing Tb(outer surface temperature) and decreasing Ta(inner surface temperature) in radial direction. Hoseini et al. [8] described a new analytical solution for the steady state creep in rotating thick cylindrical shells subjected to internal and external pressure. The creep response of the material is governed by Norton's law. An analytical method for predicting elastic–plastic stress distribution in a cylindrical pressure vessel has been presented by Kalali et al. [9]. An analytic solution for functionally graded rotating disc using stress function was developed by Sahni et al. [10] in 2014. The study was to determine deformations and stresses (radial and circumferential) made of functionally graded materials subjected to internal pressure. The paper published by Bohidar et al. [11] concentrated on the literature of FGMs includes historical background of FGMs, its areas of application, processing techniques and future scopes of FGMs. Sharma et al. [12] has solved problem on creep of a thin rotating disc with variable thickness and density with edge loading using transition theory. Using transition theory, the problem of a thin rotating disc made of transversely isotropic material is solved [13].

In this paper, elastic – plastic stresses have been calculated for thick–walled rotating cylinder under internal pressure using classical theory. Results obtained are depicted graphically and compared with the known solution without variation in rotation and external pressure by Chen et al. [14].

2   Governing Equations

Consider a thick–walled axis-symmetric cylinder having internal and external radii as 'a' and 'b' respectively. It is assumed that the graded properties of the cylinder are along the radial direction only. The cylinder is rotating with an angular velocity 'ω'. The Hooke's law (stress – strain relationship) in the elastic region can be expressed as

$$\varepsilon_r = \left(\frac{1-v^2}{E}\right)\left(\sigma_r - \frac{v}{1-v}\sigma_\theta\right) \tag{1}$$

$$\varepsilon_\theta = \left(\frac{1-v^2}{E}\right)\left(\sigma_\theta - \frac{v}{1-v}\sigma_r\right) \tag{2}$$

or

$$\sigma_r = \frac{E(1-v)}{(1-2v)(1+v)}\left(\varepsilon_r + \frac{v}{1-v}\varepsilon_\theta\right) \tag{3}$$

$$\sigma_\theta = \frac{E(1-v)}{(1-2v)(1+v)}\left(\varepsilon_\theta + \frac{v}{1-v}\varepsilon_r\right), \tag{4}$$

and

$$\sigma_z = v(\sigma_r + \sigma_\theta), \tag{5}$$

where E is the Young's modulus and v is the Poisson's ratio.

Here σ_r, σ_θ and σ_z are the radial, circumferential and axial stresses, respectively.

The strain – displacement relation following the geometry of the problem is defined as [14]

$$\varepsilon_r = \frac{du}{dr}, \varepsilon_\theta = \frac{u}{r}, \varepsilon_{r\theta} = 0, \tag{6}$$

where u is the displacement component in the radial (r) direction and ε_r and ε_θ are the strains in radial and tangential directions, respectively.

Using Eq. (6) and eliminating 'u' from them, the compatibility equation is formed as

$$r\frac{d\varepsilon_\theta}{dr} + \varepsilon_\theta - \varepsilon_r = 0. \tag{7}$$

As the problem is axis-symmetric, the stress equilibrium equation inside the FGM cylinder is

$$\frac{d\sigma_r}{dr} + \frac{\sigma_r - \sigma_\theta}{r} + \rho r \omega^2 = 0. \tag{8}$$

From Eq. (8), substituting σ_θ into Eqs. (1) and (2), we get the strains (radial and circumferential) in terms of radial stress as

$$\varepsilon_r = \left(\frac{1+v}{E}\right)\left((1-2v)\sigma_r - vr\frac{d\sigma_r}{dr} - v\rho r^2\omega^2\right) \tag{9}$$

$$\varepsilon_\theta = \left(\frac{1+v}{Ev}\right)\left((1-v)r\frac{d\sigma_r}{dr} + (1-v)\rho r^2\omega^2 + (1-2v)\sigma_r\right). \tag{10}$$

Inserting Eqs. (9) and (10) into compatibility Eq. (7), we get

$$\frac{d^2\sigma_r}{dr^2} + \frac{d\sigma_r}{dr}\left(\frac{3}{r} - \frac{1}{E}\frac{dE}{dr}\right) - \frac{(1-2v)}{(1-v)rE}\frac{dE}{dr}\sigma_r = \frac{\rho r \omega^2}{E}\frac{dE}{dr} - \frac{(3-2v)}{(1-v)}\rho\omega^2 \tag{11}$$

Let

$$E = E_0 r^\phi, \tag{12}$$

where ϕ is the gradation parameter.

Then Eq. (11) reduces to

$$r^2\frac{d^2\sigma_r}{dr^2} + r\frac{d\sigma_r}{dr}(3-\phi) - C\emptyset\sigma_r = \left(\rho\omega^2\phi - \frac{(3-2v)}{(1-v)}\rho\omega^2\right)r^2 \tag{13}$$

The above differential equation is a non-homogeneous Cauchy's differential equation of second order.

The total solution is given as

$$\sigma_r = C_1 r^{\lambda_1} + C_2 r^{\lambda_2} + \left(\frac{\rho\omega^2\phi - \frac{(3-2v)}{(1-v)}\rho\omega^2}{8 - (2+C)\phi}\right)r^2, \tag{14}$$

where $\lambda_1 = \frac{(\phi-2)+\sqrt{(2-\phi)^2+4C\phi}}{2}$, $\lambda_2 = \frac{(\phi-2)-\sqrt{(2-\phi)^2+4C\phi}}{2}$ and $C = \frac{(1-2v)}{(1-v)}$.

The circumferential and axial stresses are calculated from Eqs. (8) and (5), respectively

$$\sigma_\theta = C_1(1+\lambda_1)r^{\lambda_1} + C_2(1+\lambda_2)r^{\lambda_2} + 3r^2\left(\frac{\rho\omega^2\phi - \frac{(3-2v)}{(1-v)}\rho\omega^2}{8-(2+C)\phi}\right) + \rho r^2\omega^2 \tag{15}$$

and

$$\sigma_z = v\left[C_1(2+\lambda_1)r^{\lambda_1} + C_2(2+\lambda_2)r^{\lambda_2} + 4r^2\left(\frac{\rho\omega^2\phi - \frac{(3-2v)}{(1-v)}\rho\omega^2}{8-(2+C)\phi}\right)\right] + v\rho r^2\omega^2. \quad (16)$$

The arbitrary constants C_1 and C_2 can be calculated from the boundary conditions.

$$\begin{aligned} \sigma_r &= -p_i \quad \text{at} \quad r = a \\ \sigma_r &= 0 \quad \text{at} \quad r = b \end{aligned} \quad (17)$$

Using the boundary conditions (17) C_1 and C_2 are calculated as

$$C_1 = \frac{p_i b^{\lambda_2}}{b^{\lambda_1}a^{\lambda_2} - b^{\lambda_2}a^{\lambda_1}} + \left(\frac{a^2 b^{\lambda_2} - b^2 a^{\lambda_2}}{b^{\lambda_1}a^{\lambda_2} - b^{\lambda_2}a^{\lambda_1}}\right)\left(\frac{\rho\omega^2\phi - \frac{(3-2v)}{(1-v)}\rho\omega^2}{8-(2+C)\phi}\right),$$

$$C_2 = \frac{p_i b^{\lambda_1}}{b^{\lambda_2}a^{\lambda_1} - b^{\lambda_1}a^{\lambda_2}} + \left(\frac{a^2 b^{\lambda_1} - b^2 a^{\lambda_1}}{b^{\lambda_2}a^{\lambda_1} - b^{\lambda_1}a^{\lambda_2}}\right)\left(\frac{\rho\omega^2\phi - \frac{(3-2v)}{(1-v)}\rho\omega^2}{8-(2+C)\phi}\right).$$

3   Fully Plastic Stresses (C → 0, v → 1/2)

The fully plastic stresses becomes

$$\sigma_r = C_3 r^{\lambda_3} + C_4 r^{\lambda_4} + \left(\frac{\rho\omega^2\phi - 4\rho\omega^2}{8-2\phi}\right)r^2 \quad (18)$$

$$\sigma_\theta = C_3(1+\lambda_3)r^{\lambda_3} + C_4(1+\lambda_4)r^{\lambda_4} + 3r^2\left(\frac{\rho\omega^2\phi - 4\rho\omega^2}{8-2\phi}\right) + \rho r^2\omega^2 \quad (19)$$

$$\sigma_z = \frac{1}{2}\left[C_3(2+\lambda_3)r^{\lambda_3} + C_4(2+\lambda_4)r^{\lambda_4} + 4r^2\left(\frac{\rho\omega^2\phi - 4\rho\omega^2}{8-2\phi}\right)\right] + v\rho r^2\omega^2, \quad (20)$$

where
$$\lambda_3 = 0, \quad \lambda_4 = \phi - 2,$$

$$C_3 = \frac{p_i b^{\lambda_4}}{b^{\lambda_3}a^{\lambda_4} - b^{\lambda_4}a^{\lambda_3}} + \left(\frac{a^2 b^{\lambda_4} - b^2 a^{\lambda_4}}{b^{\lambda_3}a^{\lambda_4} - b^{\lambda_4}a^{\lambda_3}}\right)\left(\frac{\rho\omega^2\phi - 4\rho\omega^2}{8-2\phi}\right),$$

$$C_4 = \frac{p_i b^{\lambda_3}}{b^{\lambda_4}a^{\lambda_3} - b^{\lambda_3}a^{\lambda_4}} + \left(\frac{a^2 b^{\lambda_3} - b^2 a^{\lambda_3}}{b^{\lambda_4}a^{\lambda_3} - b^{\lambda_3}a^{\lambda_4}}\right)\left(\frac{\rho\omega^2\phi - 4\rho\omega^2}{8-2\phi}\right).$$

4   Numerical Discussion and Illustration

In this paper, stresses – radial and circumferential are calculated by varying geometric parameter $\emptyset$. For a numerical illustration, the following values are taken for calculation:
  a = 0.4 m,   b = 1 m,   v = 0.33, ω = 250, 500 rad/s, $\emptyset$ = 0, 0.25, 0.5, 0.75, 0.9, $\rho = 2000$ Kg/m^3, p = 5, 20 N/m^2.

  Graphs are drawn using MATHEMATICA software. In Figs. 1 and 2, graphs are drawn for stresses – radial and circumferential against radii under internal pressure. It has been observed from Fig. 1 that with the increase in internal pressure the radial stress increases and is zero at the external radii and maximum at the internal radii. With the increase in geometric parameter '$\emptyset$', i.e. from 0 to 0.9, the radial stress increases and is compressive. Young's modulus increases with the increase in value of '$\emptyset$'. In Fig. 2, graph is drawn between circumferential stress and radii. It is seen that circumferential

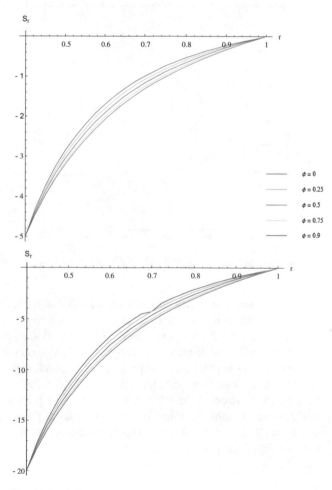

Fig. 1. Radial stress against radii for $p_i = 5, 20$ without rotation.

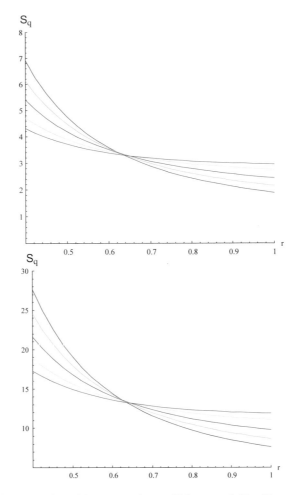

Fig. 2. Circumferential stress against radii for $p_i = 5, 20$ without rotation

stress is maximum at the internal radii and minimum at the external radii. For $\emptyset = 0$ the tangential stress is maximum at the internal radius as compared to other parameters where as reverse behavior is seen at the external radii, i.e. for $\emptyset = 0.9$, it is maximum at the external radii as compared to other values. This behavior change occurs at the point $r = 0.63$ as we can see from the graph. The importance of circumferential stress is to counter the effect of radial stress. For parameter $\emptyset = 0, 0.25, 0.5$ the circumferential stress is more than the radial stress at the internal radii, but as the parameter increases the circumferential stresses becomes less than the radial stresses at the internal radii as we compare Figs. 1 and 2. But at $r = 0.6$ the circumferential stress is more than the radial stress for all values of the geometric parameter '$\emptyset$'.

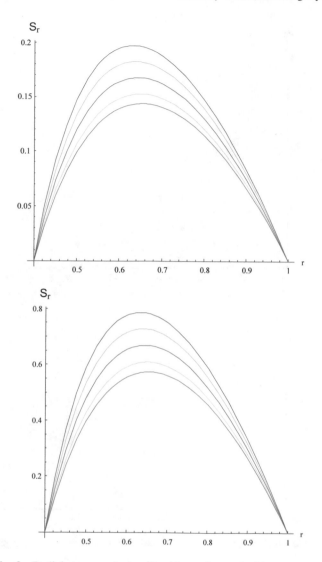

Fig. 3. Radial stress against radii with rotation and without pressure

In Figs. 3 and 4, the effect of rotation is seen against radii keeping pressure as zero. In Fig. 3, the radial stress is zero at the internal and external radii and maximum in between. With the increase in angular speed, the radial and circumferential stresses increases and it is tensile. With the increase in speed of rotation, a significant increase is seen, but circumferential stress is always more than the radial stress. The functionally graded materials with $\emptyset \leq 0.5$ is preferred as radial stress is always less than the circumferential stress.

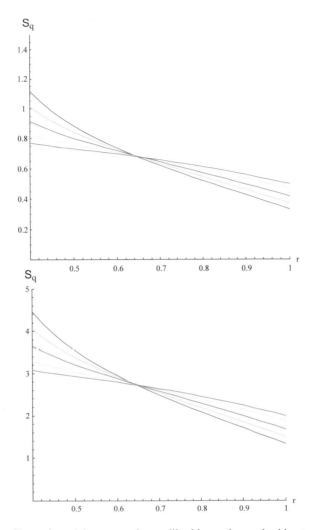

Fig. 4. Circumferential stress against radii with rotation and without pressure

In Figs. 5 and 6, fully plastic stresses are calculated with pressure and without rotation and in Figs. 7 and 8, fully plastic stresses are calculated without pressure and with rotation. The plastic stresses so obtained are more than the elastic stresses as we observed from graphs 5 to 8.

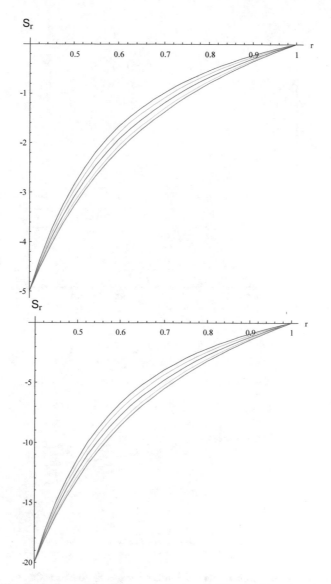

Fig. 5. Fully plastic radial stress for p = 5, 20 and without rotation

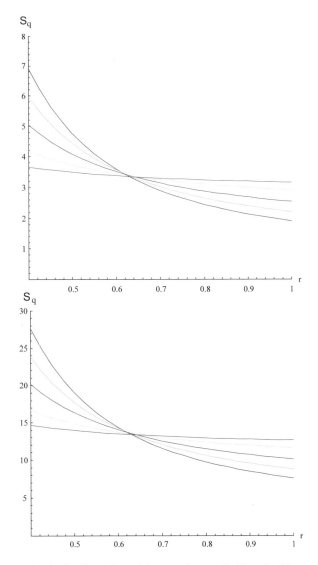

Fig. 6. Fully plastic circumferential stress for p = 5, 20 and without rotation

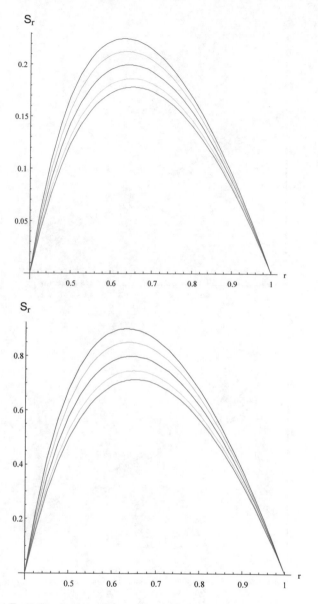

Fig. 7. Fully plastic radial stresses with rotation and without pressure

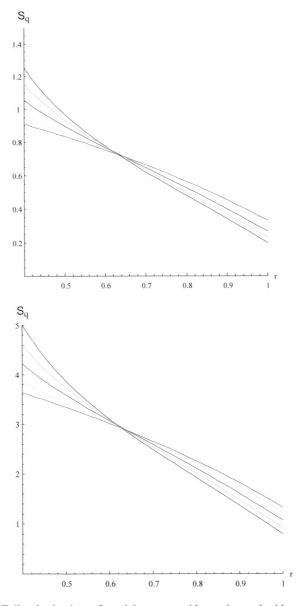

Fig. 8. Fully plastic circumferential stresses with rotation and without pressure

5   Conclusions

With the increase in internal pressure, both the radial and circumferential stresses increases. The increase in parameter '∅' decreases both the stresses. The initial yielding will start from internal radii as the difference between the two stresses is maximum at the internal radii. The circumferential stress is higher for '$\emptyset = 0$' and lower for other values at the internal radii.

References

1. Reddy, J.N., Wang, C.M., Kitipornchai, S.: Axisymetric bending of functionally graded circular and annular plate. Eur. J. Mech. A Solids **18**, 185–199 (1999)
2. Horgan, C.O., Chan, A.M.: The pressurized hollow cylinder or disk problem for functionally graded isotropic linearly elastic materials. J. Elast. **55**, 43–59 (1999)
3. Chakraborty, A., Gopalakrishnan, S., Reddy, J.N.: A new beam finite element for the analysis of functionally graded materials. Int. J. Mech. Sci. **45**, 519–539 (2003)
4. You, L.H., Ou, H., Zheng, Z.Y.: Creep deformations and stresses in thick-walled cylindrical vessels of functionally graded materials subjected to internal pressure. Compos. Struct. **78**, 285–291 (2007)
5. Bayat, M., Saleem, M., Sahari, B.B., Hamouda, A.M.S., Mahdi, E.: Thermo elastic analysis of a functionally graded rotating disk with small and large deflections. Thin-Walled Struct. **45**, 677–691 (2007)
6. Nejad, M.Z., Rahimi, G.H.: Deformation and stresses in rotating FGM pressurized thick hollow cylinder under thermal load. Sci. Res. Essay **4**(3), 131–140 (2009)
7. Callioglu, H., Sayer, M., Demir, E.: Stress analysis of functionally graded discs under mechanical and thermal loads. Indian J. Eng. Mater. Sci. **18**, 111–118 (2011)
8. Hoseini, Z., Nejad, M.Z., Niknejad, A., Ghannad, M.: New exact solution for creep behavior of rotating thick-walled cylinders. J. Basic Appl. Sci. Res. **1**(10), 1704–1708 (2011)
9. Kalali, A.T., Hadidi-Moud, S.: A semi-analytical approach to elastic-plastic stress analysis of FGM pressure vessels. J. Solid Mech. **5**(1), 63–73 (2013)
10. Sahni, M., Sahni, R.: Functionally graded rotating discs with internal pressure. Eng. Autom. Probl. **3**, 125–128 (2014)
11. Bohidar, S.K., Sharma, R., Mishra, P.R.: Functionally graded materials: a critical review. Int. J. Res. **1**(7), 289–301 (2014)
12. Sharma, S., Sahni, M.: Creep analysis of thin rotating disc having variable thickness and variable density with edge loading. Ann. Fac. Eng. Hunedoara Int. J. Eng. **11**, 289–296 (2013)
13. Sharma, S., Sahni, M.: Elastic-plastic transition of transversely isotropic thin rotating disc. Contemp. Eng. Sci. **2**(9), 433–440 (2009)
14. Chen, J.J., Tu, S.T., Xuan, F.Z., Wang, Z.D.: Creep analysis for a functionally graded cylinder subjected to internal and external pressure. J. Strain Anal. **42**, 69–77 (2007)

Mathematical Model of Magnetic Field Penetration for Applied Tasks of Electromagnetic Driver and Ferromagnetic Layer Interaction

Yuriy M. Zaporozhets[1], Yuriy P. Kondratenko[2](✉),
and Volodymyr Y. Kondratenko[3]

[1] Institute of Renewable Energy, National Academy of Sciences of Ukraine,
20A Krasnogvardejskaya Str., Kiev 02094, Ukraine
[2] Department of Intelligent Information Systems, Petro Mohyla Black Sea State
University, 68-th Desantnykiv Str. 10, Mykolaiv 54003, Ukraine
y_kondrat2002@yahoo.com
[3] Department of Mathematical and Statistical Sciences,
University of Colorado Denver, Denver, USA

Abstract. This paper deals with the investigations of an interaction between magnetic driver and ferromagnetic surface based on the calculation of magnetic field parameters in conditions of various thicknesses of layer. Special attention is paid for development of mathematical model of magnetic field penetration through flat soft magnetic layer. Applied aspects of developed mathematical model implementation in robotics, automation of different technological processes, renewable energy equipment and other industrial devices are discussed in the paper in details.

Keywords: Mathematical model · Magnetic field · Penetration
Layer · Thicknesses · Ferromagnetic surface · Electromagnetic driver
Mobile robot

1 Introduction

In many technical applications, the problem arises of calculating the electro-magnetic fields and their interaction with certain objects in the regions of space separated by layers with different permeability of materials. As applied to electrostatics such flat problem was solved by Greenberg [7], by the method of integral equations for the charge density σ induced by exciting field on the media interface ("secondary sources"), using Fourier integral transform. However, its general solution in the form of Fourier representation of functions is unsuitable for direct computation, besides the construction of the direct transformation of the Fourier transform for non-trivial configurations of the exciting field is rather difficult, and often even impossible; especially as it relates to reverse transformation, which, in fact is the real (final) solution. In [7] the final solution is obtained only for a particular case when the external field is created by a solitary charge (thread), and the format of the field is mounted in this solution,

© Springer International Publishing AG, part of Springer Nature 2018
A. M. Gil-Lafuente et al. (Eds.): FIM 2015, AISC 730, pp. 40–53, 2018.
https://doi.org/10.1007/978-3-319-75792-6_4

therefore it cannot be extended to arbitrary configuration of the exciting field, so these results may have only limited application.

At the same time, in the real technical systems and devices to perform a variety of manufacturing operations there are magnetic fields mainly used. Therefore an actual problem is the investigation of regularities of magnetic field penetration through magnetically soft layer and interaction of electromagnetic driver with ferromagnetic surface.

For this purpose, in this paper on the analogy of above-mentioned electrostatic problem it is built the mathematical model for calculation of static magnetic field in unlimited 2-dimensional space divided by infinitely extended layer of magnetically soft material in two parts, in one or both of which it is specified the external exciting field H^0 or disposed in arbitrary way the primary sources of magnetic field, specified by its spatial and magnetic parameters. The subject to determine is resultant field H_p, formed by the interaction of the primary field with magnetically soft layer.

The mathematical model is based on the "secondary sources" method [7, 25], according to which from exposure to exciting field the "secondary sources" – surface "magnetic charges" with density σ, emerge on all surface of any body, in our case – a layer. These conventional "magnetic charges" form the secondary (induced) field H_σ and together with the external one create the resultant field $H_p = H^0 + H_\sigma$. The magnetostatic task consists in finding of charges σ density distribution on the surface S which in 2-dimensional space is defined by the integral equation [7, 19]

$$\sigma(P) = 2\lambda \left[H_n^0(P) + \frac{1}{2\pi} \oint_S \sigma(M) \frac{\partial}{\partial n_P} \left(\ln \frac{1}{r_{PM}} \right) ds_M \right],$$

or, that's the same,

$$\sigma(P) = 2\lambda H_n^0(P) + \frac{\lambda}{\pi} \oint_S \sigma(M) \frac{\cos \psi_P}{r_{PM}} ds_M, \tag{1}$$

where $r_{PM} = \sqrt{(x_P - x_M)^2 + (y_P - y_M)^2}$ – radius vector, directed from integration point M to watch point P on the surface S;

n_P – normal to the surface S at point P;

$\cos \psi_P = \cos(\bar{n}_P, \bar{r}_{PM})$, where $(\bar{n}_P, \bar{r}_{PM})$ – the angle between normal vector $\bar{n}_P$ to the surface S and radius vector $\bar{r}_{PM}$ at watch point;

H_n^0 – normal to the surface S component of external magnetic field strength;

λ – factor of magnetic permeability; $\lambda = (\mu_i - \mu_e)/(\mu_i + \mu_e)$, where μ_i – relative permeability of material in the interior of layer, and μ_e – in the exterior space (for air $\mu_e = 1$).

Accordingly, the secondary field from the charge σ at arbitrary point in space Q is calculated in the same way as a field (potential gradient) of a simple layer of usual charges [25]:

$$H_\sigma(Q) = -\text{grad}_Q \frac{1}{4\pi} \oint_S \sigma(M) \left(\ln \frac{1}{r_{QM}} \right) ds_M. \tag{2}$$

2 Mathematical Model of Magnetic Field Penetration Through Magnetically Soft Layer

For infinitely extended flat layer the closed surface of S is split on two parallel horizontal sites: upper side of layer S_1 and the bottom side S_2 and also two side segments of size h as a layer thickness. However the contribution to integral from infinitely far side segments tends to zero since for magnetic field that has mainly dipole character, value of σ together with a kernel of integral decreases not slower than $1/r^2$.

The settlement model of such task is shown in Fig. 1.

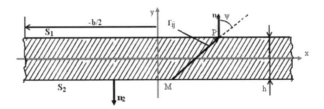

Fig. 1. Settlement model of linear magnetic system ($b \to \infty$)

According to this model, the normals n_1 and n_2 have opposite directions, and the value of $\cos \psi_P = \cos(\bar{n}_P, \bar{r}_{PM})$ when the points P and M are disposed on the same side S_1 or S_2, becomes zero, since in this case, the angle $\psi = \pi/2$; otherwise, the kernel of the integral

$$G_{PM} = \frac{\cos \psi_P}{r_{PM}} = \frac{h}{(x - \xi)^2 + h^2}, \tag{3}$$

where x and ξ – coordinates of points P and M on the surfaces S_1 and S_2 respectively or vice versa.

Then the Eq. (1) is modified in a system of two integrated equations

$$
\begin{cases}
\sigma_1(x) = 2\lambda H_1(x) + \dfrac{\lambda}{\pi} \displaystyle\int_{-\infty}^{\infty} \sigma_2(\xi)\dfrac{h}{(x-\xi)^2 + h^2}\,d\xi \\[3mm]
\sigma_2(x) = 2\lambda H_2(x) + \dfrac{\lambda}{\pi} \displaystyle\int_{-\infty}^{\infty} \sigma_1(\xi)\dfrac{h}{(x-\xi)^2 + h^2}\,d\xi
\end{cases},
\tag{4}
$$

where the indices "1" and "2" correspond to the surfaces of integration S_1 and S_2, and in the notation of functions H additional index n for simplicity is omitted.

Unlike [7] in the presented model for the decision of system (4) we use a method of iterations [18, 27].

For this purpose let us put in consideration the integral operator of such look, as in (4), which is acting on arbitrary function φ that satisfies the Dirichlet conditions [23]:

$$
\frac{1}{\pi}\int_{-\infty}^{\infty}\varphi(\xi)\frac{h}{(x-\xi)^2+h^2}\,d\xi = \frac{1}{\pi}\int_{-\infty}^{\infty}\varphi_j\frac{h}{(x_i-x_j)^2+h^2}\,dx_j = K_{i,j}\varphi_j,
\tag{5}
$$

where evidently $K_{i,j} = K_{j,i} = K$.

Then the system (4) may be represented as follows:

$$
\begin{cases}
\sigma_1 = 2\lambda H_1 + \lambda K_{12}\sigma_2 \\
\sigma_2 = 2\lambda H_2 + \lambda K_{21}\sigma_1
\end{cases}.
\tag{6}
$$

Considering that the choice of initial approach when performing iterations doesn't matter [17, 27] we'll put by turns on each step the value of σ_2 in the first equation, and nextly the value of σ_1 in the second equation of system (6), then for σ_1 will get the following sequence:

$$
\begin{aligned}
\sigma_1 = {} & 2\lambda\Big[H_1 + \lambda^2 K^{(2)}H_1 + \lambda^4 K^{(4)}H_1 + \ldots + \lambda^{2m}K^{(2m)}H_1 + \ldots\Big] \\
& + 2\Big[\lambda^2 K^{(1)}H_2 + \lambda^4 K^{(3)}H_2 \ldots + \lambda^{2m}K^{(2m+1)}H_2 + \ldots\Big],
\end{aligned}
\tag{7}
$$

where $K^{(m)}$ denotes a multiple (m-tuple) application of operator K to appropriate functions

$$
K^{(m)}\varphi = \frac{1}{\pi^m}\int_0^{\infty}\int_0^{\infty}\ldots\int_0^{\infty}\frac{\varphi_j h^m\,dx_{j,1}dx_{j,2}\ldots dx_{j,m}}{\Big[(x_i-x_{j,1})^2+h^2\Big]\Big[(x_i-x_{j,2})^2+h^2\Big]\ldots\Big[(x_i-x_{j,m})^2+h^2\Big]}.
\tag{8}
$$

For σ_2 in the same way we'll get the similar back-symmetric to (7) sequence.

As a result, the equations of system (6) in folded mode considering the expression (7) for functions σ assume such a form

$$
\begin{cases}
\sigma_1 = 2\lambda \displaystyle\sum_{m=0}^{\infty} (\lambda)^{2m} \left[K^{(2m)} H_1 + \lambda K^{(2m+1)} H_2 \right] \\
\sigma_2 = 2\lambda \displaystyle\sum_{m=0}^{\infty} (\lambda)^{2m} \left[K^{(2m)} H_2 + \lambda K^{(2m+1)} H_1 \right]
\end{cases}. \tag{9}
$$

Now let us consider the action of operator (5) to function $\varphi(x)$. As it meets the conditions stated above, it may be presented, according to [23] by the Fourier integral:

$$
\varphi(x) = \frac{1}{\pi} \int_0^{\infty} d\alpha \int_{-\infty}^{\infty} \varphi(t) \cos \alpha(t - x) dt, \tag{10}
$$

Changing in (10) the order of integration, from (5) we get:

$$
I(x) = K\varphi(x) = \frac{1}{\pi} \int_0^{\infty} d\alpha \int_{-\infty}^{\infty} \varphi(t) dt \int_{-\infty}^{\infty} \cos \alpha(t - y) \frac{h}{(x - y)^2 + h^2} dy. \tag{11}
$$

Making the change of variables $x - y = -s$ and expanding the cosine of the difference of arguments $(t - x) - s$, and taking into account that on an infinite interval, the integral over s of the odd function $(\sin \alpha s)$ equal to zero, we find the inner integral in (11), using tabular formula from [6]:

$$
\int_{-\infty}^{\infty} \cos \alpha(t - y) \frac{h}{(x - y)^2 + h^2} dy = \frac{\pi}{h} e^{-\alpha h} \cos \alpha(t - x). \tag{12}
$$

As a result, we get:

$$
I(x) = K\varphi(x) = \frac{1}{\pi} \int_{-\infty}^{\infty} \varphi(t) \int_0^{\infty} e^{-\alpha h} \cos \alpha(t - x) d\alpha dt. \tag{13}
$$

Repeated application of the operator K to received double integral, or equivalently, applying the operator $K^{(2)}$ to function $\varphi(x)$ leads to such expression:

$$
KI(x) = K^{(2)}\varphi(x) = \frac{1}{\pi} \int_{-\infty}^{\infty} \varphi(t) \int_0^{\infty} e^{-2\alpha h} \cos \alpha(t - x) d\alpha dt. \tag{14}
$$

The further continuing of this procedure gives a multiple operator $K^{(m)}$ as follows:

$$K^{(m)}\varphi = \int\limits_0^\infty e^{-m\alpha h}d\alpha \int\limits_{-\infty}^\infty \varphi(t)\cos\alpha(t-x)dt.\tag{15}$$

Then from (7) we get for σ_1 in developed form:

$$\sigma_1(x) = 2\lambda\sum\limits_{m=0}^\infty \lambda^{2m}\left[\begin{array}{l}\int\limits_0^\infty e^{-2m\alpha h}d\alpha \int\limits_{-\infty}^\infty H_1(t)\cos\alpha(t-x)dt+ \\ +\int\limits_0^\infty \lambda e^{-\alpha h}e^{-2m\alpha h}d\alpha \int\limits_{-\infty}^\infty H_2(t)\cos\alpha(t-x)dt\end{array}\right].\tag{16}$$

Since the functions H_1 and H_2 as the components of the magnetic field, in terms of the tasks are harmonic, and therefore, absolutely integrable, then the inner integrals in (16) converge uniformly for all values of the parameter α. Then they can be integrated over this parameter under the integral sign [23], which gives

$$\sigma_1(x) = 2\lambda\int\limits_{-\infty}^\infty dt \int\limits_0^\infty \left[H_1(t) + \lambda e^{-\alpha h}H_2(t)\right]\cos\alpha(t-x)\sum\limits_{m=0}^\infty \lambda^{2m}e^{-2m\alpha h}d\alpha.\tag{17}$$

The expression for density of "magnetic charge" on the reverse side of a flat layer will be analogous – back-symmetrical relatively components of magnetic field:

$$\sigma_2(x) = 2\lambda\int\limits_{-\infty}^\infty dt \int\limits_0^\infty \left[H_2(t) + \lambda e^{-\alpha h}H_1(t)\right]\cos\alpha(t-x)\sum\limits_{m=0}^\infty \lambda^{2m}e^{-2m\alpha h}d\alpha.\tag{18}$$

Finally, the sums in these expressions present no other than a geometric progression, which may be added in known way. Then we obtain:

$$\sigma_1(x) = 2\lambda\int\limits_{-\infty}^\infty dt \int\limits_0^\infty \left[H_1(t) + \lambda e^{-\alpha h}H_2(t)\right]\frac{\cos\alpha(t-x)}{1 - \beta e^{-2\alpha h}}d\alpha;\tag{19}$$

$$\sigma_2(x) = 2\lambda\int\limits_{-\infty}^\infty dt \int\limits_0^\infty \left[H_2(t) + \lambda e^{-\alpha h}H_1(t)\right]\frac{\cos\alpha(t-x)}{1 - \beta e^{-2\alpha h}}d\alpha,\tag{20}$$

where it's denoted $\beta = \lambda^2$.

The set of formulas (19) and (20) is the solution of assigned task. Its feature is that $\sigma_1(x)$ and $\sigma_2(x)$ functions formally turn out mutually independent and are connected just through external exciting field.

Unlike the solution presented in [7, 19], the system (19) and (20) is applicable for any external field, the components of which are included in the formulas in their natural form, and improper integrals in these expressions converge absolutely. Kernel of integral operators in these expressions, in fact, are the resolvents of the original Eq. (4), however, the inner integrals over parameter α of given species by direct integration cannot be expressed in a finite combination of elementary functions [6] and lead to functional series. However, obtained expressions are quite suitable to build computational algorithms, although the calculation of double integrals in infinite limits, as well as the summation of infinite series, is a rather cumbersome process that stretches in time.

Taking into account these circumstances, this paper offers a different approach to solving the problem, which does not require double integration or summation of a long series.

3   Derivation of Versatile Method for Solution of Magnetic Layer Equations in External Magnetic Field

Consider the most common case where a flat layer is ferromagnetic with $\mu \to \infty$ and, respectively, λ and $\beta \to 1$. Then the resolvents in (19) and (20) will be the following integrals:

$$R_1(t,x) = R_1 = \int_0^\infty \frac{\cos \alpha(t-x)}{1 - e^{-2\alpha h}} d\alpha; \quad R_2(t,x) = R_2 = \int_0^\infty \frac{e^{-\alpha h}\cos \alpha(t-x)}{1 - e^{-2\alpha h}} d\alpha. \quad (21)$$

For convenience of further consideration let's designate:

$$(t-x)/2h = \omega; \quad 2h\alpha = s. \quad (22)$$

As known, the function $\cos \varphi$ is the real part of a complex number $z = e^{i\varphi}$: $\cos \varphi = \mathrm{Re}(e^{i\varphi})$, therefore one may write down

$$R_1 = \frac{1}{2h} \int_0^\infty \frac{\cos \omega s}{1 - e^{-s}} ds = \frac{1}{2h} \mathrm{Re} \int_0^\infty \frac{e^{-i\omega s}}{1 - e^{-s}} ds. \quad (23)$$

This integral, as well as previous one cannot be taken in the finite species, but it is associated with special features what we will use. In particular, it is known that the logarithmic derivative of Γ-function (gamma function) in the Gaussian representation (so-called digamma- or just ψ-function) has the form [16, 28]:

$$\psi(z) = \int_0^\infty \frac{e^{-\tau}}{\tau} d\tau - \int_0^\infty \frac{e^{-z\tau}}{1 - e^{-\tau}} d\tau. \quad (24)$$

Then the desired integral in (23) at $z = i\omega$ will be equal to

$$\int\limits_0^\infty \frac{e^{-z\tau}}{1 - e^{-\tau}} d\tau = \int\limits_0^\infty \frac{e^{-\tau}}{\tau} d\tau - \psi(z). \tag{25}$$

The integral in the right part of this expression is no other than limit value of the integral exponential function [1] $\lim\limits_{\xi \to 0} \mathrm{Ei}(-\xi) = \lim\limits_{\xi \to 0} \int\limits_\xi^\infty \frac{e^{-\tau}}{\tau} d\tau$, which does not depend on z (i.e. ω and, respectively, $t - x$), therefore when it's being put in the integrals in (19) and (20), it features as a numeric constant, say, Θ. Then the above integrals, despite the extending of Θ to infinity, turn in the following:

$$\int\limits_{-\infty}^\infty H_1(t)\Theta dt = \Theta \int\limits_{-\infty}^\infty H_1(t)dt = 0; \quad \int\limits_{-\infty}^\infty H_2(t)\Theta dt = \Theta \int\limits_{-\infty}^\infty H_2(t)dt = 0. \tag{26}$$

Zero value of integrals in this case is determined by Gauss's theorem for a magnetic field – the full magnetic flux through infinitely extended surface (H_1 and H_2 are the field components, normal to this surface) equals to zero. But actually from infinity in expression (25), i.e. addend $\mathrm{Ei}(0)$, it is easy to get rid – enough to take away of it the function similar (23), with an exponent indicator which does not depend on x and t (that is ω), in particular, $\zeta = 1$. In this case, according to [15] we get:

$$\int\limits_0^\infty \frac{e^{-z\tau}}{1 - e^{-\tau}} d\tau - \int\limits_0^\infty \frac{e^{-\tau}}{1 - e^{-\tau}} d\tau = \int\limits_0^\infty \frac{e^{-\tau}}{\tau} d\tau - \psi(z) - \int\limits_0^\infty \frac{e^{-\tau}}{\tau} d\tau + \psi(1) = -\psi(z) - C,$$

$$\tag{27}$$

where C – known constant of Euler-Mascheroni; C = 0,577215665 [1, 29].

This result coincides with the integral given in [6, p. 318, p. 3.311 (6)]. Thus, the desired function is determined by the following expression:

$$R_1(t, x) = -\frac{1}{2h} \mathrm{Re}\, \psi(i\omega) - \frac{C}{2h}, \tag{28}$$

but after the substitution of expression (28) in external integrals at system (19) and (20) the constant C, as well as in (26), will give zero therefore in a resolvent formula it may be ignored.

Thus, the first component of the integral operator from (21)

$$R_1(t, x) = R_1(\omega) = -\frac{1}{2h} \mathrm{Re}\, \psi(i\omega). \tag{29}$$

In [1, 22, 29] an explicit expression of the real part ψ-functions is given, that is:

$$\text{Re}\,\psi(iy) = -C + y^2 \sum_{k=1}^{\infty} \frac{1}{k(k^2 + y^2)}, \tag{30}$$

where the sum of series has a finite appearance [21]:

$$\sum_{k=1}^{\infty} \frac{1}{k(k^2 + y^2)} = \frac{1}{2y^4} - \frac{\pi}{2y^3}\,\text{cth}\,\pi y + \frac{\pi^2}{6y^2}.$$

Considering that the last summand in this formula, together with the constant C when integrating in (19)–(20) will be zero, according to (30) we write down the final expression for the first component of the resolvent:

$$R_1(t,x) = R_1(\omega) = -\frac{1}{2h}\left(\frac{1}{2\omega^2} - \frac{\pi}{2\omega}\,\text{cth}\,\pi\omega\right)$$
$$= -\frac{1}{2h}\left[\frac{(2h)^2}{2(t-x)^2} - \frac{2\pi h}{2(t-x)}\,\text{cth}\,\pi\frac{t-x}{2h}\right],$$

that is

$$R_1(t,x) = \left[\frac{\pi}{2(t-x)}\,\text{cth}\,\pi\frac{t-x}{2h} - \frac{h}{(t-x)^2}\right]. \tag{31}$$

The second component of the resolvent R_2 according to (21) differs from the first one by presence of additional multiplicand e^{-zh} under the integral, thereby taking into account (22) the function z in the exponent indicator in (23) becomes a fractional:

$$R_2 = \frac{1}{2h}\,\text{Re}\int_0^{\infty} \frac{e^{-(\frac{1}{2}+i\omega)s}}{1 - e^{-s}}\,ds, \tag{32}$$

what ensues that we obtain, similar to (27)–(29), ψ-function with fractional argument. As regards to such function the transformations (30)–(31) won't take place therefore it is expedient to make permutation in expressions (7) in sequences with odd indices of operator K, namely: to perform the first operation $K^{(1)}H_i = H_i^{(1)}$ and to produce the subsequent actions already over function $H_i^{(1)}$ (i takes values 1 and 2). Then the system (9) gains another form:

$$\sigma_1 = 2\lambda \sum_{m=0}^{\infty} \lambda^{2m} K^{(2m)}\left[H_1 + \lambda H_2^{(1)}\right];$$
$$\sigma_2 = 2\lambda \sum_{m=0}^{\infty} \lambda^{2m} K^{(2m)}\left[H_2 + \lambda H_1^{(1)}\right]. \tag{33}$$

As a result the integral form of equations in system (31)–(32), also transforms and according to (33) becomes like this:

$$\sigma_1(x) = 2\lambda \int\limits_{-\infty}^{\infty} dt \int\limits_0^{\infty} \left[H_1(t) + \lambda H_2^{(1)}(t)\right] \frac{\cos \alpha(t-x)}{1 - \beta e^{-2\alpha h}} d\alpha;$$

$$\sigma_2(x) = 2\lambda \int\limits_{-\infty}^{\infty} dt \int\limits_0^{\infty} \left[H_2(t) + \lambda H_1^{(1)}(t)\right] \frac{\cos \alpha(t-x)}{1 - \beta e^{-2\alpha h}} d\alpha.$$

(34)

Eventually, as the system (34) contains a uniform kernel, then the resolvent for it will be shared – one that is defined by the expression (31): $R = R_1$.

If now to denote:

$$H_1 + \lambda H_2^{(1)} = H_{1,2} \text{ and } H_2 + \lambda H_1^{(1)} = H_{2,1},$$

(35)

(taking into account that H_2 projection on the normal n_2 has the inverse sign in relation to H_1), finally, the solution of considered magnetostatic problem in which the magnetic permeability of magnetically soft layer approaches to 1 ($\lambda = 1$), gets the fully completed form:

$$\sigma_1(x) = 2\lambda_1 \int\limits_{-\infty}^{\infty} H_{1,2}(t) \left[\frac{\pi}{2(t-x)} \operatorname{cth} \pi \frac{t-x}{2h} - \frac{h}{(t-x)^2}\right] dt;$$

$$\sigma_2(x) = 2\lambda_2 \int\limits_{-\infty}^{\infty} H_{2,1}(t) \left[\frac{\pi}{2(t-x)} \operatorname{cth} \pi \frac{t-x}{2h} - \frac{h}{(t-x)^2}\right] dt.$$

(36)

The kernels of integral operators in (36) appear as singular (discontinuous), but actually they have no singularity, as it's easily to make certain by expanding the function cth $\pi\omega$ ($\omega = (t - x)/2\,h$ – see (22)) in the partial fractions [6]

$$\operatorname{cth} \pi\omega = \frac{1}{\pi\omega} + \frac{2\omega}{\pi} \sum\limits_{k=1}^{\infty} \frac{1}{\omega^2 + k^2},$$

as a result from (36) we get:

$$\sigma_1(x) = 2\lambda \int\limits_{-\infty}^{\infty} H_{1,2}(t) \sum\limits_{k=1}^{\infty} \frac{2h}{(2kh)^2 + (t-x)^2} dt;$$

$$\sigma_2(x) = 2\lambda \int\limits_{-\infty}^{\infty} H_{2,1}(t) \sum\limits_{k=1}^{\infty} \frac{2h}{(kh)^2 + (t-x)^2} dt.$$

(37)

As one can see, at coincidence of coordinates t and x, none of series members does undergo discontinuity, and the value of the kernel in this case is determined by the sum of the series [21]

$$\left[\frac{\pi}{2(t-x)}\operatorname{cth}\pi\frac{t-x}{2h}-\frac{h}{(t-x)^2}\right]\Bigg|_{t=x}=\frac{1}{2h}\sum_{k=1}^{\infty}\frac{1}{k^2}=\frac{\pi^2}{12h}. \tag{38}$$

Therefore, when performing numerical integration using Eq. (36), the computational procedure has to be provided the process branching at $t = x$ on transition to formula (38). To calculate the kernel by the formula (37) is hardly expedient, because a convergence of the series essentially depends on the ratio of $(t - x)/2\ h$.

Thus, the algorithm of calculation of the magnetic field in an unbounded space in the presence of separating flat magnetically soft layer for an arbitrary configuration of the exciting field H^0 is reduced to calculation according to specified parameters by means of operator (14) and ratio (35) of combinations of the normal component double-side distribution of this field on surfaces S_1 and S_2, the calculation of density distribution of secondary sources σ on these surfaces and the calculation of induced field H_σ based on (6), and then the resulting field $H_p = H^0 + H_\sigma$ in any point of space. All the above calculations are quite affordable for well-known programs and software, such for example, as Excel, MathCad, Mathematica and so on [5, 26, 32].

4 Applied Aspects of Developed Mathematical Model Implementation

The calculations of desired magnetic field parameters based on the developed (36) and (37) mathematical model (MM) deals with the design and optimization of magnetic systems for different electromagnet drivers which can be used for automation of various technological processes. In particular, such electromagnet drivers and their components can be used: in robotics [9] for design of robot's sensor systems [10], control systems [24], grippers [30], arms and moving bases [11]; in applied technological systems based on the electromagnetic vibration-damper, vibration-exciter or magnetically supporting plates [8]; in equipment for applied use in the field of renewable energy [33] and so on.

As example of developed MM's implementation it is possible to consider modern mobile robot with magnetically operated propulsion wheels [12, 13, 32]. The wheels should have high controlling characteristics for operating on the ferromagnetic surfaces with different positions (horizontal, vertical, under various corners, etc.). Problems of control over magnetically operated drivers of mobile robots can be successfully solved by sequent using developed (36) and (37) mathematical model, methods for calculation and modeling of power components of electromagnetic wheel-drivers [32] and formation of efficient control signals for mobile robot's movement and maneuvering [2, 4].

Embedded control systems for such kind of mobile robots can be designed using FPGA-platform [3, 14] and reconfiguration technology [20]. Modeling results [30, 31] for magnetic sensors and trajectory control of mobile robot with electromagnetic driver confirm the efficiency of MM (36) and (37) application for investigations of interaction between wheel-drive and ferromagnetic surface with different parameter h of the magnetic layer.

5 Conclusions

Mathematical model (36) and (37), synthesized in this article, is efficient for investigation of parameters, characteristics and peculiarities of magnetic interaction between magnet driver and working surface with different thicknesses. The usage of the developed analytical models (36) and (37) has significant advantage for accuracy of calculations, time of modeling and program implementation of the formed models in comparison with models with layer parameter $h \rightarrow \infty$.

Suggested approach can be used for design of different types of electromagnetic drivers in robotics, renewable energy systems, automation of production and technological processes, etc.

Modeling results confirm the efficiency of proposed universal analytic models for solving research problems, partly presented in Sect. 4.

References

1. Abramovich, M., Stegun, I. (eds.): Handbook of Mathematical Functions with Formulas, Graphs and Mathematical Tables (1979). Ed. Nauka, Moscow (In Russian)
2. Collon, C., Rudolph, J.: Invariant feedback control for the kinematic car on the sphere. Syst. Control Lett. **61**(10), 967–972 (2012)
3. Drozd, J., Drozd, A.: Models, methods and means as resources for solving challenges in co-design and testing of computer systems and their components. In: Proceedings of the 9th International Conference on Digital Technologies 2013, Zhilina, Slovak Republic, 29–31 May, pp. 225–230 (2013)
4. Franke, M., Rudolph, J., Woittennek, F.: Motion planning and feedback control of a planar robotic unicycle model. Meth. Models Autom. Robot. **14**(1), 501–506 (2009)
5. Gerdt, V.P., Prokopenya, A.N.: Simulation of quantum error correction with Mathematica. In: Lecture Notes in Computer Science, vol. 8136, pp. 116–129. Springer, Berlin (2013)
6. Gradshtein, I.S., Ryzhik, I.M.: Tables of Integrals, Series and Products. Academic press, Cambridge (1963). Fizmatgiz, Moscow (In Russian)
7. Greenberg, G.A.: Selected Topics of Mathematical Theory of Electrical and Magnetic Phenomena. AS of USSR, Moscow (1948). (In Russian)
8. Kiltz, L., Join, C., Mboup, M., Rudolph, J.: Fault-tolerant control based on algebraic derivative estimation applied on a magnetically supported plate. Control Eng. Pract. **26**, 107–115 (2014)
9. Kondratenko, Y.P.: Robotics, automation and information systems: future perspectives and correlation with culture, sport and life science. In: Lecture Notes in Economics and Mathematical Systems, vol. 675, pp. 43–56 (2015)
10. Kondratenko, Y., Shvets, E., Shyshkin, O.: Modern sensor systems of intelligent robots based on the slip displacement signal detection. In: Annals of DAAAM for 2007 and Proceedings of the 18th International DAAAM Symposium on Intelligent Manufacturing and Automation, Vienna, pp. 381–382 (2007)

11. Kondratenko, Y., Gordienko, E.: Neural networks for adaptive control system of caterpillar turn. In: Annals of DAAAM for 2011 and Proceedings of the 22nd International DAAAM Symposium on Intelligent Manufacturing and Automation, Vienna, Austria, 20–23 October 2011, pp. 305–306 (2011)
12. Kondratenko, Y.P., Zaporozhets Y.M.: Propulsion wheel of mobile robot, UA Patent 45369 U, Ukraine, Bulletin 21 (2009). (In Ukrainian)
13. Kondratenko, Y.P., Zaporozhets, Y.M., Kondratenko, V.Y.: Method of magnetically operated displacement of mobile robot, UA Patent 47369 U, Ukraine, Bulletin 2 (2010). (In Ukrainian)
14. Kondratenko, Y., Gordienko, E.: Implementation of the neural networks for adaptive control system on FPGA. In: Annals of DAAAM for 2012 and Proceeding of the 23th International DAAAM Symposium on Intelligent Manufacturing and Automation 23(1), 389–392 (2012)
15. Kratzer, A., Franz, V.: Transcendental functions. Publ. House of foreign literature, Moscow (1963). (In Russian)
16. Lebedev, N.N.: Special functions and their applications. Fizmatgiz, Moscow (1963). (In Russian)
17. Lizorkin, P.I.: A course of differential and integral equations with additional chapters of analysis. Nauka, Moscow (1981). (In Russian)
18. Manzhirov, A.V., Polyanin, A.D.: Handbook of Integral Equations: Methods of Solution. Publishing House, Moscow (2000). (In Russian)
19. Mirolyubov, N.N., Kostenko, M.V., Levinshtein, M.L., Tikhodeev, N.N.: Methods for calculating the electrostatic fields. Higher School, Moscow (1963). (In Russian)
20. Palagin, A., Opanasenko, V.: Reconfigurable computing technology. J. Cybernetics and Systems Analysis 43(5), 675–686 (2007)
21. Prudnikov, A.P., Brychkov, Y.A., Marichev, O.I.: Integrals and Series. Elementary Functions. Fizmatlit, Moscow (2002). (In Russian)
22. Prudnikov, A.P., Brankov, Y.A., Marichev, O.I.: Integrals and Series. Special Functions. Additional Chapters, vol. 3. Fizmatlit, Moscow (2003). (In Russian)
23. Smirnov, V.I.: A Course of Higher Mathematics, vol. 2. Nauka, Moscow (1974). (In Russian)
24. Tkachenko, A.N., Brovinskaya, N.M., Kondratenko, Y.P.: Evolutionary adaptation of control processes in robots operating in non-stationary environments. Mech. Mach. Theor. 18(4), 275–278 (1983)
25. Tozoni, O.V.: Method of secondary sources in electrical engineering. Energy, Moscow (1975). (In Russian)
26. Vasiliev, A.N.: Scientific computing in Microsoft Excel. Publ. House, Moscow (2004). (In Russian)
27. Verlan, A.F., Sizikov, V.S.: Methods of Solution of Integral Equations with Computer Programs. Naukova Dumka, Kiev (1978). (In Russian)
28. Whittaker, E.T., Watson, J.N.: A Course of Modern Analysis. Part two. Transcendental Functions. Fizmatgiz, Moscow (1963). (In Russian)
29. Yanke, E., Emde, F., Lesh, F.: Special Functions: Formulas, Graphics, Tables. Nauka, Moscow (1964). (In Russian)
30. Zaporozhets, Y.M., Kondratenko, Y.P., Shyshkin, O.S.: Mathematical model of slip displacement sensor with registration of transversal constituents of magnetic field of sensing element. Tech. Electrodynamics 4, 67–72 (2012). (In Ukrainian)
31. Zaporozhets, Y.M., Kondratenko, V.Y., Kondratenko, Y.P.: Peculiarities of computer modeling of components of magnetic field near pole faces. Electron. Model. 35(2), 95–108 (2013)

32. Zaporozhets, Y.M., Kondratenko, Y.P.: Problems and features of control over magnetically operated drivers of mobile robots. Electron. Model. **35**(5), 109–122 (2013)
33. Zaporozhets, Y.M.: On the problem of magnetic field penetration through flat soft magnetic layer and its applied use in field of renewable energy. Altern. Energ. Ecol. **23**(163), 12–24 (2014)

Stress Analysis of a Pressurized Functionally Graded Rotating Discs with Variable Thickness and Poisson's Ratio

Manoj Sahni[1(✉)] and Ritu Sahni[2]

[1] Department of Mathematics, School of Technology, PDPU,
Gandhinagar 382007, Gujarat, India
manoj_sahani17@rediffmail.com
[2] Centre for Engineering and Enterprise, UIAR,
Gandhinagar 382007, Gujarat, India
ritusrivastava1981@gmail.com

Abstract. This paper deals with the analytical study of stress analysis and effect of variable thickness with variation in Young's modulus and constant Poisson ratio of Pressurized Functionally Graded Rotating Discs. In another case, the functionally graded material with constant thickness and variable Poisson ratio is studied, so that the effects of Poisson ration can be analyzed. Stress analysis has been done on the rotating discs of constant thickness as well as the varying thickness and it comes out that variable thickness discs perform better than constant thickness with varying Poisson's ration as it shows a significant decrease in stresses.

Keywords: Annular disc · Thickness · FGM · Poisson's ratio

1 Introduction

In recent years Functionally Graded composite material have been attracting many researchers as it plays very important role in many industrial applications. Basically composite materials have superior and unique properties because they are spatially constructed by suppressing least desirable properties and combining most desirable properties. Most of the materials consist of impurities, so they are not suitable for various applications purpose, hence these materials are developed by removing those impurities and adding suitable contents so that they becomes flexible and suitable for the spatial requirements. Functionally Graded composite material (FGCM) is also designed by varying the composition, microstructure from one material to another material across the volume so that the resultant material have the best of both materials. The resultant material has varying properties such as change in chemical, mechanical, thermal and electrical properties which strengthen the material and may be used to avoid corrosion, fatigue, fracture and other problems. Different FGM materials are designed depending on the areas of applications.

In this paper we are dealing with Functionally Graded Rotating Discs with variable thickness. A variable thickness rotating disk is extensively been used in many industrial

© Springer International Publishing AG, part of Springer Nature 2018
A. M. Gil-Lafuente et al. (Eds.): FIM 2015, AISC 730, pp. 54–62, 2018.
https://doi.org/10.1007/978-3-319-75792-6_5

problems of interest such as in flywheels, gears, in steam and gas turbine rotors, air cleaning machines, in the food processing equipments, in the generation of electric power, disc lasers device in medical, and many more. In the study of discs, stress analysis at high speed is very important to study the sensitivity of the disc stability under various circumstances. The literature reveals that lots of work has been done on the problems related to stress analysis of FGM discs. Further elastic – plastic and creep behaviour of circular disks (solid and annular) are investigated by many researchers [1–4, 9, 11].

Eraslan and Orcan [2] obtained analytical solution for elastic-plastic deformation and stresses of linearly hardening rotating solid disk of exponentially variable thickness. Here the annular region of the disc is divided into three regions –plastic (inner and outer) and elastic. In continuation of the work done by Eraslan in 2002, a closed-form solution is obtained by Eraslan [3] for the stress distribution in rotating hyperbolic solid disk under plane stress. In this research paper, the plastic core consists of three different plastic regions with different mathematical forms of the yield criterion. The authors Nie et al. [5] has studied the tailoring of volume fractions of constituents in functionally graded hollow cylinders and spheres. The results obtained by the authors help the structural engineers and material scientists to optimally design inhomogeneous cylinders and spheres. The creep mathematical model has been developed by Deepak et al. [1] to investigate the steady state creep under variable thickness rotating disc made of functionally graded SiCp.

The tangential and radial strain rates in FGM discs with linear and hyperbolic thickness profiles are respectively lower by about two and three orders of magnitude when compared to a constant thickness FGM disc. A disc made of Al-SiCp having different thickness profiles and reinforcements gradients varying from inner to outer radii have been studied by Khanna et al. [4].

Seth transition theory [6], a new theory developed for elastic-plastic and creep transition which is different from classical theory in 1962. This theory is applied to many research problems related to disk (solid and annular), cylinders (thin and thick), shells and plates, etc. It is different from classical theory where no incompressibility condition, no yielding criteria are assumed. Sharma et al. [7, 8] using transition theory solve problem of rotating cylinder made of transversely isotropic cylinder which was subjected to pressure at internal radii with and without temperature. A problem of creep behaviour with density and thickness variation with edge loading was solved by Sharma et al. [9]. In continuation of the above the transitional stresses are calculated for a thick-walled circular cylinder [10] under external pressure. With the variation in Young's modulus under the effect of pressure, stress analysis was done by Sahni et al. [11].

In this paper, we have studied the effects of variation of thickness and young's modulus parameter keeping Poisson's ration as constant. On the other hand, the thickness is taken constant and the Poisson ration variation with the Young's modulus is studied.

2 Mathematical Formulation

Consider an annular disc of internal and external radii 'a' and 'b' respectively. The disc is rotating with an angular speed 'ω'.

The equation of equilibrium in two dimensions are given as,

$$\frac{\partial \sigma_{rr}}{\partial r} + \frac{1}{r}\frac{\partial \tau_{r\theta}}{\partial \theta} + \frac{\sigma_{rr} - \sigma_{\theta\theta}}{r} + \rho\omega^2 r = 0, \tag{1}$$

$$\frac{\partial \tau_{r\theta}}{\partial r} + \frac{1}{r}\frac{\partial \sigma_{\theta\theta}}{\partial \theta} + 2\frac{\tau_{r\theta}}{r} = 0, \tag{2}$$

where σ_{rr} and $\sigma_{\theta\theta}$ is the radial and hoop stresses respectively. Here $\tau_{r\theta}$ is the shearing stress of the disc, ρ is the density and ω is the angular speed of the rotation.

An axis-symmetric problem is considered in which the stresses are independent of theta and the Eq. (1) reduces to Eq. (3) and Eq. (2) vanishes.

$$\frac{d\sigma_{rr}}{dr} + \frac{\sigma_{rr} - \sigma_{\theta\theta}}{r} + \rho\omega^2 r = 0. \tag{3}$$

When the thickness variation, the above Eq. (3) becomes

$$\frac{d(hr\sigma_{rr})}{dr} - h\sigma_{\theta\theta} + h\rho\omega^2 r^2 = 0, \tag{4}$$

where h is the thickness of the disc varying with position vector 'r'.

In classical theory, the compatibility condition is to be satisfied given as

$$\epsilon_r = \frac{d(r\epsilon_\theta)}{dr} \tag{5}$$

where ϵ_r and ϵ_θ is the strain along the radial and circumferential direction.

The Hooke's law for functionally graded materials is defined as

$$\epsilon_r = \frac{1}{E(r)}(\sigma_{rr} - \upsilon(r)\sigma_{\theta\theta}) \quad \text{and} \quad \epsilon_\theta = \frac{1}{E(r)}(\sigma_{\theta\theta} - \upsilon(r)\sigma_{rr}) \tag{6}$$

where $\upsilon(r)$ is the Poisson's ratio varying radially.

The stress function method is used to find the stresses along radial direction. The stress function is defined as to satisfy the equilibrium equation

$$\sigma_{rr} = \frac{F[r]}{r * h[r]}, \quad \text{and} \quad \sigma_{\theta\theta} = \frac{F'[r]}{h[r]} + \rho\omega^2 r^2 \tag{7}$$

where F[r] is the stress function.

The boundary conditions are defined as

$$\sigma_{rr} = -p_i \quad \text{at} \quad r = a$$
$$\sigma_{rr} = -p_0 \quad \text{at} \quad r = b. \tag{8}$$

where p_i and p_0 are the internal and external pressure respectively.

Case 1: Under the variation of thickness and moduli
Considering the Poisson ration as constant and the thickness and Young's modulus varying linearly as

$$E[r] = E0 * r, \ h[r] = h0 * r \tag{9}$$

Using (9), (7) and (6) into (5), we get the stress function as

$$F[r] = r\left(-\frac{h0r^3\rho(2+\upsilon)\omega^2}{7+2\upsilon} + r^{-\sqrt{2-2\upsilon}}C[1] + r^{\sqrt{2-2\upsilon}}C[2]\right).$$

Here $C[1]$ and $C[2]$ are arbitrary constants, which are calculated using boundary conditions given as

$$
\begin{aligned}
F[r] = {} & (h0r^{1-\sqrt{2-2\upsilon}}(ab(b^{2\sqrt{2-2\upsilon}}r^{3+\sqrt{2-2\upsilon}} - b^{3+\sqrt{2-2\upsilon}}r^{2\sqrt{2-2\upsilon}} + a^{2\sqrt{2-2\upsilon}}(b^{3+\sqrt{2-2\upsilon}} \\
& - r^{3+\sqrt{2-2\upsilon}}) + a^{3+\sqrt{2-2\upsilon}}(-b^{2\sqrt{2-2\upsilon}} + r^{2\sqrt{2-2\upsilon}}))\rho(2+\upsilon)\omega^2 \\
& + r^2(7+2\upsilon)(-ab^{\sqrt{2-2\upsilon}}(a^{2\sqrt{2-2\upsilon}} - r^{2\sqrt{2-2\upsilon}})p_0 \\
& + a^{\sqrt{2-2\upsilon}}b(b^{2\sqrt{2-2\upsilon}} - r^{2\sqrt{2-2\upsilon}})p_i)))/(ab(a^{2\sqrt{2-2\upsilon}} - b^{2\sqrt{2-2\upsilon}})(7+2\upsilon)).
\end{aligned}
\tag{10}
$$

If the thickness and Young's modulus varying non-linearly as

$$E[r] = E0 * r^2, \quad h[r] = h0 * r^2. \tag{11}$$

Using again (11), (7) and (6) into (5) and using the boundary condition, we get

$$
\begin{aligned}
F[r] = {} & -\frac{1}{4a^2b^2\left(a^{2\sqrt{5-4\upsilon}} - b^{2\sqrt{5-4\upsilon}}\right)}h0r^{2-\sqrt{5-4\upsilon}}\left(a^2b^2\left(a^{2\sqrt{5-4\upsilon}}\right.\right. \\
& \left.- b^{2\sqrt{5-4\upsilon}}\right)r^{3+\sqrt{5-4\upsilon}}\rho\omega^2 \\
& + r^{2\sqrt{5-4\upsilon}}\left(a^2b^{\sqrt{5-4\upsilon}}(b^5\rho\omega^2 - 4r^3p_0)\right. \\
& \left.- a^{\sqrt{5-4\upsilon}}b^2(a^5\rho\omega^2 - 4r^3p_i)\right) \\
& + a^{\sqrt{5-4\upsilon}}b^{\sqrt{5-4\upsilon}}\left(a^{2+\sqrt{5-4\upsilon}}(-b^5\rho\omega^2 + 4r^3p_0)\right. \\
& \left.\left.+ b^{2+\sqrt{5-4\upsilon}}(a^5\rho\omega^2 - 4r^3p_i)\right)\right).
\end{aligned}
\tag{12}
$$

Case 2: Under the variation of Young's modulus and Poisson ratio

Here the thickness of the disc is taken constant and Young's modulus and Poisson ratio varying linearly as

$$E[r] = E0 * r, \quad \upsilon[r] = \upsilon0 * r \tag{13}$$

Using (13), (7) and (6) into (5), and using the boundary condition (8), we get the stress function as

$$
\begin{aligned}
F[r] = {}& \frac{1}{55ab\left(a^{\sqrt5} - b^{\sqrt5}\right)} r^{-\frac{\sqrt5}{2}}\Bigg(ab\rho\Big(-22b^{\frac12(5+\sqrt5)}r_2^{\frac12+\sqrt5} \\
& + 22a^{\frac12(5+\sqrt5)}\Big(-b^{\sqrt5}\sqrt r + r^{\frac12+\sqrt5}\Big) - 5b^{\frac12(7+\sqrt5)}r^{\frac12+\sqrt5}\upsilon0 \\
& + 5a^{\frac12(7+\sqrt5)}\Big(-b^{\sqrt5}\sqrt r + r^{\frac12+\sqrt5}\Big)\upsilon0 + b^{\sqrt5}r^{3+\frac{\sqrt5}{2}}(22+5r\upsilon0) \\
& + a^{\sqrt5}\sqrt r\Big(22b^{\frac12(5+\sqrt5)} + 5b^{\frac12(7+\sqrt5)}\upsilon0 \\
& - r^{\frac12(5+\sqrt5)}(22+5r\upsilon0)\Big)\Big)\omega^2 \\
& + 55ab^{\frac12(1+\sqrt5)}r\Big(-a^{\sqrt5}\sqrt r + r^{\frac12+\sqrt5}\Big)p_0 \\
& + 55a^{\frac12(1+\sqrt5)}br\Big(b^{\sqrt5}\sqrt r - r^{\frac12+\sqrt5}\Big)p_i\Big).
\end{aligned}
\tag{14}
$$

If the Poisson's ratio and Young's modulus varying non-linearly as

$$E[r] = E0 * r^2, \quad \upsilon[r] = \upsilon0 * r^2 \tag{15}$$

The stress function using the boundary condition (8) is calculated as

$$
\begin{aligned}
F[r] = {}& \frac{1}{14ab(a^{2\sqrt2} - b^{2\sqrt2})}r^{1-\sqrt2}(-ab(a^{2\sqrt2} - b^{2\sqrt2})r^{2+\sqrt2}\rho(7+r^2\upsilon0)\omega^2 \\
& + r^{2\sqrt2}(-ab^{\sqrt2}(b^3\rho(7+b^2\upsilon0)\omega^2 - 14rp_0) + a^{\sqrt2}b(a^3\rho(7 \\
& + a^2\upsilon0)\omega^2 - 14rp_i)) + a^{\sqrt2}b^{\sqrt2}(a^{1+\sqrt2}(b^3\rho(7+a^2\upsilon0)\omega^2 \\
& - 14rp_0) + b^{1+\sqrt2}(-a^3\rho(7+a^2\upsilon0)\omega^2 + 14rp_i)))
\end{aligned}
\tag{16}
$$

3   Graphical Interpretation

In this paper, the values that are taken are given below:

$$a = 0.25\,\text{m.}, \ b - 0.5\,\text{m.}, \ \omega = 25, \ 50\,\text{rad/s}, \ p_i = 10\,\text{MPa}, \ p_0 = 10\,\text{MPa},$$
$$h0 = 0.5\,\text{m.}, \ \rho = 500\,\text{Kg/m}^3.$$

For case 1, the radial and circumferential stresses are calculated for different values of angular speed. As can be seen from the Fig. 1 that with the increase in angular speed, the radial stresses increase as the centrifugal force increases with the increase in angular speed. For linear variation the radial stresses are high, whereas for quadratic variation it is low as compared to linear variation. It is because with the non-linear variation both the thickness and Young's modulus decreases with the increase in radii.

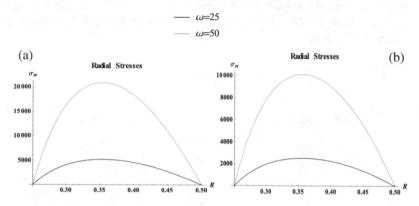

Fig. 1. Radial Stress with different angular speeds, (a) Linear Profile, (b) Quadratic Profil under pressure $p_i = p_0 = 10$ and $v = 0.2$.

The circumferential stresses are shown in Fig. 2. It is seen from Fig. 2(a) that the circumferential stresses are maximum at the internal radii and minimum at the external radii. For quadratic variation, the inverse behaviour is seen, but is observed that for both the cases, the hoop stresses are high as compared to the radial stresses. For higher angular speed, the circumferential stresses increases.

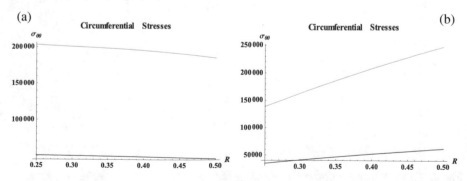

Fig. 2. Circumferential Stress with different angular speeds, (a) Linear Profile, (b) Quadratic Profile under pressure $p_i = p_0 = 10$ and $v = 0.2$.

Figures 3 and 4 are drawn for radial and hoop stresses, respectively but with different Poisson ratio and fixed angular speed $\omega = 25$ rad/s. Under equal pressure on both the boundaries, it is seen that for higher Poisson ratio, the radial stress is high as seen in Fig. 3 for both linear and quadratic profile. It is also seen for linear variation that the radial stress is high. The resisting stress is higher at the internal radius and lower at the outer as observed from Fig. 4(a). For Poisson ratio $v = 0.4$, the circumferential stress is more as compared to $v = 0.2$ at the inner radii as seen from Fig. 4(a). An inverse trend is seen at radii $r = 0.4$, where the circumferential stress is more for $v = 0.2$. With second order variation, the hoop stresses are lower at the internal surface for both Poisson's ratio.

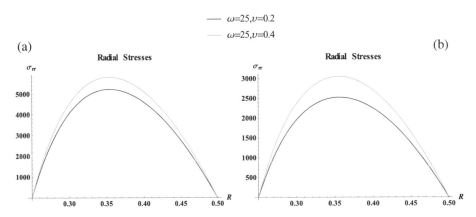

Fig. 3. Radial Stress with angular speed $\omega = 25$, (a) Linear Profile, (b) Quadratic Profile under pressure $p_i = p_0 = 10$ with different Poisson ratio.

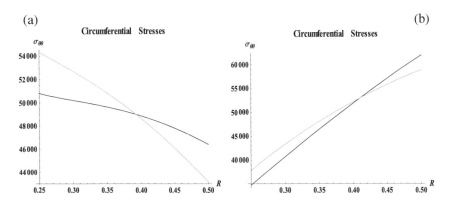

Fig. 4. Circumferential Stress with angular speed $\omega = 25$, (a) Linear Profile, (b) Quadratic Profile under pressure $p_i = p_0 = 10$ with different Poisson ratio.

Case 2: Under the variation of Young's modulus and Poisson ratio
In this case the following values are chosen for reference, i.e. $v0 = 0.05$. Figures 5 and 6 are drawn for radial and circumferential stresses for constant thickness and varying Poisson ratio. For higher angular speed the radial and circumferential stresses are high for both the cases. For quadratic profile, the radial stress decreases and circumferential stresses shows a sharp increase at the external surface. The circumferential stress is minimum at the internal radii and maximum at the external radii.

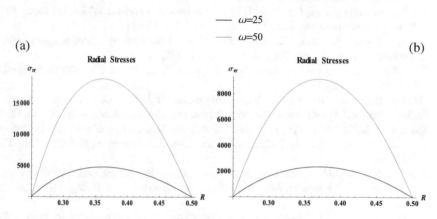

Fig. 5. Radial Stress with different angular speeds, (a) Linear Profile, (b) Quadratic Profile under pressure $p_i = p_0 = 10$.

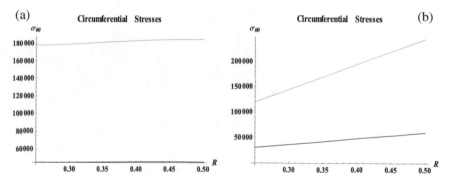

Fig. 6. Circumferential Stress with different angular speeds, (a) Linear Profile, (b) Quadratic Profile under pressure $p_i = p_0 = 10$

4 Conclusion

Stress analysis has been done on the rotating discs of constant thickness as well as the varying thickness and it comes out that variable thickness discs perform better than constant thickness with varying Poisson's ration as it shows a significant decrease in stresses.

References

1. Deepak, D., Garg, M., Gupta, V.K.: Creep behaviour of rotating FGM disc with linear and hyperbolic thickness profiles. Kragujev. J. Sci. **37**, 35–48 (2015)
2. Eraslan, A.N., Orcan, Y.: Elastic-plastic deformation of a rotating solid disk of exponentially varying thickness. Mech. Mater. **34**, 423–432 (2002)
3. Eraslan, A.N.: Tresca's yield criterion and linearly hardening rotating solid disks having hyperbolic profiles. Res. Eng. **69**(1), 17–28 (2004)
4. Khanna, K., Gupta, V.K., Nigam, S.P.: Creep analysis of a variable thickness rotating FGM disc using Tresca criterion. Def. Sci. J. **65**(2), 163–170 (2015)
5. Nie, G.J., Zhong, Z., Batra, R.C.: Material tailoring for functionally graded hollow cylinders and spheres. Composit. Sci. Technol. **71**, 666–673 (2011)
6. Seth, B.R.: Transition theory of elastic-plastic deformation, creep and relaxation. Nature **195** (4844), 896–897 (1962)
7. Sharma, S., Sahni, M., Kumar, R.: Elastic-plastic transition of transversely isotropic thick-walled rotating cylinder under internal pressure. Def. Sci. J. **59**(3), 260–264 (2009)
8. Sharma, S., Sahni, M., Kumar, R.: Thermo elastic-plastic transition of transversely isotropic thick-walled rotating cylinder under internal pressure. Adv. Theor. Appl. Mech. **2**(3), 113–122 (2009)
9. Sharma, S., Sahni, M.: Creep analysis of thin rotating disc having variable thickness and variable density with edge loading. Ann. Fac. Eng. Hunedoara Int. J. Eng. **11**(3), 289–296 (2013)
10. Sharma, S., Sahni, M.: Thermo elastic-plastic transition of a homogeneous thick-walled circular cylinder under external pressure. Struct. Integr. Life **13**(1), 3–8 (2013)
11. Sahni, M., Sahni, R.: Functionally graded rotating disc with internal pressure. Eng. Autom. Probl. **3**, 125–130 (2014)

Computational Intelligence and Optimization Techniques

Fuzzy Graph with Application to Solve
Task Scheduling Problem

Vivek Raich[1](✉), Shweta Rai[2], and D. S. Hooda[3]

[1] Government Holkar (Model Autonomous) Science College,
A.B. Road, Near Bhawarkua, Indore 452017, Madhya Pradesh, India
drvivekraich@gmail.com
[2] Christian Eminent College, F-7, Ravi Shankar Shukla Nagar, HIG Main Road,
Indore 452011, Madhya Pradesh, India
[3] Jaypee University of Engineering and Technology,
A-B Road, raghogarh, Guna 473226, Madhya Pradesh, India

Abstract. The concept of obtaining fuzzy sum of fuzzy colorings problem has a novel application in scheduling theory. The problem of scheduling N jobs on a single machine and obtaining the minimum value of the job completion times is equivalent to finding the fuzzy chromatic sum of the fuzzy graph modeled for this problem. In the present paper the task scheduling problem is solved by using fuzzy graph.

Keywords: Fuzzy graph · k-fuzzy coloring · Chromatic number
Chromatic fuzzy sum · Γ-chromatic sum

1   Introduction

The field of mathematics plays vital role in various fields. One of the important areas in Mathematics is graph theory which is used in several models. The origin of graph theory started with Konigsberg bridge problem in 1735. It was a long standing problem until solved by Leonhard Euler by means of graph. The coloring problem consists of determining the chromatic number of a graph and an associated coloring function.

Let G be a simple graph with n vertices. A coloring of the vertices of G is a mapping f: V(G) → N such that adjacent vertices are assigned different colors. The chromatic sum of a graph introduced by [4] is defined as the smallest possible total over all vertices that can occur among all colorings. Senthilraj [5] generalized these concepts to fuzzy graphs. He defined fuzzy graphs with fuzzy vertex set and fuzzy edge set. He also generalized the concept of the chromatic joins and chromatic sum of a graph to fuzzy graphs by defining the fuzzy chromatic sum of fuzzy graph.

In the present paper a problem of scheduling N jobs on a single machine is considered and the minimum value of the job completion times is obtained and that is equivalent to finding the fuzzy chromatic sum of the fuzzy graph modeled for this problem by considering the example of scheduling 6 jobs on a single machine. This result is generalized by considering the case of scheduling 8 tasks on a single machine and a minimum value of the task completion time is obtained and is also extended for *n* job.

© Springer International Publishing AG, part of Springer Nature 2018
A. M. Gil-Lafuente et al. (Eds.): FIM 2015, AISC 730, pp. 65–73, 2018.
https://doi.org/10.1007/978-3-319-75792-6_6

2 Preliminaries

Definition 2.1: Let X is a non-empty set, (called the universal set or the universe of discourse or simply domain). Then a function $\tilde{A}: X \to [0, 1]$, define fuzzy set on X where, '$\tilde{A}$' is called the membership function and $\mu_{\tilde{A}}(x)$ is called the membership grade of x. This can be written as

$$\tilde{A} = \{\langle x, \mu_{\tilde{A}}(x) \rangle : x \in X\}, \text{ where each pair } (x_1, \mu_{\tilde{A}}(x_1)) \text{ is called a singleton.}$$

In classical fuzzy set theory the set I is usually defined as the interval $[0, 1]$ such that,

$$\mu_A(x) = \begin{cases} 0 : x \notin X \\ 1 : x \in X \end{cases}$$

It takes any intermediate value between 0 and 1 represents the degree in which $x \in A$. The set I could be discrete set of the form $I = \{0, 1, \dots k\}$ where $\mu_A(x) < \mu_A(x')$ indicates that the degree of membership of x to A is lower than the degree of membership of x'.

Definition 2.2 [3]: Let V be a finite nonempty set. The triple $\hat{G} = (V, \sigma, \mu)$ is called a fuzzy graph on V where σ and μ are fuzzy sets on V and E, respectively, such that $\mu(uv) \leq \sigma(u) \wedge \sigma(v)$ for all $u, v \in V$ and $uv \in E$. For fuzzy graph $\hat{G} = (V, \sigma, \mu)$, the elements V and E are called set of vertices and set of edges of G respectively.

Definition 2.3 [3]: A fuzzy graph $\hat{G} = (V, \sigma, \mu)$ is called a complete fuzzy graph if $\mu(uv) = \sigma(u) \wedge \sigma(v)$ for all $u, v \in V$ and $uv \in E$. We denote this complete fuzzy graph by $\hat{G}k$.

Definition 2.4 [3]: Two vertices u and v in $\hat{G}$ are called adjacent if $(1/2)[\sigma(u) \wedge \sigma(v)] \leq (uv)$.

Definition 2.5 [3]: The edge uv of $\hat{G}$ is called strong if u and v are adjacent. Otherwise it is called weak.

Definition 2.6 [1]: A family $\Gamma = \{\gamma_1, \gamma_2, \dots \dots, \gamma_k\}$ of fuzzy sets on V is called a k-fuzzy coloring of $\hat{G} = (V, \sigma, \mu)$, if

a. $\wedge \Gamma = \sigma$,

b. $\gamma_i \wedge \gamma_j = 0$ and

c. For every strong edge uv of $\hat{G}$, $\gamma_i(u) \wedge \gamma_j(v) = 0$ for $1 \leq i \leq k$.

The above definition of k-fuzzy coloring was defined on fuzzy set of vertices.

Definition 2.7 [5]: The least value of k for which G has a fuzzy coloring, denoted by $x^f(G)$ is called the fuzzy chromatic number of G.

Definition 2.8 [1]: For a k-fuzzy coloring $\Gamma = \{\gamma_1, \gamma_2, \ldots, \gamma_k\}$ of a fuzzy graph of G, Γ chromatic fuzzy sum of G denoted by $\sum_\Gamma (G)$ is defined as

$$\sum_\Gamma (G) = 1 \sum_{x \in C_1} \theta_1(x) + 2 \sum_{x \in C_2} \theta_2(x) + \ldots \ldots + k \sum_{x \in C_k} \theta_k(x),$$

Where $C_i = supp\gamma_i$ and $\theta_i(x) = \max\{\sigma(x) + \mu(xy)/y \in C_i\}$.

Definition 2.9 [1]: The chromatic fuzzy sum of G denoted by $\sum (G)$ is defined as follows

$$\sum (G) = \min\{\frac{\sum_\Gamma (G)}{\Gamma} \text{is fuzzy coloring}\}.$$

The number of fuzzy coloring of G is finite and so there exist a fuzzy Γ_0 which is called minimum fuzzy coloring of G such that $\sum (G) = \sum_{\Gamma_0} (G)$.

Theorem 2.1 [2]: Let G be a fuzzy graph and $\Gamma_0 = \{\gamma_1, \gamma_2, \ldots \ldots, \gamma_k\}$ is minimum fuzzy sum coloring of G. Then $\sum_{x \in C_1} \theta_1(x) \geq \sum_{x \in C_2} \theta_2(x) \geq \ldots \ldots \geq \sum_{x \in C_k} \theta_k(x)$.

Theorem 2.2 [2]: For a fuzzy graph $\hat{G} = (V, \sigma, \mu)$, $\sum (G) \leq \frac{3}{4}[(x^f(G) + 1)h(\sigma)|V|$, where $h(\sigma)$ is height of σ and $|V|$ is cardinality of V.

Corollary 2.1: Let $\hat{G} = (V, \sigma, \mu)$ be a connected fuzzy graph with e strong edges. Then the lower und for $\sum (G)$ is $w\sqrt{8e}$, where $w = \max\{\sigma(x) + \mu(xy) > o, x \in V, (x, y)$ is weak edge of $G\}$.

Corollary 2.2: The fuzzy chromatic sum lies between $w\sqrt{8e}$ and $\frac{3}{4}[(x^f(G) + 1)h(\sigma)|V|$.

3 Main Problem and Solution

Assume that at any time the machine is capable to perform any number of tasks and these tasks are independent or conflicts between them are less than one. Consider the time consuming for task 1 and 4 is 0.4 h, for tasks 3 and 6 is 0.3 h, for tasks 2 and 5 is 1 h, for tasks 7 and 8 is 0.2 h.

And also Task $\{(2, 5), (5, 6), (6, 7)\}$ conflict together with 0.1 h, Task $\{(1, 2), (1, 4), (1, 5), (2, 4), (4,7), (4, 8)\}$ conflict together with 0.4 h and Task $\{(1, 3), (2, 8), (3, 4), (4, 5), (5, 7)\}$ conflict together with 0.3 h.

Now, we define the fuzzy graph for above problem.

Let $\hat{G} = (V, \sigma, \mu)$ where V is the set of all task, $\sigma(x)$ is the amount of consuming time of machine for each $x \in V$ and $\mu(x, y)$ is the measure of the conflict between the task x and y. Finding the minimum value of job completion time for this problem is equivalent to the chromatic sum of $\hat{G}$.

The fuzzy graph $\hat{G} = (V, \sigma, \mu)$ corresponding to our example is defined as follows:
Let $V = \{1, 2, 3, 4, 5, 6, 7, 8\}$,

$$\sigma(i) = \begin{cases} 0.4 \; for \; i = 1, 4 \\ 1.0 \; for \; i = 2, 5 \\ 0.3 \; for \; i = 3, 6 \\ 0.2 \; for \; i = 7, 8 \end{cases}$$

$$\mu(i, j) = \begin{cases} 0.1, & for \; i, j = \{(2, 5), (5, 6), (6, 7)\} \\ 0.4, & for \; i, j = \{(1, 2), (1, 4), (4, 7), (4, 8)\} \\ 0.3, & for \; i, j = \{(1, 3), (2, 8), (3, 4), (4, 5), (5, 7)\} \end{cases}$$

The fuzzy graph for above problem is (Fig. 1).
Strong edges: (1, 2), (1,3), (1, 4), (1, 5), (2, 4), (3, 4), (4, 5), (4, 7), (4, 8), (5, 7), (6, 7).
Weak edges: (1, 6), (1, 7), (1, 8), (2, 3), (2, 5), (2, 6), (2, 7), (2, 8), (3, 5), (3, 6), (3, 7), (3, 8), (4, 6), (5, 6), (5, 8), (6, 8), (7, 8).

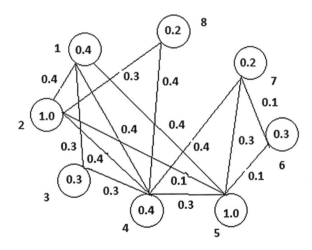

Fig. 1. Fuzzy graph for problem

Let $\Gamma_1 = \{\gamma_1, \gamma_2, \dots, \gamma_8\}$ be a family of fuzzy set defined on V where,

$$\gamma_1(i) = \begin{cases} 1.0, & for \; i = 2 \\ o, & otherwise \end{cases},$$

$$\gamma_2(i) = \begin{cases} 1.0, & for\ i = 5 \\ o, & otherwise \end{cases},$$

$$\gamma_3(i) = \begin{cases} 0.4, & for\ i = 1 \\ o, & otherwise \end{cases},$$

$$\gamma_4(i) = \begin{cases} 0.4, & for\ i = 4 \\ o, & otherwise \end{cases},$$

$$\gamma_5(i) = \begin{cases} 0.3, & for\ i = 3 \\ o, & otherwise \end{cases},$$

$$\gamma_6(i) = \begin{cases} 0.3, & for\ i = 6 \\ o, & otherwise \end{cases}$$

$$\gamma_7(i) = \begin{cases} 0.2, & for\ i = 7 \\ o, & otherwise \end{cases},$$

$$\gamma_8(i) = \begin{cases} 0.2, & for\ i = 8 \\ o, & otherwise \end{cases}$$

Table 1. k- fuzzy coloring

V	γ_1	γ_2	γ_3	γ_4	γ_5	γ_6	γ_7	γ_8	$max\ \gamma_1 = \sigma(i)$
1	0	0	0.4	0	0	0	0	0	0.4
2	1.0	0	0	0	0	0	0	0	1.0
3	0	0	0	0	0.3	0	0	0	0.3
4	0	0	0	0.4	0	0	0	0	0.4
5	0	1.0	0	0	0	0	0	0	1.0
6	0	0	0	0	0	0.3	0	0	0.3
7	0	0	0	0	0	0	0.2	0	0.2
8	0	0	0	0	0	0	0	0.2	0.2

From the Table 1, we can see that Γ_1 satisfied all the properties of k-fuzzy coloring.

Therefore G has 8-Coloring and $x^f(G) = 8$. For this 8-Coloring, Γ_1 chromatic number can be calculated as follows:

$$C_1 = \{2\}, C_2 = \{5\}, C_3 = \{1\}, C_4 = \{4\}, C_5 = \{3\}, C_6 = \{6\}, C_7 = \{7\}, C_8 = \{8\}$$

$$\theta_1(2) = \max\{1 + 0\} = 1, \theta_2(5) = \max\{1 + 0\} = 1, \theta_3(1) = \max\{0.4 + 0\} = 0.4,$$
$$\theta_4(4) = \max\{0.4 + 0\} = 0.4, \theta_5(3) = \max\{0.3 + 0\} = 0.4,$$
$$\theta_6(6) = \max\{0.3 + 0\} = 0.3$$
$$\theta_7(7) = \max\{0.2 + 0\} = 0.2, \theta_8(8) = \max\{0.2 + 0\} = 0.2$$

The Γ_1 chromatic fuzzy sum of G

$$\sum_{\Gamma_1} (G) = 1(1) + 2(1) + 3(0.4) + 4(0.4) + 5(0.3) + 6(0.3) + 7(0.2) + 8(0.2)$$
$$= 12.1.$$

Let $\Gamma_2 = \{\gamma_1, \gamma_2, \ldots, \gamma_5\}$ be a family of fuzzy set defined on V where

$$\gamma_1(i) = \begin{cases} 1.0, & for\ i = 5 \\ 0.3, & for\ i = 3 \\ o, & otherwise \end{cases},$$

$$\gamma_2(i) = \begin{cases} 1.0, & for\ i = 2 \\ o, & otherwise \end{cases},$$

$$\gamma_3(i) = \begin{cases} 0.4, & for\ i = 1 \\ 0.2, & for\ i = 7, 8 \\ o, & otherwise \end{cases}$$

$$\gamma_4(i) = \begin{cases} 0.1, & for\ i = 4 \\ o, & otherwise \end{cases},$$

$$\gamma_5(i) = \begin{cases} 0.3, & for\ i = 6 \\ o, & otherwise \end{cases}$$

Table 2. k- fuzzy coloring

V	γ_1	γ_2	γ_3	γ_4	γ_5	$max\ \gamma_1 = \sigma(i)$
1	0	0	0.4	0	0	0.4
2	0	1.0	0	0	0	1.0
3	0.3	0	0	0	0.3	0.3
4	0	0	0	0.4	0	0.4
5	1.0	0	0	0	0	1.0
6	0	0	0	0	0	0.3
7	0	0	0.2	0	0	0.2
8	0	0	0.2	0	0	0.2

Again from the (Table 2), we can see that Γ_2 satisfied all the properties of k- fuzzy coloring.

Therefore G has 5-Coloring and $x^f(G) = 5$. For this 5-Coloring, Γ_2 chromatic number can be calculated as follows:

$$C_1 = \{3,5\}, C_2 = \{2\}, C_3 = \{1,7,8\}, C_4 = \{4\}, C_5 = \{3\},$$

$$\theta_1(3) = \max\{0.3 + 0, 0.3 + 0\} = 0.3,$$
$$\theta_1(5) = \max\{1 + 0, 1 + 0\} = 1,$$
$$\theta_2(2) = \max\{1 + 0\} = 1,$$
$$\theta_3(1) = \max\{0.4 + 0, 0.4 + 0, 0.4 + 0\} = 0.4,$$
$$\theta_3(7) = \max\{0.2 + 0, 0.2 + 0, 0.2 + 0\} = 0.2,$$
$$\theta_3(8) = \max\{0.2 + 0, 0.2 + 0, 0.2 + 0\} = 0.2,$$
$$\theta_4(4) = \max\{0.4 + 0\} = 0.4,$$
$$\theta_5(6) = \max\{0.3 + 0\} = 0.3.$$

The Γ_2 chromatic fuzzy sum of G

$$\sum_{\Gamma_2}(G) = 1(1 + 0.3) + 2(1) + 3(0.4 + 0.2 + 0.2) + 4(0.4) + 5(0.3) = 8.4.$$

Let $\Gamma_3 = \{\gamma_1, \gamma_2, \gamma_3\}$ be a family of fuzzy set defined on V where

$$\gamma_1(i) = \begin{cases} 1.0, \text{ for } i = 2, 5 \\ 0.3, \text{ for } i = 3, 6, \\ o, \quad otherwise \end{cases}$$

$$\gamma_2(i) = \begin{cases} 0.4, \quad \text{for } i = 1 \\ 0.2, \text{ for } i = 7, 8, \\ o, \quad otherwise \end{cases}$$

$$\gamma_3(i) = \begin{cases} 0.4, \text{ for } i = 4 \\ o, \quad otherwise \end{cases}.$$

Table 3. k- fuzzy coloring

V	γ_1	γ_2	γ_3	$\max \gamma_1 = \sigma(i)$
1	0	0.4	0	0.4
2	1.0	0	0	1.0
3	0.3	0	0	0.3
4	0	0	0.4	0.4
5	1.0	0	0	1.0
6	0.3	0	0	0.3
7	0	0.2	0	0.2
8	0	0.2	0	0.2

Again from the Table 3, we can see that Γ_3 satisfied all the properties of k- fuzzy coloring.

Therefore G has 3-Coloring and $x^f(G) = 5$. For this 3-Coloring, Γ_3 chromatic number can be calculated as follows:

$$C_1 = \{2, 3, 5, 6\}, C_2 = \{1, 7, 8\}, C_3 = \{4\}$$

$$\theta_1(3) = \max\{0.3 + 0, 0.3 + 0, 0.3 + 0, 0.3 + 0\} = 0.3,$$
$$\theta_1(5) = \max\{1 + 0.1, 1 + 0, 1 + 0, 1 + 0.1\} = 1.1,$$
$$\theta_1(2) = \max\{1 + 0, 1 + 0, 1 + 0.1, 1 + 0\} = 1.1,$$
$$\theta_1(6) = \max\{0.4 + 0, 0.4 + 0, 0.4 + 0.1, 0.4 + 0\} = 0.5,$$
$$\theta_2(7) = \max\{0.2 + 0, 0.2 + 0, 0.2 + 0\} = 0.2,$$
$$\theta_2(8) = \max\{0.2 + 0, 0.2 + 0, 0.2 + 0\} = 0.2,$$
$$\theta_2(1) = \max\{0.4 + 0, 0.4 + 0, 0.4 + 0\} = 0.4, \theta_3(6) = \max\{0.3 + 0\} = 0.3.$$

The Γ_3 chromatic fuzzy sum of G

$$\sum\nolimits_{\Gamma_3}(G) = 1(1.1 + 1.1 + 0.5 + 0.3) + 2(0.2 + 0.2 + 0.4) + (0.3) = 5.5.$$

Therefore the fuzzy chromatic sum of G is

$$\sum(G) = \min\{\Gamma_1, \Gamma_2, \Gamma_3\} = \min\{12.1, 8.4, 5.5\} = 5.5.$$

Calculation for w:

$$w = \min\left\{\begin{matrix} 0.4 + 0, 0.4 + 0, 0.4 + 0, 1 + 0, 1 + 0.1, 1 + 0, 1 + 0, 0.3 + 0, 0.3 + 0, 0.3 \\ + 0, 0.3 + 0, 0.4 + 0, 1 + 0.1, 1 + 0, 0.3 + 0, 0.2 + 0 \end{matrix}\right\} = 0.2.$$

Lower bound of $\sum(G)$ is $w\sqrt{8e} = 0.2\sqrt{8} \times 12 = 0.2 \times 4 \times \sqrt{6} = 1.9592$.
Now,

$$\frac{3}{4}[(x^f(G) + 1)h(\sigma)|V| = \frac{3}{4}[(4 + 1) \times 1 \times 8] = 30.$$

4   Conclusion

The fuzzy chromatic number lies between 30 and 1.9592. In our problem, $\sum(G) = 5.5$. Therefore the minimum time of task completion of our problem is 5.5 h.

References

1. Eslahchi, C., Onagh, B.N.: Vertex strength of fuzzy graphs. Int. J. Math. Math. Sci. (2006). https://doi.org/10.1155/IJMMS/2006/43614
2. Munoz, S., Ortuno, T., Ramirez, J., Yanez, J.: Coloring fuzzy graphs. Omega **32**, 211–221 (2005)
3. Nivethan, V., Parvathi, A.: Fuzzy total coloring and chromatic number of a complete fuzzy graph. Int. J. Emerg. Trends Eng. Dev. **3**(6), 377–384 (2013)
4. Senthilraj, S.: On the matrix of chromatic joins. Int. J. Appl. Theor. Inf. Technol. **4**(3), 106–110 (2008)
5. Senthilraj, S.: Fuzzy graph application of job allocation. Int. J. Eng. Innovative Technol. **1**(2), 7–10 (2012)

SmartMonkey: A Web Browser Tool for Solving Combinatorial Optimization Problems in Real Time

Xavier Ruiz[1], Laura Calvet[2(✉)], Jaume Ferrarons[2], and Angel Juan[2]

[1] Incubio, Barcelona, Spain
[2] Computer Science Department, IN3-Open University of Catalonia, Barcelona, Spain
`lcalvetl@uoc.edu`

Abstract. This paper introduces SmartMonkey, a novel web-browser approach for solving NP-hard combinatorial optimization problems in "real time" (usually a few seconds). Our approach makes use of randomized algorithms that are run in parallel on a set of independent machines available on the Internet. These machines do not need to be configured, and no client application needs to be installed on them. Instead, just by opening a web page in a web browser, the computational resources of the machine become available for the algorithms to be executed. Being a configuration-free approach, it offers a great advantage to end users, since they are relieved from the usually complex and time-consuming configuration tasks that characterize other distributed-computing approaches. Computational tests have been carried out using different algorithms for solving NP-hard combinatorial optimization problems in transportation and production scheduling. The results show that our approach allows obtaining near-optimal solutions in real time, which can be especially interesting for supporting decision-making processes, especially those in small and medium enterprises, in a wide range of application fields including logistics, transportation, smart cities, and manufacturing.

Keywords: Logistics & transportation · Production
Combinatorial optimization · Parallel and distributed computing
Randomized algorithms · Metaheuristics

1 Introduction

Internet plays an essential role in our society, permitting us not only to communicate and obtain information in 'real-time' but, as it will be discussed in this paper, also to solve complex decision-making problems in 'real-time'. This is achieved by the use of Distributed and Parallel Computing Systems (DPCS), which allow the aggregation of multiple autonomous computing resources interacting to achieve a common goal [9]. Under this general concept, there are a number of different paradigms. Grid Computing [11] appeared at the beginning of the 1990s decade, aiming to combine dispersed computational resources. It was initially used in scientific and research fields. Another relevant paradigm sharing the same goal is Cloud Computing [2]. It emerged at the beginning of this century, focusing on a provider-client model in which users do not

© Springer International Publishing AG, part of Springer Nature 2018
A. M. Gil-Lafuente et al. (Eds.): FIM 2015, AISC 730, pp. 74–86, 2018.
https://doi.org/10.1007/978-3-319-75792-6_7

belong to any particular organization, only pay for the resources they use and can consume them at any time, without having to forecast their computing needs in a mid-large term. This paradigm eases Small and Medium Enterprises (SMEs) the acquisition of computer resources, since they do not longer need up-front investments or over-provisioning to face peak loads. More recent paradigms are Volunteer Computing [21] and Desktop Grids [6]. Both attempt to gather surplus or idle computing resources. While the first focuses on end-users that voluntarily make available their resources for a third entity or project, the second utilizes resources from a given organization network. A successful project in the Volunteer Computing paradigm is SETI@home (http://setia-thome.berkeley.edu) of BOINC [1]. However, it is usually a complex task to gather a considerable number of volunteers. Besides the savings, a particular interesting feature of Desktop Grids is that SMEs avoid sending private data to an external provider, and become more environmentally friendly.

The main contribution of this work is the description and testing of an efficient, flexible, and browser-based framework to facilitate access to computational resources [3] and, ultimately, solve Combinatorial Optimization Problems (COPs) in 'real time' (a few seconds). This framework allows the employment of new versions of web browsers (such as Google Chrome, Firefox, and Internet Explorer) as nodes in a cluster. The only required step is to visit a website. The embedded JavaScript code into this website enables the communication with the job dispatcher service. It may be considered a more scalable paradigm than traditional grid computing, since the connection of people is boosted by the fact that no third party software installation is required. Due to the relevance of COPs for SMEs and the amount of academic works proposing the implementation of DPCS for addressing them [23, 24], we illustrate the working and the potential benefits of our approach by solving two classic NP-hard COPs in the fields of transportation (vehicle routing) and production (scheduling).

The rest of the paper is organized as follows: Sect. 2 provides an overview of Distributed and Parallel Computing Systems. Afterwards, Sect. 3 discusses the potential of DPCS-based approaches in solving real-life SMEs problems. The description of our browser-based platform is included in Sect. 4. The web-based approach developed to tackle COPs is explained in Sect. 5. Section 6 contains the numerical experiments carried out to analyse the efficiency of our approach, while Sect. 7 discusses the results of these experiments. Finally, Sect. 8 summarizes the main findings of this work.

2   Distributed and Parallel Computing Systems

Desktop computers have become affordable machines that most people use every day for both work and leisure. Despite their current capacity, numerous institutions and individuals require more computational resources to execute intensive problem-solving processes. In these cases, Distributed and Parallel Computing Systems constitute a useful approach. Multi-processors and/or multi-computers paradigms may be employed. A multi-processors schema refers to a set of physical processing units sharing a machine (mono-core CPU, multi-core CPU, or a combination of both). In this schema, a task represents a logical concept including instructions to be executed by an algorithm or application, while a process can be defined as a running instance of a computer program. A process might consist in different threads implementing one or more tasks. Tasks, threads or processes share a global memory system, on which their communications rely. On the other hand, a multi-computer schema presents a set of physical machines linked via network connections. These machines can be coupled geographically (in a supercomputing environment, for example) or in a more distributed environment (as in Cloud Computing). The main parallel paradigm is message passing, in which tasks and processes of different machines interchange data packets by sending and receiving messages to communicate.

Nowadays, there is no need for a user of designing and building a new computing infrastructure, since there are organizations which satisfy scalable computational demands at a reasonable price. Grids are mainly required in high-performance-computing scientific projects. [4, 10] provide an overview of this paradigm analysing several projects. In contrast, Cloud platforms constitute a solution for enterprises to obtain additional resources (public clouds), or to manage the resources owned (private cloud). Some examples of platforms providing these services are Amazon's Elastic Compute Cloud (http://aws.amazon.com/ec2) and Microsoft's Azure Services Platform (www.microsoft.com/windowsazure). Volunteer Computing models are mostly used for scientific and academic projects. Highly popular implementations are BOINC [1], Condor [17] and Entropia [7]. Examples of communities created to aggregate computational resources are Seti@HOME (http://setiathome.berkeley.edu) and Distributed.net (http://www.distributed.net). Finally, Desktop computing is employed by private organizations [19].

3   Relevance of DPCS for SMEs

SMEs are responsible for a significant part of the wealth generated in all developed economies. Often, they do neither possess advanced technical knowledge nor modern computational resources. However, a number of them could benefit from having more resources, for example to speed up intensive-computation processes or to obtain a higher performance. In order to access them, DPCS offer two alternatives: *(a)* to pay for using resources from an external provider (a cloud platform, for instance), which can be presented as virtual machines; and *(b)* to employ underutilized computer resources owned by the SME. This idea of aggregating idle or unused resources characterizes also Volunteer Computing Systems. The main difference between both paradigms is that while the latter is usually associated to dynamic (any user can freely enter and leave) and heterogeneous environments, an SME knows the characteristics and the availability of its machines. Obviously, their scalability is also more limited.

The alternative of using SME's underutilized resources presents several advantages. Firstly, SMEs do not have to send private information to servers of an external enterprise. Secondly, it is a cheaper solution since the SME does already have the resources. Finally, the energy consumption is reduced by seizing these resources, which could be still consuming otherwise [5]. In this same direction, it can be argued that the environmental footprint of this alternative may be lower than that of a large digital warehouse, because the heat concentration is lower. These Desktop Grids Systems may be formed by personal computers with more computing capabilities than the required (standard computers in which employees mainly use word processors and spreadsheets, for instance) or that are not used during some specific days (weekends, holidays, etc.) or hours (night, midday, etc.). Moreover, resources from several SMEs may be gathered [14] to build a larger computing system (Fig. 1). They can rely on a directory-of-resources service that keeps updated information of available computing resources. Once a user requires executing a process, he sends a query to this directory to select the resources and organize the tasks to perform. Once these tasks have been completed, the result is sent back to the user.

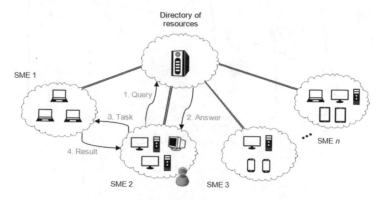

Fig. 1. Aggregating resources from one or several SMEs.

4   Description of the SmartMonkey Platform

SmartMonkey aims at facilitating the aggregation of a high number of computational resources -without any additional cost for the company- by seizing underutilized or idle resources. It is based on software already installed in most computers, web browsers. Using a modern version of some of the commonest (Google Chrome, Firefox, or Internet Explorer), it may integrate a computer into the computing network. The only action required is to visit a website with an embedded JavaScript code that enables the communication in real-time with a job dispatcher service. Each one of the jobs includes a piece of data and the computing task to perform. Additional steps such as downloading, installing, or setting up additional software are not required, which makes this option a very attractive one for most SMEs. Since the ease to add new resources, this approach can be considered highly scalable. As other Desktop Computing Systems, it has the advantage that does not require sending private data to an external enterprise or third party. Therefore, the described platform constitutes a flexible, simple, and scalable approach with multiple applications in SMEs, which reduces the cost of acquiring additional computational capabilities. Figure 2 shows the major difference between the BOINC framework and our web-based framework in terms of engagement. While our framework is just one step away to start processing, BOINC networks need a more complicated sequence of steps. Therefore, our configuration-free approach can turn out to be much more appealing to SMEs because its ease of use and fast scalability.

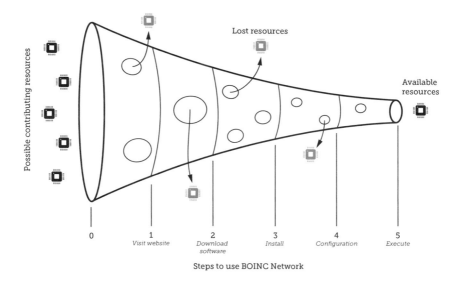

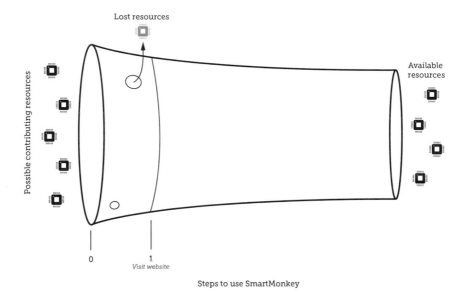

Fig. 2. Comparison between BOINC and SmartMonkey engagement process.

The platform architecture is the typical of a master-slave cluster. The system has been designed to free the master from computationally expensive tasks. For the experiments described later in this paper, a single master has been sufficient to handle all the workload. In a production environment, the system could easily scale to thousands of slaves or even further when considering other architectures like a multi-master environment, etc. In our case, the master was placed on a dedicated server located on a cloud provider (Softlayer). The slaves were located over 2 different locations: The UOC's Lab and the Incubio's offices. The execution process goes as follows. First, the end-user submits the task to be executed to the master. This task consists mainly in a set of Map and Reduce functions written in JavaScript, as well as their input dataset. The master is responsible of creating a list of jobs. Each job is composed of a chunk from the dataset and the source of code that has to be executed over each piece of data. The master delivers and ensures that jobs are evenly distributed. After each job is completed, the master receives the results and stores them in a file or a database depending on the execution flow given by the user. The master keeps track of the jobs that have been assigned and processed. Different measures handle unfinished jobs, errors or exceptions that could appear unexpectedly by either rescheduling the jobs or stopping the execution and reporting the error.

Some of the main benefits of this paradigm are cross-platform support and cluster scalability. Adding a slave to the system is as easy as adding a JavaScript snippet to any website and opening it with a modern browser. The loaded code is the responsible to communicate with the master and ask for new jobs to execute. As the code of each job is provided by the master, the code of the slave is in charge of receiving a new job, executing it, and then sending the results back to the master. This process is entirely written on JavaScript code, avoiding specificities of that language from specific browsers in such a way that it can run on a heterogeneous pool of different web browsers (such as Internet Explorer, Google Chrome and Firefox) as well as on different operating systems.

5 The Multi-agent Solving Approach

In sectors such as logistics and transportation, telecommunication, healthcare, finance, and production, a huge number of real-life decision-making processes can be modeled as COPs. Frequently, managers are required to take crucial decisions in real time. Designing routes to quickly provide medical assistance after a natural disaster, rescheduling flights because of some delays caused by unexpected circumstances, or reducing the risk of some savings in the stock market being affected by unpredicted events are a few examples. Despite the fact that exact methods exist for addressing COPs, they usually require a high amount of time to solve real-sized problem instances. As a consequence, approximate methods are widely implemented. Most of them are probabilistic, which means that their solution depends on the seed used for a pseudo-random number generator. It has been proved that the execution time that an algorithm needs to report high-quality solutions can be reduced depending on this seed [16].

According to [23, 24], DPCS are commonly employed to solve COPs. The typical approach in the related literature applies a master-slave scheme, in which a master or coordinator processor sends tasks to a set of slave processors in order to execute an intensive-computing process. Each slave is responsible for solving the same problem instance considering a different scenario, each one formed by a set of parameters and/or a seed. Once a slave has completed its task, it sends the solution to the master that stores it. In the simplest version, there is no communication between slaves. Following this approach, we aim to execute multiple instances of the algorithm at the same time, each with a different seed. As shown in Fig. 3, each of these instances can be considered a cloned agent that is searching the solution space.

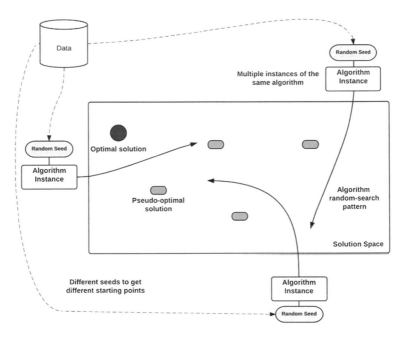

Fig. 3. A multi-agent approach for solving COPs.

6 Computational Experiments

In order to illustrate the benefits of the presented approach, two relevant NP-hard COPs have been addressed: the Capacitated Vehicle Routing Problem (CVRP) and the Permutation Flowshop Sequencing Problem (PFSP). For each problem, an efficient algorithm has been employed. The CVRP (Fig. 4) is a classical routing problem in which a fleet of capacitated vehicles must visit a number of customers and satisfy their demands. The routes start and finish at a specific depot, where all goods are initially held. Moving between nodes (customers or depot) has associated a distance-based cost. Typically, the aim is to minimize the total cost of the designed routes. Despite this simple description, the CVRP is a NP-hard problem with an extraordinary number of real-life applications in the transportation field.

The randomized version of the popular Clarke and Wright savings (CWS) heuristic [8], described in [13], has been chosen to solve the CVRP. Initially, this heuristic computes a dummy solution in which there is a route for each customer. Afterwards, a list of all edges connecting two customers is created. It is sorted according to the savings

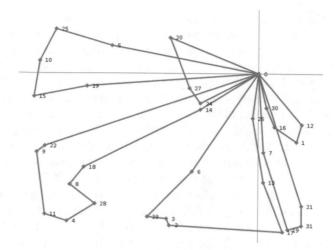

Fig. 4. A CVRP instance with customers being served from a central depot.

that would be obtained if the dummy solution was modified to include an edge by merging the corresponding routes. This is an iterative process that considers all edges and carries out merges only if no constraint is violated. The randomized version of the heuristic introduces variability in the sort process, erasing the greedy behavior of the classic version. Particularly, it assigns a given probability to each edge to be selected in first place, which is coherent with the savings (i.e., edges with a highest saving will have associated a higher probability). Then, the list and the probabilities are updated, and the process is repeated until all edges have been studied. The Kelly instances are used to test our approach [12].

The PFSP (Fig. 5) is a well-known scheduling problem. An instance is characterized by a number of independent jobs that have to be processed on a set of independent machines. Each machine can execute at most one job at a time, and all jobs must be processed in the same order in each machine. The aim is to minimize the maximum completion time, so called makespan.

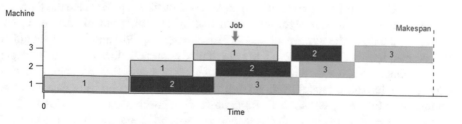

Fig. 5. A PFSP instance with 3 machines and 3 jobs.

The ILS-ESP algorithm [16] has been employed to address the PFSP. It relies on the Iterated Local Search (ILS) metaheuristic [18]. In this algorithm, a biased-randomized version of the NEH heuristic [20] is used to construct an initial solution. The original version computes the total processing time required for each job and creates a solution by iteratively selecting the remaining job with the highest value. The new version is a non-deterministic heuristic capable of generating a set of high-quality solutions without losing the logic behind the original version. Once the initial solution is created, a classical local search is applied which attempts to find a better solution in its neighborhood. The resulting solution, called *baseSol*, is stored. Then, a set of instructions are repeated until a stopping condition (number of iterations or limit of time, for instance) is met. The *baseSol* is perturbed and a local search is applied to the new solution (*newSol*). This solution is stored if it is better than the best solution found so far (*bestSol*). Moreover, the *baseSol* is set to the *newSol* if this solution is better or passes a Demon-like acceptance criterion [24]. Finally, the best solution found is returned. We have tested our approach on the 120 Taillard's benchmark instances [22]. They are grouped in 12 sets of 10, which are characterized by the following pairs of numbers of jobs and machines: $20 \times 5, 20 \times 10, 20 \times 20, 50 \times 5, 50 \times 10, 50 \times 20, 100 \times 5, 100 \times 10, 100 \times 20, 200 \times 10,$ 200×20, and 500×20. The instance resolution has been performed considering a specific combination of the parameters 'limit of time' (1, 5, 10, 15, 20, and 30 s) and 'number of agents' running in parallel (1, 4, 8, 16, 32, and 64).

All the experiments have been carried out using 64 slaves and 1 master. The master specifications are 3.5 GHz Intel Xeon-IvyBridge with 8 GB of RAM. The slaves are a heterogeneous set of desktop computers not having more than 8 GB of RAM and up to 8 cores each. The machines were connected to the parallel computing environment using one of the following browsers: Microsoft Internet Explorer, Google Chrome or Mozilla Firefox, all of them with JavaScript enabled. The slaves were connected over a usual shared internet connection. For this reason, latencies or high speed connections were considered negligible.

7   Analysis of Results

Figure 6 shows the results obtained after running the algorithm for solving the Kelly instances during 20 s of clock time per instance. Considering all instances, the first boxplot shows the gaps between the best known solution (BKS) and the solution generated by the CWS heuristic. The remaining boxplots show the gaps between the BKS and different executions of our algorithm, each one using a different number of agents running in parallel. The number of agents tested were: 64, 128, and 256. It should be noticed that, for the 20 s considered, the distributed approach allows to reduce the gap down to almost 5% even for a reasonably low number of agents (i.e., 64). Note we were using a simple biased-randomized version of the CWS heuristic, this gap could be reduced even further by employing a more powerful algorithm such as the SR-GCWS-CS [15]. Therefore, the approach can provide reasonably good solutions in just a few seconds without any algorithm fine-tuning or software installation/configuration effort.

Regarding the PFSP instances, Table 1 summarizes the results of our computational experiments using a maximum time of 5 s. Each row refers to a different set of instances. Each column shows the gap between the BKS and our solution for different numbers of

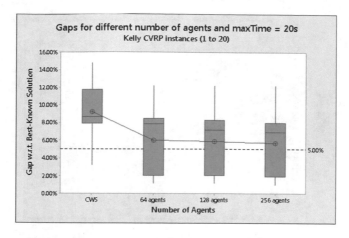

Fig. 6. Results for the CVRP using the Kelly instances.

agents (1, 4, 8, 16, 32, and 64). Notice that the gaps shrink as the number of agents working in parallel is increased.

Table 1. Results for the PFSP considering the Taillard instances. Gaps for different number of agents and a maximum tome of 5 s.

Taillard set	BKS −1A (%)	BKS −4A (%)	BKS −8A (%)	BKS −16A (%)	BKS −32A (%)	BKS −64A (%)
20 × 5	0.00	0.00	0.00	0.00	0.00	0.00
20 × 10	0.08	0.08	0.04	0.00	0.00	0.00
20 × 20	0.06	0.02	0.01	0.00	0.00	0.00
50 × 5	0.01	0.01	0.01	0.01	0.00	0.00
50 × 10	0.84	0.74	0.72	0.65	0.61	0.54
50 × 20	3.21	2.85	2.76	2.68	2.65	2.58
100 × 5	0.05	0.02	0.00	0.00	0.00	0.00
100 × 10	0.53	0.33	0.25	0.22	0.21	0.18
100 × 20	3.14	2.67	2.66	2.63	2.56	2.46
200 × 10	0.40	0.26	0.24	0.23	0.20	0.14
200 × 20	2.36	2.20	2.17	2.15	2.06	2.00
500 × 20	1.88	1.53	1.42	1.32	1.32	1.26
Averages	1.05	0.89	0.86	0.82	0.80	0.76

Figure 7 summarizes similar results for different values of the maximum clock time. It can be observed that, as time increases or as the number of agents increases, the average gap (for the entire set of benchmark instances) decreases. A detailed case is illustrated in Fig. 8, which displays the scatterplot of costs versus limit of time and number of agents for a given instance.

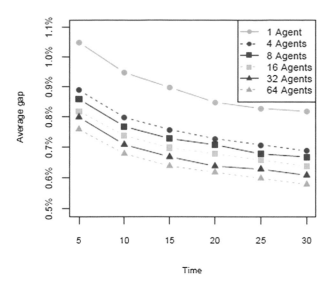

Fig. 7. Average gaps for different numbers of agents and limits of time.

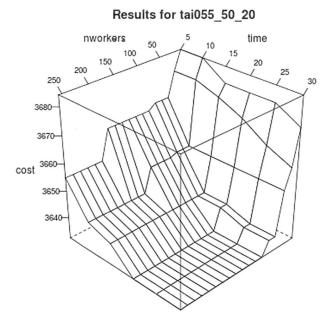

Fig. 8. Objective solutions for different numbers of agents and limits of time.

8 Conclusions

This paper has discussed the benefits of Distributed and Parallel Computing Systems (DPCS) for Small and Medium Enterprises (SMEs), which often lack advanced technical skills and modern equipment. In a globalized and dynamic environment, SMEs may become more competitive by obtaining a higher number of computational resources. DPCS offer two possible solutions: the first is to acquire them from an external provider (e.g., a public cloud), and the second consists in seizing the underutilized resources, which is the same aim that characterizes Volunteer Computing but focusing on owned resources. The most relevant advantages are related to data-privacy, costs, and environmental footprint.

We have presented SmartMonkey, a browser-based platform for Distributed Computing. It constitutes a flexible, cheap and simple approach, since it relies on available resources, do not require download/install software, and can be initialized just by accessing a website from a web browser. A JavaScript code embedded in the website allows the communication between the networked resources. Note that this technology is easily scalable due to its simplicity and the possibilities it offers for aggregating new resources.

The potential of DPCS in Mathematical Optimization has already been highlighted in the related literature. Numerous SMEs frequently face complex Combinatorial Optimization Problems (COPs) that require a real-time solution (e.g., flight rescheduling or online routing). Since exact methods are usually incapable of addressing real-sized problem instances in short computing times, approximate methods are widely used. In particular, randomized algorithms may provide a number of high-quality solutions by being repeatedly executed with a different seed. This property suggests an easy way to parallelize the resolution of a problem instance in which each aggregated machine solves the problem instance with a given seed and only the best found solution is returned.

The applicability of the described platform is tested on two classic COPs in the fields of transportation and production. The proposed approach relies on the master-slave architecture and analyzes how solutions vary with time and as the number of available agents is also modified. For the instances analyzed, our approach is able to provide high-quality solutions in real-time.

Acknowledgments. This work has been partially supported by the Spanish Ministry of Economy and Competitiveness (grant TRA2013-48180-C3-P), FEDER, and the Department of Universities, Research & Information Society of the Catalan Government (Grant 2014-CTP-00001). We are also grateful to Incubio for their support during the development of this research work.

References

1. Anderson, D.: BOINC: a system for public-resource computing and storage. In: Proceedings of the 5th IEEE/ACM International Workshop on Grid Computing, pp. 4–10 (2004)
2. Armbrust, M., Fox, A., Griffith, R., Joseph, A., Katz, R., Konwinski, A., Lee, G., Patterson, D., Rabkin, A., Stoica, I., Zaharia, M.: Above the Clouds: A Berkeley View of Cloud Computing. Technical report No. UCB/EECS200928 (2009)

3. Berry, K.: Distributed and Grid Computing via the Browser. http://www.csc.villanova.edu/~tway/courses/csc3990/f2009/csrs2009/Kevin_Berry_Grid_Computing_CSRS_2009.pdf. Accessed 14 Sept 2017
4. Buyya, R., Venugopal, S.: A gentle introduction to grid computing and technologies. CSI Commun. 9–19 (2005)
5. Cabrera, G.: Service allocation methodologies for contributory computing environments. Doctoral dissertation. Open University of Catalonia, Spain (2014)
6. Cerin, C., Fedak, G.: Desktop Grid Computing. CRC Press, Boca Raton (2012)
7. Chien, A., Calder, B., Elbert, S., Bhatia, K.: Entropia: architecture and performance of an enterprise desktop grid system. J. Parallel Distrib. Comput. 63(5), 597–610 (2003)
8. Clarke, G., Wright, J.: Scheduling of vehicles from a central depot to a number of delivering points. Oper. Res. 12, 568–581 (1964)
9. Coulouris, G., Dollimore, J., Kindberg, T.: Distributed Systems: Concepts and Design, 4th edn. Pearson Education, Harlow (2005)
10. Foster, I.: What is the Grid? A Three Point Checklist (2002). http://dlib.cs.odu.edu/WhatIsTheGrid.pdf. Accessed 14 Sept 2017
11. Foster, I., Kesselman, C.: The Grid 2: Blueprint for a New Computing Infrastructure. Elsevier, Amsterdam (2003)
12. Golden, B., Raghavan, S., Wasil, E. (eds.): The Vehicle Routing Problem: Latest Advances and New Challenges. Springer, New York (2008)
13. Juan, A., Cáceres, J., González, S., Riera, D., Barrios, B.: Biased randomization of classical heuristics. Encycl. Bus. Anal. Optim. 1, 314–324 (2014)
14. Juan, A., Faulin, J., Jorba, J., Cáceres, J., Marquès, J.: Using parallel & distributed computing for solving real-time vehicle routing problems with stochastic demands. Ann. Oper. Res. 207, 43–65 (2013)
15. Juan, A., Faulin, J., Jorba, J., Riera, D., Masip, D., Barrios, B.: On the use of Monte Carlo simulation, cache and splitting techniques to improve the Clarke and Wright savings heuristics. J. Oper. Res. Soc. 62(6), 1085–1097 (2010)
16. Juan, A., Lourenço, H.R., Mateo, M., Luo, R., Castella, Q.: Using iterated local search for solving the flow-shop problem: parallelization, parametrization, and randomization issues. Int. Trans. Oper. Res. 21(1), 103–126 (2014)
17. Litzkow, M., Livny, M., Mutka, M.: Condor-a hunter of idle workstations. In: Proceedings of 8th International Conference on Distributed Computing Systems, pp. 104–111 (1998)
18. Lourenço H.R., Martin O., Stützle T.: Iterated local search: framework and applications. In: Gendreau, M., Potvin, J.Y. (eds.) Handbook of Metaheuristics. 2nd edn. Springer, Boston (2010)
19. Nadiminti, K., Buyya, R.: Enterprise Grid computing: State-of-the-Art (2005). http://www.cloudbus.org/reports/EOSJArticleTR05.pdf Accessed 14 Sept 2017
20. Nawaz, M., Enscore, E., Ham, I.: A heuristic algorithm for the m-machine, n-job flowshop sequencing problem. OMEGA 11, 91–95 (1983)
21. Sarmenta, L.F.G.: Volunteer Computing. Doctoral dissertation. Massachusetts Institute of Technology (2001)
22. Taillard, E.: Benchmark for basic scheduling problems. Eur. J. Oper. Res. 64, 278–285 (1993)
23. Talbi, E. (ed.): Parallel Combinatorial Optimization. Wiley, Chichester (2006)
24. Talbi, E. (ed.): Metaheuristics: From Design to Implementation. Wiley, Chichester (2009)

Synthesis of Analytic Models for Subtraction of Fuzzy Numbers with Various Membership Function's Shapes

Yuriy P. Kondratenko[1($\boxtimes$)] and Nina Y. Kondratenko[2]

[1] Department of Intelligent Information Systems, Petro Mohyla Black Sea State University, 68-th Desantnykiv Street 10, Mykolaiv 54003, Ukraine
y_kondrat2002@yahoo.com
[2] Darla Moore School of Business, University of South Carolina, 901 Sumter Street, Byrnes Suite 301, Columbia, SC 29208, USA
nykondratenko@gmail.com

Abstract. In this paper authors present new universally applicable analytical models of the result's MFs with the description of synthesis procedures for the subtraction operation with triangular fuzzy numbers and various shapes of MFs. The general soft computing analytic models are given based on the developed library consisting of the 16 general resulting models. Specific properties of the developed soft computing models are discussed with interpretation to ship bunkering problem.

Keywords: Fuzzy number · Membership Function · Shape · Subtraction Library of models

1 Introduction

For mathematical formalization of various processes and systems in uncertainty it is feasible to use the theory of fuzzy sets and fuzzy logic [16]. The specialists have a great interest in such intelligent approach in terms of practical applications of its mathematical methods in different fields: business process management, engineering, economics, finances as well as science and technology. There are multiple fundamental theoretical contributions to the developments of fuzzy sets and fuzzy logic theory made by scientists all over the world [1, 3, 5, 7–10, 13].

Fuzzy sets and fuzzy logic are used for the tasks of decision making in uncertainty, in particular for problems of routes and trajectory optimization, implementation of model-based approach to modeling and evaluation of collaborative processes, decision-making in medical diagnostics and forecasting epidemic processes in different regions of the world [9, 11, 15]. It is also widely used for improving efficiency of investment in uncertainty [1], modeling and decision-making in business process management, financial analysis [3] and engineering [6, 14], increasing efficiency of sport management and so on. To solve the abovementioned problems in most cases it is necessary to fulfill the fuzzy arithmetic operations with corresponding fuzzy sets, including such operations as addition, subtraction, multiplication and division [4, 7, 9, 12].

© Springer International Publishing AG, part of Springer Nature 2018
A. M. Gil-Lafuente et al. (Eds.): FIM 2015, AISC 730, pp. 87–100, 2018.
https://doi.org/10.1007/978-3-319-75792-6_8

Let us consider a fuzzy set $\underset{\sim}{A}$ as pairs $\left(x, \mu_{\underset{\sim}{A}}(x)\right)$, that is specified on the universal set E [12, 16] and any element $x, x \in E$ of the fuzzy set $\underset{\sim}{A}$ corresponds to a specific value of the membership function (MF) $\mu_{\underset{\sim}{A}}(x) \in [0, 1]$.

The inverse models of MFs of resulting fuzzy sets acquired after implementation of any arithmetic operations using α-cuts approach [4]

$$A_{\alpha} = \{x \big| \mu_{\underset{\sim}{A}}(x) \geq \alpha\}, \ \alpha \in [0, 1], x \in R,$$

do not always provide high performance of such soft computing operations. It also often leads to complications in solving real-time decision-making and control problems.

It is well-known that the computational algorithms for the arithmetic operations based on using α-cuts of the relevant fuzzy sets [4, 7, 9, 12] (inverse approach) have a high computational complexity, as it is performed in turn for all α-levels ($\alpha_i \in [0, 1]$, $i = 0, 1, 2, \ldots, r$, $\alpha_0 = 0$, $\alpha_r = 1$) with the step of discreteness $\Delta\alpha$. The value of $\Delta\alpha$ taking into consideration that $\alpha_{i+1} = \alpha_i + \Delta\alpha$ significantly affects the accuracy and operating speed of the computational procedures performance [7, 12].

Thus, the development of corresponding analytic models based on generalized direct approach allows formalizing the procedures of fuzzy arithmetic operations and improving such parameters as dependability, operating speed and accuracy of their realizations [7, 9].

The aim of this paper is to synthesize the library of analytical models of the resulting MFs for arithmetic operation (subtraction) with various MF's shape of fuzzy sets providing an opportunity to significantly reduce the volume and complexity, improve its operating speed as well as the accuracy of the fuzzy information processing in computer decision support systems (CDSS).

2 Inverse and Direct Models for Subtraction of TrFNs

The triangular fuzzy number (TrFN) $\underset{\sim}{A} = (a_1, a_0, a_2)$ is called fuzzy number $\underset{\sim}{A}$ that has a MF $\mu_{\underset{\sim}{A}}(x)$ of triangular shape, where $\mu_{\underset{\sim}{A}}(a_1) = 0$; $\mu_{\underset{\sim}{A}}(a_0) = 1$; $\mu_{\underset{\sim}{A}}(a_2) = 0$. Generalized model A_{α}, synthesized on the basis of inverse approach, and direct model in a form of triangular MF $\mu_{\underset{\sim}{A}}(x)$ of the TrFN $\underset{\sim}{A}$ are determined [4] by the appropriate relevant dependencies (1) and (2):

$$A_{\alpha} = [a_1(\alpha), a_2(\alpha)] = [a_1 + \alpha(a_0 - a_1), a_2 - \alpha(a_2 - a_0)], \tag{1}$$

$$\mu_{\underset{\sim}{A}}(x) = \begin{cases} 0, \forall(x \leq a_1) \cup (x \geq a_2) \\ (x - a_1)/(a_0 - a_1), \forall(a_1 < x \leq a_0) \\ (a_2 - x)/(a_2 - a_0), \forall(a_0 < x < a_2) \end{cases}. \tag{2}$$

Subsets A_α and B_α of fuzzy sets $\underset{\sim}{A}$ and $\underset{\sim}{B}$ determine the appropriate α-cuts, which can be written as follows: $A_\alpha = [a_1(\alpha), a_2(\alpha)]$, $B_\alpha = [b_1(\alpha), b_2(\alpha)]$, where $\alpha \in [0, 1]$, $\underset{\sim}{A}, \underset{\sim}{B} \in R^+$, and inverse models for arithmetic operations of addition, subtraction, multiplication and division can be written as [4, 7, 9, 12]

$$A_\alpha(+)B_\alpha = [a_1(\alpha) + b_1(\alpha), \, a_2(\alpha) + b_2(\alpha)]. \tag{3}$$

$$A_\alpha(-)B_\alpha = [a_1(\alpha) - b_2(\alpha), \, a_2(\alpha) - b_1(\alpha)]. \tag{4}$$

$$A_\alpha(\cdot)B_\alpha = [a_1(\alpha)b_1(\alpha), \, a_2(\alpha)b_2(\alpha)]. \tag{5}$$

$$A_\alpha(:)B_\alpha = [a_1(\alpha)/b_2(\alpha), \, a_2(\alpha)/b_1(\alpha)]. \tag{6}$$

Direct analytic model $\mu_{\underset{\sim}{C}}(x)$ of the resulting MF, for example for subtraction $\underset{\sim}{C} = \underset{\sim}{A}(-)\underset{\sim}{B}$, synthesized using [4, 7], can be presented as follows:

$$\mu_{\underset{\sim}{C}}(x) = \begin{cases} \forall x \in R: \\ 0, \, \forall (x \le a_1 - b_2) \cup (x \ge a_2 - b_1) \\ (x - a_1 + b_2)/(a_0 - b_0 - a_1 + b_2), \forall (a_1 - b_2 < x \le a_0 + b_0) \\ (a_2 - b_1 - x)/(a_2 - b_1 - a_0 + b_0), \forall (a_0 - b_0 < x < a_2 + b_1) \end{cases} ; \tag{7}$$

In addition the calculations based on the mentioned above α-cuts [4, 12] for implementation of fuzzy arithmetic operations where computational algorithms are often used and realized through the use of max-min or min-max convolutions [12] in some cases lead to increased complexity and reduced operating speed of performance or to the moment of obtaining the resulting MFs that does not meet the requirements of convexity and normality of fuzzy sets.

3 Synthesis of Masks of TrFNs with Various Shapes of MFs

The direct models $\mu_{\underset{\sim}{C}}(x)$ represented by Eqs. (7)–(10) for different arithmetic operations are validated only for TrFNs $\underset{\sim}{A} = (a_1, a_0, a_2)$ and $\underset{\sim}{B} = (b_1, b_0, b_2)$ under the following conditions (Fig. 1a): $a_1 < a_0 < a_2, b_1 < b_0 < b_2$.

At the same time a lot of real input values for decision-making processes can be presented as TrFNs with different shapes of MF, in particular, for $\underset{\sim}{B} = (b_1, b_0, b_0)$ as presented in Fig. 1b, etc. So, for each concrete case a decision-maker should develop the analytic model of resulting fuzzy set for implementation of corresponding arithmetic operation if the TrFNs, for example, $\left(\underset{\sim}{A}, \underset{\sim}{B}\right)$ have different shapes of MFs. The

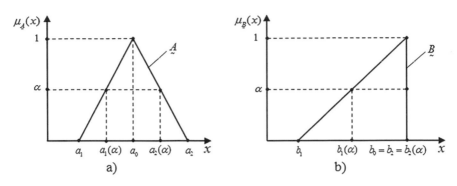

Fig. 1. TrFNs with different shape of MF: (a) $a_1 < a_0 < a_2$; (b) $b_1 < b_0 = b_2$

main idea of this research is to develop a library of inverse and direct analytic models of the resulting fuzzy sets $\tilde{C}$, for realization of subtraction as arithmetic operation with TrFNs A and B with various combinations of MF's shapes.

Proposition 1. It is a proposition to form a masks [9, 11] of TrFNs $\underset{\sim}{A}$ and $\underset{\sim}{B}$ with corresponding shape of MF in the following way

$$\text{mask } \underset{\sim}{A} = \{s,n\}, \text{ mask } \underset{\sim}{B} - \{m,r\},$$

where indicators n, s, r and m are defined as

$$n = \begin{cases} 0, & \text{if} \quad a_0 < a_2 \\ 1, & \text{if} \quad a_0 = a_2 \end{cases}; s = \begin{cases} 0, & \text{if} \quad a_0 > a_1 \\ 1, & \text{if} \quad a_0 = a_1 \end{cases}; r = \begin{cases} 0, & \text{if} \quad b_0 < b_2 \\ 1, & \text{if} \quad b_0 = b_2 \end{cases};$$

$$m = \begin{cases} 0, & \text{if} \quad b_0 > b_1 \\ 1, & \text{if} \quad b_0 = b_1 \end{cases}.$$

Proposition 2. It is a proposition to represent each pair $\left(\underset{\sim}{A}, \underset{\sim}{B}\right)$ [9, 11] by mask $\left(\underset{\sim}{A}, \underset{\sim}{B}\right)$, which can be defined as combination of mask $\underset{\sim}{A} = \{s,n\}$ and mask $\underset{\sim}{B} = \{m,r\}$ in the following way

$$\text{mask}\left(\underset{\sim}{A}, \underset{\sim}{B}\right) = \{s,n,m,r\}. \tag{8}$$

Let us denote and form a 16-component's library of resulting mathematical models $\{M_1 \ldots M_{16}\}$ for subtraction of all possible combinations of TrFN $\left(\underset{\sim}{A}, \underset{\sim}{B}\right)$ with different shapes of MFs according to corresponding mask $\left(\underset{\sim}{A}, \underset{\sim}{B}\right) = \{s, n, m, r\}$:

Model M_1 : mask $\left(\underset{\sim}{A}, \underset{\sim}{B}\right) = \{s, n, m, r\} = \{0, 0, 0, 0\}; a_1 < a_0 < a_2; b_1 < b_0 < b_2;$

Model M_2 : mask $\left(\underset{\sim}{A}, \underset{\sim}{B}\right) = \{s, n, m, r\} = \{0, 0, 0, 1\}; a_1 < a_0 < a_2; b_1 < b_0 = b_2;$

Model M_3 : mask $\left(\underset{\sim}{A}, \underset{\sim}{B}\right) = \{s, n, m, r\} = \{0, 0, 1, 0\}; a_1 < a_0 < a_2; b_1 = b_0 < b_2;$

Model M_4 : mask $\left(\underset{\sim}{A}, \underset{\sim}{B}\right) = \{s, n, m, r\} = \{0, 0, 1, 1\}; a_1 < a_0 < a_2; b_1 = b_0 = b_2;$

Model M_5 : mask $\left(\underset{\sim}{A}, \underset{\sim}{B}\right) = \{s, n, m, r\} = \{0, 1, 0, 0\}; a_1 < a_0 = a_2; b_1 < b_0 < b_2;$

Model M_6 : mask $\left(\underset{\sim}{A}, \underset{\sim}{B}\right) = \{s, n, m, r\} = \{0, 1, 0, 1\}; a_1 < a_0 = a_2; b_1 < b_0 = b_2;$

Model M_7 : mask $\left(\underset{\sim}{A}, \underset{\sim}{B}\right) = \{s, n, m, r\} = \{0, 1, 1, 0\}; a_1 < a_0 = a_2; b_1 = b_0 < b_2;$

Model M_8 : mask $\left(\underset{\sim}{A}, \underset{\sim}{B}\right) = \{s, n, m, r\} = \{0, 1, 1, 1\}; a_1 < a_0 = a_2; b_1 = b_0 = b_2;$

Model M_9 : mask $\left(\underset{\sim}{A}, \underset{\sim}{B}\right) = \{s, n, m, r\} = \{1, 0, 0, 0\}; a_1 = a_0 < a_2; b_1 < b_0 < b_2;$

Model M_{10} : mask $\left(\underset{\sim}{A}, \underset{\sim}{B}\right) = \{s, n, m, r\} = \{1, 0, 0, 1\}; a_1 = a_0 < a_2; b_1 < b_0 = b_2;$

Model M_{11} : mask $\left(\underset{\sim}{A}, \underset{\sim}{B}\right) = \{s, n, m, r\} = \{1, 0, 1, 0\}; a_1 = a_0 < a_2; b_1 = b_0 < b_2;$

Model M_{12} : mask $\left(\underset{\sim}{A}, \underset{\sim}{B}\right) = \{s, n, m, r\} = \{1, 0, 1, 1\}; a_1 = a_0 < a_2; b_1 = b_0 = b_2;$

Model M_{13} : mask $\left(\underset{\sim}{A}, \underset{\sim}{B}\right) = \{s, n, m, r\} = \{1, 1, 0, 0\}; a_1 = a_0 = a_2; b_1 < b_0 < b_2;$

Model M_{14} : mask $\left(\underset{\sim}{A}, \underset{\sim}{B}\right) = \{s, n, m, r\} = \{1, 1, 0, 1\}; a_1 = a_0 = a_2; b_1 < b_0 = b_2;$

Model M_{15} : mask $\left(\underset{\sim}{A}, \underset{\sim}{B}\right) = \{s, n, m, r\} = \{1, 1, 1, 0\}; a_1 = a_0 = a_2; b_1 = b_0 < b_2;$

Model M_{16} : mask $\left(\underset{\sim}{A}, \underset{\sim}{B}\right) = \{s, n, m, r\} = \{1, 1, 1, 1\}; a_1 = a_0 = a_2; b_1 = b_0 = b_2.$

The authors illustrate the method [10] of forming the inverse $C_\alpha = [c_1(\alpha), c_2(\alpha)]$ and direct $\mu_C(x)$ generalized analytical models of the resulting fuzzy sets $\underset{\sim}{C} = \underset{\sim}{A}(*)\underset{\sim}{B}$ for any arithmetic operation $(*) \in \{(+), (-), (\times), (:)\}$ with a pair $\left(\underset{\sim}{A}, \underset{\sim}{B}\right)$ of TrFNs using, as an example, the arithmetic operation "subtraction": $\underset{\sim}{C} = \underset{\sim}{A}(-)\underset{\sim}{B}.$

Preliminary, we form the inverse generalized models $A_\alpha = [a_1(\alpha), a_2(\alpha)]$ and $B_\alpha = [b_1(\alpha), b_2(\alpha)]$ for the TrFNs $\underset{\sim}{A} = (a_1, a_0, a_2)$ and $\underset{\sim}{B} = (b_1, b_0, b_2)$ in the set of non-negative real numbers R^+, $\underset{\sim}{A}, \underset{\sim}{B} \subset R^+$:

$$A_\alpha = [a_1(\alpha), a_2(\alpha)] = [a_1 + \alpha(a_0 - a_1), a_2 - \alpha(a_2 - a_0)], \qquad (9)$$

$$B_\alpha = [b_1(\alpha), b_2(\alpha)] = [b_1 + \alpha(b_0 - b_1), b_2 - \alpha(b_2 - b_0)]. \tag{10}$$

4 Synthesis of Inverse and Direct Analytic Models for Subtraction of TrFNs with Various MF's Shapes

The procedure of the inverse and direct analytic models synthesis [10] for the arithmetic operation "subtraction" with TrFNs contains several steps.

Step 1. Based on (4) and (5) we can receive an inverse model C_α for $\alpha-$ cut of a fuzzy set $\underset{\sim}{C} = \underset{\sim}{A}(-)\underset{\sim}{B} = (c_1, c_0, c_2)$ as TrFN:

$$C_\alpha = A_\alpha(-)B_\alpha = [a_1(\alpha) - b_2(\alpha), a_2(\alpha) - b_1(\alpha)] = [c_1(\alpha), c_2(\alpha)], \tag{11}$$

where $c_1(\alpha) = a_1(\alpha) - b_2(\alpha); c_2(\alpha) = a_2(\alpha) - b_1(\alpha)$.

Step 2. Using the Eq. (11) for the inverse submodel $c_1(\alpha)$ of the left branch of the resulting MF $\underset{\sim}{C} = \underset{\sim}{A}(-)\underset{\sim}{B}$ we can find the parameter α, which will be a function $f_L(\cdot)$ of such known arguments as $c_1(\alpha), a_1, a_0, b_1, b_0$:

$$\alpha = f_L[c_1(\alpha), a_1, a_0, b_1, b_0].$$

Step 3. The transition from inverse to direct approach [4, 7, 9, 10] shows that x is a parameter of the function $\alpha = f_L[c_1(\alpha), a_1, a_0, b_1, b_0]$, that is

$$\alpha = f_L[x, a_1, a_0, b_1, b_0], \ x = c_1(\alpha) \in [c_1, c_0].$$

Substituting $\mu_{\underset{\sim}{C}}(x)$ for α allows us to get a direct model of the left branch of the resulting MF represented as

$$\forall x \in [c_1, c_0] : \ \mu_{\underset{\sim}{C}}(x) = f_L[x, a_1, a_0, b_1, b_0].$$

Step 4. Using the Eq. (11) for the inverse submodel $c_2(\alpha)$ of right branch of resulting MF $\underset{\sim}{C} = \underset{\sim}{A}(-)\underset{\sim}{B}$ we can find the parameter α, which will be a function $f_R(\cdot)$ of such known arguments as $c_2(\alpha), a_2, a_0, b_2, b_0$:

$$\alpha = f_R[c_2(\alpha), a_2, a_0, b_2, b_0].$$

Step 5. The transition from the inverse to direct approach [4, 7, 9, 10] shows that x is a parameter of the function $\alpha = f_R[c_2(\alpha), a_2, a_0, b_2, b_0]$, that is

$$\alpha = f_R[x, a_2, a_0, b_2, b_0],$$

where $x = c_2(\alpha) \in [c_0, c_2]$.

Substituting $\mu_C(x)$ instead of α allows us to get a direct model of the right branch of the resulting MF represented as

$$\forall x \in [c_0, c_2] : \quad \mu_C(x) = f_R[x, a_2, a_0, b_2, b_0].$$

Step 6. Using the results of *Step 3* and *Step 5* we can get a full direct analytical model for resulting MF of fuzzy set C after subtraction $A(-)B$ of two TrFNs:

$$\mu_C(x) = \begin{cases} 0, & \forall(x \le c_1) \cup (x \ge (c_2)) \\ f_L[x, a_1, a_0, b_1, b_0], & \forall(c_1 < x \le c_0) \\ f_R[x, a_2, a_0, b_2, b_0], & \forall(c_0 < x \le c_2) \end{cases}. \tag{12}$$

The authors represent the library of developed inverse $C_\alpha = [c_1(\alpha), c_2(\alpha)]$ and direct $\mu_C(x)$ models $\{M_1, M_2, \ldots, M_{16}\}$, which were synthesized based on the abovementioned 6-steps procedure for different shapes of TrFNs A and B.

Library's model M_1:

$$\begin{aligned} C_\alpha = A_\alpha(-)B_\alpha &= [a_1(\alpha), a_2(\alpha)](-)[b_1(\alpha), b_2(\alpha)] \\ &= [(a_1 - b_2) + \alpha((a_0 + b_2) - (a_1 + b_0)), (a_2 - b_1) - \alpha((a_2 + b_0) - (a_0 + b_1))]; \end{aligned}$$
$$(13)$$

$$\mu_C(x) = \begin{cases} 0, & \forall(x \le a_1 - b_2) \cup (x \ge a_2 - b_1) \\ (x - a_1 + b_2)/(a_0 + b_2 - a_1 - b_0), & \forall(a_1 - b_2 < x \le a_0 - b_0) \\ (a_2 - b_1 - x)/(a_2 + b_0 - a_0 - b_1), & \forall(a_0 - b_0 < x < a_2 - b_1) \end{cases}.$$
$$(14)$$

Library's model M_2:

$$\begin{aligned} C_\alpha = A_\alpha(-)B_\alpha &= [a_1(\alpha), a_2(\alpha)](-)[b_1(\alpha), b_0] \\ &= [(a_1 - b_0) + \alpha(a_0 - a_1), (a_2 - b_1) - \alpha((a_2 + b_0) - (a_0 + b_1))]; \end{aligned} \tag{15}$$

$$\mu_C(x) = \begin{cases} 0, & \forall(x \le a_1 - b_0) \cup (x \ge a_2 - b_1) \\ (x - a_1 + b_0)/(a_0 - a_1), & \forall(a_1 - b_0 < x \le a_0 - b_0) \\ (a_2 - b_1 - x)/(a_2 + b_0 - a_0 - b_1), & \forall(a_0 - b_0 < x < a_2 - b_1) \end{cases}.$$
$$(16)$$

Library's model M_3:

$$\begin{aligned} C_\alpha = A_\alpha(-)B_\alpha &= [a_1(\alpha), a_2(\alpha)](-)[b_0, b_2(\alpha)] \\ &= [(a_1 - b_2) + \alpha((a_0 + b_2) - (a_1 + b_0)), (a_2 - b_0) - \alpha(a_2 - a_0)]; \end{aligned} \tag{17}$$

$$\mu_{\underset{\sim}{C}}(x) = \begin{cases} 0, & \forall(x \le a_1 - b_2) \cup (x \ge a_2 - b_0) \\ (x - a_1 + b_2)/(a_0 + b_2 - a_1 - b_0), & \forall(a_1 - b_2 < x \le a_0 - b_0) \\ (a_2 - b_0 - x)/(a_2 - a_0), & \forall(a_0 - b_0 < x < a_2 - b_0) \end{cases}.$$

(18)

Library's model M_4:

$$\begin{aligned} C_\alpha = A_\alpha(+)B_\alpha &= [a_1(\alpha), a_2(\alpha)](-)[b_0, b_0] \\ &= [(a_1 - b_0) + \alpha(a_0 - a_1), (a_2 - b_0) - \alpha(a_2 - a_0)]; \end{aligned}$$

(19)

$$\mu_{\underset{\sim}{C}}(x) = \begin{cases} 0, & \forall(x \le a_1 - b_0) \cup (x \ge a_2 - b_0) \\ (x - a_1 + b_0)/(a_0 - a_1), & \forall(a_1 - b_0 < x \le a_0 - b_0) \\ (a_2 - b_0 - x)/(a_2 - a_0), & \forall(a_0 - b_0 < x < a_2 - b_0) \end{cases}.$$

(20)

Library's model M_5:

$$\begin{aligned} C_\alpha = A_\alpha(+)B_\alpha &= [a_1(\alpha), a_0](-)[b_1(\alpha), b_2(\alpha)] \\ &= [(a_1 - b_2) + \alpha((a_0 + b_2) - (a_1 + b_0)), (a_0 - b_1) - \alpha(b_0 - b_1)]; \end{aligned}$$

(21)

$$\mu_{\underset{\sim}{C}}(x) = \begin{cases} 0, & \forall(x \le a_1 - b_2) \cup (x \ge a_0 - b_1) \\ (x - a_1 + b_2)/(a_0 + b_2 - a_1 - b_0), & \forall(a_1 - b_2 < x \le a_0 - b_0) \\ (a_0 - b_1 - x)/(b_0 - b_1), & \forall(a_0 - b_0 < x < a_0 - b_1) \end{cases}.$$

(22)

Library's model M_6:

$$\begin{aligned} C_\alpha = A_\alpha(-)B_\alpha &= [a_1(\alpha), a_0](-)[b_1(\alpha), b_0] \\ &= [(a_1 - b_0) + \alpha(a_0 - a_1), (a_0 - b_1) - \alpha(b_0 - b_1)]; \end{aligned}$$

(23)

$$\mu_{\underset{\sim}{C}}(x) = \begin{cases} 0, & \forall(x \le a_1 - b_0) \cup (x \ge a_0 - b_1) \\ (x - a_1 + b_0)/(a_0 - a_1), & \forall(a_1 - b_0 < x \le a_0 - b_0) \\ (a_0 - b_1 - x)/(b_0 - b_1), & \forall(a_0 - b_0 < x < a_0 - b_1) \end{cases}.$$

(24)

Library's model M_7:

$$\begin{aligned} C_\alpha = A_\alpha(-)B_\alpha &= [a_1(\alpha), a_0](-)[b_0, b_2(\alpha)] \\ &= [(a_1 - b_2) + \alpha((a_0 + b_2) - (a_1 + b_0)), a_0 - b_0]; \end{aligned}$$

(25)

$$\mu_{\underset{\sim}{C}}(x) = \begin{cases} 0, & \forall(x \le a_1 - b_2) \cup (x \ge a_0 - b_0) \\ (x - a_1 + b_2)/(a_0 + b_2 - a_1 - b_0), & \forall(a_1 - b_2 < x < a_0 - b_0) \\ 1, & \forall(x = a_0 - b_0) \end{cases}.$$

(26)

Library's model M_8:

$$C_\alpha = A_\alpha(-)B_\alpha = [a_1(\alpha), a_0](-)[b_0, b_0] = [(a_1 - b_0) + \alpha(a_0 - a_1), a_0 - b_0]; \quad (27)$$

$$\mu_{\underset{\sim}{C}}(x) = \begin{cases} 0, & \forall(x \leq a_1 - b_0) \cup (x \geq a_0 - b_0) \\ (x - a_1 + b_0)/(a_0 - a_1), & \forall(a_1 - b_0 < x < a_0 - b_0) \\ 1, & \forall(x = a_0 - b_0) \end{cases}. \quad (28)$$

Library's model M_9:

$$\begin{aligned} C_\alpha = A_\alpha(-)B_\alpha &= [a_0, a_2(\alpha)](-)[b_1(\alpha), b_2(\alpha)] \\ &= [(a_0 - b_2) + \alpha(b_2 - b_0), (a_2 - b_1) - \alpha((a_2 + b_0) - (a_0 + b_1))]; \end{aligned} \quad (29)$$

$$\mu_{\underset{\sim}{C}}(x) = \begin{cases} 0, & \forall(x \leq a_0 - b_2) \cup (x \geq a_2 - b_1) \\ (x - a_0 + b_2)/(b_2 - b_0), & \forall(a_0 - b_2 < x \leq a_0 - b_0) \\ (a_2 - b_1 - x)/(a_2 + b_0 - a_0 - b_1), & \forall(a_0 - b_0 < x < a_2 - b_1) \end{cases}. \quad (30)$$

Library's model M_{10}:

$$\begin{aligned} C_\alpha = A_\alpha(-)B_\alpha &= [a_0, a_2(\alpha)](-)[b_1(\alpha), b_0] \\ &= [a_0 - b_0, (a_2 - b_1) - \alpha((a_2 + b_0) - (a_0 + b_1))]; \end{aligned} \quad (31)$$

$$\mu_{\underset{\sim}{C}}(x) = \begin{cases} 0, & \forall(x < a_0 - b_0) \cup (x \geq a_2 - b_1) \\ 1, & \forall(x = a_0 - b_0) \\ (a_2 - b_1 - x)/(a_2 + b_0 - a_0 - b_1), & \forall(a_0 - b_0 < x < a_2 - b_1) \end{cases}. \quad (32)$$

Library's model M_{11}:

$$\begin{aligned} C_\alpha = A_\alpha(-)B_\alpha &= [a_0, a_2(\alpha)](-)[b_0, b_2(\alpha)] \\ &= [(a_0 - b_2) + \alpha(b_2 - b_0), (a_2 - b_0) - \alpha(a_2 - a_0)]; \end{aligned} \quad (33)$$

$$\mu_{\underset{\sim}{C}}(x) = \begin{cases} 0, & \forall(x \leq a_0 - b_2) \cup (x \geq a_2 - b_0) \\ (x - a_0 + b_2)/(b_2 - b_0), & \forall(a_0 - b_2 < x \leq a_0 - b_0) \\ (a_2 - b_0 - x)/(a_2 - a_0), & \forall(a_0 - b_0 < x < a_2 - b_0) \end{cases}. \quad (34)$$

Library's model M_{12}:

$$C_\alpha = A_\alpha(-)B_\alpha = [a_0, a_2(\alpha)](-)[b_0, b_0] = [a_0 - b_0, (a_2 - b_0) - \alpha(a_2 - a_0)]; \quad (35)$$

$$\mu_{\underset{\sim}{C}}(x) = \begin{cases} 0, & \forall(x < a_0 - b_0) \cup (x \geq a_2 - b_0) \\ 1, & \forall(x = a_0 - b_0) \\ (a_2 - b_0 - x)/(a_2 - a_0), & \forall(a_0 - b_0 < x < a_2 - b_0) \end{cases}. \quad (36)$$

Library's model M_{13}:

$$\begin{aligned} C_\alpha &= A_\alpha(-)B_\alpha = [a_0, a_0](-)[b_1(\alpha), b_2(\alpha)] \\ &= [(a_0 - b_2) + \alpha(b_2 - b_0), (a_0 - b_1) - \alpha(b_0 - b_1)]; \end{aligned} \quad (37)$$

$$\mu_{\underset{\sim}{C}}(x) = \begin{cases} 0, & \forall(x \leq a_0 - b_2) \cup (x \geq a_0 - b_1) \\ (x - a_0 + b_2)/(b_2 - b_0), & \forall(a_0 - b_2 < x \leq a_0 - b_0) \\ (a_0 - b_1 - x)/(b_0 - b_1) & \forall(a_0 - b_0 < x < a_0 - b_1) \end{cases}. \quad (38)$$

Library's model M_{14}:

$$C_\alpha = A_\alpha(-)B_\alpha = [a_0, a_0](-)[b_1(\alpha), b_0] = [a_0 - b_0, (a_0 - b_1) - \alpha(b_0 - b_1)]; \quad (39)$$

$$\mu_{\underset{\sim}{C}}(x) = \begin{cases} 0, & \forall(x < a_0 - b_0) \cup (x \geq u_0 - b_1) \\ 1, & \forall(x = a_0 - b_0) \\ (a_0 - b_1 - x)/(b_0 - b_1), & \forall(a_0 - b_0 < x < a_0 - b_1) \end{cases}. \quad (40)$$

Library's model M_{15}:

$$C_\alpha = A_\alpha(-)B_\alpha = [a_0, a_0](-)[b_0, b_2(\alpha)] = [(a_0 - b_2) + \alpha(b_2 - b_0), a_0 - b_0]; \quad (41)$$

$$\mu_{\underset{\sim}{C}}(x) = \begin{cases} 0, & \forall(x \leq a_0 - b_2) \cup (x > a_0 - b_0) \\ (x - a_0 + b_2)/(b_2 - b_0), & \forall(a_1 - b_2 < x < a_0 - b_0) \\ 1, & \forall(x = a_0 - b_0) \end{cases}. \quad (42)$$

Library's model M_{16}:

$$C_\alpha = A_\alpha(-)B_\alpha = [a_0, a_0](-)[b_0, b_0] = [a_0 - b_0, a_0 - b_0]; \quad (43)$$

$$\mu_{\underset{\sim}{C}}(x) = \begin{cases} 0, & \forall(x < a_0 - b_0) \cup (x > a_0 - b_0) \\ 1, & \forall(x = a_0 - b_0) \end{cases}. \quad (44)$$

The main components [9] of the proposed algorithm for subtraction of TrFNs based on the library M1-M16 of analytic models (13)–(44) for different MF's shape are:

Step 1. Evaluation of mask $\{s, n, m, r\}$ of the TrFNs $\left(\underset{\sim}{A}, \underset{\sim}{B}\right)$ based on the masks $\{s, n\}$ for TrFN $\underset{\sim}{A} = (a_1, a_0, a_2)$ and $\{m, r\}$ for TrFN $\underset{\sim}{B} = (b_1, b_0, b_2)$.

Step 2. The choice of the corresponding analytic model $M^* \in \{M_1 \ldots M_{16}\}$ from the developed library of the models $\{M_1, M_2, \ldots, M_{16}\}$, according to mask $\{s, n, m, r\}$ (8).

Step 3. The choice of the direct model $\mu_C(x)$ of the resulting TrFN $C = A(-)B$.

Step 4. The calculation of the value of confidence level $\mu_C(x^*)$ for the desired argument x^* using the synthesized direct model $\mu_C(x)$.

Let us consider an example for realization of arithmetic operation "subtraction" for pair $\left(A, B\right)$ with the different MF's shape of TrFNs $A = (3, 3, 7)$, $B = (4, 9, 9)$.

Using the proposed algorithm we can determine in an automatic mode the corresponding mask $A = \{1, 0\}$, mask $B = \{0, 1\}$, mask of the pair $\left(A, B\right)$

$$\text{mask}\left(A, B\right) = \{s, n, m, r\} = \{1, 0, 0, 1\}$$

and the corresponding model M_{10} from the library of models $\{M_1, M_2, \ldots, M_{16}\}$, which includes preliminary synthesized inverse (31) and direct (32) models. The MF $\mu_C(x)$ of the resulting TrFN $C = A(-)B$ is presented as follows

$$\mu_C(x) = \begin{cases} 0, & \forall(x < -6) \cup (x \geq 3) \\ 1, & \forall(x = -6) \\ (3 - x)/9, & \forall(-6 < x < 3) \end{cases}.$$

Modeling results for different pairs of the TrFNs with various MF's shape confirm the high efficiency of the developed algorithm and library of models.

5 Implementation of Universal Analytic Models in Solving Real-Life Capacitated Vehicle Routing Problem with Fuzzy Demands

The implementation of developed direct analytic models (14), (16), (18), (20), (22), (24), (26), (28), (28), (32), (34), (36), (38), (40), (42), (44) for calculation of the resulting MFs $\mu_C(x)$, according to TrFNs with various shape of MF (Sect. 3), allows using one step automation mode for operation $C = A(-)B$. In many practical and theoretical cases, such direct analytic models $\mu_C(x) = \mu_{A(-)B}(x)$ may have efficient introducing to evaluation and decision support processes.

Let's consider an example with application of developed library of analytic models (13)–(44) in solving real-life capacitive vehicle routing problem with fuzzy demands [5, 13, 15].

One of the most important vehicle routing problems (VRP) [14] is a routing problem for bunkering tankers [5, 15]. Such kind of tankers should provide bunkering (transportation and unloading) operations for various served (ordered) ships located in different marine ports and open sea points. Marine practice shows that very often the information about fuel demands of served ships and ports is uncertain. Typically ship sends the order for quantity of fuel supply to bunkering company as an approximate value. It is possible to represent such kind of orders as fuzzy demands, for example, as fuzzy numbers with triangular membership function with various MF's shape. Taking into account the restricted fuel capacity D_{max} of each tanker and fuzzy demands of ships' orders, the well-known VRP will be converted to capacitate vehicle routing problem (CVRP) in uncertainty. The efficiency of preliminary bunkering operations planning can be evaluated by possibility to serve all ships' orders with maximum of possible quantity of unloaded fuel and with minimum length of the total tankers' routes.

The CVRP with fuzzy demands $q_j = \left(\underset{\sim}{q_j}, \hat{q}_j, \bar{q}_j \right)$ in nodes is considered in [5, 13, 15], where $\hat{q}_j$ is a value of membership function of triangular fuzzy number $\tilde{q}_j$ with $\mu(\hat{q}_i) = 1$; $\underset{\sim}{q_j}$ and $\bar{q}_j$ are the lowest and highest possible values of demand, respectively, $\mu\left(\underset{\sim}{q_j} \right) = 0$, $\mu(\bar{q}_j) = 0$.

Solving such kind of decision-making problems deals with implementation of corresponding arithmetic operation "subtraction" for calculating the remain cargo $\Delta\underset{\sim}{D}_i$ at tanker-refueller after serving each i-th point of destinations with fuzzy demands $\underset{\sim}{q_1}, \underset{\sim}{q_2}, \ldots, \underset{\sim}{q_i}, \ldots, \underset{\sim}{q_r}, (i = 1, \ldots, r)$ [5, 15]:

$$\Delta\underset{\sim}{D}_i = D_{max} - \underset{\sim}{q_1} - \underset{\sim}{q_2} - \ldots - \underset{\sim}{q_i} - \ldots - \underset{\sim}{q_r}, (i = 1, \ldots, r).$$

All these demands $\underset{\sim}{q_1}, \underset{\sim}{q_2}, \ldots, \underset{\sim}{q_r}$ may have various shapes of MFs for their representation as triangular fuzzy models [9, 15]. The usage of the developed in Sect. 4 fuzzy models (13)–(44) and algorithm allows increasing the efficiency of decision-making processes for planning and optimization of tanker's routes [5, 15] according to sequenced calculation procedure:

$$\Delta\underset{\sim}{D}_1 = D_{max} - \underset{\sim}{q_1};$$

$$\Delta\underset{\sim}{D}_2 = \Delta\underset{\sim}{D}_1 - \underset{\sim}{q_2} = D_{max} - \underset{\sim}{q_1} - \underset{\sim}{q_2};$$

$$\Delta\underset{\sim}{D}_i = \Delta\underset{\sim}{D}_{i-1} - \underset{\sim}{q_i} = D_{max} - \underset{\sim}{q_1} - \underset{\sim}{q_2} - \ldots - \underset{\sim}{q_i}; (i = 3, \ldots, r).$$

6 Conclusions

The resulting fuzzy models and algorithm developed and outlined in this paper may be successfully implemented for the arithmetic operations with fuzzy input data that can be presented as TrFNs with various shapes of MFs. The usage of the developed analytical models (13)–(44) has a significant advantage for accuracy of calculations, time of modeling and program implementation of the formed models in comparison with step by step models of arithmetic operations with triangular fuzzy numbers based on the algorithms of sorting and max-min convolutions [4, 12]. Modeling results for subtraction of different fuzzy numbers with TrFNs and various shapes of MFs confirm the efficiency of proposed universal calculation algorithm based on the library of 16 analytic models for different applications. Among the perspectives for developed library application are CDSS for decision making in business process management, estimation of software development costs, medical diagnostics, practical project planning, investment in uncertainty [1], decision making in political [2], sport and cultural management [6], financial analysis [3] and others.

References

1. Gil-Aluja, J.: Investment in Uncertainty. Kluwer Academic Publishers, Dordrecht (1999)
2. Gil-Lafuente, A.M., Merigo J.M.: Decision making techniques in political management. In: Lodwick, W.A., Kacprzhyk, J. (eds.) Fuzzy Optimization. Studies in Fuzziness and Soft Computing, vol. 254, pp. 389–405. Springer-Verlag, Heidelberg (2010)
3. Gil-Lafuente, A.M.: Fuzzy Logic in Financial Analysis. Studies in Fuzziness and Soft Computing, vol. 175. Springer, Berlin (2005)
4. Kaufmann, A., Gupta, M.: Introduction to Fuzzy Arithmetic: Theory and Applications. Van Nostrand Reinhold Company, New York (1985)
5. Kondratenko, G.V., Kondratenko, Y.P., Romanov, D.O.: Fuzzy models for capacitive vehicle routing problem in uncertainty. In: Proceedings of 17th International DAAAM Symposium "Intelligent Manufacturing and Automation: Focus on Mechatronics & Robotics", pp. 205–206. Vienna, Austria (2006)
6. Kondratenko, Y.P.: Robotics, automation and information systems: future perspectives and correlation with culture, sport and life science. In: Gil-Lafuente, A.M., Zopounidis, C. (eds.) Decision Making and Knowledge Decision Support Systems. Lecture Notes in Economics and Mathematical Systems, vol. 675, pp. 43–56. Springer International Publishing, Switzerland (2015)
7. Kondratenko, Y., Kondratenko, V.: Soft computing algorithm for arithmetic multiplication of fuzzy sets based on universal analytic models. In: Ermolayev, V. et al. (eds.) Information and Communication Technologies in Education, Research, and Industrial Application. ICTERI'2014. Communications in Computer and Information Science, vol. 469, pp. 49–77. Springer International Publishing Switzerland (2014)
8. Kondratenko, Y.P., Sidenko, I.V.: Decision-making based on fuzzy estimation of quality level for cargo delivery. In: Zadeh, L.A. et al. (eds.) Recent Developments and New Directions in Soft Computing. Studies in Fuzziness and Soft Computing, vol. 317, pp. 331–344. Springer International Publishing, Switzerland (2014)

9. Kondratenko, Y.P., Kondratenko, N.Y. Soft computing analytic models for increasing efficiency of fuzzy information processing in decision support systems. In: Hudson, R. (ed.) Decision Making: Processes, Behavioral Influences and Role in Business Management, pp. 41–78. Nova Science Publishers, New York (2015)
10. Kondratenko, Y.P., Kondratenko, V.Y., Kondratenko, N.Y.: The Method of Analytic Models Synthesis for Resulting Fuzzy Sets in Realization of Fuzzy Arithmetic Operations. Ukraine Patent for Utility Model №. 68118, Bulletin №. 5 (2012). (in Ukrainian)
11. Kotov, Y.B.: New Mathematical Approaches to the Problems of Medical Diagnostics. Editorial EPCC, Moscow (2004). (in Russian)
12. Piegat, A.: Fuzzy Modeling and Control. Springer, Heidelberg (2001)
13. Teodorovic, D., Pavkovich, G.: The fuzzy set theory approach to the vehicle routing problem when demand at nodes is uncertain. Fuzzy Sets Syst. **82**, 307–317 (1996)
14. Toth, P., Vigo, D. (eds.): The Vehicle Routing Problem. SIAM, Philadelphia (2002)
15. Werners, B., Kondratenko, Y.P.: Fuzzy multi-criteria optimisation for vehicle routing with capacity constraints and uncertain demands. In: Proceedings of the International Congress on Cost Control, pp. 145–159, 17–18 March 2011, Barcelona, Spain. CCID/ASEPUC, Barcelona (2011)
16. Zadeh, L.A.: Fuzzy sets. Inf. Control **8**, 338–353 (1965)

Knowledge-Based Decision Support System with Reconfiguration of Fuzzy Rule Base for Model-Oriented Academic-Industry Interaction

Yuriy P. Kondratenko$^{(\boxtimes)}$, Galyna V. Kondratenko, and Ievgen V. Sidenko

Department of Intelligent Information Systems, Petro Mohyla Black Sea National University, 68-th Desantnykiv Str. 10, Mykolaiv 54003, Ukraine
y_kondrat2002@yahoo.com, galvlad09@rambler.ru,
emoty@mail.ru

Abstract. In this work the current state of the problem, which consists in choice the rational model of academic-industry interaction such as "University – IT-company" is analyzed. To solve this problem it is developed and researched the intelligent decision support system (DSS) based on fuzzy logic for multi-criterion evaluation the most rational model of academic-industry inter-action such as "University – IT-company" in case of changing dimension of input coordinates vector.

Keywords: Decision support system · Fuzzy logic · Membership function
Linguistic term · Rule base · Reconfiguration · Academic-industry interaction

1 Introduction

The essential influence on the general development and integration level of informational technologies into Ukrainian or any national economy and into world market's segments is done by results of high-efficiency and mutually profitable interaction of universities and IT-companies. Herein implementation of the new models of interaction requires consideration and preliminary processing of large amount of input data, in particular, based on analysis of preliminary interaction experience of involved parties, their main achievements, competitiveness, advantages and directions for the development, scientific and educational levels of participants of future academic-industrial consortium, employment level of students, university professors and IT-companies, etc. [1, 10]. Incorrectly chosen model of interaction as well as non-observation of relevant conditions of collaboration within the interaction such as "University – IT-companies" can lead to undesired and unexpected consequences, including the loss of significant amount of intellectual and/or material resources, lowering educational-qualification level of specialists, appearing of limitation in education and development of ability to creative thinking [6, 7, 11, 13].

© Springer International Publishing AG, part of Springer Nature 2018
A. M. Gil-Lafuente et al. (Eds.): FIM 2015, AISC 730, pp. 101–112, 2018.
https://doi.org/10.1007/978-3-319-75792-6_9

2   The Statement of Researched Problem

Increase of interaction efficiency can be influenced by decision support systems (DSS) for model-oriented academic-industry interaction, which is developed on the basis of the latest methods, technologies, and approaches of system analysis, forecasting, fuzzy logics, neural networks, artificial intelligence, etc. [2, 5, 20, 21, 24, 32]. Usage of the above mentioned methods when designing modern DSS allows to process the essential amount of different-type information on a new level of intellectual interaction of a decision maker (DM) and computer system [26, 27]. Nowadays there is still an unsolved question of selecting partnership models based on developing the system of multi-criterion assessment of possible level of interaction between universities and IT-companies. Usage of such class DSS in some specific practical cases makes it possible to select the best variant of the model of interaction development such as "University – IT-company" [6, 13, 22, 28, 29].

The aim of this work is development and research DSS based on fuzzy logic to increase the efficiency of multi-criterion decision making processes for model-oriented interaction such as "University – IT-company" in case of changing dimension of input coordinates vector.

Preliminary researches and analysis of successful interaction experience within different-type consortia prove that nowadays solving the task of estimating the level of interaction between universities and IT-companies involves the selection of one of the four formed alternative models [6, 7] as alternative decision variants E_i, $(i = 1...4)$, where decision variant E_1 corresponds with the model A1 (interaction between university and IT-company in the sphere of education and study organization, knowledge sharing, targeted personnel training for IT-industry); variant E_2 – model A2 (organization and support of certification processes of interaction results); variant E_3 – model B (creating collective center of scientific researches, developing collective scientific projects); variant E_4 – model C (creation of student research groups with business orientation and realization of startups). Herein the efficiency of process of selecting interaction model essentially depends on chosen criterion x_j, $(j = 1, 2, ..., n)$, which characterizes each partner of the relevant future consortium such as "University – IT-company". Usage of fuzzy logics and hierarchical structure of input data (coordinates) when developing model-oriented DSS of such type allows to increase efficiency of multi-criterion selection of interaction model between universities and IT-companies, which is achieved by simplifying the process of formation and processing knowledge, taking into account significant amount of quality indicators and selection of optimal solution for a large amount of input expert information [26, 27, 32].

3   Structure and Rule Base Reconfiguration of Fuzzy DSS with Hierarchically-Organized Structure

In this paper there is considered an approach of designing fuzzy hierarchical DSS with the exception of defuzzyfication procedure on each previous hierarchical level and fuzzyfication procedure - on the next relevant hierarchical level. Herein the result of fuzzy logical conclusion of each subsystem of previous level of hierarchy directly (without additional processing) is passed to a block of fuzzy logical conclusion of relevant subsystem of each DSS hierarchy level [26, 27]. Informational technologies which realize DSS of such type, provide support to decision-making with high performance. Herein the quantity and complexity of computing operations significantly are reduced by eliminating intermediate "defuzzyfication-fuzzyfication" procedures between nearest hierarchical levels [14, 16–18, 23].

Let's consider a fuzzy one-level system (Fig. 1) that simulates dependency

$$y = F(x_1, x_2, \ldots, x_9),$$

where x_i, $i = 1, 2, \ldots, 9$ − input linguistic variables; y − output variable.

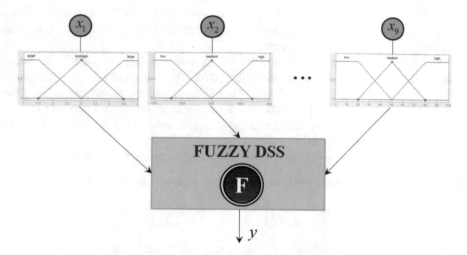

Fig. 1. The structure of one-level fuzzy DSS

Connections between input variables and output value are described with the help of fuzzy rules of one knowledge base f_1.

One of the possible variants of hierarchically-organized structure of fuzzy DSS we will design (Fig. 2) on the basis of decomposition of input coordinates vector $(x_1, x_2, \ldots, x_9)$ by joining them in the next i three-component group combination E_i:

$$E_i = \{\{x_1, x_2, x_3, x_4\}, \{x_5, x_6\}, \{x_7, x_8, x_9\}\}.$$

Herein relevant subsystems of one of the alternative structures of fuzzy DSS (Fig. 2), for example j, realize next functional dependencies

$$St_j = \{y_1, y_2, \ldots, y_4, y\} = \left\{ \begin{array}{l} y_1 = f_1(x_1, x_2, x_3, x_4), y_2 = f_2(x_5, x_6), \\ y_3 = f_3(x_7, x_8, x_9), y_4 = f_4(y_1, y_2), y = f_5(y_3, y_4) \end{array} \right\}.$$

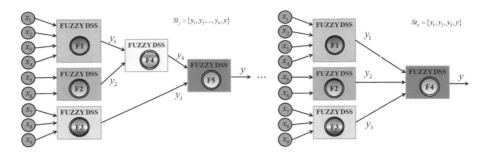

Fig. 2. The alternative hierarchical structures $St_j, \ldots, St_n$ of multi-level fuzzy DSS

The results of investigating [16–18, 26, 27] fuzzy hierarchically-organized systems prove that number of production rules significantly decreases in the process of the fuzzy DSS synthesis based on hierarchical decomposition of input $(x_1, x_2, \ldots, x_9)$ and intermediate (y_1, y_2, y_3, y_4) parameters.

Nowadays there is enough amount of software means to synthesize fuzzy hierarchically-organized DSS, in particular, MATLAB, CubiCalc, FuzzyTECH, etc. [27]. Alternative variants of DSS structures (Fig. 2), which are implemented with exception of intermediate "defuzzyfication-fuzzyfication" procedures between nearest hierarchical levels, allow on the designing stage to optimize the fuzzy DSS by its structural reconfiguration [25].

The main stages of structural reconfiguration procedure are as follows:

1. Synthesis of alternative variants of group combinations of input parameters of fuzzy DSS

$$\bar{E} = \{E_1, E_2, \ldots, E_i, \ldots, E_k\},$$

   where k - number of alternative variants of group combinations.
2. Synthesis of alternative variants of structural organization of fuzzy DSS

$$\bar{St} = \{St_1(E_1), St_2(E_2), \ldots, St_i(E_i), \ldots, St_k(E_k)\}.$$

3. Assessment of each alternative variant of structures $St_i(E_i)$ according to criterion of adequacy $K(St_i), i = \{1, \ldots, k\}$ and indicators of decision making quality.
4. Selection of optimal configuration of hierarchically-organized structure of the fuzzy DSS

$$St_{opt} = Arg \underset{i}{Max} K(St_i), i \in \{1, \ldots, k\}.$$

In most practical cases reconfiguration of hierarchical structure of the fuzzy DSS according to subparagraphs 1–4 is also appropriate when targets, indicators (criteria) of assessments, preferences system and DM priorities, etc. are changing.

In this study there is considered the developed by authors model-oriented DSS for selecting model $(m = 4)$ of interaction between universities and IT-companies according to preliminary proposed and defined criteria $(n = 27)$. The experience of professionals in the sphere of designing specialized fuzzy system of different purpose shows that with one-level structure of DSS in cases of large dimension of input coordinates vector $X = \{x_j\}, j = 1\ldots n$ sensitivity of their fuzzy rule bases to changes of input coordinates (criterion) values reduces $x_j, (j = 1, 2, \ldots, n)$ [15, 17, 18]. This is primarily due to complexity of creating relevant fuzzy rules to realize all possible dependences between input and output parameters of the system $y_k = f(x_1, x_2, \ldots, x_{27}), k = 1\ldots K$.

There is shown (Fig. 3) the variant of proposed by authors hierarchically-organized DSS St_s to select the best model $E^*, (E^* \in E, E = \{E_1, E_2, E_3, E_4\})$ of interaction between universities and IT-companies, which is created on the basis of input coordinates vector decomposition $X = \{x_j\}, j = 1\ldots 27$ with their association in the next s-group combination:

$$X_s = \left\{ \begin{array}{l} \{x_1, x_2, x_3\}, \{x_4, x_5, x_6, x_7\}, \{x_8, x_9, \ldots, x_{13}\}, \{x_{14}, x_{15}, x_{16}, x_{17}\}, \\ \{x_6, x_{18}, x_{19}\}, \{x_{18}, x_{19}, \ldots, x_{23}\}, \{x_{24}, x_{25}, x_{26}, x_{27}\} \end{array} \right\}.$$

Herein, corresponding subsystems of DSS (Fig. 3), among them $\{FES_1, FES_2, \ldots, FES_{10}, FES_{11}\}$, realize next functional dependencies for s alternative structure $St_s = \{y_1, y_2, \ldots, y_{10}, y\}$ of DSS:

$$St_s = \left\{ \begin{array}{l} y_1 = f_1(x_1, x_2, x_3), y_2 = f_2(x_4, x_5, x_6, x_7), y_3 = f_3(x_8, x_9, \ldots, x_{13}), \\ y_4 = f_4(x_{14}, x_{15}, x_{16}, x_{17}), y_5 = f_5(x_6, x_{18}, x_{19}), y_6 = f_6(x_{18}, x_{19}, \ldots, x_{23}), \\ y_7 = f_7(x_{24}, x_{25}, x_{26}, x_{27}), y_8 = f_8(y_1, y_2), y_9 = f_9(y_3, y_4), \\ y_{10} = f_{10}(y_5, y_6), y = f_{11}(y_7, y_8, y_9, y_{10}) \end{array} \right\}.$$

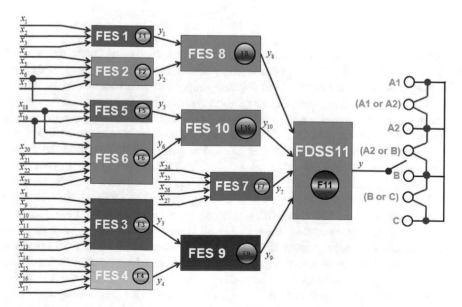

Fig. 3. The structure of knowledge-based fuzzy DSS for selecting model of interaction within consortia "University – IT-company"

So, for example, the fourth subsystem $y_4 = f_4(x_{14}, x_{15}, x_{16}, x_{17})$ for assessment of level of business orientation of department in relevant university is being created (Fig. 3) on the basis of four input coordinates (x_{14} – IT-certification of department teachers, x_{15} – number of business courses, x_{16} – experience in organizing student companies, x_{17} – experience in organizing mixed creative teams for execution and realization of IT-projects), which are combined according to common abilities, and one output coordinate (y_4 – level of business orientation of university department) with realization of relevant knowledge base, which includes 81 fuzzy rules of production type (Table 1). To design fuzzy rule bases for developed structure of DSS (Fig. 3) there are used linguistic terms $\{L, M, H\}$ for input coordinates, $\{L, LM, M, MH, H\}$ for intermediate and output coordinates with triangular shape of membership function [26, 27, 31, 32].

Table 1. Selective rule-set of knowledge base of the fourth subsystem

Number of rule	2	3	4		43	44	45	46		77	78	79
x_{14}	L	L	L	...	M	M	M	M	...	H	H	H
x_{15}	L	L	L	...	M	M	M	H	...	H	H	H
x_{16}	L	L	M		H	H	H	L		M	M	H
x_{17}	M	H	L	...	L	M	H	L	...	M	H	L
y_4	L	LM	L		M	M	MH	M		MH	H	MH

In the process of decision making using the fuzzy hierarchically-organized DSS with variable structure of input coordinates vector there appears a need to develop effective approaches to the reduction (reconfiguration) of rule bases of fuzzy models [16–18]. The necessity of relevant reconfiguration including input coordinates, which are at the option of DM excluded from the vector of input coordinates, appears in case of interaction between DM and DSS in interactive modes. In such modes DM can reduce the dimension of DSS vector of input coordinates, excluding from the further consideration those coordinates, the values of which DM does not know or cannot get accurately [27]. For example, synthesized DSS operates effectively with 15 input coordinates ($N = 15$), but in some cases DM can accurately estimate only 11 input signals ($N_E = 11$) and other 4 input signals ($N_{NE} = 4$) DM excludes from consideration, as they are not evaluated, $N_E + N_{NE} = N$. Herein the dimension of input coordinates vector X is reduced from 15 to 11. In the process of the fuzzy DSS operation with the fixed structure of knowledge bases and under variable structure of input data vector ($N_E < N$) the results of decision making y get deformed. This is due to the fact that meaning of input coordinates ($N_{NE} = 4$), which do not take part in designing fuzzy DSS and are equal to zero, through corresponding fuzzy rules influence negatively on the result y [17, 18].

One of the approaches to reconfiguration of DSS with relevant reduction of rule bases with variable structure of input coordinates vector is the approach that consists in identification of minor model parameters [26]. Herein the number of rules essentially decreases, that allows increasing sensitivity of the system to changes of input coordinates vector's values. There are also methods of assessing the importance of input indicators, among which there are: trial and error method, method of fuzzy average curves, and method of expert ranking [27]. Herein the relevant methods are dependent enough from dimension of the vector of input coordinates. When the number of system inputs increases the complexity of their assessment increases too. Using the known method of weighting coefficients of fuzzy rules there is a necessity in their additional configuration at every structural reconfiguration of input data [26].

Limited features of the considered approaches and methods of DSS reconfiguration with relevant reduction of rule bases do not allow using them directly for optimization of the fuzzy hierarchical DSS with variable structure of vector of input data.

With the implementation of the first stage (Fig. 4) antecedents of all 81 rules of the fourth subsystem $y_4 = f_4(x_{14}, x_{15}, x_{16}, x_{17})$ are processed. Herein singular signal is sent to informational inputs of keys Key 1,...,Key 4. If, for example, two input signals $x_{16} = NE$ and $x_{17} = NE$ are not interesting for DM or they cannot be evaluated, then Key 1 and Key 2 are locked, and to the control inputs of Key 5, Key 6 and integrator $\sum 1$ there come relevant singular signals. Herein Key 5 and Key 6 are locked, that provides reduction of rules antecedents (Fig. 4) by means of automatic exclusion of all components with input coordinates x_{16}, x_{17} as Key 7 and Key 8 stay unlocked.

Modified antecedents contain only input coordinates x_{14} and x_{15}, as to informational inputs of Key 5 and Key 6 from rule base comes information about quantitative characteristics $\{L = 1, M = 2, H = 3\}$ of relevant LT of activated rules of the fourth DSS subsystem. Herein reduced rule base for realization of modified dependency $y_4 = f_4(x_{14}, x_{15})$ is reduced from 81 to 9 rules [17, 26].

With the implementation of the second stage (Fig. 4) correction of consequents of components y_i of output coordinate y_4 is done according to reduced rule base, that is automatically designed after realization of the first stage (numerical values $\{L = 1, M = 2, H = 3\}$ of relevant LT of reduced rules come to inputs of integrator $\sum 2$, Fig. 4).

To illustrate in details the method of two-stage reconfiguration of fuzzy DSS rule bases let's consider, for example, fuzzy rule № 37 (Fig. 4):

$$\text{IF } x_{14} = M \text{ AND } x_{15} = M \text{ AND } x_{16} = L \text{ AND } x_{17} = L \text{ THEN } y_4 = LM. \qquad (1)$$

In the process of the first stage reconfiguration the antecedent of this rule (1) automatically is adjusted and newly formed antecedent is transformed to the form:

$$\text{IF } x_{14} = M \text{ AND } x_{15} = M. \qquad (2)$$

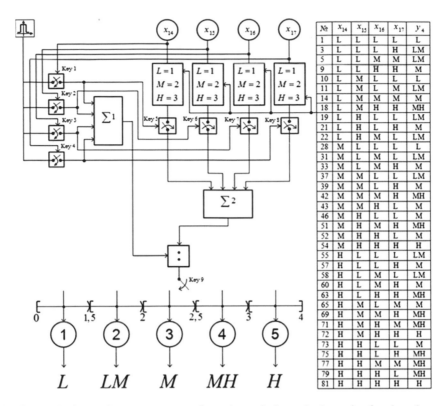

№	x_{14}	x_{15}	x_{16}	x_{17}	y_4
1	L	L	L	L	L
3	L	L	L	H	LM
5	L	L	M	M	LM
9	L	L	H	H	M
10	L	M	L	L	L
11	L	M	L	M	LM
14	L	M	M	M	M
18	L	M	H	H	MH
19	L	H	L	L	LM
21	L	H	L	H	M
22	L	H	M	L	LM
28	M	L	L	L	L
31	M	L	L	M	LM
33	M	L	M	H	M
37	M	M	L	L	LM
39	M	M	L	H	M
42	M	M	M	H	MH
43	M	M	H	L	M
46	M	H	L	L	M
51	M	H	M	H	MH
52	M	H	H	L	M
54	M	H	H	H	H
55	H	L	L	L	LM
57	H	L	L	H	M
58	H	L	M	L	LM
60	H	L	M	H	M
63	H	L	H	H	MH
65	H	M	L	M	M
69	H	M	M	H	MH
71	H	M	M	M	MH
72	H	M	H	H	H
73	H	H	L	L	M
75	H	H	L	H	MH
77	H	H	M	M	MH
79	H	H	H	L	MH
81	H	H	H	H	H

Fig. 4. Mechanism of two-stage reconfiguration of the rule base in fourth subsystem $y_4 = f_4(x_{14}, x_{15}, x_{16}, x_{17})$

With implementation of the second stage reconfiguration by calculating ratio $U_{out}(\sum 2)/U_{out}(\sum 1)$ of output signals of integrators $\sum 1$ and $\sum 2$ there is formed an assessment:

$$\text{Result} \in \{[0, 1.5), [1.5, 2), [2, 2.5), [2.5, 3), [3, 4]\}. \qquad (3)$$

Depending on the size of signal Result (3) the form of consequent with the help of Key 9 is transformed to the one of relevant linguistic terms $LT_{Result} = \{L, LM, M, HM, H\}$ (Fig. 4). For the rule № 37 the signal Result = 2 and respectively (Fig. 4) Result $\in [2, 2.5)$, that provides automatic correction of consequent and modified rule № 37 takes its final form (4):

$$\text{IF } x_{14} = M \text{ AND } x_{15} = M \text{ THEN } y_4 = M. \qquad (4)$$

Characteristic surfaces of the fourth subsystem $y_4 = f_4(x_{14}, x_{15}, x_{16}, x_{17})$ are shown on the Fig. 5.

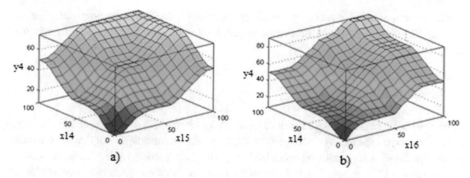

Fig. 5. Characteristic surfaces of the fourth subsystem: (a) $y_4 = f_4(x_{14}, x_{15}); x_{16}, x_{17} - const$; (b) $y_4 = f_4(x_{14}, x_{16}); x_{15}, x_{17} - const$

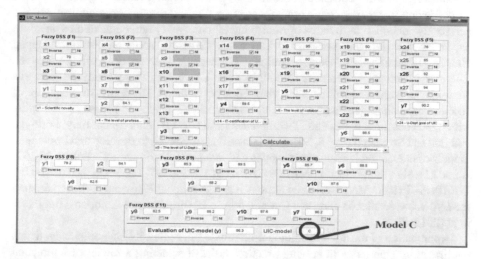

Fig. 6. The interface of developed knowledge-based DSS for selecting the model of academic-industry interaction for consortia "University – IT-company"

Human-computer interface, program realization and results of DSS work for selecting the model of interaction within consortia such as "University – IT-company" are shown on Fig. 6 ($x_5, x_9, x_{10}, x_{14}, x_{15}$ are not interesting).

For the presented on Fig. 6 set of input data $X = \{x_j\}$, $j = 1 \ldots 27$ developed DSS on fuzzy logic (Fig. 3) creates on its output consolidated signal, which recommends corresponding future partners for interaction (specific University and specific IT-company) to choose as optimal model E^* the model of interaction C: $E^* = C, (E^* \in E, E = \{E_1 = A1, A2, B, C\})$. As input data for modular DSS different-type input variables $X = \{x_j\}$, $j = 1 \ldots 27$ are used, which characterize performance indicators of university (of relevant IT-department) and IT-companies, which are part of academic-industrial interaction. Some of input data are quantitative, and some – qualitative. Quantitative input indicators can be created on the basis of results of statistical information processing, and qualitative – on the basis of results of expert evaluations (using individual and group assessments) [3, 4, 15, 20].

4 Conclusions

In this paper there are shown results of developing hierarchically-organized DSS, which is synthesized on the basis of using fuzzy logic, to increase efficiency of decision-making processes for selecting optimal model E^* of partner interaction under consortia such as "University – IT-company". Made by authors analysis of samples of successful innovative interaction of academic institutions and IT-companies [7, 9–11, 13] proves that creation of different groups, consortia, associations and alliances such as "University – IT-company" to solve current and future problems in higher education sphere based on mutual working experience in computer science area and internet-communications is a perspective direction in the area of improving efficiency of higher education system. In particular, the National Aerocosmic University "Kharkiv Aviation Institute" named after M. E. Zhukovskiy, Odessa National Polytechnic University, Yuriy Fedkovych Chernivtsi National University, Chernihiv State University, Petro Mohyla Black Sea National University, Institute of Cybernetics of National Academy of Sciences of Ukraine and others are members of such international academic-industrial consortia, which includes universities and IT-companies from Great Britain, Spain, Italy, Portugal, Ukraine and Sweden [6, 7, 19, 30]. This consortium is created to develop and implement models of interaction between universities and industry (IT-companies) such as A1, A2, B and C within the project TEMPUS- CABRIOLET 544497-TEMPUS-1-2013-1-UK-TEMPUS-JPHES "Model-oriented approach and Intelligent Knowledge–Based System for Evolvable Academia-Industry Cooperation in Electronics and Computer Engineering" (2013–2016).

Approbation of the developed model-oriented DSS proves its high efficiency, that is confirmed by authors as in solving practical tasks of selecting a model of interaction within consortia such as "University – IT-company", and in solving different-type tasks of transport logistics [8, 12, 14, 16, 18] in particular when selecting the best transport company from the set of existing alternative variants, etc.

References

1. Bogel, S., Stieglitz, S., Meske, C.: A role model-based approach for modelling collaborative processes. Bus. Process Manage. J. **20**(4), 598–614 (2014)
2. Drozd, J., Drozd, A., Maevsky, D., Shapa, L.: The levels of target resources development in computer systems. In: Proceedings of IEEE East-West Design & Test Symposium, Kiev, Ukraine, pp. 185–189 (2014)
3. Gil-Aluja, J.: Investment in Uncertainty. Kluwer Academic Publishers, Dordrecht (1999)
4. Gil-Lafuente, A.M.: Fuzzy Logic in Financial Analysis. StudFuzz, vol. 175. Springer, Berlin (2005)

5. Kazymyr, V.V., Sklyar, V.V., Lytvyn, S.V., Lytvynov, V.V.: Communications management for academia-industry cooperation in IT-engineering: training. In: Kharchenko, V.S. (ed.). MESU, ChNTU, NASU "KhAI", Chernigiv, Kharkiv (2015)
6. Kharchenko, V.S., Sklyar, V.V.: Conception and models of interaction between University Science and IT-Industry: S2B-B2S. J. Kartblansh **8–9**, 170–174 (2012). (in Russian)
7. Kharchenko, V.S., Sklyar, V.V.: Cooperation between Universities and IT-Industry: some problems and solutions. J. Kartblansh **3–4**, 43–50 (2014). (in Russian)
8. Kondratenko, G.V., Kondratenko, Y.P., Romanov, D.O.: Fuzzy models for capacitive vehicle routing problem in uncertainty. In: Proceediongs of 17th International DAAAM Symposium "Intelligent Manufacturing and Automation: Focus on Mechatronics & Robotics", Vienna, Austria, pp. 205–206 (2006)
9. Kondratenko, Y.P.: Application of active-IIDL in leading Universities of Mykolaiv region of Ukraine. In: Proceedings of International Active-HDL Conference, Kharkiv, Ukraine, pp. 60–66 (2001)
10. Kondratenko, Y.P.: Future perspectives of inter-institutional cooperation on international and regional level. In: Anales del Curso Academico 2006–2007, vol. XXIX, pp. 34–41. Real Academia de Ciencias Economicas y Financieras, Tomo, Barcelona (2008)
11. Kondratenko, Y.P.: The role of inter-university consortia for improving higher education system. In: Proceedings of Phi Beta Delta (Ed. by Michael Smithee), pp. 26–27. Honor Society for International Scholars, USA (2011)
12. Kondratenko, Y.P., Encheva, S.B., Sidenko, E.V.: Synthesis of intelligent decision support systems for transport logistic. In: Proceeding of the 6th IEEE International Conference on Intelligent Data Acquisition and Advanced Computing Systems: Technology and Applications, IDAACS 2011, Prague, Czech Republic, vol. 2, pp. 642–646, 15–17 September 2011. https://doi.org/10.1109/IDAACS.2011.6072847
13. Kondratenko, Yu., Kharchenko, V.: Analysis of features of innovative collaboration of Academic Institutions and IT-Companies in Areas S2B and B2S. Tech. News **1**(39), 15–19 (2014). (in Ukrainian)
14. Kondratenko, Y.P., Klymenko, L.P., Sidenko, I.V.: Comparative analysis of evaluation algorithms for decision-making in transport logistics. In: Jamshidi, M., Kreinovich, V., Kazprzyk, J. (eds.) Advance Trends in Soft Computing. Studies in Fuzziness and Soft Computing, vol. 312, pp. 203–217. Springer, Cham (2014). https://doi.org/10.1007/978-3-319-03674-8_20
15. Kondratenko, Y., Kondratenko, V.: Soft computing algorithm for arithmetic multiplication of fuzzy sets based on universal analytic models. In: Ermolayev , V., Mayr, H., Nikitchenko, M., Spivakovsky, A., Zholtkevych, G. (eds.) Information and Communication Technologies in Education, Research, and Industrial Application, ICTERI 2014. Communications in Computer and Information Science, vol. 469, pp. 49–77. Springer International Publishing, Switzerland (2014). https://doi.org/10.1007/978-3-319-13206-8_3
16. Kondratenko, Y.P., Sidenko, I.V.: Design and reconfiguration of intelligent knowledge-based system for fuzzy multi-criterion decision making in transport logistics. J. Comput. Optim. Econ. Finance **6**(3), 229–242 (2014)
17. Kondratenko, Y.P., Sidenko, I.V.: Method of actual correction of the knowledge database of fuzzy decision support system with flexible hierarchical structure. J. Comput. Optim. Econ. Finance **4**(2–3), 57–76 (2012)
18. Kondratenko, Y.P., Sidenko, I.V.: Decision-making based on fuzzy estimation of quality level for Cargo delivery. In: Zadeh, L., Abbasov, A., Yager, R., Shahbazova, S., Reformat, M. (eds.) Recent Developments and New Directions in Soft Computing. Studies in Fuzziness and Soft Computing , vol. 317, pp. 331–344. Springer International Publishing, Switzerland (2014).https://doi.org/10.1007/978-3-319-06323-2_21

19. Kondratenko, Y., Sydorenko, S.: Cooperation between Ukrainian Universities and Aldec Inc. (USA) in the field of VHDL and Verilog introduction to design of digital devices. In: Proceedings of Intern. Conf. "Higher Education Perspectives: The Role of Inter-University Consortia", Mykolaiv, Ukraine, pp. 150–153 (2004)
20. Lodwick, W.A., Untiedt, E. Introduction to fuzzy possibilistic optimization: In: Lodwick, W.A., Kacprzhyk, J. (eds.) Fuzzy Optimization. Studies in Fuzziness and Soft Computing, vol. 254, pp. 33–62. Springer, Heidelberg (2010)
21. Lytvynov, V.V., Kharchenko, V.S., Lytvyn, S.V., Saveliev, M.V., Trunova, E.V., Skiter, I. S.: Tool-based support of university-industry cooperation in IT-engineering. ChNTU, Chernigiv (2015)
22. Meerman, A., Kliewe, T. (eds.): Fostering University-Industry Relationships, Entrepreneurial Universities and Collaborative Innovations. Good Practice Series (2013)
23. Mendel, J.M.: Uncertain Rule-Based Fuzzy Logic Systems: Introduction and New Directions. Prentice Hall PTR, Upper Saddle river (2001)
24. Merigo, J.M., Gil-Lafuente, A.M., Yager, R.R.: An overview of fuzzy research with bibliometric indicators. Appl. Soft Comput. **27**, 420–433 (2015)
25. Palagin, A.V., Opanasenko, V.N.: Reconfigurable computing technology. Cybern. Syst. Anal. **43**(5), 675–686 (2007)
26. Piegat, A.: Fuzzy Modeling and Control. Springer, Heidelberg (2001)
27. Rotshtein, A.P.: Intelligent Technologies of Identification: Fuzzy Logic, Genetic Algorithms. Neural Networks. Universum Press, Vinnitsya (1999). (in Russian)
28. Starov, O., Kharchenko, V., Sklyar, V., Khokhlenkov, N.: Startup company and spin-off advanced partnership via web-based networking. In: Proceedings of the University-Industry Interaction Conference, Amsterdam, Netherlands, pp. 294–310 (2013)
29. Starov, O., Sklyar, V., Kharchenko, V., Boyarchuk, A., Phillips, C.: A student-in-the-middle approach for successful university and business cooperation in IT. In: Proceedings of the University-Industry Interaction Conference, Barcelona, Spain, pp. 193–207 (2014)
30. Vom Brocke, J., Schmiedel, T., Recker, J., Trkman, P., Mertens, W., Viaene, S.: Ten principles of good business process management. Bus. Process Manage. J. **20**(4), 530–548 (2014)
31. Whalen, T.: Decision making under uncertainty with ordinal linguistic data. In: Ruan, D., Kacprzyk, J., Fedrizzi, M. (eds.) Soft Computing for Risk Evaluation and Management. Applications in Technology, Environment and Finance, pp. 3–16. Physica-Verlag, Heidelberg (2001)
32. Zadeh, L.A.: Fuzzy sets. Inf. Control **8**(3), 338–353 (1965)

Multi-capacity, Multi-depot, Multi-product VRP with Heterogeneous Fleets and Demand Exceeding Depot Capacity

Gabriel Alemany[1], Angel A. Juan[1(✉)], Roberto Garcia[2], Alvaro Garcia[2], and Miguel Ortega[2]

[1] IN3 - Computer Science Department,
Open University of Catalonia, Barcelona, Spain
ajuanp@gmail.com
[2] Industrial Engineering Department,
Universidad Politecnica de Madrid, Madrid, Spain

Abstract. This paper presents a four-step metaheuristic for addressing a rich and real-life vehicle routing problem. A set of customers request several products that must be delivered using a heterogeneous fleet of trucks with different compartments (one per product). These vehicles depart from a set of depots, which do not have enough capacity for meeting the aggregated customers' demand of products. Therefore, some vehicles must visit an external facility at the beginning of their routes in order to obtain the necessary products to deliver. A real-world case has been solved, providing savings in reduced computing times.

Keywords: Logistics and transportation
Combinatorial optimization · Randomized algorithms · Metaheuristics

1 Introduction

Distributing goods to end customers, providing high levels of service at a reasonable cost, may be of paramount importance. Sometimes, this distribution must be done by road, and several products that cannot be mixed are involved. When products can be easily stored but transportation is costly, having several distribution depots may be an appropriate approach for reducing transportation costs. Each depot contains a number of vehicles for delivering products to the end customers. In addition to that, multi-compartment vehicles are usually required in those scenarios with multiple-products that cannot be mixed in the same compartment. All these characteristics increase the complexity of the distribution logistics. Finally, due to the evolution of the automotive industry, trucks of different capacities are acquired over time, so that in most cases fleets are heterogeneous. As result of the previous considerations, obtaining routes on a daily basis is not done in an effective manner, leading to far-from-optimal performances of the distribution process.

© Springer International Publishing AG, part of Springer Nature 2018
A. M. Gil-Lafuente et al. (Eds.): FIM 2015, AISC 730, pp. 113–123, 2018.
https://doi.org/10.1007/978-3-319-75792-6_10

Being a rich and real-life version of the well-known vehicle routing problem (VRP) [1, 10, 13], the problem considered in this paper is also NP-hard. Thus, it cannot be efficiently addressed using exact methods. In fact, for medium- and large-sized instances heuristic methods are much more appropriate for finding high-quality solutions in reduced computing times. In this paper, we propose a four-step heuristic for solving the aforementioned problem, which includes some specific features not found in the existing literature. The rest of the paper is structured as follows: Sect. 2 offers a more detailed description of the considered problem; Sect. 3 provides a short review of related work; Sect. 4 describes the heuristic approach proposed for solving the problem; computational experiments are analyzed in Sect. 5; finally, Sect. 6 highlights the main findings of this work.

2   Problem Description

The problem consists in minimizing the cost of serving the daily demand of a set of customers. Each of the customers may request different quantities of various products. The customers are served from a set of depots that are owned by the company (Fig. 1). At each depot, a fixed fleet of heterogeneous vehicles is based. Each type of vehicle offers a specific loading capacity. All vehicles have three compartments, which implies that a vehicle can transport no more than three different products at once since the products cannot be mixed in the same compartment. Every vehicle has a maximum driving time, which imposes a maximum route length for every vehicle. After completing their assigned route, vehicles must return to their base depot. At the beginning of the day, each depot has a known amount of each product at the beginning of the day. Each customers has to be fully served from a single depot, although it is possible to use different vehicles from the same depot to serve a customer –each of these vehicles serving a different product. In addition to the in-house depots –the ones owned by the company–, there are also outsourcing facilities that offer extra capacity to the logistics system. These external facilities hold large amounts of all products. They belong to the upstream logistics company and provide products to smaller and medium distributors as the one considered here. Vehicles can access the external facilities only if they are completely empty. Once at the external facility, vehicles can load their tanks with different products, which are either delivered to the end customers or used to replenish the company's depots. The cost function of this problem is obtained by aggregating both fixed and variable costs. Thus, there is a fixed cost for using each vehicle. In addition to that, there is a variable cost which is proportional to the traveled distance.

Therefore, the problem consists in minimizing the distribution cost when assigning customers to depots and, for the vehicles in each depot, creating routes and defining the amount of each product these vehicles will deliver to the end customers. The cost associated with visiting an external facility is simply a distance-based cost.

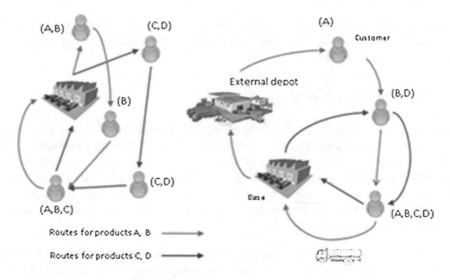

Fig. 1. Scheme of distribution system configuration.

3   Related Work

The optimization problem considered in this paper is a combination of the following ones: the multi-depot VRP (MDVRP), the multi-product VRP (MPVRP), and the VRP with heterogeneous fleets (HeVRP). Regarding the MDVRP, contributions date back to 40 years ago. For example, the authors in [15] discuss a VRP variant with more than one depot. Their algorithm first assign customers to depots and then combines routes trying to maximize savings. More recently, other authors have approached the problem combining biased-randomization techniques and iterated local search [9]. Within the context of the MPVRP, literature can be found related to the distribution of oil-derivatives [12]. These authors formulate a multi-product VRP with a single depot and a heterogeneous fleet of multi-compartment trucks. In order to solve the problem they use branch-and-price techniques. Also including a multi-period problem, authors in [3] propose a heuristic to optimize the petrol station replenishment problem over a time horizon. Moreover, we also found similar studies related to selective garbage collection. In these studies, different types of products cannot be mixed in the vehicle. On the contrary, they have to be stored in different compartments. Thus, in [5] a glass waste collection is addressed, where glasses of different colors must be kept in separated compartments inside the truck. A first approximation to the HeVRP is provided in [14]. In this paper, a column generation heuristic is employed to deal with the different vehicle capacities. Likewise, in [11] the authors propose a method based on a variant of their "record-to-record travel" algorithm for the basic VRP. In this case, however, the basic algorithm is extended to account for the heterogeneous fleet. Other recent approaches for solving the HeVRP, considering both fixed and variable costs, can be found in

[16] and in [6]. The first authors use a reactive tabu search metaheuristic, while the second authors use a successive approximation method based on biased-randomization techniques.

4   Solving Approach

The problem entails three different partitioning decisions. First, partitioning customers by assigning each one to a depot. Secondly, splitting customers into subcustomers, i.e.: a customer requesting n different products is split into n subcostomers, each of these requesting a single product and located at the same position as the original customer. Thirdly, partitioning all routes corresponding to every depot into subsets of routes, one per type of vehicle. Finally, a four-step approach has been adopted:

1. Customers are assigned to depots, generating an assignment map for each depot (notice that the union of all these maps covers all the customers).
2. Routing plans are generated for every customers-depot assignment map.
3. Visits to the system storage are inserted (if convenient) into the routing plans.
4. Routing plans associated with every map are improved.

In order to generate near-optimal solutions, a biased-randomized heuristic is applied [8]. In a biased-randomized heuristic, the possible constructive movements are sorted according to a quality criterion and then a probability is assigned to each of these movements based on the selected criterion [7]. In our case, the probability given to each option is defined by a Geometric probability distribution with parameter β, with $0 < \beta < 1$, as discussed in [4]. Figure 2 shows the general flowchart of our algorithm. In the following subsections, each of the previous steps are described in more detail.

4.1   Assigning Customers to Depots

The result of this step is a set of as many maps as depots, where a map includes all the customers to be covered by the corresponding depot. The classical Clarke and Wrights savings heuristic [2] is used in this step. Also, for each depot d customers are sorted according to their marginal savings, where the marginal savings of assigning a customer i to a depot d are defined as the difference between the cost of serving customer i from depot d and the cost of serving customer i from the closest depot other than d.

Once customers are sorted as described, the assignment of customers is done in to stages (Fig. 3):

- Stage 1: Selecting the depot on a round-robin basis, until every depot has an amount of assigned demands that does not exceed the total demand that can be served by its fleet of vehicles.
- Stage 2: Choosing the customer for the depot selected in Stage 1 according to the surplus capacity and the marginal savings value.

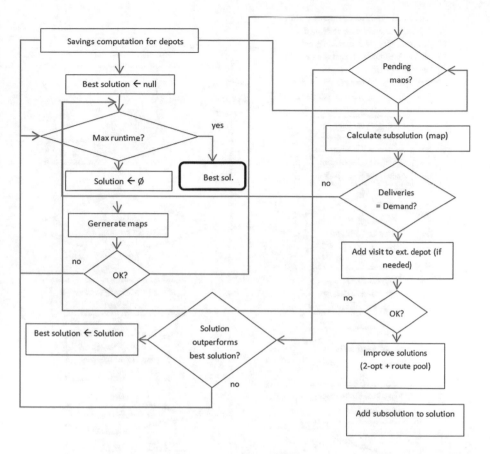

Fig. 2. General flowchart of the proposed algorithm.

Notice that the assignment of customers to depots is based on the aforementioned biased-randomization process. This first source of 'oriented' randomness allows obtaining a wide variety of potentially interesting maps. This is a key step in order to find a near-optimal solution after the routing process is applied to the right map. In any case, a customer is assigned to the depot only if this has the necessary capacity available. Since the detailed routes for trucks are not considered here, a map may not lead to feasible routes. This will be identified in the following steps. In the case this issue appears, new maps will have to be generated. Finally, inventory levels are disregarded at this stage, since some vehicle might go to the external facility (assuming the corresponding cost). The process finishes when all customers have been assigned to a depot. If at some point there is not enough capacity for the non-yet-assigned customers –due to a poor choice of the first assignments–, then the process starts all over again.

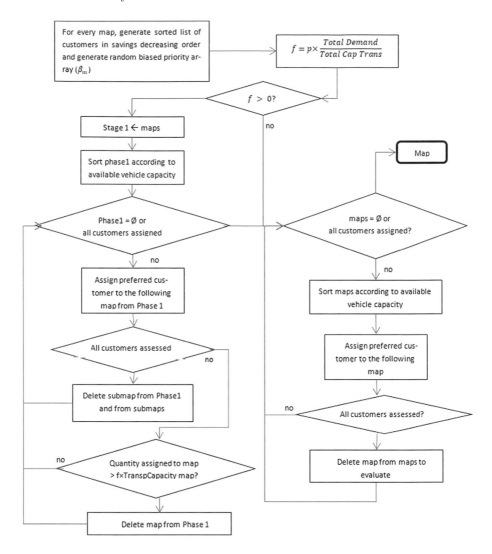

Fig. 3. Flowchart for the map generation process.

4.2   Generating Routes

First, customers are disaggregated in subcustomers, one for each requested product. Again, for every map the well-known savings-based routing heuristic is applied [2]. The biased randomization is applied again at this point, providing a large number of routes for every map. The routes to be selected depend on the number of vehicles of each type. Finally, routes for subcustomers are combined, so that a complete routing plan is obtained (eliminating any loops corresponding to visiting twice the same customer). When this process has been completed for all types of vehicles, the resulting solution is expected to be complete (i.e., it should cover all customers). Otherwise, the routing process is interrupted and a new iteration begins.

4.3 Visiting the External Facility

When the total demand assigned to a depot exceeds its initial inventory, it becomes necessary to visit the external facility to acquire the associated product(s). In particular, visits to the external facility are always performed at the very beginning of a route. Also, among the routes that may visit the external facility, the ones chosen to do it are those with customers in the neighborhood of the external facility –in order to avoid costly detours.

4.4 Improving the Routes

First, a fast 2-opt algorithm is applied to every route generated. After that, alternative routes are evaluated among those stored in a pool (typically a quick-access hash map). Thus, if a previously existing route visited the very same customers at a lower cost, the new route is substituted by the one already in the pool. If not, the new route is stored in the pool for future route improvements.

5   Computational Experiments

After developing the algorithm, some tests were run with the aim of both evaluating its performance under several scenarios and tuning the parameters. All the experiments were conducted on a computer with an Intel Core 2 Duo T6600 processor at 2.20 GHz and 4 GB RAM. The experiments allowed us to get some insight on how the algorithm performed for a series of configurations and for instances with different features. Thus, for example, it was possible to solve instances with 120 customers in about 15 s. With the algorithm tuned, a case study based on a real-world context was carried out. Data was provided by a company that delivers oil-derivatives to different types of customers, including households -mainly for heating– and companies that require these products for their activity. This company has three types of vehicles, which are trucks of small, medium, and large sizes. Each of these vehicles has two or three compartments, as detailed in Table 1. In addition, small-size trucks cannot access the external facility.

Table 1. Vehicles characteristics

Vehicle type	Compartment 1 capacity (liters)	Compartment 2 capacity (liters)	Compartment 3 capacity (liters)
1 (small)	2000	2000	N/A
2 (medium)	1800	3150	4050
3 (large)	6000	10500	13500

Fig. 4. Location of customers in a real-life case study.

For each route in the distribution plan, there is a fixed cost associated with the employed vehicle as well as a variable cost that depends on the route length. The total transportation capacity is 78,000 liters/day, which corresponds to the fleet size of the company. Thus, a real company instance was solved and compared

Table 2. Distance matrix for every pair of node in the instance

Customers	C0	C1	C2	C3	C4	C5	C6	C7	C8	C9	C10	C11	C12	C13	C14	C15
Depot	10.0	11.3	25.0	20.0	40.0	10.0	5.0	24.0	10.0	2.5	47.0	9.0	2.5	30.0	28.8	21.0
C0		10.0	16.0	11.0	41.0	18.0	8.5	30.0	20.0	5.2	53.5	20.9	16.6	20.7	21.5	20.7
C1			21.0	17.0	47.0	29.0	17.5	38.0	28.0	11.0	58.6	24.9	20.7	24.6	27.3	26.4
C2				17.0	51.0	21.0	18.7	23.0	18.0	29.1	53.8	19.8	31.0	16.2	12.9	14.7
C3					31.0	27.5	16.5	28.0	27.0	18.1	54.9	28.0	29.5	10.8	13.5	12.5
C4						7.0	35.5	49.5	31.0	36.9	69.6	28.4	31.7	53.7	56.4	55.5
C5							11.0	14.5	3.5	14.2	43.2	3.1	12.7	32.6	35.3	35.9
C6								31.0	13.0	2.6	45.3	13.4	9.3	24.5	27.1	26.1
C7									15.0	27.3	34.0	15.6	29.2	44.1	34.1	50.7
C8										14.6	37.5	2.6	13.1	33.1	35.7	36.3
C9											51.5	18.0	11.8	25.6	28.3	27.3
C10												40.0	49.9	64.8	54.8	60.0
C11													14.4	33.9	36.6	37.1
C12														38.7	34.1	38.7
C13															5.0	20.3
C14																22.0

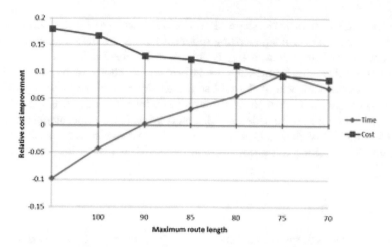

Fig. 5. Cost improvement vs. maximum route distance.

with the actual solution that the company implemented on a specific day. The location of the customers is given in Fig. 4.

Table 2 shows the distance matrix (in km.) between each pair of nodes, including the depot and the customers.

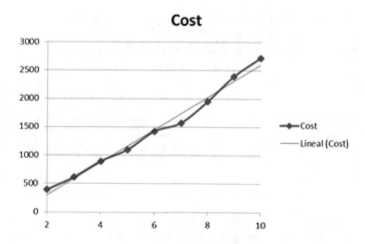

Fig. 6. Cost vs. instance size.

For this particular case, the company had to deliver 4 products to 16 customers from a single depot, using two types of vehicles: 3 small-size ones and 2 large-size ones. The company routes involved 7 routes and 366.3 km. When solving the problem without imposing any upper bound on the distance of any route, the algorithm provided a solution in 0.23 s, with 4 routes, and a cost of 300.4.

Notice that this cost is 18% better than the one provided by the company. The algorithm was run several times using different values for the maximum length of any route. Results are given in Fig. 5. As it can be observed, there is a trade-off between the route length and the cost improvement that can be obtained.

Finally, a first test was performed to assess the behavior of the algorithm for different problem sizes. A total of 9 instances where solved, with bases ranging from 2 (instance 1) to 10 (instance 9), and with a number of customers ten times the number of bases. Figure 6 shows the increase in cost as the problem size increases. In these experiments, computing times were limited to a maximum of 15 s.

6 Conclusions

In this paper we addressed the specific needs that arise in some distribution contexts, as that of the company that inspired this work, where various products must be delivered to several customers using multi-compartment heterogeneous trucks. After describing the problem, we have designed and implemented an algorithm that combines several techniques to solve the optimization problem in an efficient way. We have applied the algorithm to a real case, obtaining a solution that clearly outperforms the one currently employed by the company. In addition, we have analyzed how the quality of the solution varies as the maximum route length is limited by a parametric threshold.

Acknowledgments. This work has been partially supported by the Spanish Ministry of Economy and Competitiveness (TRA2013-48180-C3-P, TRA2015-71883-REDT), FEDER, and the Erasmus+ Program (2016-1-ES01-KA108-023465).

References

1. Caceres-Cruz, J., Arias, P., Guimarans, D., Riera, D., Juan, A.A.: Rich vehicle routing problem: survey. ACM Comput. Surv. (CSUR) **47**(2), 32 (2015)
2. Clarke, G., Wright, J.W.: Scheduling of vehicles from a central depot to a number of delivery points. Oper. Res. **12**(4), 568–581 (1964)
3. Cornillier, F., Boctor, F.F., Laporte, G., Renaud, J.: A heuristic for the multi-period petrol station replenishment problem. Eur. J. Oper. Res. **191**(2), 295–305 (2008)
4. Grasas, A., Juan, A.A., Faulin, J., De Armas, J., Ramalhinho, H.: Biased randomization of heuristics using skewed probability distributions: a survey and some applications. Comput. Ind. Eng. **110**, 216–228 (2017)
5. Henke, T., Speranza, M.G., Wäscher, G.: The multi-compartment vehicle routing problem with flexible compartment sizes. Eur. Oper. Res. **246**(3), 730–743 (2015)
6. Juan, A.A., Faulin, J., Caceres-Cruz, J., Barrios, B.B., Martinez, E.: A successive approximations method for the heterogeneous vehicle routing problem: analyzing different fleet configurations. Eur. J. Ind. Eng. **8**(6), 762–788 (2014)
7. Juan, A.A., Faulin, J., Ferrer, A., Lourenço, H.R., Barrios, B.: MIRHA: multi-start biased randomization of heuristics with adaptive local search for solving non-smooth routing problems. Top **21**(1), 109–132 (2013)

8. Juan, A.A., Faulin, J., Ruiz, R., Barrios, B., Gilibert, M., Vilajosana, X.: Using oriented random search to provide a set of alternative solutions to the capacitated vehicle routing problem. In: Operations Research and Cyber-Infrastructure, pp. 331–345. Springer, Boston (2009)
9. Juan, A.A., Pascual, I., Guimarans, D., Barrios, B.: Combining biased randomization with iterated local search for solving the multidepot vehicle routing problem. Int. Trans. Oper. Res. **22**(4), 647–667 (2015)
10. Laporte, G.: What you should know about the vehicle routing problem. Naval Res. Logistics **54**(8), 811–819 (2007)
11. Li, F., Golden, B., Wasil, E.: A record-to-record travel algorithm for solving the heterogeneous fleet vehicle routing problem. Comput. Oper. Res. **34**(9), 2734–2742 (2007)
12. Pasquale, A., Maurizio, B., Antonio, S.: Solving a fuel delivery problem by heuristic and exact approaches. Eur. J. Oper. Res. **152**(1), 170–179 (2004)
13. Prins, C.: A simple and effective evolutionary algorithm for the vehicle routing problem. Comput. Oper. Res. **31**(12), 1985–2002 (2004)
14. Taillard, É.: A heuristic column generation method for the heterogeneous fleet VRP. RAIRO Oper. Res. **33**(1), 1–14 (1999)
15. Tillman, F.A., Cain, T.M.: An upperbound algorithm for the single and multiple terminal delivery problem. Manag. Sci. **18**(11), 664–682 (1972)
16. Wang, Q., Ji, Q., Chiu, C.H.: Optimal routing for heterogeneous fixed fleets of multicompartment vehicles. Math. Prob. Eng. **2013**, 13 (2014)

Generalized OWA-TOPSIS Model Based on the Concept of Majority Opinion for Group Decision Making

Binyamin Yusoff[1(✉)], José M. Merigó[2,3],
and David Ceballos Hornero[1]

[1] Department of Mathematical Economics and Finance,
University of Barcelona, Barcelona, Spain
binyaminy@yahoo.com
[2] Department of Management Control and Information Systems,
University of Chile, Santiago, Chile
[3] Risk Center, University of Barcelona, Barcelona, Spain

Abstract. In this paper, an extension of OWA-TOPSIS model by inclusion of a concept of majority opinion and generalized aggregation operators for group decision making is proposed. To achieve this objective, two fusion schemes in TOPSIS model are designed. First, an external fusion scheme to aggregate the experts' judgments with respect to the concept of majority opinion on each criterion is proposed. Then, an internal fusion scheme of ideal and anti ideal solutions that represents the majority of experts is proposed using the Minkowski OWA distance with the inclusion of relative importances of criteria. The advantages of the proposed model include, a consideration of soft majority concept as a group aggregator and a flexibility in applying the decision strategies for analyzing the decision making process. In addition, instead of calculate the majority opinion with respect to the individual experts' judgments on each alternative, the proposed method takes into account the majority of experts on each criterion, in which reflects the specificity on criteria for overall decision. A numerical example is provided to demonstrate the applicability of the proposed method and comparisons are made between some aggregation operators and distance measures.

Keywords: TOPSIS · Group decision making · OWA operator
IOWA operator · Minkowski distance · Majority opinion

1 Introduction

Multi-criteria decision analysis (MCDA) is one of the active topics in the field of operations research. MCDA deals with the problem to select, prioritize or rank a finite number or discrete set of courses of action. There are a number of techniques in the literature which developed to deal with different types of MCDA problems, see [5] for the state-of-art surveys of MCDA techniques. The technique for order performance by similarity to ideal solution (TOPSIS) is one of the mentioned methods and was proposed at first by Hwang and Yoon [7] based on the concept of ideal distances of

© Springer International Publishing AG, part of Springer Nature 2018
A. M. Gil-Lafuente et al. (Eds.): FIM 2015, AISC 730, pp. 124–139, 2018.
https://doi.org/10.1007/978-3-319-75792-6_11

alternatives. The ranking of the alternatives are referred to the shortest distance from positive ideal solution and the farthest from negative ideal solution.

The TOPSIS method is flexible to be integrated with other methods as extension model, for example in a case of group decision making [13, 14]. In complex problems, such in group decision environment, much attention have been focused on the aggregation of the preferences among the group. The ordered weighted averaging (OWA) aggregation operator as proposed by Yager [17] is one of the approaches normally used for the aggregation of preferences in MCDA techniques. The OWA operator provides a parameterized class of mean-type aggregation operators, such as the max, arithmetic average and min, among others, with flexibility for inclusion of linguistic quantifier in the aggregation process [16]. With such characteristic, the OWA operator can be interpreted as a generalization of the original decision making process suggested by Bellman and Zadeh [1]. In addition, Yager and Filev [18] have proposed the induced OWA (IOWA) aggregation operator as an extension of the OWA operator with an additional feature for the complex decision making. Subsequently, Merigo and Gil-Lafuente [9] generalized the IOWA operator to include other mean operators such as the induced ordered weighted geometric average (IOWGA), the induced ordered weighted harmonic average (IOWHA), the ordered weighted quadratic average (IOWQA) operators, etc. In the literature, the OWA and IOWA based operators have been successfully applied in some of the MCDA techniques, for instance see [4, 8, 19].

There are some studies that have been done on the TOPSIS under group decision making. Shih et al. [13] for example, have provided an analysis on the TOPSIS model under group environment. The aggregations of the individual experts' judgments as overall group decision are calculated using the arithmetic and geometric means under some distance normalization methods, e.g., Manhattan distance and Euclidean distance. Afterward, Chen et al. [4] extended the TOPSIS under group decision making by inclusion of the OWA operator. Three fusion schemes were suggested where one deals with the local judgment (or internal aggregation) of each expert, providing flexibility for decision strategies on criteria (e.g., either total or partial compensation of criteria) and the rest are defined as global judgments (or external aggregations) which deal with the fusion of individual experts' judgments as the collective judgment of experts or group. The external aggregation based-OWA provides a flexibility or tolerance in selection of individual experts' judgments in a similar way as the aggregation of criteria (e.g., either to consider all or partial compensation of experts). As in the application side, Shafiqul et al. [12] have proposed an integrated approach of fuzzy TOPSIS-OWA and geographic information system (GIS) for evaluating water quality problems. However, for the above-mentioned models, a consensus measure or agreement of experts is not explicitly considered.

Pasi and Yager [11] have proposed the concept of majority opinion for group decision making by using the OWA and IOWA aggregation operators based on the linguistic quantifiers. In this approach, the quantifier-guided aggregation function generating the group decision making is defined by an IOWA operator. Hence, the opinions of the experts supporting each other on each alternative are taken into account in deriving the group decision making. Subsequently, based on Pasi and Yager's method, Hajimirsadeghi and Lucas [6] have proposed the inclusion of majority concept in the TOPSIS model under group environment where the consensus measure is explicitly included. In addition, Boroushaki and Maczewski [2] utilized the concept of fuzzy majority for GIS-based multi-criteria group decision making.

Nevertheless, the method as proposed in [11] is simply focused on the similarity values on each alternative with disregard the conflicts or incoherence between the performance judgments of the single criterion. In this case, two experts may produce the same performance judgment for an alternative even their single performance judgments of the criteria are completely different. In consequence, Bordogna and Sterlacchini [3] have proposed an extension of the Pasi and Yager's method by considering the consensus on each criterion instead of on each alternative and calculate the consensus by the Minkowski OWA on coherent evaluation based on similarity measure. This alternative approach has some advantages include providing the uniformity in considering the behavior of majority of experts regarding the proportion of criteria to consider and a robust decision by determine the performance judgments on each criterion of the experts [3].

In this paper, the integration of TOPSIS under group decision making with the concept of majority opinion and the generalize aggregation operators is proposed as a modified model. This model is an extension of the methods proposed in [4, 6] by considering the majority concept or consensus on each criterion instead of consensus on each alternative. Some modifications on the concept of majority opinion introduced in [11] with the idea proposed in [3] is put forward to be applied in the TOPSIS model under group decision making problems. However, the focus in this paper is limited to the case of homogeneous group decision making where each expert associated with identical degree of importance for each criterion. The rest of the paper is structured as follows: in Sect. 2, some preliminaries related to the definitions and concepts used in this paper are presented. In Sect. 3, the general framework of the classical TOPSIS method is given in comparison with the proposed model; Sect. 4, the proposed model on generalized OWA-TOPSIS based on majority concept is explained. In Sect. 5, a numerical example is provided and comparisons are made. Finally, the conclusions are drawn in Sect. 6.

2   Preliminaries

In the following, the basic aggregation operators that are used in this paper are briefly discussed.

2.1   OWA, IOWA and IGOWA Operators

Definition 1 [17]. An OWA operator of dimension n is a mapping $OWA : R^n \rightarrow R$ that has an associated weighting vector W of dimension n such that $\sum_{j=1}^{n} w_j = 1$ and $w_j \in [0, 1]$, then:

$$OWA(a_1, a_2, \ldots, a_n) = \sum_{j=1}^{n} w_j b_j,$$ (1)

Where b_j is the jth largest a_i and R is the set of positive real numbers.

Definition 2 [18]. An IOWA operator of dimension n is a mapping $IOWA : R^n \to R$ that has an associated weighting vector W of dimension n such that $\sum_{j=1}^{n} w_j = 1$ and $w_j \in [0, 1]$, then:

$$IOWA(\langle z_1, a_1 \rangle, \langle z_2, a_2 \rangle, \ldots, \langle z_n, a_n \rangle) = \sum_{j=1}^{n} w_j b_j, \tag{2}$$

where b_j is the a_i value of the IOWA pair $\langle z_i, a_i \rangle$ having the jth largest z_i, z_i is the order-inducing variable and a_i is the argument variable. Note that, in case of 'ties' between argument values, the policy suggested in [16] will be implemented, in which each argument of tied IOWA pair is replaced by their average.

Definition 3 [9]. An IGOWA operator of dimension n is a mapping $IGOWA : R^n \to R$ that has an associated weighting vector W of dimension n such that $\sum_{j=1}^{n} w_j = 1$ and $w_j \in [0, 1]$, then:

$$IGOWA(\langle z_1, a_1 \rangle, \langle z_2, a_2 \rangle, \ldots, \langle z_n, a_n \rangle) = \left(\sum_{j=1}^{n} w_j b_j^{\lambda} \right)^{1/\lambda}, \tag{3}$$

where b_j is the a_i value of the IGOWA pair $\langle z_i a_i \rangle$ having the jth largest z_i, z_i is the order inducing variable, a_i is the argument variable and λ is a parameter such that $\lambda \in (-\infty, \infty)$. With different values of λ, various type of weighted average can be derived. For instance, when $\lambda = -1$, the IOWHA operator can be obtained, when $\lambda = 0$, then the IOWG is generated, for $\lambda = 2$, the IOWQA operator is derived, etc. The OWA, the IOWA, and the IGOWA operators are all commutative, monotonic, bounded and idempotent [9, 17, 18].

2.2   Minkowski Distance OWA

Definition 4 [10]. A Minkowski OWAD operator of dimension n is a mapping $MOWAD : R^n \times R^n \to R$ that has an associated weighting vector W of dimension n such that $\sum_{j=1}^{n} w_j = 1$ with $w_j \in [0, 1]$ and the distance between two sets A and B is given as follows:

$$MOWAD(d_1, d_2, \ldots, d_n) = \left(\sum_{j=1}^{n} w_j e_j^{\lambda} \right)^{1/\lambda}, \tag{4}$$

where e_j is the jth largest of the d_i and d_i is the individual distance between A and B, such that $d_i = |a_i - b_i|$ with λ is a parameter in a range $\lambda \in (-\infty, \infty)$. By setting different values for the norm parameter λ, some special distance measures can be derived. For example, if $\lambda = 1$, then the Manhattan OWA distance can be obtained, $\lambda = 2$ then the Euclidean OWA distance can be acquired, $\lambda = \infty$ then Tchebycheff OWA is derived, etc.

2.3   OWA Operators with Inclusion of Linguistic Quantifiers

The linguistic quantifier was introduced by Zadeh [20] as a generalization of the existential (at least one) and universal (all) quantifiers of classical logic. Linguistic quantifiers are expressed by the terms, for example, *most, many, half, some, few* to indicate an approximate way a quantity of the elements belonging to a reference set (or the universe of discourse). There are two types of quantifiers, in which absolute and proportional [20]. Throughout this paper only relative quantifier is used. A relative quantifier can be denoted as a function as follows:

$$Q : I \rightarrow I \; satisfies \; Q(0) = 0, \exists x \in I \ni Q(x) = 1. \tag{5}$$

If Q is a fuzzy subset corresponding to a relative linguistic quantifier, then for any value x in the unit interval the membership grade $Q(x)$ corresponds to the compatibility of the value x with the concept in which Q is representing [20].

There are some categories exist for quantifiers, see [15]. But in the context of MCDA, regular increasing monotone (RIM) quantifiers are sufficient, as one wants to represent the fact that the larger the number of satisfied criteria the more satisfied the solution is [15]. RIM quantifier is defined such that:

$$Q(0) = 0, \; Q(1) = 1 \; and \; Q(x) \geq Q(y) \; if \; x > y. \tag{6}$$

In addition, in [15, 20] two kinds of fuzzy quantified propositions are defined. First, "$Q X \; are \; Y$", i.e., Q elements of set X satisfy the fuzzy predicate Y. The other proposition is "$Q B X \; are \; Y$", i.e., Q elements of set X which satisfy the fuzzy predicate B also satisfy the fuzzy predicate Y. The linguistic quantifiers Q can be represented in the forms (Yager [17]):

$$Q(r) = r^{\alpha}, \alpha > 0, \tag{7}$$

where parameter α indicate the degree of inclusion for different elements. By changing the α values, different decision strategies can be derived. For example, the linguistic quantifier "*most of the criteria*" is represented as ($\alpha = 2$). Yager [17] defined the OWA aggregation from Q by defining the weights in the following way:

$$w_j(x) = Q\left(\frac{j}{n}\right) - Q\left(\frac{j-1}{n}\right), \quad j = 1, 2, \ldots, n, \tag{8}$$

where w_j represents the increase of satisfaction in getting j with respect to $j - 1$ criteria satisfied. In the case where the criteria c_j to be aggregated have relative importances v_j associated with them $(v_j, c_j(x))$, the inclusion of importances in OWA operators from Q can be defined as follows (Yager [15]):

$$w_j(x) = Q\left(\frac{\sum_{k=1}^{j} u_k}{T}\right) - Q\left(\frac{\sum_{k=1}^{j-1} u_k}{T}\right), \tag{9}$$

where u_k denote the importances associated with the criteria that has the jth largest satisfaction b_j to x, such as (u_j, b_j) and $T = \sum_{j=1}^{n} u_k$, the total sum of importances. For example, if $c_3(x) = j$th largest of the $c_j(x)$, then $b_1 = c_3(x)$ and $u_1 = v_3$, then $(v_3, c_3(x)) = (u_1, b_1)$.

2.4   The Concept of Majority Opinion in Group Decision Making

Normally, in group decision making the unanimous agreement is not easy to achieve due to conflict interest among the experts. Hence, the concept of majority is crucial as it is required to find a soft agreement that satisfies the opinions of the majority of the experts, e.g., "most" of experts. OWA aggregator with regular linguistic quantifiers is not ideal for modeling the concept of majority, e.g., OWA aggregation with the quantifier "most of" produces a value that is reflect the satisfaction of the proposition "most of the criteria have to be satisfied" instead of "satisfaction value of most of the criteria" [11]. Therefore, a mechanism based on IOWA operators with an inducing ordering variable which denotes the similarity of the elements to be aggregated is proposed in [11] to model the majority opinion.

The methodology used to obtain the majority opinion is described as follows: The quantifier based on membership function for the "most" of experts is given as:

$$Q(r) = \begin{cases} 1 & \text{if } r \geq 0.9, \\ 2r - 0.8 & \text{if } 0.4 < r < 0.9, \\ 0 & \text{if } r \leq 0.4, \end{cases} \tag{10}$$

and is used throughout this paper. The values of the inducing variable are obtained by means of a function of the supports or similarities between pairs of the values to be aggregated. A support function is a binary function that used to compute a value $supp(x_i, x_j)$, which expresses support from x_j to x_i. In this case, the more similar or close the two values then the more they support to each other. The support function can be given as follow:

$$supp(x_i, x_j) = \begin{cases} 1 & \text{if } |x_i - x_j| < \beta, \\ 0 & \text{otherwise.} \end{cases} \tag{11}$$

This support function then is used to measure the support of each expert's opinion with respect to the other experts' opinions and the overall support is calculated by the sum of all supports. By the same process, the overall support values for the other experts can be derived. These values are used as the values of order inducing variable and they are reordered in non-decreasing order. The overall support values of x_i with respect to $x_j, j = (1, 2, \ldots, n)$ can be given as:

$$s_i = \sum_{j=1}^{n} supp(x_i, x_j), \tag{12}$$

for $i \neq j$, with $U = (s_1, \ldots, s_n)$ as the order inducing variable and $s_1 \leq \ldots \leq s_n$. To compute the non-decreasing weights of the weighting vector, define the values t_i based on a modification of the s_i values: $t_i = s_i + 1$ (the similarity of a_i with itself, similarity value equal to one). The t_i values are in increasing order, $t_1 \leq \ldots \leq t_n$. On the basis of t_i values, the weights of the weighting vector are computed as follows:

$$w_i = \frac{Q(t_i/n)}{\sum_{j=1}^{n} Q(t_i/n)}. \tag{13}$$

The value $Q(t_i/n)$ denotes the degree to which a given member of the considered set of values represents the majority, referring to "the satisfaction value of Q of the criteria" instead of "Q of the criteria have to be satisfied" such in generalized quantifiers. Then, the final evaluation derived using the IOWA operators.

As previously mentioned in [3], the extension of the Pasi and Yager's method as the alternative approach is proposed. Instead of concentrated on single alternative to calculate the majority opinion of individual experts, they focused on each criterion as to represent the majority aggregation. The proposal of this paper is to deal with Pasi and Yager's method with a slight modification made on the majority opinion concept as suggested in [3] and then applied the method to the TOPSIS model under homogeneous group decision making.

3 The TOPSIS Model

In this section, the classical TOPSIS is presented prior to the extension model proposed in this paper.

The classical TOPSIS analysis procedure [7, 13] can be summarized as the following steps: First, a decision matrix for each expert $D^h, h = 1, 2, \ldots, k$, is constructed as follows:

$$D^h = \begin{array}{c} \\ A_1 \\ \vdots \\ A_m \end{array} \begin{pmatrix} C_1 & \cdots & C_n \\ a_{11}^h & \cdots & a_{1n}^h \\ \vdots & \ddots & \vdots \\ a_{m1}^h & \cdots & a_{mn}^h \end{pmatrix}, \tag{14}$$

where A_i indicates the alternative $i(i = 1, 2, \ldots, m)$ and C_j denotes the criterion $j(j = 1, 2, \ldots, n)$, and a_{ij}^h denotes the preferences on consequence data space (original and raw information) for alternative A_i with respect to criterion C_j.

Then, each decision matrix which represents each expert $D^h, h = 1, 2, \ldots, k$ is normalized to $R^h, h = 1, 2, \ldots, k$, using the vector normalization:

$$x_{ij}^h = \frac{a_{ij}^h}{\sqrt{\sum_{i=1}^{m} \left(a_{ij}^h\right)^2}}, \tag{15}$$

Determine the ideal and anti-ideal solutions V^{h+} and $V^{h-}, h = 1, 2, \ldots, k$, for each decision maker. The ideal and anti-ideal are obtained as follows:

$$V^{h+} = \{x_1^{h+}, \ldots, x_n^{h+}\}$$
$$= \{(max_i x_{ij}^h | j \in J), (min_i x_{ij}^h | j \in J')\}, \tag{16}$$

$$V^{h-} = \{x_1^{h-}, \ldots, x_n^{h-}\}$$
$$= \{(min_i x_{ij}^h | j \in J), (max_i x_{ij}^h | j \in J')\}, \tag{17}$$

where J is associated with the set of benefit criteria and J' is associated with the set of cost criteria.

Next, calculate the separation measures from the ideal and anti-ideal solutions for the group. This step can be divided in the following steps:

In this step, the separation measure from ideal and anti-ideal solutions are computed with a distance metric. The manipulation for Minkowski's L_p metric as the distance measure is described as follows:

$$D_i^{h+} = \left\{ \sum_{j=1}^n w_j^h \left(v_{ij}^h - v_j^{h+} \right)^p \right\}^{1/p}, \tag{18}$$
$$\text{for } i = 1, 2, \ldots, m,$$

$$D_i^{h-} = \left\{ \sum_{j=1}^n w_j^h \left(v_{ij}^h - v_j^{h-} \right)^p \right\}^{1/p}, \tag{19}$$
$$\text{for } i = 1, 2, \ldots, m,$$

where $p \geq 1$ and integer, w_j^h is the weight for the criterion j and decision maker h, and $\sum_{j=1}^n w_j^h = 1$, $h = 1, 2, \ldots, k$. Note that with $p = 2$, then D_i^{h+} and D_i^{h-} provide the Euclidean distance, and the metric with $p = 1$ is the Manhattan distance. At this stage, each expert derived the ranking of all alternative individually (i.e., internal aggregation).

Then in the next stage, the consensus or final ranking of all alternatives (i.e., external aggregation) are calculated using the group aggregators such as arithmetic mean or geometric mean such that:

$$D_i^{G+} = D_i^{1+} \oplus \ldots \oplus D_i^{k+}, \text{ for alternative } i, \tag{20}$$

$$D_i^{G-} = D_i^{1-} \oplus \ldots \oplus D_i^{k-}, \text{ for alternative } i, \tag{21}$$

The operators $\oplus$ are either geometric mean or arithmetic mean.

Finally, calculate the relative closeness RC_i^G to the ideal solution for the group. The relative closeness can be calculated according to Eq. (22) and the alternatives are ranked in descending order.

$$RC_i^G = \frac{D_i^{G-}}{D_i^{G-} + D_i^{G+}}, i = 1, 2, \ldots m. \tag{22}$$

Note that the larger the value of RC_i^G denotes the better performance of the alternative.

4 Generalized OWA-TOPSIS Based on Majority Concept

In this section, the algorithm for the proposed model is explained. Two stages of external and internal fusion schemes are presented, where external fusion scheme deals with the aggregation of majority opinion of experts and internal fusion scheme deals with the implementation of decision strategies, the proportion of criteria to consider.

4.1 External Fusion Scheme: Inclusion of Majority Concept of Experts

Step 1: Construct decision matrix $D^h, h = 1, 2, \ldots, k$, for each expert as in Eq. (14).

Step 2: Convert the consequence data a_{ij}^h to the value data x_{ij}^h using the vector normalization function, $x_{ij}^k = f\left(a_{ij}^k\right)$ for the decision matrix $D^h, h = 1, 2, \ldots, k$, of each expert as in the Eq. (15).

Step 3: At this stage, instead of directly determine the ideal and anti-ideal solutions for each expert such in classical TOPSIS, this step deals with the aggregation (i.e., group aggregation) of the normalized experts' judgments $D^h = \left[x_{ij}^h\right]$ with respect to each criterion C_j using the concept of majority opinion. First, calculate the support of each expert, l, $\forall l = 1, 2, \ldots, k$ with respect to all the other experts in evaluating each alternative A_i with respect to each single criterion C_j such in Eq. (12). But, since the normalized values x_{ij}^h are very close to each other, then the support function in this paper is modified as follow: $\left(|x_i - x_j| \times 10\right)$. In this case, the supported values are the inducing variables and these values are reordered in increasing order, with $(\beta = 0.2)$ is used in throughout this paper:

$$s^h = \left(s_{ij}^1, \ldots, s_{ij}^k\right) = \left(Supp_i\left(x_j^l, x_j^1\right), \ldots, Supp_i\left(x_j^l, x_j^k\right)\right), \tag{23}$$
$$\forall l = 1, 2, \ldots, k.$$

Step 4: To compute the non-decreasing weights of the weighting vector, then define the values $t^h, h = 1, 2, \ldots, k$ based on a modification of the s^h support values: $t^h = s^h + 1$ (the similarity of x_j^l with itself, similarity value is equal to one). The t^h values are in increasing order. On the basis of t^h values, the weights of the weighting vector are computed as follows:

$$w^h = \frac{Q(t^h/k)}{\sum_{h=1}^{k} Q(t^h/k)}. \tag{24}$$

The value $Q(t^h/k)$ denotes the degree to which a given member of the considered set of experts represents the majority, which is the quantifier "*most*" such in Eq. (10).

Then, the aggregation of the experts' judgments $D^h = \left[x_{ij}^h\right]$ with respect to each criterion C_j using the concept of majority opinion can be given as follows:

$$GIOWA\left(\left\langle s_1, x_j^1 \right\rangle, \ldots, \left\langle s_k, x_j^k \right\rangle\right) = \left(\sum_{j=1}^{n} w_{(j)} b_{(j)}^{\lambda}\right)^{1/\lambda} \tag{25}$$

where λ is parameter for the generalization of aggregation operator. At this stage, a new decision matrix which represent the majority of experts on all criteria is derived as follows:

$$D^{Maj} = \begin{array}{c} \\ A_1 \\ \vdots \\ A_m \end{array} \begin{array}{c} C_1 \quad \cdots \quad C_n \\ \begin{pmatrix} x_{11}^{Maj} & \cdots & x_{1n}^{Maj} \\ \vdots & \ddots & \vdots \\ x_{m1}^{Maj} & \cdots & x_{mn}^{Maj} \end{pmatrix} \end{array}, \tag{26}$$

where x_{ij}^{Maj} represent the majority opinion of experts with respect to each criterion C_j for the alternative A_i.

4.2 Internal Fusion Scheme: Inclusion of Decision Strategies on Criteria

Step 5: Determine the ideal and anti-ideal solutions V^+ and V^- for the majority of experts such in Eqs. (16) and (17).

Step 6: Calculate the separation measures from the ideal and anti-ideal solutions using the Minkowski OWA distance such in Eq. (4). This step can be divided in the following steps:

Compute the separation measures from ideal solution, where the argument variables as distance measure are reordered in ascending order; the shortest distance to the ideal solution is the best as follows:

$$D_i^+ = MOWAD(d_1, \ldots, d_n)^+ = \left\{\sum_{j=1}^{n} w_j^+ \left(e_j^+\right)^p\right\}^{1/p}, \tag{27}$$
$$\text{for } j = 1, 2, \ldots, n,$$

where e_j^+ is the jth smallest of the d_i and d_i is the individual distance between x_j and x_j^+, such that $d_i = \left|x_j - x_j^+\right|$.

By the similar way, compute the separation measure for anti-ideal solution, where the argument variables as distance measure are ordered in descending order; as the farthest distance to the anti ideal solution is the best.

$$D_i^- = MOWAD(d_1, \ldots, d_n)^- = \left\{ \sum_{j=1}^n w_j^- \left(e_j^- \right)^p \right\}^{1/p},$$

for $j = 1, 2, \ldots, n$, (28)

where e_j^- is the jth largest of the d_i and d_i is the individual distance between x_j and x_j^-, such that $d_i = \left| x_j - x_j^- \right|$.

Step 7: Derive the weighting vectors for ideal and anti-ideal solutions using RIM (quantifier guided aggregation) for criteria $\{c_1, \ldots, c_n\}$, where the $v_j, j = 1, 2, \ldots, n$ is relative importance of each criterion associated with criterion c_j such in Eq. (9). The weights of ideal solution can be calculated as follow:

$$w_j^+ = Q\left(\sum_{k=1}^j u_k \right) - Q\left(\sum_{k=1}^{j-1} u_k \right),$$ (29)

where u_k denote the importance associated with the criterion that has the jth smallest e_j^+, such as (u_j, b_j). The integer w_j^+ is the inclusion of relative importance associated with linguistic quantifier for the criterion C_j, and $\sum_{j=1}^n w_j^+ = 1$.

By the similar way, the weights of anti-ideal solution can be calculated as follow:

$$w_j^- = Q\left(\sum_{k=1}^j u_k \right) - Q\left(\sum_{k=1}^{j-1} u_k \right),$$ (30)

where u_k denote the importance associated with the criterion that has the jth greatest e_j^-, such as (u_j, b_j). The integer w_j^- is the weights associated with anti-ideal solutions and $\sum_{j=1}^n w_j^- = 1$.

Step 8: Calculate the relative closeness RC_i^{Maj} to the ideal solution for the group. The relative closeness can be calculated as follows:

$$RC_i^{Maj} = \frac{D(a^i)^-}{D(a^i)^- + D(a^i)^+}, i = 1, 2, \ldots m,$$ (31)

where the alternatives are ranked in descending order. Note that the larger the value of RC_i^{Maj} denotes the better performance of the alternative.

5 Illustrative Example

In this section, the case study of human resource selection problem for a local chemical company as described in [13] is used. There are 17 candidates (alternatives) and 4 decision makers considered, and the candidates' qualification is measured through a number of objective and subjective tests. The basic data for this experiment is demonstrated in Tables 1 and 2. However, in order to consort with the context of this paper, a slight modification is made on the original data regarding the weights associated to experts on each criterion. In this case, a homogenous type of problem is considered by attaching identic weights of criteria for each expert as follows: language test (C_1), professional test (C_2), safety rule test (C_3), professional skills (C_4), computer skills (C_5), panel interview (C_6) and 1-on-1 interview (C_7) as 0.066, 0.196, 0.066, 0.130, 0.130, 0.216 and 0.196, respectively.

Table 1. Decision matrix of human resource selection problem – objective attribute

No.	Knowledge test			Skill tests	
	Language test	Professional test	Safety rule test	Professional skills	Computer skills
1	80	70	87	77	76
2	85	65	76	80	75
3	78	90	72	80	85
4	75	84	69	85	65
5	84	67	60	75	85
6	85	78	82	81	79
7	77	83	74	70	71
8	78	82	72	80	78
9	85	90	80	88	90
10	89	75	79	67	77
11	65	55	68	62	70
12	70	64	65	65	60
13	95	80	70	75	70
14	70	80	79	80	85
15	60	78	87	70	66
16	92	85	88	90	85
17	86	87	80	70	72

The comparison between the classical TOPSIS [13], the modified Hajimirsadeghi and Lucas's model and the proposed model, specifically with respect to both distance measures (i.e., Manhattan and Euclidean) are made and shown in Table 3. As can be seen, the rankings resulted from all methods show slightly different results, especially

Table 2. Decision matrix of human resource selection problem – subjective attribute

No.	DM 1		DM 2		DM 3		DM 4	
	Panel	1-on-1	Panel	1-on-1	Panel	1-on-1	Panel	1-on-1
1	80	75	85	80	75	70	90	85
2	65	75	60	70	70	77	60	70
3	90	85	80	85	80	90	90	95
4	65	70	55	60	68	72	62	72
5	75	80	75	80	50	55	70	75
6	80	80	75	85	77	82	75	75
7	65	70	70	60	65	72	67	75
8	70	60	75	65	75	67	82	85
9	80	85	95	85	90	85	90	92
10	70	75	75	80	68	78	65	70
11	50	60	62	65	60	65	65	70
12	60	65	65	75	50	60	45	50
13	75	75	80	80	65	75	70	75
14	80	70	75	72	80	70	75	75
15	70	65	75	70	65	70	60	65
16	90	95	92	90	85	80	88	90
17	80	85	70	75	75	80	70	75

Note: (Panel = Panel interview, 1-on-1 = One-on-one interview)

on the first two candidates: a_9 and a_{16}. The classical TOPSIS method for both distance measures ranked a_{16} as the best and then a_9. On contrary, Hajimirsadeghi and Lucas's model ranked a_9 as the top ranking and then followed by a_{16} for both distance measures. The proposed method ranked different results for each of distance measures, refer Table 3. In Table 4, the confidence measures for all methods are provided, include (1) the sum of absolute difference between relative closeness of the consecutive alternatives, (2) the minimum of the absolute difference between relative closeness of the consecutive alternatives, and (3) the range of calculated relative closeness for the alternatives. Based on the results of measure 2, the proposed method under Euclidean distance indicates the highest difference (0.0042) between relative closeness of the consecutive alternatives compared to the others. Hence, this imply that, the experts can be more convincing to make the decision when the distinction values between each alternative is significant.In addition, the proposed method can represent the uniformity of the decision strategies compared to method based on majority opinion of experts on each alternative. Therefore, the aggregation on specific criteria can be easily adjusted which represent the majority of experts.

Table 3. Final distance performance and rankings of the aggregation

Classical TOPSIS-GDM with arithmetic mean				Hajimirsadeghi and Lucas [6]				The proposed model			
Manhattan distance		Euclidean distance		Manhattan OWAD		Euclidean OWAD		Manhattan OWAD		Euclidean OWAD	
RC	R	RC	R	RC	R	RC	R	RC	R	RC	R
0.6169	7	0.6122	5	0.6065	7	0.5916	7	0.4981	5	0.4949	5
0.4393	14	0.4338	14	0.4703	13	0.4587	14	0.2106	15	0.2538	15
0.8212	3	0.7767	3	0.8830	3	0.8278	3	0.7306	3	0.7008	3
0.4603	12	0.4645	13	0.5036	11	0.4722	13	0.2548	14	0.2978	14
0.4559	13	0.4651	12	0.4695	14	0.4746	12	0.3771	10	0.3890	11
0.6677	4	0.6596	4	0.6884	4	0.6812	4	0.6024	4	0.5971	4
0.4655	11	0.4732	11	0.4958	12	0.4764	11	0.3106	13	0.3247	12
0.5779	8	0.5755	8	0.5715	8	0.5857	8	0.3755	11	0.4102	9
0.9103	2	0.8729	2	0.9531	1	0.9150	1	0.8821	1	0.8459	2
0.5187	10	0.5167	10	0.5259	10	0.5276	10	0.3809	9	0.3932	10
0.1715	16	0.2145	16	0.1436	16	0.2078	16	0.0459	17	0.0757	17
0.1561	17	0.1838	17	0.1396	17	0.1549	17	0.0483	16	0.0876	16
0.5654	9	0.5626	9	0.5687	9	0.5581	9	0.4173	8	0.4225	8
0.6195	5	0.6063	7	0.6521	5	0.6429	5	0.4779	6	0.4845	6
0.4079	15	0.4223	15	0.3784	15	0.3964	15	0.3583	12	0.3146	13
0.9104	1	0.8899	1	0.9277	2	0.9128	2	0.8817	2	0.8709	1
0.6190	6	0.6081	6	0.6498	6	0.6135	6	0.4351	7	0.4445	7

Note: (RC = Relative Closeness, R = Ranking)

Table 4. Confidence measures for different methods

Measure	Classical TOPSIS-GDM with arithmetic mean		Hajimirsadeghi and Lucas [6]		The proposed method	
	Manhattan distance	Euclidean distance	Manhattan OWAD	Euclidean OWAD	Manhattan OWAD	Euclidean OWAD
1	0.7543	3.6597	0.8136	0.7601	0.8362	0.7952
2	0.0001	0.0006	0.0009	0.0018	0.0004	**0.0042**
3	[0.1561, 0.9104]	[0.1838, 0.8899]	[0.1396, 0.9531]	[0.1549, 0.9150]	[0.0459, 0.8821]	[0.0757, 0.8709]

6 Conclusions

In this paper, the extension of OWA-TOPSIS model under group decision making is presented. The emphasis is focused on the integration of majority concept as group aggregator for the experts' judgments, where the consensus among the experts on each criterion is concentrated. Then, the generalized aggregation operators are incorporated to provide more flexibility in the fusion process as to present the majority of experts. Some modifications on separation measures are made using the Minkowski OWA distance with inclusion of relative importances associated to the criteria. The analysis on human resource selection problem is implemented to test the reliability of the proposed model and some comparisons are made between other OWA-TOPSIS models proposed in the literature. The advantage of the proposed model includes, provide the uniformity in modeling the behavior of majority experts regarding the proportion of criteria to consider or decision strategies to analyze. The model also provides a robust decision by taking into account the consensual judgments on each criterion instead of on each alternative in which derived from the individual judgments of each expert. For future research, the extension of the model to the case of heterogeneous group decision making will be proposed and the application on the selection of financial products will be implemented.

References

1. Bellman, R.E., Zadeh, L.A.: Decision making in a fuzzy environment. Manag. Sci. **17**, 141–164 (1970)
2. Boroushaki, S., Maczewski, J.: Using the fuzzy majority approach for GIS-based multicriteria group decision-making. Comput. Geosci. **36**, 302–312 (2010)
3. Bordogna, G., Sterlacchini, S.: A multi criteria group decision making process based on the soft fusion of coherent evaluations of spatial alternatives. In: Zadeh, L.A., et al. (eds.) Recent Developments and New Directions on Soft Computing, Studies in Fuzziness and Soft Computing, vol. 317 (2014)
4. Chen, Y., Li, K.W., Liu, S.F.: An OWA-TOPSIS method for multiple criteria decision analysis. Expert Syst. Appl. **38**, 5205–5211 (2011)
5. Figueira, J., Greco, S., Ehrgott, M.: Multiple Criteria Decision Analysis: State of the Art Surveys. Springer, New York (2005)
6. Hajimirsadeghi, H., Lucas, C.: Extended TOPSIS for multi-criteria group decision making with linguistic quantification and concept of majority opinion, Final project for "Fuzzy Systems" course, School of Electrical and Computer Engineering, University of Tehran, Spring (2009)
7. Hwang, C.L., Yoon, K.: Multiple Attribute Decision Making. Springer, Berlin (1981)
8. Kacprzyk, J., Zadrozny, S., Nurmi, H.: The role of the OWA operators as a unification tool for the representation of collective choice sets. In: Yager, R.R., et al. (eds.) Recent Development in the OWA Operators, STUDFUZZ, vol. 265, pp. 149–166 (2011)
9. Merigó, J.M., Gil-Lafuente, A.N.: The induced generalized OWA operator. Inf. Sci. **179**, 729–741 (2009)
10. Merigó, J.M., Gil-Lafuente, A.M.: Using OWA operator in the Minkowski distance. Int. J. Soc. Humanit. Sci. **2**, 564–572 (2008)
11. Pasi, G., Yager, R.R.: Modelling the concept of majority opinion in group decision making. Inf. Sci. **176**, 390–414 (2006)
12. Shafiqul, M.I., Sadiq, R., Rodriguez, M.J., Najjaran, H., Francisque, A., Hoorfar, M.: Evaluating water quality failure potential in water distribution systems: a Fuzzy-TOPSIS-OWA based methodology. Water Resour. Manag. **27**, 2195–2216 (2013)
13. Shih, H.S., Shyur, H.J., Lee, E.S.: An extension of TOPSIS for group decision making. Math. Comput. Model. **45**, 801–813 (2007)

14. Taib, C.M.I.C., Yusoff, B., Abdullah, M.L., Wahab, A.F.: Conflicting bi-fuzzy multi-attribute group decision making model with application to flood control project. Group Decis. Negot. **25**(1), 157–180 (2016)
15. Yager, R.R.: Quantifier guided aggregation using OWA operators. Int. J. Intell. Syst. **11**, 49–73 (1996)
16. Yager, R.R., Kacprzyk, J.: The Ordered Weighted Averaging Operators: Theory and Applications. Kluwer, Norwell (1997)
17. Yager, R.R.: On ordered weighted averaging aggregation operators in multi-criteria decision making. IEEE Trans. Syst. Man Cybern. **18**, 183–190 (1988)
18. Yager, R.R., Filev, D.P.: Induced ordered weighted averaging operators. IEEE Trans. Syst. Man Cybern. Part B Cybern. **29**, 141–150 (1999)
19. Yusoff, B., Merigó, J.M.: Analytic hierarchy process under group decision making with some induced aggregation operators. In: IPMU 2014, Montpellier, France (2014)
20. Zadeh, L.A.: A computational approach to fuzzy quantifiers in natural languages. Comput. Math. Appl. **9**, 149–184 (1983)

Fuzzy Logic Approach Applied into Balanced Scorecard

Carolina Nicolás[1], Jaume Gil-Lafuente[2],
Angélica Urrutia Sepúlveda[3], and Leslier Valenzuela Fernández[4(✉)]

[1] School of Economics and Business, University Santo Tomás, Valdivia, Chile
[2] School of Economics and Business, University of Barcelona, Barcelona, Spain
[3] Facultad de Ingeniería, University Católica del Maule, Talca, Chile
[4] School of Economics and Business, University of Chile, Santiago, Chile
lvalenzu@fen.uchile.cl

Abstract. In this paper we propose to apply fuzzy logic methodologies for measuring key performance indicator into Balanced Scorecard (BSC) Customer Perspective. This study provide fuzzy key performance indicator (FKPI) to the customer experience management for uncertainty measures. The proposal is a step forward in terms of methodologies for creation of performance indicators in the field of marketing, where you have the opportunity to work with uncertain or vague data (human language). The significant contribution of the research is include qualitative data indicator, result from analysis the information that expresses the client in text format, to BSC. The methodology used is the model proposed by Mamdani fuzzy inference which is based on the fuzzy logic theory and fuzzy subsets. Software Matlab was used for the analysis.

Keywords: Fuzzy Key Performance Indicator · Fuzzy inference
Customer experience management · Balanced Scorecard
Mamdani inference method

1 Introduction

The management of a company needs to develop key performance indicators integrated in balanced scorecard, which considers financial measures that tell executives the results of actions already taken. And complements the financial measures with three other sets of operational measures related to the customer, internal processes, and the ability of the organization to learn and improve [20]. Thus, the availability of a consistent and coherent set of indicators is a prerequisite to make informed decisions, aligned with the goals of the organization [9, 11, 19].

Non-financial indicators that form an integral measuring system performance and associated with the BSC customer perspective are usually created based on quantitative data, the product of transactional relationships with customers. It is considered that mathematical approaches of the calculation of quality evaluation of the systems involving human factors do not reflect a significant feature of the processed data due to their natural vagueness. The appropriate theoretical background for the formalization of data vagueness is the fuzzy set and fuzzy logic theory [30].

© Springer International Publishing AG, part of Springer Nature 2018
A. M. Gil-Lafuente et al. (Eds.): FIM 2015, AISC 730, pp. 140–151, 2018.
https://doi.org/10.1007/978-3-319-75792-6_12

The study on "Fuzzy key performance indicator" (FKPI) is limited in the literature [24, 30, 47, 49].

Schmitt [44] is recognized by his contributions to customer experience management. In his proposed model, Customer Experience Management (CEM), he shows how to use the power of an experimental approach so as to be able to connect with the customer in each interaction or contact point. An agreement between many authors can be observed in terms of how customer experience management should include all of the contact points between the customer and the company [12, 14, 18, 32, 38, 44, 50].

Considering the importance that today reaches managing the customer experience, and provision for consumers to point at any social network their experiences with companies is that the empirical study focused on the analysis of the text given freely by the customers in a survey that evaluates this variable.

The aim of this paper is to advance within the Fuzzy Key Performance Indicators in customer experience theory. Thus, proposed to apply fuzzy logic methodologies for measuring indicators into Balanced Scorecard (BSC) Customer Perspective.

To clarify, "classic data" in this study refers to "numerical" format information that is traditionally analyzed by companies and the academic world, while "fuzzy data" refers to all the information in "linguistic" format.

In the following sections of this study, the theoretical framework and the methodology applied are explained, ending with the Fuzzy Key Performance Indicators (FKPI) for the Customer Experience Management model.

2   Conceptual Framework and Development of Hypotheses

In this section, we briefly review the Balanced Scorecard, key performance indicator; fuzzy set and fuzzy number.

2.1   Balanced Scorecard (BSC)

The balanced scorecard (Balanced Score Card), hereinafter BSC is a system of strategic planning and management that has been used extensively in business and industry, government, and nonprofit organizations worldwide in order to align business activities to the vision and strategy of the organization, improve internal and external communications, and monitor organization performance against strategic goals. It was developed by Kaplan and Norton [19] as a framework for performance measurement that adds non-financial strategic measures to traditional financial metrics, and delivery to managers and executives a more 'balanced' view of organizational performance [19, 20]. The BSC focuses primarily on the two major problems of modern organizations: the proper performance measurement and evaluation of the successful implementation of the strategy of the organization. Overall, the BSC system is considered to be simultaneously a performance measurement system, a system of valuation of the strategy and a communication tool [19], defined by four perspectives: financial, customer, process domestic business and, finally, learning and growth. For many researchers the BSC is not just a measurement system, but it is a management system that enables organizations to clarify their vision and strategy and transform them into action [34, 36, 39].

The BSC has been widely applied in the private sector. The "Gartner Group" estimated that at least half of the Fortune 1000 companies use BSC [36] methodology.

2.2 Key Performance Indicator

KPIs are meters that reflect the health of an organization [8]. A key performance indicator (KPI) is a performance measurement that evaluates the success of a particular activity. Success can be either the achievement of an operational goal or the progress toward strategic goals [5].

The goals to be measured in the realm of organizational performance are latent u objective constructs (e.g. customer satisfaction, employee loyalty, customer experience, profit, among others.). Similar to an instrument of mediation, a performance indicator measuring addressed some aspect of organizational performance with respect to a reference object. Therefore, a performance indicator is based on the hypothesis that properly represents a particular aspect of organizational performance [40]. Moreover, an indicator involves an epistemological principle only if their validity is successful it will serve its purpose.

An indicator is valid if it really measures the manifestations appearance to be measured; in other words, the validity implies objectivity and reliability [7]. An indicator is measurement target if not depends on the judgment of a particular person. And it will be reliable if repeat measurements produce the same result. Although the question of how to judge the validity of a measure is well known in the scientific philosophy, he has never been, nevertheless, properly answered. The validity of a performance indicator depends on the truth of the hypothesis after the indicator or in other words, how well it can be justified [7].

2.3 Fuzzy Set and Fuzzy Number

The origins of fuzzy logic are associated with investigations by Zadeh [51], professor at the University of California, USA, disseminated through his article titled "Fuzzy sets": where he presented a few sets with no clear or precise limits, playing an important role in pattern recognition, interpretation of meanings, and abstraction [48]. It is widely used to handle the imprecise and uncertain information in the real-world problems [52]. The main contribution of this theory is that it allows study of ambiguity, allowing more akin to express human expression, "natural language" logical relationships.

Thus, fuzzy logic suggests that in the field of social sciences provide technical management in an uncertain environment, which is based on the concept "gradual simultaneity", which can handle vague or difficult to specify, and to change the operation or status of a specific system [13].

Definition 1. Fuzzy set: $\tilde{A}$ in a universe of discourse X is characterized by a membership function $\mu_{\tilde{A}}(x)$ which associates with each element x in X a real number in the interval [0, 1]. The function value $\mu_{\tilde{A}}(x)$ is termed the degree of membership of x in $\tilde{A}$.

Definition 2. Triangular fuzzy number: A fuzzy number is a subset fuzzy, A, normal, convex and whose reference is the real numbers. It is the instrument par excellence of the Fuzzy Logic to represent amounts estimated or observed fuzzy.

In change, for unordered referential domains can be defined labels or scalar. These labels can be a similarity function is defined for each pair of values of the domain D and establishes a relationship of similarity or proximity to measure the similarity or resemblance between two elements of the domain. Generally, the similarity values are normalized in the interval [0, 1], corresponding to "0" to the meaning such as "totally different" and "1" meaning "totally like," Urrutia [48].

A positive triangular fuzzy number $\tilde{n}$ can be defined as $\tilde{n} = (n1, n2, n3)$, where $n1 \leq n2 \leq n3$ and $n1 > 0$.

The membership function $\mu_{\tilde{n}}(x)$ of triangular fuzzy [52].

$$\mu_{\tilde{n}}(x) = \begin{cases} \frac{x-n1}{n1-n2} & n_1 < x < n_2 \\ \frac{n3-x}{n3-n2} & n_2 < x < n_3 \\ 0, & Otherwise \end{cases}$$

Definition 4 [46]: Linguistic Variable is a variable whose values are expressed in text terms. Linguistic variables are very useful in dealing with situations which are too complex or not well defined to be reasonably described in conventional quantitative expressions.

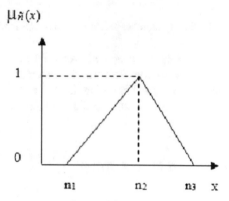

3   Methodology

This paper presents the first phase, exploratory character, of the study on performance indicators on the interactive customer experience. An online style of structured questionnaire was used as an information gathering tool. It was defined to be able to measure the interactive experience of customers with the company being studied, geared towards listening to the customer's voice, thus empirically proving the Customer Interactive Experience the analysis model proposed in this investigation with the gathered information.

The universe studied was made up of active company customers that have flown in the last year. The type of sampling was probabilistic, simple random sample, obtaining a final valid sample of 960 surveyed. Geographical Area: Spain.

So as to be able to solve the problem, a methodological sequence that covered from the choosing of the indicators for each experience, passing through the design of the general architecture of the system, to the fuzzy model with the help of the MATLAB fuzzy logic toolbox, was used. In this case, the centroid in the Defuzzyfication method was used.

The proposed for measuring the fuzzy customer experience indicators are performed using the free text format, gathered with an open question on the survey applied to customers of airline company. Is defined as a grouping of graduated responses; fuzzy inference using the Mamdani model, its rules are of the IF – THEN kind. As observed, a rule from the set of rules or knowledge base has two parts: the antecedent and the conclusion, to see Fig. 1, software used MATLAB.

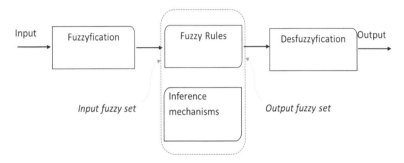

Fig. 1. Fuzzy inference using the Mamdani model

The analysis considered: data storage, with KNIME perform the extraction and transformation of data from the source, data filtering, storage of relevant words, This step is of great importance, here are beginning to think about the stage of opinion mining. Then JAVA relevant words are analyzed and finally developed in MATLAB a fuzzy indicator.

4   Empirical Results and Discussion

Fuzzy Indicators – Free Text

This stage of the investigation analyses free text, gathered with an open question on the survey applied to customers of the airline company. The text mining is done with the Knime software which takes every word and sees the number of coincidences it has within the text, thus "rescuing" those words that were more frequently mentioned in the comments. This information is organized for the purpose of knowing which words present a higher degree of frequency; these words will be used as search and analysis patterns later on. The analysis methodology used assigns a degree of belonging to the significant words found by the system, be it by positive or negative real numbers. This lets us infer how positive or negative customer experiences with the company actually are. See Fig. 2. At this stage of analysis we used JAVA software.

Figure 2 shows that the application requires three sources of information: the analysis of words of higher usage, the comments that contain these words and our dictionary made up of the adjectives. The analysis carried out with the help of these information entries results in the approval indicator, which is our main goal at this stage.

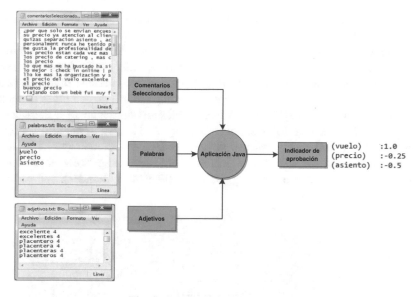

Fig. 2. Diagram: opinion mining

The indicators obtained, associated to the words obtained, are: flights (1.0), price (−0.25), and seat (−0.5). These results respond to the purpose of the investigation because the fact that they produce indicators that report on customer perceptions regarding their experiences with the company. The value of the indicators can be in the [−2, 2] interval, assigned values according to the degree of presence of an adjective in a word. This means that, for example, −0.25 is not a very critical indicator. However, it is a concrete value and more closely reflects the actual conditions of the company in terms of this indicator, supplying greater knowledge regarding the situation of the company.

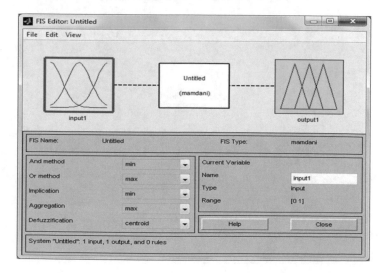

Fig. 3. Fuzzy interface module belonging to MATLAB

Next, once we already have the experience we have indicators represent those values in an indicator, to see Fig. 3.

The previous results revealed valuable information to the company, but this information can be "encapsulated", delivering a single guiding indicator in terms of the current company situation. The Matlab tool was incorporated because of this. This tools takes three significant words with their corresponding indicators, and gives us a global view regarding customer experience; a fuzzy indicator. Inference with the Mamdani (1975) process is carried out for this. This is a fuzzy, also known as linguistic, inference method that has a wider range of use than other fuzzy methodologies. The authors based themselves on the research of Zadeh (1965) regarding fuzzy algorithms for complex systems and decision making processes (Figs. 4, 5, 6 and 7).

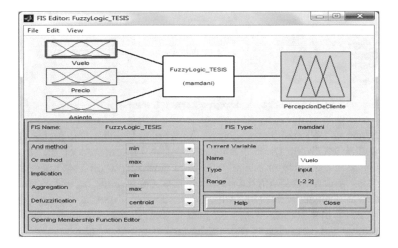

Fig. 4. Creation of linguistic variables

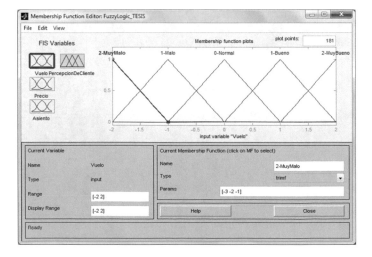

Fig. 5. Definition of linguistic labels and membership function to output variable

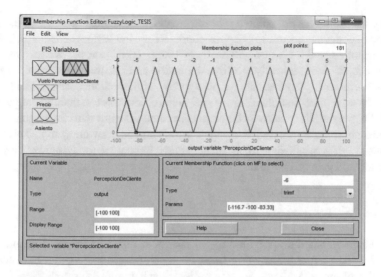

Fig. 6. Definition of linguistic labels and membership function for input variables

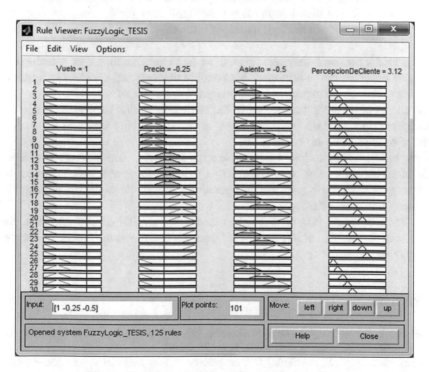

Fig. 7. Final fuzzy indicator of customer experience

Once the indicators are entered into the created model, they tell us that the level of customer experience regarding services rendered by the airline reaches 3.12%. This final indicator can take a value within the [−100%, 100%] range. We are therefore in the presence of a numerically positive indicator and, at the same time, a positive indicator for the company because people approve of their service. It's not an incredible number, but the fact that there are more positive comments than negative ones is evident. Or, maybe, the positive comments are more relevant than the negative ones in the comments done by customers, which is relevant for the company. This analysis paved the way for proposing fuzzy management indicators for customer experiences through the analysis of free text.

5 Conclusions

The organization has data provided by its clients in different formats (quantitative or qualitative). To extract information from the data and get knowledge is necessary to apply different methodologies. So, one option to achieve a better understanding of the linguistic data is through fuzzy logic methodologies.

After comparing the different methodologies used for the analysis of uncertain data, it was possible to see that defining key fuzzy indicators for the management of customer experience, formed through linguistic information contributed by the customers, is possible.

The results obtained, therefore, also allow us to indicate that after listening to and analyzing the voice of the customer, it's possible to create management indicators that form the customer's perspective in an Integral Control Panel.

The areas of marketing work a lot of linguistic information to measure the customer experience. Thus, methodologies for the management of uncertainty in the marketing areas of organizations are contributed.

The business implications presented are related to decision making. The development of an indicator relative to customer experience, to measure performance of the progress toward strategic goals and/or operational goals; allows the organization to be informed of customer feedback.

In future research, it is considered to create an indicator of customer experience, considering the analysis of free text, as for example comments on Twitter or Facebook, among other social networks.

References

1. Åberg, A., Shahmehri, N.: The role of human Web assistants in e-commerce: an analysis and a usability study. Internet Res. **10**(2), 114–125 (2000)
2. Ananiadou, S., Kell, D., Tsujii, J.: Text mining and its potential applications in systems biology. Trends Biotechnol. **24**(12), 571–579 (2000)
3. Berland, G.K., Elliott, M.N., Morales, L.S., Algazy, J.I., et al.: Health information on the internet accessibility, quality, and readability in English and Spanish. J. Am. Med. Assoc. **285**(20), 2612–2621 (2001)

4. Berry, L.L., Carbone, L.P., Haeckel, S.H.: Managing the total customer experience. MIT Sloan Manag. Rev. **43**(3), 85–89 (2002)
5. Cabeza, L., Galindo, E., Prieto, C., Barreneche, C., Fernández, E.: Key performance indicators in thermal energy storage: survey and assessment. Renew. Energy **83**, 820–827 (2015)
6. Carbone, L.P., Haeckel, S.H.: Engineering customer experiences. Mark. Manag. **3**(3), 8–19 (1994)
7. Edwards, J.R., Bagozzi, R.P.: On the nature and direction of the relationship between constructs and measures. Psychol. Meth. **5**, 155–174 (2000)
8. Enns, B.: Key performance indicators for new business development. Critical Briefings for the Business of Persuasion, pp. 2–6 (2005)
9. Epstein, M.J., Manzoni, J.F.: The balanced scorecard and tableau de bord: a global perspective on translating strategy into action. In: Working Paper, 97/82/IC/SM, INSEAD, Paris (1998)
10. Flynn, B.B., Sakakibara, S., Schroeder, R.G., Bates, K.A., Flynn, E.J.: Empirical research methods in operations management. J. Oper. Manag. **9**(2), 250–284 (1990)
11. Fortuin, L.: Performance indicators - Why, where and how? Eur. J. Oper. Res. **34**(1), 1–9 (1988)
12. Gentile, C., Spiller, N., Noci, G.: How to sustain the customer experience: an overview of experience components that co-create value with the customer. Eur. Manag. J. **25**(5), 395–410 (2007)
13. Gil-Aluja, J.: Elements for a Theory of Decision in Uncertainty, pp. 16–18. Kluwer Academic Publishers, Boston (1999)
14. Grewal, D., Michael, M.L., Kumarc, V.: Customer experience management in retailing: an organizing framework. J. Retail. **85**(1), 1–14 (2009)
15. Han, K.H., Gu, K.J., Minseok, S.S.: A process-based performance measurement framework for continuous process improvement. Int. J. Ind. Eng. **14**(3), 220–228 (2007)
16. Hird, W.: Recycled water-case study: BlueScope steel, Port Kembla steelworks. Desalination **188**(3), 97–103 (2006)
17. Hoffman, D., Novak, T.P.: Flow online: lessons learned and future prospects. J. Interact. Mark. **23**, 23–34 (2009)
18. Holbrook, M.B., Hirschman, E.C.: The experiential aspects of consumption: consumer fantasies, feelings and fun. J. Consum. Res. **9**(2), 132–140 (1982)
19. Kaplan, R.S., Norton, D.: The balanced scorecard measures that drives performance. Harvard Bus. Rev. **70**(1), 71–79 (1992)
20. Kaplan, R.S., Norton, D.: Using the balanced scorecard as a strategic management system. Harvard Bus. Rev. **74**(1), 75–85 (1996)
21. Kaplan, R.S., Norton, D.: Measuring the strategic readiness of intangible assets. Harvard Bus. Rev. **82**(2), 52–63 (2004)
22. Kerin, R.A., Ambuj, J., Daniel, J.H.: Store shopping experience and consumer price-quality-value perceptions. J. Retail. **68**(4), 376–397 (1992)
23. Kim, S.H., Cha, J., Knutson, B.J., Jeffrey, A.B.: Development and testing of the Consumer Experience Index (CEI). Manag. Serv. Qual. **21**(2), 112–132 (2011)
24. Khireldin, A., Zaher, H.M., Elmoneim, A.M.: A fuzzy approach for evaluating the performance and service quality of airport, Egyptian Aviation Academy, Cairo University, Egypt (2011)
25. Kotler, P., Armstrong, G.: Fundamentos de Marketing, 8th edn. Pearson Education, México (2008)
26. Kumar, V.: Customer Relationship Management. In: Wiley International Encyclopedia of Marketing. Wiley, Hoboken (2010)

27. Kumar, V., Venkatesan, R., Reinartz, W.: Knowing what to sell, when, and to whom. Hardvard Bus. Rev. **84**, 1–9 (2006)
28. Lynch, R.L., Cross, K.F.: Measure Up - The Essential Guide to Measuring Business Performance. Mandarin, London (1991)
29. Mamdani, E.H., Assilian, S.: An experiment in linguistic synthesis with a fuzzy logic controller. Int. J. Man Mach. Stud. **7**(1), 1–13 (1975)
30. Mensik, M., Miroslav, P.: Fuzzy approaches applied into balanced scorecard customer perspective. In: Proceedings of the International Conference on Advanced ICT and Education, pp. 848–856 (2013)
31. McKenzie, S.B., Podsakoff, P.M., Jarvis, C.B.: The problem of measurement model misspecification in behavioral and organizational research and some recommended solutions. J. Appl. Psychol. **90**, 710–730 (2005)
32. Meyer, C., Schwager, A.: Understanding customer experience. Harvard Bus. Rev. **85**, 116–126 (2007)
33. Morley, L., Leonard, D., David, M.: Variations in Vivas: quality and equality in British PhD assessments. J. Stud. High. Educ. **27**(3), 263–273 (2002)
34. Nair, M.: Essentials of Balanced Scorecard. Emerge Inc., Wiley, Hoboken (2004)
35. Nambisan, P., Watt, J.H.: Managing customer experiences in online product communities. J. Bus. Res. **64**, 889–895 (2011)
36. Niven, P.R.: Balanced Scorecard Step-by-Step: Maximizing Performance and Maintaining Results. Wiley, Hoboken (2015)
37. Novak, T.P., Hoffman, D.L., Yung, Y.F.: Measuring the customer experience in online environments: a structural modeling approach. Mark. Sci. **19**(1), 22–42 (2002)
38. Otnes, C.C., Ilhan, B.E., Kulkarni, A.: The Language of Marketplace Rituals: Implications for Customer Experience Management. J. Retail. **88**(3), 367–383 (2012)
39. Olsona, E.M., Slater, S.F.: The balanced scorecard, competitive strategy, and performance. Bus. Horiz. **45**(3), 11–16 (2002)
40. Peng, T.J.A., Pike, S., Roos, G.: Intellectual capital and performance indicators: Taiwanese healthcare sector. J. Intellect. Capital **8**(3), 538–556 (2007)
41. Pine, B.J., Gilmore, J.H.: The Experience Economy: Work is Theatre and Every Business a Stage. Harvard Business School Press, Boston (1999)
42. Rose, S., Clark, M., Samouel, P., Hair, N.: Online customer experience in e-retailing: an empirical model of antecedents and outcomes. J. Retail. **88**(2), 308–322 (2012)
43. Schmitt, B.H., Rogers, D.L.: Handbook on Brand and Experience Management. Edward Elgar Publishing, Northampton (2009)
44. Schmitt, B.H.: Customer Experience Management: A Revolutionary Approach to Connecting with Your Customers. Mc Graw Hill Publication, New York (2003)
45. Schmitt, B.H.: Big Think Strategy: How to Leverage Bold Ideas and Leave Small Thinking Behind. Harvard Business School Press, Boston (2007)
46. Tavana, N.: A fuzzy-QFD approach to balanced scorecard using an analytic network process. Int. J. Inf. Decis. Sci. **5**(4), 331–362 (2013)
47. Tchesmedjiev, P., Vassilev, P.: Determining intuitionistic fuzzy index for overall employees' performance based on objectives and KPIs. In Proceedings of International Conference on IFSs, Sofia, vol. 14(2), pp. 88–90 (2008)
48. Urrutia, A., Galindo, J., Piattini, M.: Modeling data using fuzzy attributes. In: 22nd International Conference of the Chilean Computer Science Society (SCCC-2002), 6–8 November 2002, Copiapo, Chile, pp. 117–123 (2002)
49. Wang, M.L.: Lin, and Y.H.: To construct a monitoring mechanism of production loss by using fuzzy Delphi method and fuzzy regression technique -A case study of IC package testing company. Expert Syst. Appl. **35**(3), 1156–1165 (2008)

50. Yang, C.L., Chuang, S.P., Huang, R.H.: Manufacturing evaluation system based on AHP/ANP approach for wafer fabricating industry. Expert Syst. Appl. **36**(8), 11369–11377 (2009)
51. Zadeh, L.A.: Fuzzy sets. Inf. Control **8**(3), 338–353 (1965)
52. Zimmermann, H.J.: Fuzzy Set Theory and its Applications, 2nd edn. Kluwer Academic Publishers, Boston (1991)

Role of Octagonal Fuzzy Numbers in Solving Some Special Fuzzy Linear Programming Problems

Felbin C. Kennedy[1(✉)] and S. U. Malini[2]

[1] Stella Maris College (Autonomous), Chennai, Tamil Nadu, India
felbinckennedy@gmail.com
[2] D.G. Vaishnav College (Autonomous), Chennai, Tamil Nadu, India
malinisul4@gmail.com

Abstract. In the area of Fuzzy Operational Research, fuzzy transportation problem and fuzzy assignment problem were dealt with by several authors using a variety of fuzzy real numbers that are listed in the literature survey in [4, 5], wherein the problems are handled by converting them to be crisp case. Unlike in the case of real numbers, fuzzy numbers have no natural order. As a consequence, there are several ranking methods in literature [1, 2, 7] to compare or/and rank fuzzy numbers introduced by various authors since 1976. Based on the context of the application, some methods seem to be more appropriate than others.

The concept of octagonal fuzzy numbers and a ranking procedure using their α-cuts was introduced by the authors in an earlier paper [4]. Also the concept on symmetric octagonal fuzzy numbers has been introduced by the authors in [3]. The parameter 'k', $0 \leq k \leq 1$ involved in the definition of octagonal fuzzy numbers may be chosen appropriately to suit the situation/problem.

In this paper, symmetric octagonal fuzzy numbers are used to solve fuzzy transportation problem and fuzzy assignment problem, as distributive property holds only in the case of symmetric octagonal fuzzy numbers. New methods to solve the fuzzy transportation problem and fuzzy assignment problem are proposed in Sects. 4 and 5 respectively. Both the problems are solved without converting them to crisp problems. The solution procedures are illustrated with numerical examples.

Keywords: Symmetric octagonal fuzzy numbers
Fuzzy transportation problem · Fuzzy assignment problem

1 Introduction

The transportation problem and assignment problem are special cases of linear programming problems. Using transportation models we can find out how many units are to be transported from which origin to which destination so that the cost incurred is least, as well the availability at the origin and the requirements at the destination are satisfied. Using assignment models one can determine the optimum allotment of jobs to the persons in such a way that the cost incurred is least or the profit attained is maximum. The uncertainty in the real world is inevitable owing to some unexpected

© Springer International Publishing AG, part of Springer Nature 2018
A. M. Gil-Lafuente et al. (Eds.): FIM 2015, AISC 730, pp. 152–163, 2018.
https://doi.org/10.1007/978-3-319-75792-6_13

situations. It may so happen that the cost coefficients and the supply and demand quantities may be uncertain due to uncontrollable factors. In order to quantify the uncertain information in making decisions Zadeh [8] introduced the notion of fuzziness.

In this paper, a new method is proposed for finding a fuzzy optimal solution for a fuzzy transportation problem where the cost coefficients, supply and demand are special octagonal fuzzy numbers. Also a fuzzy analogue of Hungarian method is used for finding a fuzzy optimal solution for a fuzzy assignment problem where the cost, time or profit is special octagonal fuzzy numbers.

2 Octagonal Fuzzy Numbers

For the sake of completeness we recall from [3, 4], the required definitions and results.

Definition 1. A generalized octagonal fuzzy number denoted by $\tilde{A}_\omega$ is defined to be the ordered quadruple $\tilde{A}_\omega = (l_1(r), s_1(t), s_2(t), l_2(r))$, for $r \in [0, k]$, and $t \in [k, \omega]$ where

  (i) $l_1(r)$ is a bounded right continuous non decreasing function over $[0, \omega_1]$, $[0 \leq \omega_1 \leq k]$
  (ii) $s_1(t)$ is a bounded right continuous non decreasing function over $[k, \omega_2]$, $[k \leq \omega_2 \leq \omega]$
  (iii) $s_2(t)$ is a bounded right continuous non increasing function over $[k, \omega_2]$, $[k \leq \omega_2 \leq \omega]$
  (iv) $l_2(r)$ is a bounded right continuous non increasing function over $[0, \omega_2]$, $[0 \leq \omega_1 \leq k]$.

If $\omega = 1$, then the above-defined number is called a normal octagonal fuzzy number.

The octagonal fuzzy numbers we consider for our study is a subclass of the generalized octagonal fuzzy numbers (Definition 1) defined as follows:

Definition 2. A fuzzy set $(\tilde{A}, \mu_{\tilde{A}})$ or simply $\tilde{A}$ on $\mathbb{R}^1$ is said to be a symmetric octagonal fuzzy number if there exist real numbers $a_1, a_2, a_1 < a_2$, and $h > s > g > 0$ such that

$$\mu_{\tilde{A}}(x) = \begin{cases} k\left[\frac{x}{h-s} + \frac{h-a_1}{h-s}\right], & x \in [a_1 - h, a_1 - s] \\ k, & x \in [a_1 - s, a_1 - g] \\ k + (1-k)\left[\frac{x}{g} + \frac{g-a_1}{g}\right], & x \in [a_1 - g, a_1] \\ 1, & x \in [a_1, a_2] \\ k + (1-k)\left[\frac{a_2+g}{g} - \frac{x}{g}\right], & x \in [a_2, a_2 + g] \\ k, & x \in [a_2 + g, a_2 + s] \\ k\left[\frac{a_2+h}{h-s} - \frac{x}{h-s}\right], & x \in [a_2 + s, a_2 + h] \\ 0, & \text{otherwise} \end{cases}$$

We denote it by $\tilde{A} \approx (a_1 - h, a_1 - s, a_1 - g, a_1, a_2, a_2 + g, a_2 + s, a_2 + h; k, 1)$. When $h = s = g = 0$; $\tilde{A} \approx (a_1, a_1, a_1, a_1, a_2, a_2, a_2, a_2; k, 1)$ is a degenerate form. We use $\mathscr{F}(S)$ to denote the set of all symmetric octagonal fuzzy numbers.

Definition 3 (Arithmetic Operations). Let $\tilde{A} \approx (a_1, a_2, g, g, s, s, h, h; k, 1)$ and $\tilde{B} \approx (b_1, b_2, f, f, l, l, m, m; k, 1)$ be two symmetrical octagonal fuzzy numbers then

(i) **Addition** of $\tilde{A}$ and $\tilde{B}$ denoted as $A + \tilde{B}$ is defined as

$$\tilde{A} + \tilde{B} \approx (a_1 + b_1 - (h+m), a_1 + b_1 - (s+l), a_1 + b_1 - (g+f), a_1 + b_1, a_2 + b_2, a_2 + b_2 + (g+f), a_2 + b_2 + (s+l), a_2 + b_2 + (h+m); k, 1)$$

(ii) **Subtraction** of $\tilde{A}$ and $\tilde{B}$ denoted as $\tilde{A} - \tilde{B}$ is defined as

$$\tilde{A} - \tilde{B} \approx (a_1 - b_2 - (h+m), a_1 - b_2 - (s+l), a_1 - b_2 - (g+f), a_1 - b_2, a_2 - b_1, a_2 - b_1 + (g+f), a_2 - b_1 + (s+l), a_2 - b_1 + (h+m); k, 1)$$

(iii) **Scalar Multiplication** of $\tilde{A}$ with a scalar λ is given by

$$\lambda \tilde{A} \approx \begin{cases} (\lambda a_1, \lambda a_2, \lambda g, \lambda g, \lambda s, \lambda s, \lambda h, \lambda h; k, 1), & \lambda \geq 0 \\ (\lambda a_2, \lambda a_1, \lambda g, \lambda g, \lambda s, \lambda s, \lambda h, \lambda h; k, 1) & \lambda < 0 \end{cases}$$

(iv) **Multiplication** Define a new multiplication operation (*) as

$$\tilde{A} * \tilde{B} \approx \begin{pmatrix} \left(\frac{a_1+a_2}{2}\right)\left(\frac{b_1+b_2}{2}\right) - w - \left(\left|\frac{a_1+a_2}{2}\right|m + \left|\frac{b_1+b_2}{2}\right|h\right), \left(\frac{a_1+a_2}{2}\right)\left(\frac{b_1+b_2}{2}\right) - w - \left(\left|\frac{a_1+a_2}{2}\right|l + \left|\frac{b_1+b_2}{2}\right|s\right) \\ \left(\frac{a_1+a_2}{2}\right)\left(\frac{b_1+b_2}{2}\right) - w - \left(\left|\frac{a_1+a_2}{2}\right|f + \left|\frac{b_1+b_2}{2}\right|g\right), \left(\frac{a_1+a_2}{2}\right)\left(\frac{b_1+b_2}{2}\right) - w, \\ \left(\frac{a_1+a_2}{2}\right)\left(\frac{b_1+b_2}{2}\right) + w, \left(\frac{a_1+a_2}{2}\right)\left(\frac{b_1+b_2}{2}\right) + w + \left(\left|\frac{a_1+a_2}{2}\right|f + \left|\frac{b_1+b_2}{2}\right|g\right), \\ \left(\frac{a_1+a_2}{2}\right)\left(\frac{b_1+b_2}{2}\right) + w + \left(\left|\frac{a_1+a_2}{2}\right|l + \left|\frac{b_1+b_2}{2}\right|s\right), \left(\frac{a_1+a_2}{2}\right)\left(\frac{b_1+b_2}{2}\right) + w + \left(\left|\frac{a_1+a_2}{2}\right|m + \left|\frac{b_1+b_2}{2}\right|h\right); k, 1 \end{pmatrix}$$

Using $*$ multiplication defined in Definition 3 the straightforward computation yields the following proposition.

Proposition 4. The $*$ multiplication defined in Definition 3 satisfies the distributive property

(i) $\tilde{c} * \left(\tilde{a} + \tilde{b}\right) \approx \left(\tilde{c} * \tilde{a} + \tilde{c} * \tilde{b}\right)$ and

(ii) $\tilde{c} * \left(\tilde{a} - \tilde{b}\right) \approx \left(\tilde{c} * \tilde{a} - \tilde{c} * \tilde{b}\right)$ for symmetric octagonal fuzzy numbers.

Definition 5. For any symmetric octagonal fuzzy number $\tilde{A}$ given in Definition 2,

$$\tilde{A} - \tilde{A} \not\approx (0, 0, 0, 0, 0, 0, 0, 0; k, 1)$$

But

$$\tilde{A} - \tilde{A} \approx (-(a_8 - a_1), -(a_7 - a_2), -(a_6 - a_3), -(a_5 - a_4), (a_5 - a_4), (a_6 - a_3), (a_7 - a_2), (a_8 - a_1); k, 1$$

and octagonal fuzzy number of the form

$(-(a_8 - a_1), -(a_7 - a_2), -(a_6 - a_3), -(a_5 - a_4), (a_5 - a_4), (a_6 - a_3), (a_7 - a_2),$
$(a_8 - a_1); k, 1)$ are called octagonal fuzzy number equivalent to 0, denoted by $\tilde{0}$.

3 Ranking of Octagonal Fuzzy Numbers

In this paper, ranking of octagonal fuzzy numbers introduced by the authors in [4] is used to solve fuzzy transportation problem, wherein the cost, supply and demand are octagonal fuzzy numbers.

Definition 6. Let $\tilde{A}$ be an octagonal fuzzy number. The value $M_0^{Oct}(\tilde{A})$, called the measure of $\tilde{A}$ is calculated as follows:

$$M_0^{Oct}(\tilde{A}) = \frac{1}{2} \int_0^k (l_1(r) + l_2(r)) dr + \frac{1}{2} \int_k^1 (s_1(t) + s_2(t)) dt \text{ where } 0 < k < 1 \tag{1}$$
$$= \frac{1}{4}[(a_1 + a_2 + a_7 + a_8)k + (a_3 + a_4 + a_5 + a_6)(1 - k)]$$

In particular if $\tilde{A}$ is a symmetric octagonal fuzzy number given by $\tilde{A} \approx (a_1 - h, a_1 - s, a_1 - g, a_1, a_2, a_2 + g, a_2 + s, a_2 + h; k, 1)$ then its measure is given by $M_0^{Oct}(\tilde{A}) = \frac{a_1 + a_2}{2}$. Note that in this case the measure is independent of k.

Remark 7. If $\tilde{A}$ and $\tilde{B}$ are two octagonal fuzzy numbers, then we have:

  I. $M_0^{Oct}(\tilde{A}) \leq M_0^{Oct}(\tilde{B}) \Rightarrow \tilde{A} \preccurlyeq \tilde{B}$
  II. $M_0^{Oct}(\tilde{A}) = M_0^{Oct}(\tilde{B}) \Rightarrow \tilde{A} \approx \tilde{B}$
  III. $M_0^{Oct}(\tilde{A}) \geq M_0^{Oct}(B) \Rightarrow \tilde{A} \succcurlyeq \tilde{B}$
  IV. $\max_{i \in \mathbb{N}_n} \tilde{A}_i$ and $\min_{i \in \mathbb{N}_n} \tilde{A}_i$ are determined using the measure defined in Definition 6.

4 Fuzzy Transportation Problems

In this section, a new algorithm is proposed to solve fuzzy transportation problem where the transportation cost, supply and demand are symmetric octagonal fuzzy numbers and a real life example is cited.

4.1 Mathematical Formulation of a Fuzzy Transportation Problem

Consider the following fuzzy transportation problem (FTP) having fuzzy costs, fuzzy sources and fuzzy demands,

(FTP) Minimize $\tilde{z} \approx \sum_{i=1}^m \sum_{j=1}^n \tilde{c}_{ij} * \tilde{x}_{ij}$

Subject to

$$\sum_{j=1}^{n} \tilde{x}_{ij} \approx \tilde{a}_i, \text{ for } i = 1, 2, \ldots m \tag{2}$$

$$\sum_{i=1}^{m} \tilde{x}_{ij} \approx \tilde{b}_j, \text{ for } j = 1, 2, \ldots n \tag{3}$$

$$\tilde{X}_{ij} \succcurlyeq \tilde{0} \text{ for } i = 1, 2, \ldots m \text{ and } j = 1, 2, \ldots n \tag{4}$$

where m = the number of supply points; n = the number of demand points;

$\tilde{x}_{ij} \approx \left(x_{ij}^1, x_{ij}^2, x_{ij}^3, x_{ij}^4, x_{ij}^5, x_{ij}^6, x_{ij}^7, x_{ij}^8; k, 1\right)$ is the uncertain number of units shipped from supply point i to demand point j;

$\tilde{c}_{ij} \approx \left(c_{ij}^1, c_{ij}^2, c_{ij}^3, c_{ij}^4, c_{ij}^5, c_{ij}^6, c_{ij}^7, c_{ij}^8; k, 1\right)$ is the uncertain cost of shipping one unit from supply point i to the demand point j;

$\tilde{a}_i \approx \left(a_i^1, a_i^2, a_i^3, a_i^4, a_i^5, a_i^6, a_i^7, a_i^8; k, 1\right)$ is the uncertain supply at supply point i;

$\tilde{b}_j \approx \left(b_j^1, b_j^2, b_j^3, b_j^4, b_j^5, b_j^6, b_j^7, b_j^8; k, 1\right)$ is the uncertain demand at demand point j.

The necessary and sufficient condition for the linear programming problem given above to have a solution is that

$$\sum_{i-1}^{m} \tilde{a}_i \approx \sum_{j-1}^{n} \tilde{b}_j.$$

The above problem can be put in table namely fuzzy transportation table given below:

			Supply
$\tilde{c}_{11}$	....	$\tilde{c}_{1n}$	$\tilde{a}_1$
.	.	.	.
.	.	.	.
.	.	.	.
$\tilde{c}_{m1}$	....	$\tilde{c}_{mn}$	$\tilde{a}_m$
Demand $\tilde{b}_1$	....	$.\tilde{b}_n$	

4.2   A New Algorithm to Solve Fuzzy Transportation Problem

A new algorithm for finding a fuzzy optimal solution for fuzzy transportation problem in single stage is discussed here and the procedure is adopted to solve fuzzy transportation problem where the supply, demand and transportation costs are octagonal fuzzy numbers.

Step 1:   Construct the fuzzy transportation table for the given fuzzy transportation problem and then, convert it into a balanced one if it is not.

Step 2:   Subtract from each row entries of the fuzzy transportation table the row minimum.

Step 3: Subtract from each column entries of the fuzzy transportation table the column minimum.

Step 4: Check if each column fuzzy demand is less than to the sum of the fuzzy supplies whose reduced costs in that column are fuzzy zeros (octagonal fuzzy numbers equivalent to zero). Also, check if each row fuzzy supply is less than to the sum of the column fuzzy demands whose reduced costs in that row are fuzzy zeros. If so, go to step 7. (Such reduced table is called the allotment table). If not, go to step 5.

Step 5: Draw the minimum number of horizontal lines and vertical lines to cover all the fuzzy zeros of the reduced fuzzy transportation table such that some entries of row(s) or/and column(s) which do not satisfy the condition of the Step 4 are not covered.

Step 6: Develop the new revised reduced fuzzy transportation table as follows:

  i. Find the smallest entry of the reduced fuzzy cost matrix not covered by any lines.
  ii. Subtract this entry from all the uncovered entries and add the same to all entries lying at the intersection of any two lines, and then, go to Step 4.

Step 7: Select a cell in the reduced fuzzy transportation table whose reduced cost is the maximum cost, say (α, β). If there is more than one, then select anyone.

Step 8: Select a cell in the α-row or/and β-column of the reduced fuzzy transportation table which is the only cell whose reduced cost is fuzzy zero and then allot the maximum possible to that cell. If such cell does not occur for the maximum value, find the next maximum so that such a cell occurs. If such cell does not occur for any value, we select any cell in the reduced fuzzy transportation table whose reduced cost is fuzzy zero.

Step 9: Reform the reduced fuzzy transportation table after deleting the fully used fuzzy supply points and the fully received fuzzy demand points and also, modify it to include the not fully used fuzzy supply points and the not fully received fuzzy demand points.

Step 10: Repeat Step 7 to Step 9 until all fuzzy supply points are fully used and all fuzzy demand points are fully received.

Step 11: This allotment yields a fuzzy solution to the given fuzzy transportation problem.

Now, the following theorems justifies that the solution obtained by the above algorithm is the optimal solution to the fuzzy transportation problem.

Theorem 8. Any optimal solution to the fuzzy problem (P_1) where

  (P_1) Minimize $\widetilde{z}^0 \approx \sum_{i=1}^{m} \sum_{j=1}^{n} (\tilde{c}_{ij} - \tilde{u}_i - \tilde{v}_j) * \tilde{x}_{ij}$
subject to (2) to (4) are satisfied,
where $\tilde{u}_i$ and $\tilde{v}_j$ are some symmetric octagonal fuzzy numbers, is an optimal solution to the problem (P) where
  (P) Minimize $\widetilde{z} \approx \sum_{i=1}^{m} \sum_{j=1}^{n} \tilde{c}_{ij} * \tilde{x}_{ij}$
subject to (2) to (4) are satisfied.

Proof. Now, $\tilde{z}^o \approx \sum_{i=1}^{m} \sum_{j=1}^{n} \tilde{c}_{ij} * \tilde{x}_{ij} - \sum_{i=1}^{m} \sum_{j=1}^{n} \tilde{u}_i * \tilde{x}_{ij} - \sum_{i=1}^{m} \sum_{j=1}^{n} \tilde{v}_j * \tilde{x}_{ij}$

$\approx \tilde{z} - \sum_{i=1}^{m} \tilde{u}_i * \tilde{b}_i - \sum_{j=1}^{n} \tilde{v}_j * a_j$, (from (2) and (3)).

Since $\sum_{i=1}^{m} \dot{u}_i * \tilde{b}_i$ and $\sum_{j=1}^{n} v_j * a_j$ are independent of $\ddot{x}_{ij}$ for all i and j, we can conclude that any optimal solution to the problem (P$_1$) is also a fuzzy optimal solution to the problem (P). Hence the theorem.

Theorem 9. If $\{\tilde{x}_{ij}^0, i = 1, 2, \ldots, m$ and $j = 1, 2, \ldots, n\}$ is a feasible solution to the problem (P) and $\left(\tilde{c}_{ij} - \tilde{u}_i - \tilde{v}_j\right) \succcurlyeq \tilde{0}$, for all i and j where $\tilde{u}_i$ and $\tilde{v}_j$ are some real symmetric octagonal fuzzy numbers, such that the minimum $\sum_{i=1}^{m} \sum_{j=1}^{n} \left(\tilde{c}_{ij} - \tilde{u}_i - \tilde{v}_j\right) * \tilde{x}_{ij}$ subject to (2) to (4) are satisfied, is fuzzy zero, then $\{\tilde{x}_{ij}^0, i = 1, 2, \ldots, m$ and $j = 1, 2, \ldots, n\}$ is a fuzzy optimal solution to the problem (P).

Proof. From the Theorem 8, the result follows. Hence the theorem.

Theorem 10. A solution obtained by the proposed method for a fuzzy transportation problem with equality constraints (P) is a fuzzy optimal solution for the fuzzy transportation problem (P).

Proof. To prove this it is required to describe the method in detail.

The fuzzy transportation table $\left[\tilde{c}_{ij}\right]$ for the given fuzzy transportation problem should be constructed and then the problem should be converted to a balanced one if it is not balanced.

Let $\tilde{u}_i$ be the minimum of i-th row of the table $\left[\tilde{c}_{ij}\right]$. Now, subtract $\tilde{u}_i$ from the i-th row entries so that the resulting table is $\left[\tilde{c}_{ij} - \tilde{u}_i\right]$.

Let $\tilde{v}_j$ be the minimum of j-th column of the resulting table $\left[\tilde{c}_{ij} - \tilde{u}_i\right]$. Now, subtract $\tilde{v}_j$ from the j-th column entries so that the resulting table is $\left[\tilde{c}_{ij} - \tilde{u}_i - \tilde{v}_j\right]$. It may be noted that $\tilde{c}_{ij} - \tilde{u}_i - \tilde{v}_j \succcurlyeq \tilde{0}$, for all i and j and each row and each column of the resulting table $\left[\tilde{c}_{ij} - \tilde{u}_i - \tilde{v}_j\right]$ has atleast one fuzzy zero entry.

Each column fuzzy demand of t resulting table $\left[\tilde{c}_{ij} - \tilde{u}_i - \tilde{v}_j\right]$ is less than or equal to the sum of the fuzzy supply points whose reduced costs in the column are fuzzy zero. Further, each row fuzzy supply of the resulting table $\left[\tilde{c}_{ij} - \tilde{u}_i - \tilde{v}_j\right]$ is less than or equal to sum of the column fuzzy demand points whose reduced costs in the row are fuzzy zero (If not so, as per direction given in the Step 5 and 6 in the zero point method, we can make it that way). The current resulting table is the allotment table.

Now a cell in the allotment table $\left[\tilde{c}_{ij} - \tilde{u}_i - \tilde{v}_j\right]$ whose reduced cost is maximum need to be found, say (α, β). Allot the maximum possible to a cell in the α-row or/and β-column which is the only cell in the α-row or/and β-column whose reduced cost is fuzzy zero. The resulting fuzzy transportation table is reformed after deleting the fully used fuzzy supply points and the fully received fuzzy demand points. Also, the not fully used fuzzy supply points and the not fully received fuzzy demand points are modified. The above said procedure is to be repeated till the total fuzzy supply are fully used and the total fuzzy demands are fully received.

Finally, a solution $\{\tilde{x}_{ij}, i = 1, 2, \ldots, m$ and $j = 1, 2, \ldots, n\}$ is obtained for the reduced fuzzy transportation problem whose cost matrix is $\left[\tilde{c}_{ij} - \tilde{u}_i - \tilde{v}_j\right]$ such that $\tilde{x}_{ij} \approx \tilde{0}$ for $\left[\tilde{c}_{ij} - \tilde{u}_i - \tilde{v}_j\right] \succ \tilde{0}$ and $\tilde{x}_{ij} \succ \tilde{0}$ for $\left[\tilde{c}_{ij} - \tilde{u}_i - \tilde{v}_j\right] \approx \tilde{0}$.

Therefore, the minimum $\sum_{i=1}^m \sum_{j=1}^n \left(\tilde{c}_{ij} - \tilde{u}_i - \tilde{v}_j\right) * \tilde{x}_{ij}$ subject to (2) to (4) are satisfied, is fuzzy zero. Thus, by the Theorem 9, the solution $\{\tilde{x}_{ij}, i = 1, 2, \ldots, m$ and $j = 1, 2, \ldots, n\}$ is a fuzzy optimal solution to the fuzzy transportation problem (P). Hence the theorem.

Remark 11. Distributive property cited in Proposition 4 does not hold for general trapezoidal or octagonal fuzzy numbers under usual multiplication. Hence in this paper we are considering symmetric octagonal fuzzy numbers and the new $*$ multiplication (Definition 3) to ensure Proposition 4 holds. And the Theorems 8, 9 and 10 hold only for symmetric octagonal fuzzy numbers or symmetric trapezoidal fuzzy numbers.

Remark 12. Theorems 8, 9 and 10 has been considered for trapezoidal fuzzy numbers by Pandian et al. in [6] is incorrect.

4.3 Numerical Example

A company has three plants S_1, S_2 and S_3 which supply to warehouses located at D_1, D_2, D_3 and D_4. Monthly plant capacities, monthly warehouse requirements and the cost of transporting one unit from the three plants to the four warehouses are given in the table below:

Source	Destination				Supply
	D_1	D_2	D_3	D_4	
S_1	(−1,0,1,2,3,4,5,6;0.4,1)	(0,1,2,3,4,5,6,7;0.4,1)	(8,9,10,11,12,13,14,15;0.4,1)	(4,5,6,7,8,9,10,11;0.4,1)	(1,3,5,6,7,8,10,12;0.4,1)
S_2	(−2, −1,0,1,2,3,4,5;0.4,1)	(−3,−2, −1,0,1,2,3,4;0.4,1)	(2,4,5,6,7,8,9,11;0.4,1)	(−3,−2,0,1,2,3,5,6;0.4,1)	(−2,−1,0,1,2,3,4,5;0.4,1)
S_3	(2,3,4,5,6,7,8,9;0.4,1)	(3,6,7,8,9,10,11,14;0.4,1)	(11,12,14,15,16,17,19,20;0.4,1)	(5,6,8,9,10,11,13,14;0.4, 1)	(5,7,8,10,12,14,15,17;0.4,1)
Demand	(4,5,6,7,8,9,10,11;0.4,1)	(1,2,3,5,6,8,9,10;0.4,1)	(0,1,2,3,4,5,6,7;0.4,1)	(−1,0,1,2,3,4,5,6;0.4,1)	

Determine an optimum distribution for the company in order to minimize the total transportation cost.

Now, the total fuzzy supply, $\tilde{S}$ = (4, 9, 13, 17, 21, 25, 29, 34; 0.4, 1) and the total fuzzy demand, $\tilde{D} \approx$ (4, 8, 12, 17, 21, 26, 30, 34; 0.4, 1). Since $M_0^{oct}(\tilde{S}) = M_0^{oct}(\tilde{D})$, the given problem is a balanced one.

Now, using the procedure given above we get the optimal solution as

Source	Destination				Supply
	(−45,32,−17,−5,7,19,3 4,47;0.4,1)	$\tilde{0}$	$\tilde{0}$	(−89,−64,−34, −10,14,38,68,93;0.4,1)	(1,3,5,6,7,8,10,12;0.4,1)
	(−78,−55,−28, −6,16,38,65,88;0.4,1)	(−21,−15,−7,−1,5, ll,19,25;44,63;0.4,l)	$\tilde{0}$	(−46,−30,−18, −6,6,19,32,45;0.4,1)	(−2,−1,0,1,2,3,4,5;0.4,1)
	$\tilde{0}$	(−61,−42,−23, −7,9,25,44,63;0.4.1)	$\tilde{0}$	$\tilde{0}$	(5.7,8,10,12,14,15,17;0.4,1)
Demand	(4,5,6,7,8,9,10,11;0.4,1)	(1,2,3,5,6,8,9,10;0.4,1)	(0,1,2,3,4,5,6,7;0.4,1)	(−1,0,1,2,3,4,5,6;0.4,1)	

This table satisfies the optimality conditions. Now, using the allotment rules of the proposed algorithm, we have the allotment table,

Source	Destination				Supply
		(1,2,3,5,6,8,9,10;0.4,1)	(−9,−6, −3,0,2,5,8,11;0.4,1)		(1,3,5,6,7,8,10,12;0.4,1)
			(−2,−1,0,1,2,3,0.4,1)		(−2,−1,0,1,2,3,4,5;0.4,1)
	(4,5,6,7,8,9,10,11;0.4.1)		(−12,−8,−5, −1,3,7,10,14;0.4,1)	(−1,0,1,2,3,4,5,6;0.4,1)	(5,7,8,10,12,14,15,17;0.4,1)
Demand	(4,5,6,7,8, 9,10,11,0.4,1)	(1,2,3,5,6,8,9,10;0.4,1)	(0,1,2,3,4,5,6,7;0.4,1)	(−1,0,1,2,3,4,5,6;0.4,1)	

Therefore, the fuzzy optimal solution for the fuzzy transportation problem is

$$\tilde{x}_{12} \approx (1, 2, 3, 5, 6, 8, 9, 10; 0.4, 1), \ \tilde{x}_{13} \approx (−9, −6, −3, 0, 2, 5, 8, 11; 0.4, 1),$$
$$\tilde{x}_{23} \approx (−2, −1, 0, 1, 2, 3, 4, 5; 0.4, 1), \ \tilde{x}_{31} \approx (4, 5, 6, 7, 8, 9, 10, 11; 0.4, 1),$$
$$\tilde{x}_{33} \approx (−12, −8, −5, −1, 37, 10, 14; 0, 4, 1), \ \tilde{x}_{34} \approx (−1, 0, 1, 2, 3, 4, 5, 6; 0.4, 1),$$

and the fuzzy optimal value of

$$\tilde{z} \approx (−358.5, −216.5, −88, 56, 186, 330, 458.5, 600.5; 0.4, 1).$$

5   Fuzzy Assignment Problems

In this section, a fuzzy analogue of Hungarian method is used to solve fuzzy assignment problem wherein the assigned cost, time or profit are symmetric octagonal fuzzy numbers and a real life example is cited.

5.1   Mathematical Formulation of a Fuzzy Assignment Problem

Consider the following fuzzy assignment problem (FAP) having fuzzy costs.
   (FAP) Minimize $\tilde{z} \approx \sum_{i=1}^{m} \sum_{j=1}^{n} x_{ij} \tilde{c}_{ij}$
   Subject to

$$\sum_{j=1}^{n} x_{ij} \approx 1, \ \text{for } i = 1, 2, \ldots m \tag{5}$$

$$\sum_{i=1}^{m} x_{ij} \approx 1, \text{for } j = 1, 2, \ldots n \tag{6}$$

$$x_{ij} \in \{0, 1\}, \ \text{where } x_{ij} = \begin{cases} 1, \text{if the i}^{\text{th}} \text{ person is assigned j}^{\text{th}} \text{ job} \\ 0, \text{otherwise} \end{cases}$$

is the decision variable denoting the assignment of the person i to job j. $\tilde{c}_{ij}$ is the cost of assigning the jth job to the ith person. The objective is to minimize the total cost of assigning all the jobs to the available persons. (One job to one person).

5.2   Procedure for Solving Fuzzy Assignment Problem

We shall present a solution to fuzzy assignment problem in which the cost or profit are symmetric octagonal fuzzy numbers.

Step 1:   Determine the cost table from the given problem. If the cost table is a square matrix then goes to step 3, otherwise go to step 2.

Step 2:   Add a dummy row or column, so that the cost table becomes a square matrix. The cost entries of dummy row or column are always zero.

Step 3:   Choose the least from each row and subtract it from all the elements of that row. The least number is chosen using the Definition 6.

Step 4:   In the reduced matrix obtained in step 3, choose the least from each column and subtract it from all the elements of that column. Again here the least number is chosen using the Definition 6.

Step 5:   In the reduced matrix obtained from step 4, optimal assignment can be obtained as follows:

   a. The rows are to be examined successively until a row with single zero (or equivalent to zero). Put a box ($\square$) around this zero (or equivalent to zero) and cross off ($\times$) all other zeros in its column. Continue the same until all the rows are exhausted.
   b. Repeat the same column-wise.
   c. if a row and/or column has more than one zero, then assign any one zero arbitrarily and cross of all other zeros of that row/column.
   d. Repeat (a) through (c) above until all the zeros (or equivalent to zero) are assigned either ($\square$) or ($\times$).

Step 6:   If the number of assignments ($\square$) is equal to the order of the cost matrix then optimum solution is reached. If it is less then go to step 7.

Step 7:   Now draw minimum number of vertical and/or horizontal lines to cover all zeros (or all equivalent to zero). This can be done in the following way:

   (i)   Mark ($\checkmark$) rows that do not have assigned zero.
   (ii)   Mark ($\checkmark$) columns that have zeros in the marked rows.
   (iii) Mark ($\checkmark$) rows which have assignment in the marked column.
   (iv)   Repeat steps (ii) and (iii) until the marking chain ends.
   (v)   Draw lines through unmarked rows and marked columns.

   In this way minimum number of lines can be drawn to cover all zeros.

Step 8:   Obtain a modified cost matrix as follows:

   (i)   Choose the least element from the uncovered elements.
   (ii)   Add the least value to the elements which are lying in the intersection of two lines and subtract it from all the uncovered elements.

Step 9:   Now go to step 6 and repeat the procedure until optimum solution is attained.

Remark 13. Theorems 8, 9 and 10 also holds good for fuzzy assignment problem solved in this paper. But in FAP the cost matrix need to be a square matrix (i.e., $m = n$) and the $\tilde{a}_i$'s and $\tilde{b}_j$'s are all equivalent to either $\tilde{0}$ or $\tilde{1}$.

5.3 Numerical Example

A departmental head has four subordinates, and four tasks to be performed. The subordinates differ in efficiency, and the tasks differ in their intrinsic difficulty. His estimate of time each man would take to perform each task, is given in the matrix below:

		Job			
		A	B	C	D
Person	1	(1,2,3,5,6,8,9,10;0.4,1)	(4,5,7,8,11,12,14,15;0.4,1)	(6,7,9,10,11,12,14,15;0.4,1)	(4,5,7,8,10,11,13,14;0.4,1)
	2	(4,5,6,8,10,12,13,14;0.4,1)	(1,2,3,4,7,8,9,10;0.4,1)	(1,3,4,8,10,14,15,17;0.4,1)	(1,4,5,8,9,12,13,16;0.4,1)
	3	(1,2,3,4,5,6,7,8;0 4,1)	(3,4,6,7,10,11,13,14;0.4,1	(5,7,9,10,13,14,16,18;0.4,1)	(0,3,4,6,7,9,10,13;0.4,1)
	4	(3,5,6,8,10,12,13,15;0.4,1)	(0,2,3,5,6,8,9,11;0.4,1)	(3,4,5,7,9,11,12,13;0.4,1)	(0,2,3,4,5,6,7,9;0.4,1)

Now by proceeding as in the algorithm we obtain the optimal assignment as

Person	Job	Fuzzy cost
1	3	(6,7,9,10,11,12,14,15;0.4,1)
2	2	(1,2,3,4,7,8,9,10;0.4,1)
3	1	(1,2,3,4,5,6,7,8;0.4,1)
4	4	(0,2,3,4,5,6,7,9;0.4,1)
Total		(8,13,18,22,28,32,37,42;0.4,1)

6   Conclusion

Some previous studies have provided an optimal solution for fuzzy transportation problem and fuzzy assignment problem wherein the problems were solved after converting them to crisp problems. Also the obtained results being crisp values, tend to lose some intermediate information. To capture the problem and its solution in a more natural way we use symmetric octagonal fuzzy numbers to solve FTP and FAP and solutions are also obtained as symmetric octagonal fuzzy numbers. Here, by using the proposed method the optimal value of the fuzzy transportation problem whose unit shipping costs, the supply quantities and the demand quantities are symmetric octagonal fuzzy numbers is obtained as symmetric octagonal fuzzy number. Also the optimum allocation of jobs to persons or machines wherein the cost or profit are taken as symmetric octagonal fuzzy numbers is obtained. Hence algorithms proposed for FTP and FAP can serve as an important tool for the decision makers when they are handling various types of logistic and job allocation problems having fuzzy parameters.

References

1. Basirzadeh, H., Abbasi, R.: A new approach for ranking fuzzy numbers based on α cuts. J. Appl. Math. Inf. **26**, 767–778 (2008)
2. Cheng, C.H.: A new approach for ranking fuzzy numbers by distance method. Fuzzy Sets Syst. **95**, 307–317 (1998)
3. Kennedy, F.C., Malini, S.U.: Fuzzy linear programs with octagonal fuzzy numbers. Scientiae Mathematicae Japonicae (2016). http://www.jams.or.jp/scm/abstract/e-2015/abst2015-26.pdf
4. Malini, S.U., Kennedy, F.C.: An approach for solving fuzzy transportation problem using octagonal fuzzy numbers. Appl. Math. Sci. **7**(54), 2661–2673 (2013)
5. Malini, S.U., Kennedy, F.C.: An approach for solving fuzzy assignment problem using octagonal fuzzy numbers. Int. J. Math. Sci. Eng. Appl. **7**(3), 441–449 (2013)
6. Pandian, P., Natarajan, G.: A new algorithm for finding a fuzzy optimal solution for fuzzy transportation problems. Appl. Math. Sci. **4**(2), 79–90 (2010)
7. Mukherjee, S., Basu, K.: Application of fuzzy ranking method for solving assignment problems with fuzzy costs. Int. J. Comput. Appl. Math. **5**(3), 359–368 (2010)
8. Zadeh, L.A.: Fuzzy sets. Inf. Control **8**, 338–353 (1965)

Solution of the Portfolio Optimization Model as a Fuzzy Bilevel Programming Problem

Vyacheslav Kalashnikov[1,2,3]($\boxtimes$) (iD), Nataliya Kalashnykova[4](iD),
and José G. Flores-Muñiz[4](iD)

[1] Tecnológico de Monterrey (ITESM), 64849 Monterrey, NL, Mexico
kalash@itesm.mx
[2] Central Economics and Mathematics Institute (CEMI),
Moscow 117418, Russia
[3] Sumy State University, Sumy 40007, Ukraine
[4] Universidad Autónoma de Nuevo León (UANL),
66455 San Nicolás de los Garza, NL, Mexico
nkalash2009@gmail.com

Abstract. In this chapter, we consider a mixed-integer bilevel linear programming problem with one parameter in the right-hand side of the constraints in the lower level (or, the follower's) problem. Motivated by an application to the fuzzy portfolio optimization model, we consider a particular case that consists in maximizing the investor's expected return. The functions are linear at both the upper and lower levels, and the proposed algorithm is based upon an approximation of the optimal value function using the branch-and-bound method. Therefore, at every node of this tree-type structure, we apply a new branch-and-bound procedure to deal with the integrity restriction.

Keywords: Fuzzy portfolio optimization · Integer programming
Parametric programming · Branch-and-bound approach

1 Introduction

The portfolio theory was developed to support decision making for allocation of financial assets (securities, bounds) traded at the stock exchange [30]. This allocation is known as "investment" decision making. The investor considers the asset as a matter of future income. The better combination of financial assets (securities) of the portfolio leads to better return for the investor. The portfolio contains a set of securities, and the problem of portfolio optimization targets the optimal resource allocation in investment process of trading financial assets [31]. The resource allocation means investing capital in financial assets (or goods), which gives return to the investor after certain period of time. For the investment process, the aim is to maximize the return while the investment risk has to be minimal [18].

© Springer International Publishing AG, part of Springer Nature 2018
A. M. Gil-Lafuente et al. (Eds.): FIM 2015, AISC 730, pp. 164–178, 2018.
https://doi.org/10.1007/978-3-319-75792-6_14

Harry Markowitz suggested a powerful approach for quantifying portfolio risks in 1952. According to his seminal portfolio theory [24], the analytical relations among the portfolio risk V_p, portfolio return E_p, and the values of investment x_i in the form of asset i, are described as follows:

$$E_p := \sum_{i=1}^{n} E_i x_i = E^T x; \tag{1}$$

$$V_p := \frac{1}{2} \sum_{j=1}^{n} \sum_{i=1}^{n} x_i Q_{ij} x_j = \frac{1}{2} x^T Q x, \tag{2}$$

where E_i is the average value of the return for asset i; $E^T = (E_1, \ldots, E_n) \in R^n$ is a vector of dimension $1 \times n$, and $Q \in R^{n \times n}, Q \geq 0$, is a symmetric positive semi-definite assets variance-covariance matrix. This scheme reflects the elements of fuzziness in the proposed construction.

The portfolio theory introduces the so called "standard" optimization problem as follows:

$$z := \frac{1}{2} x^T Q x - \sigma E^T x \to \min_x, \tag{3}$$

subject to

$$\sum_{i=1}^{n} x_i = 1, \tag{4}$$

and

$$x_i \geq 0, \quad i = 1, \ldots, n, \tag{5}$$

where $\sigma \geq 0$ is a parameter of the investor's risk aversion (his/her tolerance to accepting risky investments).

The numerical assessment of σ is a task for a financial analyst, and it inevitably bears some subjective footprints. This coefficient strongly influences the definition and hence, the solution of the portfolio problem. As a result, the selected value of σ affects the final investment decision as well.

Because of that, the most appropriate for the investor is the following *fuzzy bilevel programming problem* (FBLP):

$$w := E(\sigma) = E^T y \to \max_{\sigma \geq 0}, \tag{6}$$

subject to the constraint:

$$y \in \Psi(\sigma) := \{x \in R^n \ : \ x \text{ solves problem } (3)-(5)\}. \tag{7}$$

This chapter proposes an efficient numerical algorithm to solve the fuzzy bilevel programming problem (6)–(7). The method is based upon the lower level objective function optimal value techniques (*cf.*, [8]).

The rest of the chapter is organized as follows. Preliminaries, a general formulation of the problem, and the mathematical model are given in Sects. 2 and 3, respectively. The geometry of the problem is described in Sect. 4, whereas the approximation algorithm is presented in Sect. 5. Section 6 illustrates the algorithm by a numerical example, and conclusions are listed in Sect. 7, with Acknowledgments and References completing the chapter.

2   Preliminaries

Hierarchical decision making is strongly motivated by real-world applications. For example, in engineering design, the main objective of a design engineer may be constrained by the properties inherent in the process (such as minimum energy), which, in turn, may be controlled through the (parametric) decision variables chosen by the engineer. Such models can be specified within a bilevel programming (BLP) framework, where an *upper level (outer)* optimization problem is restricted by solutions of another, *lower level (inner)* optimization problem.

Hierarchical problems also arise in the (non-simultaneous) Stackelberg games [32], in which various decision makers try to maximize their utility functions with a temporal delay. Because of that, they are often not allowed to realize their decisions independently and at the same time, but are forced to act according to a certain *hierarchy*. We will consider the simplest case of such a situation, where there are only two acting decision makers. The *leader* is the one that can handle the market without any restrictions, whereas the *follower* has to act in a dependent (optimal reaction) manner.

In mathematical terms, it means that the set of variables is partitioned into two vector variables, x and y, where $y \in R^m$ are the leader's variables and $x \in R^n$ are those governed by the follower. Using y as a parameter, the follower solves a parametric optimization problem, and the values $x = x(y)$ are determined by the follower knowing the selection y of the leader. The leader has to determine the best choice of y knowing the (optimal) reaction $x = x(y)$ of the follower to the leader's decision.

Some important decision making problems may involve both discrete and continuous variables. For example, a chemical engineering design problem may contain discrete decisions regarding the existence of chemical process units *in addition to* decisions in continuous variables, such as temperature or pressure values. Problems of this class dealing with both discrete and continuous decision variables are referred to as *mixed-integer* BLPs.

A particular case of the mixed-integer bi-level programming problem is presented by the real-world problem of minimizing the cash-out penalty costs for a natural gas shipping company [9]. This problem arises when a (gas) shipper draws a contract with a pipeline company to deliver a certain amount of gas at several delivering meters. What is actually shipped may be higher or lower than the amount that had been originally agreed upon (this phenomenon is called an *imbalance*). When such an imbalance occurs, the pipeline penalizes the shipper by imposing a cash-out penalty policy. As this penalty is a function of the operating daily imbalances, an important problem for the shippers is how to carry out their daily imbalances so as to minimize the incurred penalty. On the other hand, the pipeline (the follower) tries to minimize the absolute values of the cash-outs, which produces the optimal response function taken into account by the leader in order to find the optimal imbalance operating strategy. Integer variables are involved at the lower level problem, and various algorithms solving the natural gas cash-out problem are described in [9–20].

In general, the mixed-integer BLPs can be sorted out into four classes [15]:

(I) **Integer Upper, Continuous Lower:** If the sets of inner (lower level) integer and outer (upper level) continuous variables are empty, and on the contrary, the sets of outer integer and inner continuous variables is nonempty, then the MIBLP is of Type I.

(II) **Purely Integer:** If the sets of inner and outer integer variables are nonempty, and the sets of inner and outer continuous variables are empty, then the problem is a purely integer BLP.

(III) **Continuous Upper, Integer Lower:** When the sets of inner continuous and outer integer variables are empty, and vice versa, the sets of inner integer and outer continuous variables are nonempty, then the problem is a MIBLP of Type III.

(IV) **Mixed-Integer Upper and Lower:** If the sets of both inner and outer continuous and integer variables are nonempty, then the problem is a MIBLP of Type IV.

Serious steps toward the efficient solution of the mixed-integer bilevel programming problems (MIBLP) of all four types can greatly expand the scope of decision making instances that can be modeled within the bilevel optimization framework. However, up to now, very little attention has been paid to both the solution and applications of BLPs managing discrete variables. This is mainly because these problems pose major algorithmic challenges in the development of good solution strategies.

In the literature, the methods elaborated for the solution of the MIBLPs have so far addressed a very restricted class of problems. More attention has been paid to *linear* problems. For instance, for the solution of the purely integer (Type II) linear BLP, a branch-and-bound type of enumerating technique has been proposed in [25], whereas the authors of [26] applied a kind of genetic algorithm to the same problem type. For the solution of the mixed-integer BLP of Type I, another branch-and-bounds approach has been developed in [33]. Cutting plane and parametric solution techniques have been elaborated by Dempe [3] to solve the MIBLPs, in which the lower level has only *one* upper level (outer) variable involved into the (lower level) objective function. The same author also proposed an algorithm to solve the Linear Bilevel Programming Problem (BLPP) using the simplex method with additional variables in the basis set using the theory of subgradients [2]. One of the pioneers in Bilevel Programming Jonathan Bard obtained upper bounds for the objective functions at both levels [1]. Thus he generated a non-decreasing sequence of lower bounds for the objective function at the upper level, which, under certain conditions, converges to the solution of the general BLP with continuously differentiable functions. The methods based upon decomposition techniques have been proposed in [28,36].

Mixed-integer *nonlinear* bilevel programming problems have received even less attention in the literature. The developed methods include an algorithm making use of parametric analysis to solve separable monotone nonlinear MIBLP proposed in [17], a stochastic simulated annealing method presented in [29], a global optimization approach based on parametric programming technique published in [10]. The authors of [11, 15] developed several algorithms dealing with global optimization of mixed-integer bilevel programming problems of both deterministic and stochastic nature. The sensitivity analysis for MIBLPs also was considered in [34].

In [4–7], we already started considering and solving mixed-integer linear BLPs of Type I. Sometimes, a BLP can be reduced to the solution of a multi-objective optimization problem, which is efficiently processed in [23]. Bilevel programming problems with discrete variables are also examined in [16].

The main goal of this chapter is to propose an efficient algorithm to solve the bilevel linear mixed-integer program of Type I, because in our bilevel portfolio optimization problem (6)–(7), the upper level variable σ can clearly treated as integer one. Knowing that this problem is hard to solve, we propose an algorithm generating approximations that converge to a global solution. The main novelty of the presented heuristic approach lies in the combination of branch-and-bound (B&B) technique with the simplicial subdivision algorithms. The numerical experiments demonstrate the robust performance of the developed method for the instances of a small and medium size.

2.1   General Formulation

We consider the hierarchical optimization problem in two levels: the decision making at the upper level is governed by the constraints that are (partially) defined by a lower level (parametric) optimization problem. Let the lower level problem be specified as follows:

$$\min_{x} \left\{ f(x, y) \mid g(x, y) \leq 0, h(x, y) = 0 \right\}, \tag{8}$$

where $f : R^n \times R^m \to R, g : R^n \times R^m \to R^p$ and $h : R^n \times R^m \to R^q$ with $g(x, y) = -(g_1(x, y), \ldots, g_p(x, y))^T$ and $h(x, y) = (h_1(x, y), \ldots, h_q(x, y))^T$ are given functions. Let $\psi(y)$ denote the solution set of problem (8) for a fixed parameter $y \in R^m$, which we assume to be nonempty for (at least) some values of y.

Now we can outline the bilevel programming problem as follows:

$$\min_{x, y} \left\{ F(x, y) \; : \; y \in Y, x \in \psi(y) \right\}, \tag{9}$$

where $F : R^n \times R^m \to R$ is a given function, and Y is a closed subset of R^m. This is the *upper level*, or the *leader's* problem. The whole model (8)–(9) is referred to as a *bilevel programming problem*.

In order to guarantee that the bilevel programming problem is well-defined, we assume the following:

1. The set $M = \{(x, y) \mid g(x, y) \leq 0, h(x, y) = 0\}$ is nonempty.
2. Both $F(x, y)$ and $f(x, y)$ are bounded from below on M.

Definition 1. *A pair (x, y) is said to be feasible to the linear bilevel programming problem if it satisfies $y \in Y$ and $x \in \psi(y)$.*

Definition 2. *A feasible pair (x', y') is called an optimal solution to the bilevel programming problem if $F(x', y') \leq F(x, y)$ for all (other) feasible solutions (x, y).*

3   Mathematical Model

The Mixed Integer Bilevel Linear Programming Problem with a parameter in the objective function of the lower level is stated as follows:

$$\min_{x,y} \left\{ \langle a, x \rangle + \langle b, y \rangle \mid Gy = d, x \in \psi(y), y \in Z_+^m \right\}, \tag{10}$$

which represents the upper level, where $a, x \in R^n$, $b, y \in R^m$, G is an $r \times m$ matrix, $d \in R^r$. Note that we use the optimistic version of the bilevel programming problem here, *see* [3]. From now on, $\langle \cdot, \cdot \rangle$ is the inner product. In general, $\psi(y)$ is defined as follows:

$$\psi(y) = \operatorname*{Arg\,min}_{x} \left\{ \langle c, x \rangle \mid Ax = y, x \geq 0 \right\}, \tag{11}$$

which describes the set of optimal solutions of the lower level problem (the set of rational reactions). Here $c, x \in R^n$, while A is an $m \times n$ matrix with $m \leq n$.

In our particular model, for a fixed current approximate solution $x^k \in R_+^n$ of the lower level problem, the set $\psi(\sigma)$ is defined as follows:

$$\psi(\sigma) = \operatorname*{Arg\,min}_{x} \left\{ \langle c, x \rangle + \tau \mid Ax = 1, x \geq 0 \right\}, \tag{12}$$

where

$$c^T := \left(x^k \right)^T Q - \sigma E^T, \quad \text{and } \tau := \frac{1}{2} \left(x^k \right)^T Q x^k - \sigma E^T x^k, \tag{13}$$

which describes the set of optimal solution of the approximate (linearized) lower level problem (the set of rational reactions). Here $x \in R^n$, whereas Q is an $n \times n$ matrix, A is the $1 \times n$ matrix with $a_{1i} = 1$, $i = 1, \ldots, n$.

Let us determine the optimal value function of the lower level problem as follows:

$$\varphi(y) = \min_{x} \left\{ \langle c, x \rangle \mid Ax = y, x \geq 0 \right\}. \tag{14}$$

We suppose that the feasible set of problem (11) is nonempty. Again, in the example of the portfolio optimization model, σ is the parameter that can represent the values of different degree of risk aversion on part of the investor. The lower level, depending on our objectives, may try to minimize the risk, the uncertainty of possible returns, etc.

In this chapter, we consider a reformulation of (10)–(14) based upon an approach reported in the literature (*see*, [35], or [3]) as a classical nondifferentiable optimization problem. If we take into account the lower level optimal value function (14), then problem (10)–(14) can be replaced by:

$$\min_{x,y} \left\{ \langle a, x \rangle + \langle b, y \rangle \mid Gy = d, \langle c, x \rangle \leq \varphi(y), \ Ax = y, \ x \geq 0, y \in Z_+^m \right\}. \tag{15}$$

Our work is concentrated on the lower level objective value function (14). For this reason, we show some important characteristics (*see* [14], or [5]) that will be helpful for solving problem (15).

4   The Problem's Geometry

Consider the parametric linear programming problem (14)

$$\varphi(y) = \min_{x}\left\{\langle c, x\rangle \mid Ax = y, x \geq 0\right\}.$$

In order to solve this problem, we use the dual simplex algorithm like in [5]. Let us fix a $y = y^*$, and let x^* be a basic optimal solution for $y = y^*$ with the corresponding basic matrix B, which is a quadratic submatrix of A having the same rank as A. Recall that $x^* = (x_B^*, x_N^*)^T$, with $x_B^* = B^{-1}y$ and $x_N^* = 0$. Then we can say that $x^*(y^*) = (x_B^*(y^*), x_N^*(y^*))^\top = (B^{-1}y^*, 0)^\top$ is a *basic optimal solution* of problem (14) for a *fixed parameter* y^*. Now if the following inequality holds:

$$B^{-1}y \geq 0,$$

then $x^*(y) = (x_B^*(y), x_N^*(y))^\top = (B^{-1}y, 0)^\top$ is also optimal for *another* parameter vector y.

It is possible to perturb y^* so that B remains a basic optimal matrix [14]. We denote by $\Re(B)$ the set called the *stability region* of B defined as

$$\Re(B) = \left\{y \mid B^{-1}y \geq 0\right\}.$$

For any $y \in \Re(B)$, the point $x^*(y) = (x_B^*(y), x_N^*(y))^\top = (B^{-1}y, 0)^\top$ is a basic optimal solution of problem (14). This region is nonempty because $y^* \in \Re(B)$. Furthermore, it is closed but not necessarily bounded. If $\Re(B)$ and $\Re(B')$ are two different stability regions with $B \neq B'$, then only one of the following cases is possible.

1. $\Re(B) \cap \Re(B') = \{0\}$.
2. $\Re(B) \cap \Re(B')$ is the adjacent face (the common border) of the regions $\Re(B)$ and $\Re(B')$.
3. $\Re(B) = \Re(B')$.

Moreover, $\Re(B)$ is a convex polyhedral set, on which the lower level optimal value function is finite and linear. To determine an explicit description of the function φ consider the dual problem to (14). If the value $\varphi(y)$ is finite, then

$$\varphi(y) = \max\{\langle y, u\rangle : A^\top u \leq c\}.$$

Let $u^1, u^2, \ldots, u^s$ denote the vertices of the polyhedral set $\{u : A^\top u \leq c\}$. Therefore,

$$\varphi(y) = \max\{\langle y, u^1\rangle, \langle y, u^2\rangle, \ldots, \langle y, u^s\rangle\},$$

whenever $\varphi(y)$ is finite.

By duality, for some basic matrix B_i with $y \in \Re(B_i)$ we have $B_i^\top u = c_{B_i}$ or $u = \left(B_i^\top\right)^{-1} c_{B_i}$ and, thus,

$$\langle y, u^i\rangle = \langle y, \left(B_i^\top\right)^{-1} c_{B_i}\rangle = \langle (B_i)^{-1} y, c_{B_i}\rangle.$$

Setting $x^i(y) = ((B_i)^{-1} y, 0)$ we derive

$$\varphi(y) = \max \left\{ \langle c, x^1(y) \rangle, \langle c, x^2(y) \rangle, \ldots, \langle c, x^q(y) \rangle \right\}.$$

It is easy to understand that the stability regions are represented by the segments on the y-axes. The function φ is nonsmooth, which makes this kind of problems hard to solve.

Now, we introduce the following definition from [35].

Definition 3. *Let* (x^*, y^*) *solve problem (15). Then (15) is called partially calm at* (x^*, y^*) *if there exist a constant* $\mu > 0$ *and a neighborhood* U *of* $(x^*, y^*, 0) \in R^n \times R^m \times R$, *such that for all* $(x, y, u) \in U$ *feasible to the problem:*

$$\min_{x, y, u} \left\{ \langle a, x \rangle + \langle b, y \rangle \mid Gy = d, \ \langle c, x \rangle - \varphi(y) = -u, \ Ax = y, \ x \geq 0, y \in Z_+^m \right\},$$

(16)

we have

$$\langle a, x \rangle + \langle b, y \rangle - \langle a, x^* \rangle - \langle b, y^* \rangle + \mu|u| \geq 0.$$

Here $|u|$ represents the absolute value of u.

Theorem 1. *Let* (x^*, y^*) *solve problem (10)–(14), then (15) is partially calm at* (x^*, y^*).

5   An Approximation Algorithm

The basis to start describing the algorithm is given above in this chapter. The difficulty when working with the objective value function (14) is due to the simple fact that we do not have it in an explicit form. This algorithm tries to approximate function (14) with a finite number of iterations. Also (14) is not differentiable: cf., [6,35] dealing with subdifferential calculus based upon the nonsmooth Mangasarian-Fromowitz constraint qualification.

The tools that we use in this chapter are mainly based on the fact that (14) is piecewise-linear and convex. Also, the basis for developing an efficient algorithm is given in the next theorems, important for keeping on the convexity at every level of approximation.

Definition 4. *The intersection of all the convex sets containing a given subset* W *of* R^m *is called the convex hull of* W *and is denoted by* $\operatorname{conv} W$.

Theorem 2 (Carathéodory's Theorem). *Let* W *be any set of points in* R^m, *and let* $C = \operatorname{conv} W$. *Then* $y \in C$ *if and only if* y *can be expressed as a convex combination of* $m + 1$ *(not necessarily distinct) points in* W. *In fact,* C *is the union of all the generalized d-dimensional simplices whose vertices belong to* W, *where* $d = \dim C$.

Corollary 1. *Let* $\{C_i \mid i \in I\}$ *be an arbitrary collection of convex sets in* R^m, *and let* C *be the convex hull of the union of the collection. Then every point of* C *can be expressed as a convex combination of* $m + 1$ *or fewer affinely independent points, each belonging to a different* C_i.

The details and proofs of Theorem 2 and Corollary 1 can be found in [27].

Now, we describe the proposed algorithm as follows:

Step 0. *Initialization.* Let the *list of problems* initially include only the Approximate Integer Problem (AIP) build as follows:

Consider problem (15):

$$\min_{x,y} \left\{ \langle a, x \rangle + \langle b, y \rangle : Gy = d, \langle c, x \rangle \leq \varphi(y), \ Ax = y, x \geq 0, y \in Z_+^m \right\}.$$

Now, let us generate the polytop Y composed as a convex hull of the leader's strategies at the upper level: $Y = \{y \,|\, Gy = d, y \geq 0\}$, and select $\widehat{m} + 1$ affinely independent points y^i such that $Y \subset \text{conv}\,\{y^1, \ldots, y^{\widehat{m}+1}\} \subset \{y : |\varphi(y)| < \infty\}$. Here $\widehat{m} = m - rank(G)$, and $y^2 - y^1, y^3 - y^1, \cdots, y^{\widehat{m}+1} - y^1$ form a linearly independent system. We denote this set of vertices as $V = \{y^1, \ldots, y^{\widehat{m}+1}\}$.

We also accept a tolerance parameter value $\varepsilon > 0$. Then we solve the lower level linear programming problem (14) at each vertex, i.e., find $\varphi(y^1), \ldots, \varphi(y^{\widehat{m}+1})$ and the corresponding solution vectors $\left(x^1, y^1\right), \ldots, \left(x^{\widehat{m}+1}, y^{\widehat{m}+1}\right)$.

Next, we build the first approximation of the optimal value function as follows:

$$\Phi(y) = \sum_{i=1}^{\widehat{m}+1} \lambda_i \varphi(y^i), \tag{17}$$

defined over

$$y = \sum_{i=1}^{\widehat{m}+1} \lambda_i y^i, \tag{18}$$

with $\lambda_i \geq 0$, $i = 1, \ldots, \widehat{m} + 1$, and

$$\sum_{i=1}^{\widehat{m}+1} \lambda_i = 1. \tag{19}$$

In (17), we have an expression with the variable λ related to the variable y via the links (18) and (19). Now since the function φ is convex, one has

$$\langle c, x \rangle \leq \varphi(y) \leq \Phi(y),$$

and our condition $\langle c, x \rangle \leq \varphi(y)$ in (15) can be relaxed to the following explicit inequality:

$$\langle c, x \rangle \leq \Phi(y).$$

Thus we obtain a new optimization problem, which can be solved, for example, by the branch-and-bound algorithm. The Approximate Integer Problem (AIP) is described as follows:

$$\min_{x,y} \left\{ \langle a, x \rangle + \langle b, y \rangle : Gy = d, \langle c, x \rangle \leq \Phi(y), Ax = y, x \geq 0, y \in Z_+^m \right\}. \tag{20}$$

Now let $t = 1$, and $z_t = +\infty$, where z_t is the incumbent objective value. Put this problem into the problems list. By definition, this problem corresponds to the convex polyhedron Y. Go to Step 1.

Step 1. *Termination criterion.* Stop if the list of problems is empty, or if all the current solutions of problem (5) are close enough:

$$\max_{1 \le i \neq k \le \widehat{m}+1} \left\| (x^i, y^i) - (x^k, y^k) \right\| < \varepsilon.$$

In these cases, select the point (x^r, y^r), where $\varphi(y^r) = \min\{\varphi(y^1), \ldots, \varphi(y^{\widehat{m}+1})\}$ as the best approximation to the optimal solution of the original problem.

Otherwise, arbitrarily select and remove a program from the problems list. Go to Step 2.

Step 2. Solve the problem taken from the problems list using a regular method for integer programming (e.g., like branch-and-bound) to manage the integrality restriction. Denote the set of optimal solutions as $S = \left\{ (\tilde{x}^1, \tilde{y}^1), \ldots \right\}$ and $\tilde{z}$ the objective function value. If the problem has no feasible solution, or if its objective function value is larger than z_t, then fathom this branch, let $z_{t+1} := z_t, t := t+1$, and go to Step 1. Otherwise go to Step 3.

Step 3. If the components y of all the solutions belonging to S are elements of V, then store the solutions, set $z_{t+1} := \tilde{z}, t := t+1$, and go to Step 1 (for such values of y, the point (x, y) is feasible for problem (15)). Otherwise, considering the solution $(\tilde{x}^j, \tilde{y}^j)$ from S such that the component $\tilde{y}^j$ is different from all the elements of V, we add $\tilde{y}^j$ to V, set $z_{t+1} := z_t, t := t+1$, and go to Step 4.

Step 4. *Subdivision.* Make a subdivision of the set Y corresponding to this problem. By construction, problem (20) corresponds to one set of $\widehat{m}+1$ affinely independent points, which (without any loss of generality) are assumed to be the points $y^1, \ldots, y^{\widehat{m}+1}$. Having added the point $\tilde{y}^j$ to this set, it becomes affinely dependent. Excluding one element of the resulting set, affine independence can eventually be obtained (this is guaranteed if some correct element is dropped). When one uses this approach, at most $\widehat{m}+1$ new affine independent sets arise, each corresponding to a new linear approximation of the lower level objective function on the convex hull of these points. If one such simplex T is a subset of some region of stability: $T \subset \Re(B_i)$, the feasible points (x, y) of problem (20) are also feasible for problem (15). The aim of this step is to find these simplices by subsequent subdivisions of the set Y. These problems are then added to the list of problems.

In order to calculate the new approximation of the lower level optimal value function, we proceed as follows. First, compute $\varphi(\tilde{y}^j)$, then construct one set of affinely independent points as described above, i.e., delete one of the previous points, say y^ℓ, where $\ell \in \{1, \ldots, \widehat{m}+1\}$, and calculate

$$\Phi_\ell(y) = \sum_{i=1, i \neq \ell}^{\widehat{m}+1} \lambda_i \varphi(y^i) + \mu \varphi(\tilde{y}^j),$$

defined at all

$$y = \sum_{i=1, i \neq \ell}^{\widehat{m}+1} \lambda_i y^i + \mu \tilde{y}^j, \tag{21}$$

with $\lambda_i \geq 0$, $i = 1, \ldots, \widehat{m} + 1$, $i \neq \ell$, and

$$\sum_{i=1, i \neq \ell}^{\widehat{m}+1} \lambda_i + \mu - 1. \tag{22}$$

Thus we construct at most $\widehat{m} + 1$ new problems:

$$(P^\ell) \quad \min_{x,y} \left\{ \langle a, x \rangle + \langle b, y \rangle : Gy = d, \langle c, x \rangle \leq \Phi_\ell(y), Ax = y, x \geq 0, y \in Z_+^m \right\},$$

and add them to the list of problems. Go to Step 1.

Another idea about how to solve problem (15) is to work with the exact penalty function described in [4,6,9]. Namely, one can deduce a new reformulation of (15) using the facts that the objective value function (14) is piecewise-linear, convex, and partially calm, as was mentioned in Sect. 4.

We suppose that there exists a $k_0 < \infty$ such that a point (x^0, y^0) is locally optimal for problem (15) if, and only if, it is locally optimal for the problem:

$$\min_{x,y} \left\{ \langle a, x \rangle + \langle b, y \rangle + k \left[\langle c, x \rangle - \varphi(y) \right] \mid Gy = d, Ax = y, x \geq 0, y \in Z_+^m \right\}, \tag{23}$$

for all $k \geq k_0$.

The difficulty in dealing with (23) arises from the fact that the exact penalty function:

$$\langle a, x \rangle + \langle b, y \rangle + k \left[\langle c, x \rangle - \varphi(y) \right] \tag{24}$$

is not explicit due to the nature of the lower level optimal value function (14). Moreover, the penalty function (24) is also nonconvex. For this reason, we propose to use the algorithms developed in [12,13].

6   A Numerical Example

We consider the following bilevel parametric lineal programming problem, where the upper level is described as:

$$\min_{x_1, x_2, x_3, y_1, y_2} \left\{ 3x_1 + 2x_2 + 6x_3 + 2y_1 \; : \; 4y_1 + y_2 = 10, x \in \psi(y), y_1, y_2 \in Z_+ \right\},$$

where

$$\psi(y_1, y_2) = \operatorname*{Argmin}_{x_1, x_2, x_3} \{ -5x_1 - 8x_2 - x_3 \; : \; 4x_1 + 2x_2 \leq y_1, 2x_1 + 4x_2 + x_3$$

$$\leq y_2, x_1, x_2, x_3 \geq 0 \},$$

hence the lower level optimal value is given by:

$$\varphi(y_1, y_2) = \min_{x_1, x_2, x_3} \{ -5x_1 - 8x_2 - x_3 \; : \; 4x_1 + 2x_2 \leq y_1, 2x_1 + 4x_2 + x_3$$

$$\leq y_2, x_1, x_2, x_3 \geq 0 \}.$$

The optimal solution of this problem is $(x_1^*, x_2^*, x_3^*; y_1^*, y_2^*) = (1/3, 1/3, 0; 2, 2)$. We start solving the problem with the aid of the above-described algorithm.

Step 0. We choose the vertices $y^1 = (5/2, 0)$ and $y^2 = (0, 10)$ that belong to the convex hull of the leader's strategies at the upper level (that is, $V = \{y^1, y^2\}$). Fix the tolerance value $\varepsilon = 0.1$. Now, we calculate $\varphi(y^1) = 0$ and $\varphi(y^2) = -10$, set $z_1 = +\infty$, then the first approximation is build as follows:

$$\Phi(y) = -y_2.$$

The approximate integer problem (AIP) that we add to the problems' list is given as follows:

$$\min_{x,y}\{3x_1 + 2x_2 + 6x_3 + 2y_1 \mid 4y_1 + y_2 = 10, 4x_1 + 2x_2 \le y_1$$
$$2x_1 + 4x_2 + x_3 \le y_2, -5x_1 - 8x_2 - x_3 \le -y_2$$
$$x_1, x_2, x_3 \ge 0, y_1, y_2 \in Z_+\}$$

Step 1. We select (AIP) from the list of problems.
Step 2. We solve problem (AIP) and obtain the (unique) solution $(\tilde{x}_1, \tilde{x}_2, \tilde{x}_3; \tilde{y}_1, \tilde{y}_2) = (0, 1/4, 0; 2, 2)$ with $\tilde{z} = 15/4$. Because $\tilde{z}$ is less than $+\infty$, we go to Step 3.
Step 3. As $\tilde{y} = (\tilde{y}_1, \tilde{y}_2) = (2, 2)$ is different from the elements of the set V, we add $\tilde{y} = (2, 2)$ to V, set $z_2 := +\infty$, $t := 2$, and go to Step 4.
Step 4. Make a subdivision at $\tilde{y} = (2, 2)$ thus obtaining two new problems: the first one corresponding to $conv\{y^2 = (0, 10), \tilde{y} = (2, 2)\}$, and the second one corresponding to $conv\{\tilde{y} = (2, 2), y^1 = (5/2, 0)\}$. Then we add these two new programs to the list of problems: the first one with the approximation

$$\Phi_1(y) = -17y_2/24 - 70/24,$$

and the second one with the approximation

$$\Phi_2(y) = -13y_2/6.$$

In other words, the new problems can be specified as follows:

$(P^1) \quad \min_{x_1,x_2,x_3,y_1,y_2} \{3x_1 + 2x_2 + 6x_3 + 2y_1 \; : \; 4y_1 + y_2 = 10, 4x_1 + 2x_2$

$\le y_1, 2x_1 + 4x_2 + x_3 \le y_2, -5x_1 - 8x_2 - x_3 \le \Phi_1(y), x_1, x_2, x_3 \ge 0, y_1, y_2 \in Z_+\}$.

(when removing y^1 from V), and

$(P^2) \quad \min_{x_1,x_2,x_3,y_1,y_2} \{3x_1 + 2x_2 + 6x_3 + 2y_1 \; : \; 4y_1 + y_2 = 10, 4x_1 + 2x_2$

$\le y_1, 2x_1 + 4x_2 + x_3 \le y_2, -5x_1 - 8x_2 - x_3 \le \Phi_2(y), x_1, x_2, x_3 \ge 0, y_1, y_2 \in Z_+\}$.

(when removing y^2 from V). Go to Step 1.
Step 1. We select (P^1) from the list of problems and go to Step 2.
Step 2. We solve (P^1) thus yielding the (unique) solution $(\tilde{x}_1, \tilde{x}_2, \tilde{x}_3; \tilde{y}_1, \tilde{y}_2) = (1/3, 1/3, 0; 2, 2)$ with $\tilde{z} = 17/3$. Because $\tilde{z}$ is less than z_2, then we go to Step 3.

Step 3. As $\tilde{y} = (\tilde{y}_1, \tilde{y}_2) = (2, 2)$ coincides with one of the elements of V, we store the solution $(\tilde{x}_1, \tilde{x}_2, \tilde{x}_3; \tilde{y}_1, \tilde{y}_2) = (1/3, 1/3, 0; 2, 2)$, set $z_3 := 17/3$, $t := 3$, and go to Step 1.

Step 1. Now we select (P^2) from the list of problems and go to Step 2.

Step 2. We solve (P^2) to obtain the (unique) solution $(\tilde{x}_1, \tilde{x}_2, \tilde{x}_3; \tilde{y}_1, \tilde{y}_2) = (1/3, 1/3, 0; 2, 2)$ with $\tilde{z} = 17/3$. As $\tilde{z}$ is equal to z_3, we go to Step 3.

Step 3. Because $\tilde{y} = (\tilde{y}_1, \tilde{y}_2) = (2, 2)$ coincides with one of the elements of V, we store the solution $(\tilde{x}_1, \tilde{x}_2, \tilde{x}_3; \tilde{y}_1, \tilde{y}_2) = (1/3, 1/3, 0; 2, 2)$, set $z_4 := 17/3$, $t := 4$, then go to Step 1.

Step 1. The list of problems is empty, so we finish the algorithm.

Therefore, the last stored solution $(\tilde{x}_1, \tilde{x}_2, \tilde{x}_3; \tilde{y}_1, \tilde{y}_2) = (1/3, 1/3, 0; 2, 2)$ with $z = 17/3$ is the solution obtained with our algorithm, and it coincides with the exact solution of the problem.

7   Conclusions

In this chapter, we propose an approximation algorithm to solve the Mixed-Integer Linear Programming Problem, and at the same time, using the exact penalty function, we provide the upper and lower bounds for a feasible solution. This algorithm can be applied to solve numerically the bilevel fuzzy portfolio optimization problem (6)–(7). In the latter case, $y = \sigma$ is the upper level variable, which uses to be discrete. Therefore, after having numerated various possible values of σ, the vector y can be interpreted as belonging to Z_+^m.

The work does not stop here: our goal is to analyze more alternatives such as a convexification of the exact penalty function, or making use of the subdifferential calculus as another alternative. Later, comparing the algorithms, we will choose the best one according to its performance and robustness.

Acknowledgment. The research activity of the first author was partially funded by the R&D Department of the Tecnológico de Monterrey (ITESM), Campus Monterrey, Mexico, and by the SEP-CONACYT grant CB-2013-01-221676 (Mexico), while the second and third authors were financially supported by the SEP-CONACYT grant FC-2016-01-1938 (Mexico).

References

1. Bard, J.F.: An algorithm for solving the general bilevel programming problem. Math. Oper. Res. **8**, 260–282 (1983)
2. Dempe, S.: A simple algorithm for the linear bilevel programming problem. Optimization **18**, 373–385 (1987)
3. Dempe, S.: Foundations of Bilevel Programming. Kluwer Academic Publishers, Dordrecht, London, Boston (2002)
4. Dempe, S., Kalashnikov, V.V.: Discrete Bilevel Programming with Linear Lower Level Problems. Preprint, TU Bergakademie Freiberg (2005)
5. Dempe, S., Schreier, H.: Operations Research - Deterministische Modelle und Methoden. Teubner Verlag, Wiesbaden (2006)

6. Dempe, S., Zemkoho, A.B.: A Bilevel Approach to Optimal Toll Setting in Capacitated Networks. Preprint, TU Bergakademie Freiberg (2008)
7. Dempe, S., Kalashnikov, V.V., Kalashnykova, N.I., Arévalo Franco, A.: A new approach to solving bi-level programming problems with integer upper level variables. ICIC Express Lett. **3**, 1281–1286 (2009)
8. Dempe, S., Kalashnikov, V.V., Pérez-Valdés, G.A., Kalashnykova, N.I.: Bilevel Programming Problems: Theory, Algorithms and Applications to Energy Networks. Springer, Heidelberg, New York, Dordrecht, London (2015)
9. Dempe, S., Kalashnikov, V.V., Ríos-Mercado, R.Z.: Discrete bilevel programming: application to a natural gas cash-out problem. Eur. J. Oper. Res. **166**, 469–488 (2005)
10. Faísca, N.P., Dua, V., Rustem, B., Saraiva, P.M., Pistikopoulos, E.N.: Parametric global optimization for bilevel programming. J. Glob. Optim. **38**, 609–623 (2007)
11. Floudas, C.A., Gümüş, Z.H., Ierapetritou, M.G.: Global optimization in design under uncertainty: feasibility test and flexibility index problem. Ind. Eng. Chem. Res. **40**, 4267–4282 (2001)
12. Gao, D.Y.: Canonical duality theory and solutions to constrained nonconvex quadratic programming. J. Glob. Optim. **29**, 377–399 (2004)
13. Gao, D.Y.: Solutions and optimality criteria to box constrained nonconvex minimization problems. J. Ind. Manag. Optim. **3**, 293–304 (2007)
14. Grygarová, L.: Qualitative Untersuchung des I. Optimierungsproblems in mehrparametrischer Programmierung. Appl. Math. **15**, 276–295 (1970)
15. Gümüş, Z.H., Floudas, C.A.: Global optimization of mixed-integer-bilevel programming problems. Comput. Manag. Sci. **2**, 181–212 (2005)
16. Hu, X.P., Li, Y.X., Guo, J.W., Sun, L.J., Zeng, A.Z.: A simulation optimization algorithms with heuristic transformation and its application to vehicle routing problems. Int. J. Innov. Comput. Inf. Control **4**, 1169–1182 (2008)
17. Jan, R.H., Chern, M.S.: Non-linear integer bilevel programming. Eur. J. Oper. Res. **72**, 574–587 (1994)
18. Kalashnikov, V.V., Kalashnykova, N.I., Castillo-Pérez, F.J.: Finding equilibrium in a financial model by solving a variational inequality problem. In: Le Thi, H.A., et al. (eds.) Modelling, Computation and Optimization in Information Systems and Management Sciences, Proceedings of the 3rd International Conference on Modelling Computation and Optimization in Information Systems and Management Sciences (MCO 2015), Metz, France, Part I, pp. 281–291. Springer, Cham, Heidelberg, New York, Dordrecht, London (2015)
19. Kalashnikov, V.V., Matis, T., Pérez-Valdés, G.A.: Time series analysis applied to construct US natural gas price functions for groups of states. Energy Econ. **32**, 887–900 (2010)
20. Kalashnikov, V.V., Pérez-Valdés, G.A., Kalashnykova, N.I.: A linearization approach to solve the natural gas cash-out bilevel problem. Ann. Oper. Res. **181**, 423–442 (2010)
21. Kalashnikov, V.V., Pérez-Valdés, G.A., Kalashnykova, N.I., Tomasgard, A.: Natural gas cash-out problem: bilevel stochastic optimization approach. Eur. J. Oper. Res. **206**, 18–33 (2010)
22. Kalashnikov, V.V., Ríos-Mercado, R.Z.: A natural gas cash-out problem: a bilevel programming framework and a penalty function method. Optim. Eng. **7**, 403–420 (2006)
23. Liu, C., Wang, Y.: A new evolutionary algorithm for multi-objective optimization problems. ICIC Express Lett. **1**, 93–98 (2007)

24. Markowitz, H.: Portfolio selection. J. Financ. **7**, 77–91 (1952)
25. Moore, J.T., Bard, J.F.: The mixed integer linear bilevel programming problem. Oper. Res. **38**, 911–921 (1990)
26. Nishizaki, I., Sakawa, M., Kan, T.: Computational methods through genetic algorithms for obtaining Stackelberg solutions to two-level integer programming problems. Electron. Commun. Jpn. Part 3 **86**, 1251–1257 (2003)
27. Rockafellar, R.T.: Convex Analysis. Princeton University Press, Princeton (1970)
28. Saharidis, G.K., Ierapetritou, M.G.: Resolution method for mixed integer bi-level linear problems based on decomposition techinque. J. Glob. Optim. **44**, 29–51 (2009)
29. Sahin, K.H., Ciric, A.R.: A dual temperature simulated annealing approach for solving bilevel programming problems. Comput. Chem. Eng. **23**, 11–25 (1998)
30. Sharpe, W.: Portfolio Theory and Capital Markets. McGrow Hill Book Company, New York (1970)
31. Sharpe, W., Alexander, G., Bailey, J.: Investments. Prentice Hall, England Cliffs (1999)
32. von Stackelberg, H.: Marktform und Gleichgewicht. Julius Springer, Vienna (1934.) English translation: The Theory of the Market Economy. Oxford University Press, Oxford (1952)
33. Wen, U.P., Yang, Y.H.: Algorithms for solving the mixed integer two level linear programming problem. Comput. Oper. Res. **17**, 133–142 (1990)
34. Wendell, R.E.: A preview of a tolerance approach to sensitivity analysis in linear programming. Discrete Math. **38**, 121–124 (1982)
35. Ye, J.J., Zhu, D.L.: Optimality conditions for bilevel programming problems. Optimization **33**, 9–27 (1995)
36. Zhang, R., Wu, C.: A decomposition-based optimization algorithm for scheduling large-scale job shops. Int. J. Innov. Comput. Inf. Control **5**, 2769–2780 (2009)

Analysis on Extensions of Multi-expert Decision Making Model with Respect to OWA-Based Aggregation Processes

Binyamin Yusoff[1](✉), José M. Merigó[2,3], and David Ceballos Hornero[1]

[1] Department of Mathematical Economics and Finance,
University of Barcelona, Barcelona, Spain
binyaminy@yahoo.com
[2] Department of Management Control and Information Systems,
University of Chile, Santiago, Chile
[3] Risk Center, University of Barcelona, Barcelona, Spain

Abstract. In this paper, an analysis on extensions of multi-expert decision making model based on ordered weighted averaging (OWA) operators is presented. The focus is on the aggregation of criteria and the aggregation of individual judgment of experts. First, soft majority concept based on induced OWA (IOWA) and generalized quantifiers to aggregate the experts' judgments is analyzed, in which concentrated on both classical and alternative schemes of decision making model. Secondly, analysis on the weighting methods related to unification of weighted average (WA) and OWA is conducted. An alternative weighting technique is proposed which is termed as alternative OWA-WA (AOWAWA) operator. The multi-expert decision making model then is developed based on both aggregation processes and a comparison is made to see the effect of different schemes for the fusion of soft majority opinions of experts and distinct weighting techniques in aggregating the criteria. A numerical example in the selection of investment strategy is provided for the comparison purpose.

Keywords: Multi-expert decision making · OWA operator · IOWA operator
Weighting methods · Soft majority concept

1 Introduction

In the past, various multi-criteria decision making methods have been developed as tools for modeling human decision making and reasoning, see [4, 5]. The methods have effectively used in numerous applications to deal with the rating, ranking and selection of option(s). In complex decision making, normally a group of experts or decision makers involved in which each of them offset and/or support the others for the comprehensive decision. Since then, the expansion of such models to multi-expert decision making have been extensively focused.

Central to the decision making problems, aggregation process play a crucial role in deriving the final decision, either to aggregate the criteria with respect to each option or

© Springer International Publishing AG, part of Springer Nature 2018
A. M. Gil-Lafuente et al. (Eds.): FIM 2015, AISC 730, pp. 179–196, 2018.
https://doi.org/10.1007/978-3-319-75792-6_15

to aggregate the final agreement of individual experts. Weighted average (WA) and ordered weighted averaging (OWA) operators are generally employed as aggregation processes in decision making models. OWA operators [17, 18, 19] provide a parameterized class of aggregation operators which can be ranged from minimum to maximum and average as normal case. In contrast to the WA which represents the reliability of information sources or criteria, the weights in OWA reflect the importance of values with respect to ordering. The OWA operators can be explained as applying the concept of fuzzy set theory to modify the basic aggregation process used in decision making, precisely, using generalized quantifiers [7, 8, 26] for soft aggregation processes. In addition, the induced OWA (IOWA) operators [22] as its extension deal with the problem which involve pair of values, for example, the additional parameters used to induce the argument values to be aggregated. The OWA and IOWA are useful in the case of the need to consider the attitudinal character of experts, for instance, the behavior of experts regarding the proportion of criteria to consider. Analogously, with respect to group decision making, the soft majority agreement among experts can be implemented using the IOWA operators, which synthesizes the opinions of the majority (such as semantics "most") of the experts. In this case, a majority opinion refers to consensual judgment of a majority of experts who have similar opinions.

With respect to that, the purpose of this paper is on analyzing the multi-expert decision making model based on these two aggregation processes, i.e., aggregation of criteria and aggregation of experts' judgments. At first, the soft majority concept models for aggregating the experts' judgments based on IOWA operators and linguistic quantifier are reviewed, particularly the method as proposed in [14] and its extension as proposed in [2]. The difference between the two majority concept models can be divided into: (i) on assigning the weights for the experts, (ii) the measures used in calculating the support between experts (proximity metric), and (iii) the approach in deriving the support between experts, either based on options (classical scheme) or criteria (alternative scheme). Pasi and Yager [14] proposed the method in case of weights between experts are considered as identical (homogeneous group decision making) and used the support function based on distance measure to compute the overall level of agreement between experts. Besides, the support between experts is calculated with respect to the final result of options of each expert. On the other hand, Bordogna and Sterlacchini [2] extended this idea to include the case of where the experts are assigned with different weights (heterogeneous group decision making) and utilized similarity measure based on Minkowski OWA to calculate the overall support between experts. Moreover, the approach used to calculate the support between experts is based on the similarity measure with respect to each criterion instead of on each option. In this paper, for the purpose of comparison, some modifications have been made to both methods, include an extension of the Pasi and Yager's method from classical scheme to alternative scheme. On contrary, the Bordogna and Sterlacchini's method has been modified to deal with classical scheme. Hence, two additional methods with the existing two original methods are compared as to examine the effect of the approaches on decision scheme used.

Secondly, the weighting methods which stipulate decision strategies for the compensation of criteria in making the decision are studied. Specifically, we analyze some of the methods in deriving the weights based on the unification of WA and OWA, such as, methods for including importances using combination of 'or-and' operators [18], linguistic quantifier [23], fuzzy system modeling [24], weighted OWA (WOWA) [15], OWAWA [10], hybrid WA (HWA) [16] and immediate WA (IWA) [9]. In addition, we propose an alternative OWAWA (AOWAWA) operator which combines the characteristics of IWA and OWAWA using the idea of geometric means. As comparison, the multi-expert decision making model with respect to Bordogna and Sterlacchini's approach on alternative scheme is used as to observe the results of distinct weighting techniques in aggregation of criteria.

The outline of the paper is as follows. In Sect. 2 the definitions of OWA, IOWA and Minkowski OWA distance operators are presented. In Sect. 3 the aggregation techniques for soft majority concept is discussed and then Sect. 4 reviews the weighting methods based on WA and OWA. In Sect. 5, multi-expert decision making model based on different schemes and weighting techniques of aggregation processes are outlined. A numerical example in a selection of investment strategy is provided in Sect. 6. The paper then is summed up with a conclusion in the Sect. 7.

2 Preliminaries

This section provides some definitions and basic concepts related to OWA and IOWA operators and their generalizations that will be used throughout the paper.

2.1 OWA Operator

Definition 1 [18]. An OWA operator of dimension n is mapping $OWA : R^n \to R$ that has an associated weighting vector W of dimension n, such that $w_j \in [0, 1]$ and $\sum_{j=1}^n w_j = 1$, according to the following formula:

$$OWA(a_1, \ldots, a_n) = \sum_{j=1}^n w_j a_{\sigma(j)} \tag{1}$$

where $a_{\sigma(j)}$ denotes the components of $A = (a_1, a_2, \ldots, a_n)$ being ordered in non-increasing order $a_{\sigma(1)} \geq a_{\sigma(2)} \geq \ldots \geq a_{\sigma(n)}$.

The OWA operators are all meet commutative, monotonic, bounded and idempotent properties. Given that a function $Q : [0, 1] \to [0, 1]$ as a regular monotonically non-decreasing fuzzy quantifier and it satisfies: (i) $Q(0) = 0$, (ii) $Q(1) = 1$, (iii) $a > b$ implies $Q(a) \geq Q(b)$, then the associated OWA weights can be derived using this function such in the next definition.

Definition 2 [18]. Let Q be a non-decreasing fuzzy quantifier, then a mapping $OWA : R^n \to R$ is an ordered weighted average (OWA) operator of dimension n if:

$$OWA_Q(a_1, a_2, \ldots, a_n) = \sum_{j=1}^n \omega_j a_{\sigma(j)}, \tag{2}$$

where $a_{\sigma(j)}$ denotes the components of $A = (a_1, a_2, \ldots, a_n)$ being ordered in non-increasing order $a_{\sigma(1)} \geq a_{\sigma(2)} \geq \ldots \geq a_{\sigma(n)}$ and $\omega_j = Q(\frac{i}{n}) - Q(\frac{i-1}{n})$, being a monotonic non-decreasing function.

2.2   IOWA Operator

Definition 3 [22]. An IOWA operator of dimension n is mapping $IOWA : R^n \rightarrow R$ that has an associated weighting vector W such that $w_j \in [0, 1]$ and $\sum_{j=1}^{n} w_j = 1$, according to the following formula:

$$IOWA(\langle u_1, a_1 \rangle, \langle u_2, a_2 \rangle, \ldots, \langle u_n, a_n \rangle) = \sum_{j=1}^{n} w_j a_{\sigma(j)} \tag{3}$$

where the notion $\sigma(j)$ denotes the inputs $\langle u_j, a_j \rangle$ of the order-inducing variable u_j and argument variable a_j reordered such that $u_{\sigma(1)} \geq u_{\sigma(2)} \geq \ldots \geq u_{\sigma(n)}$ and the convention that if z of the are tied, i.e., $u_{\sigma(j)} = u_{\sigma(j+1)} = \ldots = u_{\sigma(j+z-1)}$, then, the value $a_{\sigma(j)}$ is given as follow [8, 20]:

$$a_{\sigma(j)} = \frac{1}{z} \sum_{k=\sigma(j)}^{\sigma(j+z-1)} a_k \tag{4}$$

The IOWA operators are all meet commutative, monotonic, bounded and idempotent properties.

2.3   Minkowski OWA Distance

Definition 4 [11]. A Minkowski OWAD operator of dimension n is a mapping $MOWAD : R^n \times R^n \rightarrow R$ that has an associated weighting vector W of dimension n such that $\sum_{j=1}^{n} w_j = 1$ with $w_j \in [0, 1]$ and the distance between two sets A and B is given as follows:

$$MOWAD(d_1, d_2, \ldots, d_n) = \left(\sum_{j=1}^{n} w_j d_{\sigma(j)}^{\lambda} \right)^{1/\lambda}, \tag{5}$$

where $d_{\sigma(j)}$ denotes the components of $D = (d_1, d_2, \ldots, d_n)$ being ordered in non-increasing order $d_{\sigma(1)} \geq d_{\sigma(2)} \geq \ldots \geq d_{\sigma(n)}$, and d_j is the individual distance between A and B, such that $d_j = |a_j - b_j|$ with λ is a parameter in a range $\lambda \in (-\infty, \infty)$.

The MOWAD operators are all meet commutative, monotonic, bounded and idempotent properties. By setting different values for the norm parameter λ, some special distance measures can be derived. For example, if $\lambda = 1$, then the Manhattan OWA distance can be obtained, $\lambda = 2$ then the Euclidean OWA distance can be acquired, $\lambda = \infty$ then Tchebycheff OWA is derived, etc. Equivalently, OWA and IOWA can be generalized using the same formulation, see [12, 20, 25].

3 Aggregation Methods for Soft Majority Concept

In this section, the methods for aggregating the soft majority opinion of individual experts are presented. The method by Pasi and Yager [14] as well as its extension, Bordogna and Sterlacchini [2] are studied. The extension of both methods then are made and applied in a multi-expert decision making model for the analysis purpose. Before that, the general framework of decision making schemes in which the basis of Pasi and Yager's method and also Bordogna and Sterlacchini's method are presented, i.e., a classical scheme and an alternative scheme of decision making process.

3.1 Multi-expert Decision Making Schemes

In general, the method as proposed in [14] is mainly based on the classical scheme where the result of consensus measure is determined according to the support on each option of individual experts. While the method in [2] is based on the alternative scheme in which the majority opinion particularly focuses on each specific criterion.

The classical scheme of group decision making process can be divided into two stages of aggregation process, namely internal and external aggregations. The internal aggregation involves the fusion of criteria for each expert, either full or partial compensation. At this stage, the ranking of alternatives for each expert is derived. As regard to this ranking, then in the external aggregation, the soft majority concept is implemented to find the final ranking which reflects the majority opinion of individual experts. Note that the fusion of experts' judgments in this case is focused on each option as proposed in [14].

On the other hand, for the alternative approach, instead of dealing with internal aggregation at the first step, where the individual ranking of options of each expert is derived, this method initiated with the external aggregation to fuse the majority opinion with respect to each criterion. At this stage, the new decision matrix which represents the soft majority of experts is obtained. Then, the internal aggregation to fuse the criteria is performed with the flexibility to compensate the criteria for the final decision.

3.2 The Method Based on Pasi and Yager's Approach

In the following, a brief description of the aforementioned methods is conferred. Two fundamental steps in each method are on determining the inducing variable and deriving the associated weights of experts. The methodology used to obtain the majority opinion based on Pasi and Yager [14] can be expressed as follows:

Suppose that a collection of individual opinion of h experts $(h = 1, 2, \ldots, k)$ is given as the vector $P_i^h = \left(p_i^1, p_i^2, \ldots, p_i^k\right)$ with respect to each option $i, (i = 1, 2, \ldots, m)$. For a simple notation, p_h can be used instead of p_i^h since each option is evaluated independently using the same formulation. For a single option, the similarity of each expert can be calculated using the support function as follows:

$$supp(p_l, p_g) = \begin{cases} 1 & if \ |p_l - p_g| < \beta, \\ 0 & otherwise. \end{cases} \quad (6)$$

The support function represents the similarity or dissimilarity between expert l with respect to all the other experts g, such that $l, g \in h$. Then the overall support for each individual expert l can be given as:

$$u_l = \sum_{\substack{g = 1 \\ g \neq l}}^{k} supp(p_l, p_g), \quad (7)$$

where $u_l, l \in h = (1, 2, \ldots k)$ constitute the values of order inducing variable $U = (u_{\sigma(1)}, \ldots, u_{\sigma(k)})$ which ordered as $u_{\sigma(1)} \leq u_{\sigma(2)} \leq \ldots \leq u_{\sigma(k)}$. Note that, here the values of inducing variable are reordered in non-decreasing order instead of non-increasing order as in the original IOWA, such in Eq. (3). This type of ordering reflects the conformity of quantifier 'most' as to model the majority concept, see [14] for clarification.

In consequence, to compute the weights of the weighting vector, define the values t_l based on an adjustment of the values u_l, such that: $t_l = u_l + 1$ (the similarity of p_l with itself, similarity value equal to one). The t_l values are in non-decreasing order, $t_1 \leq \ldots \leq t_k$. On the basis of t_l values, the weights of the weighting vector are computed as follows:

$$w_l = \frac{Q(t_l/k)}{\sum_{l=1}^{k} Q(t_l/k)}. \quad (8)$$

The value $Q(t_l/k)$ denotes the degree to which a given member of the considered set of values represents the majority. The quantifier Q based on membership function for semantics "most" of experts can be given as follows:

$$Q(r) = \begin{cases} 1 & if \ r \geq 0.8, \\ 2r - 0.6 & if \ 0.3 < r < 0.8, \\ 0 & if \ r \leq 0.3, \end{cases} \quad (9)$$

where $r = t_l/k$. As can be seen, the weight of experts here is derived based on the arithmetic mean (AM) where each expert is considered as having an equal degree of importance or trust, e.g., reflect the average of the most of the similar values. Then, the final evaluation is determined using the IOWA operators such in Eq. (3). Note that, here the IOWA is based on the non-decreasing of inputs u_l, p_l, as well as weights w_l as to comply with the concept of majority opinions.

However, in some cases, the values of the vector $P_i^h = (p_i^1, p_i^2, \ldots, p_i^k)$ derived after the internal aggregation process are very close to each other due to, for example, the normalization process. This case then leads to the values of $|p_l - p_g|$ less differentiable and cause a difficulty in assigning a value for β. Hence, in this paper, a slight modification has been made to cope with this problem. The support function in Eq. (6) then can be modified as follows:

$$supp(p_l, p_g) = \begin{cases} 1 & if \ \frac{|p_l - p_g|}{\max_l |p_l - p_g|} < \beta, \\ 0 & otherwise. \end{cases} \tag{10}$$

3.3   The Method Based on Bordogna and Sterlacchini's Approach

In the following, the method based on Bordogna and Sterlacchini [2] is presented. Contrary to the previous method, here, the majority opinion of experts with respect to each specific criterion is conducted for every option. Suppose that a collection of individual opinion of h experts is given as vector $P_i^h = (p_i^1, p_i^2, \ldots, p_i^k)$ for each option $i, (i = 1, 2, \ldots, m)$. In this method, instead of using the support function, they used the Minkowski OWA based similarity measure to obtain the $Q_{coherence}$ for inducing variable. The $Q_{coherence}$ of each expert can be defined as follows:

$$u_l = Q_{coherence}(P_l, P_h) = MOWA(s_1, \ldots, s_k) = \left(\sum_{h=1}^{k} \omega_h s_h^\lambda \right)^{1/\lambda}, \tag{11}$$

where ω_h are the ordered weights with the inclusion of importances of experts (or trust scores of experts, $t_h, h = 1, 2, \ldots, k$), such that $\omega_h, t_h \in [0, 1]$ with $\left(\sum_{h=1}^{k} \omega_h = \sum_{h=1}^{k} t_h = 1 \right)$ and $s_l = 1 - |p_l - p_h|$ as similarity measure between expert l with respect to all the other experts h (includes itself), such that $l \in h$. The norm parameter $\lambda \in (-\infty, \infty)$ provides a generalization of the model.

Then, the order inducing vector can be given as:

$$U = (u_1, \ldots, u_k) = (Q_{coherence}(P_1, P_h), \ldots, Q_{coherence}(P_k, P_h)), \tag{12}$$

Moreover, Q as generalized quantifiers can take any semantics to modify the weights of experts (or trust degrees) for different strategies or behaviors. When $Q(x) = x$, then $Q_{coherence}$ is reduced to:

$$u_l = coherence(P_l, P_h) = \left(\sum_{h=1}^{k} t_h s_h^\lambda \right)^{1/\lambda}, \tag{13}$$

which is the weighted average of trust degrees with similarity measure of experts.

This can be explained as the generalization of trust degrees, where in [14] the trust, t_h are considered as equal, while here they can be extended to WA and OWA weights.

Subsequently, the weights of weighting vector for the IOWA operator can be deriving using the following formula:

$$m_h = \frac{argmin_h(u_1 \cdot t_1, \ldots, u_k \cdot t_k)}{\sum_{h=1}^{k} argmin_i(u_1 \cdot t_1, \ldots, u_k \cdot t_k)}, \qquad (14)$$

where m_h are ordered in non-decreasing order. Further, given the quantifier Q with semantics "most" as Eq. (9), the weighting vector $W = (\omega_1, \ldots, \omega_k)$ can computed as:

$$\omega_h = \frac{Q(m_h)}{\sum_{h=1}^{k} Q(m_h)}. \qquad (15)$$

Next, the overall aggregation process is computed using the IOWA operator with non-decreasing inputs $\langle u_l, p_l \rangle$. Similarly, here, a simple modification can be made to the similarity measure to cope with the small difference between the values as follows:

$$s(p_l, p_g) = 1 - \left(\frac{|p_l - p_g|}{\max_l |p_l - p_g|} \right). \qquad (16)$$

4   The Methods Based on Unification of WA and OWA

In this section, the method for deriving the associated weights for aggregation of criteria is discussed. In particular, the weighting methods based on unification of WA and OWA are reviewed. In addition to the previously proposed methods in the literature, an alternative weighting technique called as AOWAWA operator is suggested. The analysis on some functions that generalizes WA and OWA operators which was done in [9] i.e., WOWA, HWA, OWAWA and IWA, then is extended to include some other functions like OWA-OA, OWA-FSM, and the proposed AOWAWA.

4.1   The Existing Methods

Prior to the definition of unification of WA and OWA as weighting methods, the general definition of WA and OWA weights are given.

Definition 5. A weighting vector $V = (v_1, v_2, \ldots, v_n)$ is a weighting vector of dimension n if and only if $v_j \in [0, 1]$ and $\sum_j v_j = 1$.

Definition 6. Let P be a weighting vector of dimension n, then a mapping $WA : R^n \to R$ is a weighted average of dimension n if $WA_P(a_1, a_2, \ldots, a_n) = \sum_j p_j a_j$. The WA are monotonic, idempotent and bounded, but it is not commutative [1, 6].

Definition 7 [20]. Let W be a weighting vector of dimension n, then a mapping $OWA_W : R^n \to R$ is an ordered weighted averaging (OWA) operator of dimension n if:

$$OWA_W(a_1, a_2, \ldots, a_n) = \sum_j w_j a_{\sigma(j)}, \qquad (17)$$

where $a_{\sigma(j)}$ denotes the components of $A = (a_1, a_2, \ldots, a_n)$ being ordered in non-increasing order $a_{\sigma(1)} \geq a_{\sigma(2)} \geq \cdots \geq a_{\sigma(n)}$.

There are a number of methods proposed in the literature for obtaining the OWA weights, e.g., linguistic quantifier [18] such in Eq. (2), maximum entropy OWA [13], etc. For the overview of methods for determining OWA weights, see [26]. Next, some of the unification methods of WA and OWA are given.

Definition 8 [18]. Let P and W be two weighting vectors of dimension n., then a mapping $OWA : R^n \rightarrow R$ is an OWA operator of dimension n if:

$$OWA_{P,W}(a_1, a_2, \ldots, a_n) = \sum_j w_j a_{\sigma(j)}, \tag{18}$$

where $a_{\sigma(j)}$ denotes the components of $\check{A} = (\check{a}_1, \check{a}_2, \ldots, \check{a}_n)$ being ordered in non-increasing order $\check{a}_{\sigma(1)} \geq \check{a}_{\sigma(2)} \geq \cdots \geq \check{a}_{\sigma(n)}$ such that $\check{a}_j = H(a_j, p_j) = (p_j \vee \bar{\alpha}) \cdot (a_j)^{p_j \vee \alpha}$ and $\alpha = \sum_{j=1}^{n} \frac{n-j}{n-1} w_j$ is the orness measure and $\bar{\alpha} = 1 - \alpha$ is its complement.

This method is based on 'or-and' lattice operator and for the sake of simplicity, in this paper it can be termed as OWA-OA. Note that if $\alpha = 0$, then it is a pure 'and' operator, given as $a_j = d_j^{p_j}$. Since $w_n = 1$, then $D(x) = \begin{array}{c} Min \\ j = 1, \ldots, n \end{array} A_j(x)^{p_j}$, $A_j(x) = a_j$. Conversely, if $\alpha = 1$, then it is a pure 'or' operator, given as $a_j = p_j a_j$. Since $w_1 = 1$, then $D(x) = \begin{array}{c} Max \\ j = 1, \ldots, n \end{array} p_j A_j(x)$, $A_j(x) = a_j$. The OWA-OA operators are all meet commutative, monotonic, bounded and idempotent properties. But, OWA-OA operators do not satisfy $O_p^\eta = F_p$ and $O_\eta^w = F^w$.

Definition 9 [24]. Let P and W be two weighting vectors of dimension n, then a mapping $OWA : R^n \rightarrow R$ is an OWA operator of dimension n if:

$$OWA_{P,W}(a_1, a_2, \ldots, a_n) = \sum_i w_j a_{\sigma(j)}, \tag{19}$$

where $a_{\sigma(j)}$ denotes the components of $\hat{A} = (\hat{a}_1, \hat{a}_2, \ldots, \hat{a}_n)$ being ordered in non-increasing order $\hat{a}_{\sigma(1)} \geq \hat{a}_{\sigma(2)} \geq \cdots \geq \hat{a}_{\sigma(n)}$ given that $\hat{a}_j = H(a_j, p_j) = \bar{\alpha} \bar{p}_j + p_j a_j$ and $\bar{\alpha} = 1 - \alpha$, that is the orness measure $\alpha = \sum_{j=1}^{n} \frac{n-j}{n-1} w_j$. This method is based on fuzzy system modeling and can be termed as OWA-FSM. The OWA-FSM operators are all meet commutative, monotonic, bounded and idempotent properties. But, OWA-FSM operators do not satisfy $M_p^\eta = F_p$ and $M_\eta^w = F^w$.

Definition 10 [15]. Let P and W be two weighting vectors of dimension n, then a mapping $WOWA : R^n \rightarrow R$ is a weighted ordered weighted averaging (WOWA) operator of dimension n if:

$$WOWA_{P,W}(a_1, a_2, \ldots, a_n) = \sum_j \omega_j a_{\sigma(j)}, \tag{20}$$

where $a_{\sigma(j)}$ denotes the components of $A = (a_1, a_2, \ldots, a_n)$ being ordered in non-increasing order $a_{\sigma(1)} \geq a_{\sigma(2)} \geq \ldots \geq a_{\sigma(n)}$ and $\omega_j = w^* \left(\sum_{k \leq j} p_{\sigma(j)} \right) - w^* \left(\sum_{k \leq j} p_{\sigma(j)} \right)$ with w^* being a monotonic non-decreasing function that interpolates the points $\left((j/n), \sum_{k \leq j} w_j \right)$ together with the point $(0, 0)$. The function w^* required to be a straight line when the points can be interpolated in this way.

WOWA operators satisfy $W_p^\eta = F_p$ and $W_\eta^w = F^w$. Moreover, they are monotonic, idempotent, and bounded [16]. In a similar way that for the OWA operator, the WOWA operator can be defined using a fuzzy quantifier instead of having the weighting vector w. This definition is similar to the Yager's definition of OWA using importances [23].

Definition 11 [23]. Let Q be a non-decreasing fuzzy quantifier, let p be a weighting vector of dimension n, then a mapping $OWA : R^n \to R$ is an OWA operator of dimension n if:

$$OWA_{P,Q}(a_1, a_2, \ldots, a_n) = \sum_j \omega_j a_{\sigma(j)}, \tag{21}$$

where $a_{\sigma(j)}$ denotes the components of $A = (a_1, a_2, \ldots, a_n)$ being ordered in non-increasing order $a_{\sigma(1)} \geq a_{\sigma(2)} \geq \ldots \geq a_{\sigma(n)}$ and $\omega_j = Q \left(\sum_{k \leq j} p_{\sigma(j)} \right) - Q \left(\sum_{k \leq j} p_{\sigma(j)} \right)$.

This operator generalizes the weighted mean and the OWA operator: when $p = \left(\frac{1}{n}, \frac{1}{n}, \ldots, \frac{1}{n} \right)$ the operator reduces to the OWA operator and when $w = \left(\frac{1}{n}, \frac{1}{n}, \ldots, \frac{1}{n} \right)$ the operator reduces to the WA.

Definition 12 [16]. Let P and W be two weighting vectors of dimension n, then a mapping $HA : R^n \to R$ is a hybrid averaging (HA) operator of dimension n if:

$$HA_{P,W}(a_1, a_2, \ldots, a_n) = \sum_j w_j a_{\sigma(j)}, \tag{22}$$

where $a_{\sigma(j)}$ denotes the components of $\breve{A} = (\breve{a}_1, \breve{a}_2, \ldots, \breve{a}_n)$ being ordered in non-increasing order $\breve{a}_{\sigma(1)} \geq \breve{a}_{\sigma(2)} \geq \ldots \geq \breve{a}_{\sigma(n)}$ given that $\breve{a}_j = np_j a_j$ and n is the balancing coefficient.

HWA operator generalizes both OWA and WA operators and reflects the importance degrees of both the given argument and the ordered position of the argument. HWA operators satisfy $H_p^\eta = F_p$ and $H_\eta^w = F^w$. Moreover, they are monotonic [9].

Definition 13 [9]. Let P and W be two weighting vectors of dimension n, then a mapping $IWA : R^n \to R$ is an immediate weighted averaging (IWA) operator of dimension n if:

$$IWA_{P,W}(a_1, a_2, \ldots, a_n) = \sum_j \pi_j a_{\sigma(j)}, \tag{23}$$

where $a_{\sigma(j)}$ denotes the components of $A = (a_1, a_2, \ldots, a_n)$ being ordered in non-increasing order $a_{\sigma(1)} \geq a_{\sigma(2)} \geq \cdots \geq a_{\sigma(n)}$ and $\pi_j = w_j p_j / \sum_{j=1}^{n} w_j p_j$.

As can be seen, the IWA is a manipulation of immediate probability [3, 11, 21] by using the WA instead of probability distribution. IWA operators satisfy $I_p^{\eta} = F_p$ and $I_{\eta}^w = F^w$ [9].

Definition 14 [11]. Let P and W be two weighting vectors of dimension n, then a mapping $OWAWA : R^n \to R$ is an ordered weighted averaging-weighted average (OWAWA) operator of dimension n if:

$$OWAWA_{P,W}(a_1, a_2, \ldots, a_n) = \sum_j \varphi_j a_{\sigma(j)}, \tag{24}$$

where $a_{\sigma(j)}$ denotes the components of $A = (a_1, a_2, \ldots, a_n)$ being ordered in non-increasing order $a_{\sigma(1)} \geq a_{\sigma(2)} \geq \cdots \geq a_{\sigma(n)}$ and $\varphi_j = \beta w_j + (1 - \beta) p_{\sigma(j)}$ with $\beta \in [0, 1]$.

OWAWA operator is all meet monotonic, idempotent, bounded properties. Moreover the value returned by the OWAWA operator lies between the values returned by the WA and OWA, and coincides with them when both are equal. But, OWAWA operators do not satisfy $N_p^{\eta} = F_p$ and $N_{\eta}^w = F^w$.

In addition, by taking the advantages of IWA and OWAWA operators, a new weighting method can be derived as in the next sub-section.

4.2   The Proposed Alternative OWAWA Operator

Definition 15. Let P and W be two weighting vectors of dimension n, then a mapping $AOWAWA : R^n \to R$ is an alternative ordered weighted averaging-weighted average (AOWAWA) operator of dimension n if:

$$AOWAWA_{P,W}(a_1, a_2, \ldots, a_n) = \sum_j \hat{\varphi}_j a_{\sigma(j)}, \tag{25}$$

where $a_{\sigma(j)}$ denotes the components of A being ordered in non-increasing order $a_{\sigma(1)} \geq a_{\sigma(2)} \geq \cdots \geq a_{\sigma(n)}$ and $\hat{\varphi}_j = w_j^{\beta} * p_{\sigma(j)}^{(1-\beta)} / \sum_{j=1}^{n} w_j^{\beta} \cdot p_{\sigma(j)}^{(1-\beta)}$ with $\beta \in [0, 1]$, by convention $(0^0 = 0)$.

The AOWAWA operator is monotonic, bounded, idempotent. However, it is not commutative because the AOWAWA operator includes the WA. In addition, AOWAWA operators do not satisfy $A_p^{\eta} = F_p$ and $A_{\eta}^w = F^w$.

Theorem 1 (Monotonicity). Assume f is the AOWAWA operator, let $(a_1, a_2, \ldots, a_n)$ and $(b_1, b_2, \ldots, b_n)$ be two sets of arguments. If $a_j \geq b_j$, $\forall j \in \{1, 2, \ldots, n\}$, then:

$$f(a_1, a_2, \ldots, a_n) \geq f(b_1, b_2, \ldots, b_n).$$

Proof. It is straightforward and thus omitted.

Theorem 2 (Idempotency). Assume f is the AOWAWA operator, if $a_j = a$, $\forall j \in \{1, 2, \ldots, n\}$, then:

$$f(a_1, a_2, \ldots, a_n) = a.$$

Proof. It is straightforward and thus omitted.

Theorem 3 (Bounded). Assume f is the AOWAWA operator, then:

$$Min\{a_j\} \leq f(a_1, a_2, \ldots, a_n) \leq Max\{a_j\}.$$

Proof. It is straightforward and thus omitted.

5 Multi-expert Decision Making Model Based on Different Schemes of Aggregation Processes

In this section, some multi-expert decision making models based on classical and alternative aggregation schemes are presented. First, the majority concept of Pasi and Yager's method which is originally based on classical aggregation scheme is extended to the alternative scheme. Here, the multi-expert decision making model using Pasi and Yager's method with respect to classical scheme is stated as MEDM-PY I and for alternative scheme is denoted as MEDM-PY II. Secondly, Bordogna and Sterlacchini's method which is based on alternative scheme is modified to the case of classical method. Here, the MEDM-BS I represents decision making model using the alternative scheme and MEDM-BS II denoted as the method based on classical scheme. Moreover, for the aggregation process of criteria, each of the weighting methods based on unification of WA and OWA are implemented for comparison purpose.

5.1 The Proposed Alternative OWAWA Operator

Stage I: Internal aggregation (Local aggregation)

Step 1: First, a decision matrix for each expert $D^h, h = 1, 2, \ldots, k$, is constructed as follows:

$$D^h = \begin{array}{c} \\ A_1 \\ \vdots \\ A_m \end{array} \begin{pmatrix} C_1 & \cdots & C_n \\ a_{11}^h & \cdots & a_{1n}^h \\ \vdots & \ddots & \vdots \\ a_{m1}^h & \cdots & a_{mn}^h \end{pmatrix}, \tag{26}$$

where A_i indicates the alternative $i(i = 1, 2, \ldots, m)$ and C_j denotes the criterion $j(j = 1, 2, \ldots, n)$, and a_{ij}^h with $a_{ij}^h \in [0, 1]$ denotes the preferences for alternative A_i with respect to criterion C_j.

Step 2: Determine the weighting vector for each expert using the unification of WA and OWA. All the weighting methods can be implemented such as in Eqs. (17)–(25). In this step, the attitudinal character of experts reflects the proportion of criteria used under consideration.

Step 3: Aggregate the judgment matrix of each expert by the weighting vector in Step 2. At this stage, each expert derives the ranking/priorities of all alternatives individually (individual experts' judgments).

Stage II: External aggregation (Global aggregation)

With respect to the type of aggregation method, the consensus measure for the majority of experts can be calculated as follows:

(A) Pasi and Yager's method: MEDM-PS I (Homogeneous group decision making)

Step 4A: Determine the inducing variable using the Eqs. (6)–(7) or in case of the values are very close to each other, use the modified support function such in Eq. (10).

Step 5A: Calculate the weighting vector which represents the majority of experts using the Eq. (8) based on quantifier "most" such in Eq. (9). In this case, the weights are considered as equal for all experts.

(B) Modified version of Bordogna and Sterlacchini's method: MEDM-BS I (Heterogeneous group decision making process).

Step 4B: Determine the inducing variable using the Eq. (11) or in case of the values are very close to each other use the similarity measure such in Eq. (16).

Step 5B: Calculate the weighting vector using the Eqs. (14)–(15). In this case, the weights of experts or trust degrees are associated to each expert.

5.2   The Proposed Alternative OWAWA Operator

Stage I: External aggregation (Local aggregation)

Step 1: By the similar way, a decision matrix for each expert is constructed such in Eq. (26). Then the aggregation of majority of experts can be implemented using one of the methods as follows:

(A) Bordogna and Sterlacchini's method: MEDM-BS II

Step 2A: Determine the inducing variable such in Step 4B of classical scheme. But, instead of aggregate the opinion of experts with respect to each option, in this step, the aggregation process is conducted on each criterion.

Step 3A: Calculate the weighting vector such in Step 5B of classical scheme using the values of inducing variable in Step 2A.

(B) Extension of Pasi and Yager's method: MEDM-PS II

Step 2: Determine the inducing variable such in Step 4A of classical scheme. But, instead of aggregate the opinion of experts with respect to each option, here, the aggregation process is conducted on each criterion.

Step 3: Calculate the weighting vector such in Step 5A of classical scheme using the values of inducing variable in Step 2B.

Stage II: Internal aggregation (Global aggregation)

Step 4: Determine the weighting vector using the unification of WA and OWA such in Eqs. (17)–(25).

Step 5: Finally, aggregate the judgment matrix of majority of experts by the derived weighting vector. Here, the proportion of criteria is respected to the attitudinal character of majority of experts.

6   Numerical Example

In the following, a numerical example is presented. An investment selection problem is studied where a group of experts are assigned for the selection of an optimal strategy.

Different cases of multi-expert decision making methods are analyzed, in particular with respect to aggregation process of majority opinions of experts based on different schemes (namely classical and alternative schemes), and also on different weighting methods. Note that with this analysis, the optimal choices will be obtained depend on the scheme and aggregation operator used in each particular case. As can be seen each scheme and aggregation operator leads to different results and decisions.

Assume that a company plans to invest some money in a region. At first, they consider five possible investment options as follows: A_1 = invest in the European market, A_2 = invest in the American market, A_3 = invest in the Asian market, A_4 = invest in the African market, A_5 = do not invest money.

In order to evaluate these investments, the investor has brought together a group of experts E_k. This group considers that each investment option can be described with the following characteristics: C_1 = benefits in the short term, C_2 = benefits in the mid-term, C_3 = benefits in the long term, C_4 = risk of the investment, C_5 = other variables.

The available investment strategies, depending on the characteristic C_i and the option A_i for each expert are shown in Table 1.

The aggregated results of the different approaches are presented in the Table 2 and their rankings are given in Table 3. Should be noted that in this case, all the criteria are set to have equal degrees of importance and the experts' weights are given as 0.3, 0.1, 0.1, 0.4, 0.1 for expert E_1, E_2, E_3, E_4 and E_5, respectively for MEDM-BS I and MEDM-BS II. While for MEDM-PY I and MEDM-PY II the experts' weights are considered as equal.

As can be seen, there is a slight difference between the results which derived from both soft majority aggregation approaches with respect to different decision schemes. The majority opinion of individual experts which calculated based on the classical scheme provided A4, A2, A1, A5 and A3 as ranking for both methods. While the majority opinions computed with respect to alternative scheme gave the ranking of A4, A1, A5, A2 and A3 for both methods. Hence, the results show the effect of different decision schemes in ranking the options.

Table 1. Available investment strategies of each expert, E_h

E_1						E_2				
C1	C2	C3	C4	C5		C1	C2	C3	C4	C5
0.7	0.6	0.7	0.6	0.9		0.6	0.9	1	0.9	0.9
0.8	1	0.2	1	0.6		1	0.7	0.1	1	0.8
0.6	0.7	0.6	0.6	0.5		0.4	0.9	0.8	0.7	0.6
0.9	0.6	0.8	1	0.9		0.9	0.5	0.7	1	0.9
0.3	0.7	0.7	0.8	0.9		0.7	0.7	0.9	0.9	0.9
E_3						E_4				
C1	C2	C3	C4	C5		C1	C2	C3	C4	C5
0.5	0.7	0.9	0.8	0.9		0.4	0.7	0.9	0.8	0.8
0.9	0.9	0.2	1	0.7		0.9	0.7	0.1	0.9	0.6
0.8	0.8	0.7	0.7	0.6		0.6	0.6	0.5	0.8	0.4
0.9	0.5	0.8	1	0.7		0.7	0.5	0.7	0.7	0.9
0.8	0.7	0.8	0.9	0.8		0.4	0.6	0.7	0.8	0.9
E_5										
C1	C2	C3	C4	C5						
0.5	0.6	0.7	0.6	0.8						
0.9	0.8	0.4	0.9	0.5						
0.6	0.6	0.5	0.8	0.7						
0.8	0.7	0.6	0.9	0.8						
0.2	0.6	0.8	0.6	0.8						

Table 2. The aggregated results

	MEDM-PY I	MEDM-PY II	MEDM-BS I	MEDM-BS II
A1	0.7143	0.7726	0.7169	0.7989
A2	0.7178	0.6992	0.7200	0.6580
A3	0.6280	0.6361	0.5952	0.6057
A4	0.7886	0.8027	0.7800	0.8000
A5	0.7029	0.7225	0.6800	0.6969

Table 3. The ranking of financial strategies

Method	Ranking
MEDM-PY I	A4 > A 2 > A1 > A5 > A3
MEDM-PY II	A4 > A1 > A5 > A2 > A3
MEDM-BS I	A4 > A 2 > A1 > A5 > A3
MEDM-BS II	A4 > A1 > A5 > A2 > A3

As further analysis, we extend the method of Bordogna and Sterlacchini II to include the unification of WA and OWA weights with different criteria' weights. Tables 4 and 5 show the aggregated results and the final ordering of the financial strategies.

Table 4. The aggregated results with respect to MEDM-BS II model

	OWA (Q)	WOWA (Q)	HA	IWA	OWA WA	AOWAWA	OWA (FSM)	OWA (OA)
A1	0.880	0.764	1.193	0.872	0.845	0.851	0.355	0.255
A2	0.914	0.421	1.097	0.942	0.767	0.853	0.343	0.196
A3	0.678	0.586	0.922	0.685	0.652	0.663	0.301	0.164
A4	0.947	0.687	1.210	0.965	0.868	0.910	0.360	0.211
A5	0.838	0.657	1.066	0.806	0.778	0.785	0.330	0.234

Table 5. The ordering of financial strategies

	Ordering		Ordering
OWA (Q)	A3 > A 2 > A5 > A1 > A4	OWAWA	A2 > A4 > A5 > A1 > A3
WOWA (Q)	A1 > A5 > A4 > A2 > A3	AOWAWA	A3 > A2 > A5 > A1 > A4
HA	A2 > A3 > A5 > A1 > A4	OWA (FSM)	A2 > A3 > A5 > A1 > A4
IWA	A3 > A2 > A5 > A1 > A4	OWA (OA)	A1 > A4 > A5 > A3 > A2

The weighted average, p for each criteria is given as 0.1, 0.2, 0.3, 0.3, 0.1 and the ordering weights, w which represent 'most' of the criteria is given as 0.0044, 0.0356, 0.1200, 0.2844, 0.5556. As can be seen, the proposed AOWAWA weights with $\beta = 0.5$ provided the ranking similar to the IWA weights. While the rest weighting techniques shown slightly different results.

7   Conclusions

In this paper, the analysis on extensions of multi-expert decision making model based on ordered weighted average (OWA) operators is conducted. The focus is on the aggregation processes with respect to criteria and individual judgment of experts. First, the soft majority concept based on induced OWA (IOWA) and linguistic quantifiers to aggregate the experts' judgments is analyzed, in which concentrated on the classical and alternative schemes of decision making model. Then, analysis on the weighting methods related to integration of weighted average (WA) and OWA is conducted. The alternative weighting technique has been proposed which is termed as alternative OWA-WA (AOWAWA) operator. The multi-expert decision making model based on both aggregation processes then has been developed and a comparison is made to see the effect of different weighting techniques in aggregating the criteria and the results of using different schemes for the fusion of soft majority opinions of experts. A numerical example in the selection of investments is provided for the comparison purpose.

References

1. Beliakov, G., Pradera, A., Calvo, T.: Aggregation Functions: A Guide For Practitioners. Springer, Berlin (2007)
2. Bordogna, G., Sterlacchini, S.: A multi criteria group decision making process based on the soft fusion of coherent evaluations of spatial alternatives. In: Zadeh, L.A., Abbasov, A.M., Yager, R.R., Shahbazova, S.N., Reformat, M.Z. (eds.) Recent Developments and New Directions in Soft Computing. Studies in Fuzziness and Soft Computing. Springer, Cham (2014)
3. Engemann, K.J., Filev, D.P., Yager, R.R.: Modelling decision making using immediate probabilities. Int. J. Gen. Syst. **24**, 281–284 (1996)
4. Figueira, J., Greco, S., Ehrgott, M.: Multiple Criteria Decision Analysis: State of the Art Surveys. Springer, New York (2005)
5. Gal, T., Steward, T., Hanne, T.: Multicriteria Decision Making: Advances in MCDM Models, Algorithms, Theory and Applications. Springer Science and Business Media, Dordrecht (1999)
6. Grabisch, M., Marichal, J.L., Mesiar, R., Pap, E.: Aggregation Functions. Cambridge University Press, Cambridge (2009)
7. Kacprzyk, J.: Group decision making with a fuzzy linguistic majority. Fuzzy Sets Syst. **18**, 105–118 (1986)
8. Kacprzyk, J., Fedrizzi, M., Nurmi, H.: Group decision making and consensus under fuzzy preferences and fuzzy majority. Fuzzy Sets Syst. **49**, 21–31 (1992)
9. Llamazares, B.: An analysis of some functions that generalizes weighted means and OWA operators. Int. J. Intell. Syst. **28**(4), 380–393 (2013)
10. Merigo, J.M.: OWA operators in the weighted average and their application in decision making. Control Cybern. **41**(3), 605–643 (2012)
11. Merigo, J.M., Gil-Lafuente, A.M.: Using OWA operator in the Minkowski distance. Int. J. Soc. Hum. Sci. **2**, 564–572 (2008)
12. Merigo, J.M., Gil-Lafuente, A.M.: The induced generalized OWA operator. Inf. Sci. **179**, 729–741 (2009)
13. O'Hagan, M.: Aggregating template or rule antecedents in real-time expert systems with fuzzy set logic. In: IEEE Asilomar Conference on Signals, Systems and Computers, pp. 681–689 (1988)
14. Pasi, G., Yager, R.R.: Modelling the concept of majority opinion in group decision making. Inf. Sci. **176**, 390–414 (2006)
15. Torra, V.: The weighted OWA operator. Int. J. Intell. Syst. **12**, 153–166 (1997)
16. Xu, Z.S., Da, Q.L.: An overview of operators for aggregating information. Int. J. Intell. Syst. **18**, 953–969 (2003)
17. Xu, Z.S.: An overview of methods for determining OWA weights. Int. J. Intell. Syst. **20**, 843–865 (2005)
18. Yager, R.R.: On ordered weighted averaging aggregation operators in multi-criteria decision making. IEEE Trans. Syst. Man Cybern. **18**, 183–190 (1988)
19. Yager, R.R., Kacprzyk, J.: The Ordered Weighted Averaging Operators: Theory and Applications. Kluwer, Norwell (1997)
20. Yager, R.R.: Generalized OWA aggregation operators. Fuzzy Optim. Decis. Making **3**, 93–107 (2004)
21. Yager, R.R., Engemann, K.J., Filev, D.P.: On the concept of immediate probabilities. Int. J. Intell. Syst. **10**, 373–397 (1995)

22. Yager, R.R., Filev, D.P.: Induced ordered weighted averaging operators. IEEE Trans. Syst. Man Cybern. Part B Cybern. **29**, 141–150 (1999)
23. Yager, R.R.: Quantifier guided aggregation using OWA operators. Int. J. Intell. Syst. **11**, 49–73 (1996)
24. Yager, R.R.: Including importances in OWA aggregations using fuzzy systems modeling. IEEE Trans. Fuzzy Syst. **6**(2), 286–294 (1998)
25. Yusoff, B., Merigo, J.M.: Analytic hierarchy process under group decision making with some induced aggregation operators. In: IPMU 2014, Montpellier, France (2014)
26. Zadeh, L.A.: A computational approach to fuzzy quantifiers in natural languages. Comput. Math. Appl. **9**, 149–184 (1983)

Procedure for Staff Planning Based on the Theory of Fuzzy Subsets

Lourdes Souto Anido[1], Irene García Rondón[1(✉)],
Anna M. Gil-Lafuente[2], and Gabriela López Ruiz[1]

[1] Department of Business Administration, University of Havana, Havana, Cuba
irene@fec.uh.cu
[2] Department of Business Administration, University of Barcelona,
Barcelona, Spain

Abstract. The Staff Planning process, which has the responsibility to provide for the needs of workforce, must be borne by the companies as a key process sin the sub-heading Human Resources (HR). The Theory of Fuzzy Subsets is shown as a tool to help decision-making in this field. This paper proposes a method for Staff Planning based on the Theory of Fuzzy Subsets, to help decrease this subjectivity and ensure precise and accurate results as well assist application in the company SERVICEX.

Keywords: Human resources · Staff Planning process · Fuzzy Subsets Theory
Exponential Smoothing

1 Staff Planning Process

The success of any organization depends largely on its human resources [5, 7, 9, 11], hence the importance of planning well its needs, the same level as the technical, financial and commercial resources are planned. That's why the process of planning, which has the responsibility to provide what will be the workforce necessary for the realization of future organizational work, must be borne by the companies as a key process in the sub-heading Human Resources.

Today this function has gained prominence, being considered the base of Human Resources Management, because, as its name implies, plans and organizes the way in which the remaining functions will be carried out and feeds its results when projected.

At a time when the instability of Human Resources run against good organizational development, planning becomes more relevant. The acknowledgment of that resource's demand decreases the risk of failure in any organization, allowing building capacities for future events and conditions for achieving its goals [8].

The Planning process has been discussed and evaluated by different authors, so there are dissimilar criteria regarding its scope and concept. Some of these definitions are:

- "The process to meet the needs of Human Resources and the means to satisfy them in order to carry out the comprehensive plans of the organization, and in turn that plan involves determining the types of needs, skills or abilities and the number of people required" [10].

© Springer International Publishing AG, part of Springer Nature 2018
A. M. Gil-Lafuente et al. (Eds.): FIM 2015, AISC 730, pp. 197–217, 2018.
https://doi.org/10.1007/978-3-319-75792-6_16

- "The process through which the company expects future staff needs, while simultaneously pursuing the availability and development of individuals who are to meet these needs" [1].
- "Implementing a technique to systematically identify and forecast the demand for employees an organization will have. This allows the Personnel Department to provide the organization with the right people at the right time" [12].

According to the conceptual perspectives above, all authors emphasize that the planning process allows early identification of human resource needs presented by the company, in line with its objectives and strategies, with a view to having the required number of persons who are performing their duties at the right time efficiently and effectively for the benefit of the organization. It is valid to note that greater importance that the Cuban Standard gives to this concept, as this definition brings together the most important aspects and covers a more general and complete Human Resource Planning definition.

Among the main features this process presented can include the following [6]:

- All components and facets of human resource planning are closely interrelated in a systematic, progressive and dynamic way.
- It is a systematic process that identifies opportunities and dangers that arise in the future.
- It is modified by the introduction of new technologies. The instruments with which aims to predict future employee needs can range from very basic to very complex techniques.
- The long-term objectives must set the number of employees and the characteristics they will have.
- HR planning must be integrated both internally and externally.

Two main types of planning prevail in literature: (1) the estimated and provisional, which is performed before a company starts operating, in the same way as the rest of their sources are planned; and (2) planning that takes place once the organization is already in operation. This leads to set goals that you want to reach in a given time, assuming changes in internal and external conditions of the company that it must be able to foresee.

Human resources have two characteristics that justify their planning by themselves: the inertia of the human system and its flexibility. The first became evident on its difficult modeling over other resources, such as financial and material, since men cannot be handled at the whim of organizations. In this sense, planning intended that the input and output needs of workers in the organization are minimal or at least not have a significant impact. The second one is reflected in the slowness experienced by people to change or expand their knowledge and skills. While equipment can be acquired in a short time, knowledge and skills needed for management may be delayed. Following the issues raised, it is shown how the need to combine inertia and flexibility is an additional reason to apply the planning of human resources.

2 Staff Planning Process Based on Fuzzy Subsets Theory

Staff Planning is the process of identifying human resource needs in quantity and quality, by comparing available workers and the needs of the company to accomplish hits objectives. This action allows the company to have the necessary workers to develop service and/or productive activities.

This process implies both an economic and social cost, so it is necessary to reduce the subjectivity that is always present in the decision to develop this process. Therefore, It will make use of tools associated with the Theory of Fuzzy Subsets, such as the Fuzzy Delphi method and the Expertons, which allow to reach consensus while materialize the opinions of experts to make forecasts. Likewise, using Exponential Smoothing technique is to smooth the results of the views received.

Next, we will move to implement the Procedure for Staff Planning based on the Theory of Fuzzy Subsets, which is presented graphically (Fig. 1).

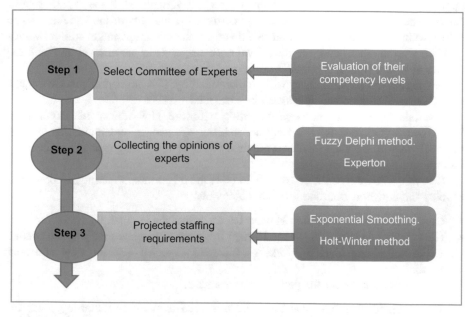

Fig. 1. Phases of Staff Planning Process based on the Theory of Fuzzy Subsets

In a first step, in order to estimate the behavior of the staffing needs, we proceed to select a Committee of Experts, with the aim that they express their views on how, in their experience, it may incur those needs. As experts, is taken to the company's HR staff and managers, but not before calculate the Competence Coefficient for each of them, in order to measure their level of competitiveness.

The Planning Process begins from the staff needs required by a company to meet its objectives. For this reason, it is requested the experts to provide an estimate of the behavior of the needs for future periods. Their views are expressed in an Expertons, which is calculated through the Fuzzy Delphi method according to the following steps [4, 11]:

1 Valuation criteria for each expert
2 Find the mean value and the Hamming distance
3 Second round of valuations
4 Determinate the competence coefficient (>0.6)
5 Construct the Expertons for each criterion
6 Determinate them arithmetical expectation of each Expertons.

Once the information provided by the experts is completed, it is proposed the use of the technique of Exponential Smoothing in order to soften the estimates and make more accurate forecasts of future HR needs.

While it is possible to quantify a variable with diffuse tools from multiple perspectives, in this case it is proposed the use of the Triangular Fuzzy Numbers (TFN), carrying three characteristic values: lower, middle and upper. Extreme (lower and upper) values demarcate where the value of the real variable, so is the experts role to determinate these characteristic values and certifies the inability of the variable to take values within the numerical range outside that form the minimum and maximum values of TFN. On the other hand, the central value corresponds to the interval at which the experts assigned maximum confidence.

Given that the actual number will be integrated by TFN instead of precise values, the values of the smoothed series also be measured in TFN.

For this case, it is proposed to work with the Holt-Winter of Exponential Smoothing model, that is both easy adaptability to changes and trends, as well as seasonal patterns. Similarly, the time required to calculate the prognosis is considerably faster.

Carrying this framework in to the field of Human Resource Planning, it intends to develop the equation from the following elements:

• The current staff representing Human Resources of the company.
• The current trend that follows the Human Resources in companies; it is, the change in the underlying level of workers which are expected between now and next month.
• The seasonal index for the period being forecast.

We conclude, therefore, that the indicators used to work as a starting point for experts to give their opinions will be real fluctuation of workers, absenteeism levels and levels of activity of the company.

3 Application of Staff Planning Process in the Company SERVICEX

SERVICEX, the Services Company of the Foreign Trade and Foreign Investment Department (MINCEX), was created with the aim of providing services to that Department and its companies, such as safety and security, general maintenance and graphic reproductions, just to mention a few.

To fulfill its mission, the company was approved its **Social objects**, through Resolution No. 304/2013, which is governed by the definitions and general principles, set out in Resolution No. 134 of April 30, 2013, being the following:

- Provide installation, maintenance and repair of buildings services, office equipment, refrigeration equipment, air conditioning, electrical, electronic and mechanical; and commissioning of the equipment.
- Provide emergency services, maintenance and repair of motor equipment, cleaning and sanitation, gardening, courier bags and artwork; as well as elevators and receptions. In the case of the courier and bag, only to MINCEX system.
- Market nonfood products.
- To provide security and protection, custody and transfer of securities and cash; as well as develop studies, advise and make recommendations on security and protection.

Similarly, the company provides a set of secondary activities derived from its social purpose, which are presented in services numbered below:

- Preparation and marketing special pins for the identification and control of staff on MINCEX entities and third parties on request.
- Researching people and determining curricula to natural and legal persons upon request.
- Transportation of personnel, sheet metal, water supply, rental of warehouses, making keys and parking services.
- Labor housing repairs under MINCEX's workers, directives and officials.
- Binding and finishing of books both on MINCEX entities and third parties.
- Gastronomy services on MINCEX's entities.
- Market food stuffs and toiletries to the employees of MINCEX.
- Preparation of opinions about basic media.
- Rental of premises, specialized equipment and qualified staff meeting.
- Scrub and grease to both MINCEX entities and third parties.

Such activities, now, are awaiting approval to be implemented and executed as an integral part of its corporate purpose.

SERVICEX proposes as a Vision:

Being a cohesive company with prestige and authority, with high political level, able to meet increasing efficiency and quality of services that customers demand, thereby contributing to the successful development of the strategic objectives of our Department.

As such, the company plays a work based on seriousness, responsibility and ethics, with the aim of achieving high levels of satisfaction in their internal and external customers, seeking to be ranked as a company of excellence, with optimal efficiency and effectiveness in business management.

The application of the proposed procedure will enable the company under study anticipate the needs of staff in 2015 in order to cope with the services demanded by MINCEX and its companies.

Step 1: Select Committee of Experts

Select the team of experts who will participate in the research is a major step in this process. According to García Rondón [3], the number of experts to select is immaterial, since the most important in the selection is not the quantity but the quality.

To select the Committee of Experts, inclusive criteria were established:

- Have experience in Human Resources
- To have years of experience with the company
- Get Competence Coefficient levels above (0.6).

The initial group was made by the Managing Director, the Director of Human Resources and three technicians in HRM. It also consulted a Health and safety technician, who has spent many years working in the company and has acquired extensive practical and theoretical knowledge on the subject of planning. Thus a total of 6 experts were completed, being as follows (Table 1):

Table 1. Committee of Expert

Experts	Charges
Expert no. 1	HRM technician
Expert no. 2	Managing Director
Expert no. 3	HRM technician
Expert no. 4	HRM technician
Expert no. 5	Human resources director
Expert no. 6	Health and safety technician

Source: Compiled from documents in the company

This group of experts will be calculated their Competence Coefficient as the Model stipulates, showing the following results (Table 2).

Table 2. Analysis of Knowledge coefficient, argumentation coefficient and Competence Coefficient

Number of experts						
Coefficients	Expert no. 1	Expert no. 1	Expert no. 1	Expert no. 1	Expert no. 1	Expert no. 1
Kc	0,7	0,8	0,9	0,4	0,9	0,6
Ka	0,8	0,9	0,8	0,5	1	0,8
K	0,75	0,85	0,85	0,45	0,95	0,7

Three of the five experts evaluated have a high level of competence, which represents 60% of the total, while the remaining two have an average level, accounting for 40%. It can be concluded that the selected team has a high preparedness reliably determine staffing needs in the company. The opinions offered by them will be concretized in Expertons, through the Fuzzy Delphi method, as will be developed below.

Step 2: Collecting the opinions of experts

After selecting the team of experts who will participate in the investigation, it was applied a first survey that should estimate what percentage should increase the service needs of the company during 2015 based on their experience. The information taken as a starting point for experts to cast their opinions was demands for monthly services in 2014, which are presented in Table 3. In addition, we also took into account the levels of absenteeism and fluctuation.

Table 3. Demand for services during 2014

Months	Service needs
January	118
February	95
March	102
April	98
May	140
June	140
July	120
August	148
September	161
October	178
November	146
December	135

Source: Compiled from documents in the company

Based on these criteria, each expert could express their opinion in a range between 1% and 10%, as they felt that this was the range of percentages in which they could move the operation to respond to staffing activity levels. This gives the possibility to collect the real opinions of people with their doubts and uncertainties, without forcing them to settle on one end of the scale. However, if the expert is certain that one of the criteria has an exact value, he may be assigned that value as an absolute range. Worth notice that if estimated a decreases in demand for services, should write the value in parentheses.

Once we have the valuations made by the experts, it is calculated the average values and Hamming distances of each criterion. The average values of the first round of surveys are shown in Table 4:

This average is calculated for each of the criteria, which represents the average evaluation of all experts. These values are used to calculate the Hamming distances, which are shown below (Table 5):

Hamming distance was calculated for each expert, and gives an idea of which one is farther from the mean. The sum of the distances of each expert describes more

Table 4. Average values. First round

Criteria	Average values	
	Min.	Max.
January	4,6	7,4
February	5,8	8,6
March	5	7,4
April	5,8	8,4
May	3	5,4
June	3,2	5,2
July	4,6	7,2
August	2,4	4,4
September	1,2	2,8
October	1	1,6
November	2,8	4,6
December	3,6	5,4

Table 5. Hamming distances

Hamming distances

Criteria	Expert no. 1	Expert no. 2	Expert no. 3	Expert no. 4	Expert no. 5
January	0,5	0,5	0,6	0,4	0,4
February	0,3	0,7	0,3	0,4	0,3
March	0,8	0,3	0,2	0,2	0,7
April	0,4	0,6	0,4	0,3	0,3
May	0,2	0,8	0,3	0,2	0,7
June	0,2	0,8	0,5	0,5	1,2
July	0,7	0,3	0,3	0,4	0,3
August	0,5	0,5	0,5	0,4	0,5
September	1	0,2	0,5	0,2	0,5
October	0,3	0,2	0,2	0,2	0,3
November	0,4	0,3	0,3	0,4	0,6
December	0,5	0,4	0,4	0,5	0,6
	5,8	5,6	4,5	4,1	6,4

accurately that the specialists who most moved away from the average of the assessments made in the first round of surveys were experts no. 5, 1 and 2, respectively. This ratio helps improve valuations for a second round, it is created for a more unified and accurate approach.

Both mean values and the distances were used in the second round of evaluation, where experts, analyzing these figures, were able to correct his previous answer according to their criteria and knowledge.

The second round of evaluation was applied using the same criteria of the first, and was recorded the final opinion of experts, which agreed that their estimates would be the same that had been offered initially after checking their answers with stockings calculated.

After being collected the information, we proceeded to perform mathematical calculations which lead to obtaining Expertons. This tool allows obtaining estimates from information obtained from experts and summarizing the main single expression.

The calculation is performed for each criterion individually and they appear in the evaluations of each of the experts who confirmed the group. It is important to note that the qualifications were offered as percentages, were taken to scale of 100.

At first, we proceeded to calculate the frequency of assessments for both the minimum and the maximum values, indicating how many times match the answers of the experts and to what scale. As this paper shows the result of this calculation is developed from the first projected month (January) (Table 6).

Table 6. Frequency on January criteria

Frequency		
0	0	0
0,01	0	0
0,02	0	0
0,03	0	0
0,04	2	0
0,05	3	0
0,06	0	0
0,07	0	3
0,08	0	2
0,09	0	0
0,1	0	0
Summation	5	5

The summation at the end of each table lets us check whether the procedure was properly applied, as long as the results match the total of experts involved in the investigation.

From this table, the observations were normalized by dividing the value of the frequency of each number on the scale between the numbers of members that make up the Committee (Table 7).

As shown, the final summation must equal one, because it represents 100% of the observations. Then it is proceeded to conform the Cumulative Frequency, calculated from the bottom up and cumulatively. As can be seen in Table 8, with this calculation it is passed to find Math Expectation. In value it is recorded specifically the opinion of all the Experton that particular criterion.

Table 7. Standardized observations of January

Standardized observations		
0	0	0
0,01	0	0
0,02	0	0
0,03	0	0
0,04	0,4	0
0,05	0,6	0
0,06	0	0
0,07	0	0,6
0,08	0	0,4
0,09	0	0
0,1	0	0
Summation	1	1

Table 8. Cumulative Frequency of January

Cumulative Frequency		
0	1	1
0,01	1	1
0,02	1	1
0,03	1	1
0,04	1	1
0,05	0,6	1
0,06	0	1
0,07	0	1
0,08	0	0,4
0,09	0	0
0,1	0	0
Expectation	0,046	0,074

To make the Expertons we took the Math expectation of all evaluated criteria. In Table 9 is displayed as it was conformed for all months of 2015:

When working with percentages, the value of Mathematics Expectation represents what percentage the service needs for 2015 should increase monthly, based on the historical record of 2014. Worth mentioning that the experts could have estimated decreases inactivity levels, which it did not happen in this case. Through Table 9 we can appreciate how in the month of January they estimated that these need should increase between 4.6% and 7.4%. Therefore, for forecasting service needs, we precede to multiply the Expectation for 2014 data.

Table 9. Expertons needs services

Months	Min.	Max.
January	0,046	0,074
February	0,058	0,086
March	0,05	0,074
April	0,058	0,084
May	0,03	0,054
June	0,032	0,052
July	0,046	0,072
August	0,024	0,044
September	0,012	0,028
October	0,01	0,016
November	0,028	0,046
December	0,036	0,054

Step 3: Projected staffing requirements

As previously explained, to carry out the proposed procedure will make use of the TFN, in order to reduce the subjectivity that may be present in the planning process. TFN are formed by three characteristic values:

- The lower ends coincide with the estimates provided by the experts in the Minimum column.
- The upper ends coincide with the estimates reported by experts in the Maximum column.
- The maximum presumption will be the value of the Mathematical Expectation of these estimates, considering the continuous case of uniform distribution; in this case it coincided with the average value of these estimates.

Finally, the service requirements estimated for 2015 were:

Table 10. Estimated service needs for 2015

Lower extreme	Maximum presumption	Upper extreme
123,428	125,08	126,732
100,51	101,84	103,17
107,1	108,324	109,548
103,684	104,958	106,232
144,2	145,88	147,56
144,48	145,88	147,28
125,52	127,08	128,64
151,552	153,032	154,512
162,932	164,22	165,508
179,78	180,314	180,848
150,088	151,402	152,716
139,86	141,075	142,29

In Table 10 is observed that the estimates given by the experts were expressed through TFN, so the demand for services in 2015 should be between these two extremes.

Although these values were estimated by a group of experts' familiar with the issue, they believed that monthly changes are too abrupt and actually cushioning of estimated figures will occur for each month by the weight of the needs of the previous month and not considered as "normal" changes from month to month, based on their knowledge, experience and intuition. Therefore, we will proceed to apply the technique of Exponential Smoothing independently of each time series, in order to soften them and create more accurate forecasts.

4   Application of Exponential Smoothing Technique

The Exponential Smoothing is a technique to smooth the data, to reduce the degree of variation and give greater importance to most recent figures. It is usually used when the forecast horizon is short term. It can be carried out through three models: Simple, non-seasonal Holt-Winter and seasonal Holt-Winter (additive and multiplicative). For these reasons, it is considered a practical and appropriate tool to smooth the estimates of the needs of services offered by the experts' and make forecasts obtained as accurate as possible. Statistical computing package Eviews 3.1 was used for the application of this technique, while it is considered easy to use econometric software.

In this research, the simple model and the Holt-Winter non-seasonal were applied in order to check which one provides a better fit to the series through the Root Mean Square Error selection criterion. Please note that the Holt-Winter method was not used on its seasonal mode because the analysis of time series corresponding to workforce does not present a seasonal pattern for the speed with which it responds to changes in the environment.

At first, the adjustment was made by the method of simple smoothing, which was proved at first glance that the series analyzed show a pattern of trend butno seasonal. Therefore, we proceeded to use the non-seasonal Holt-Winter method, because it is ideal to set Time Series presented in this pattern.

The non-seasonal Holt-Winter smoothing is a further extension of the approach to simple exponential smoothing. It is used to make predictions on the assumption of linear trend, but unlike the simple, it use two smoothing parameters α and β. It provides both the **overall level of movement** and the **future trend** in a series. The prediction equation is as follows [2]:

$$\hat{Y}_{t+T} = (a_t + b_t T),$$

Where a_t = intercept of the trend line in t, b_t = Slope of the trend line in t

This method will be applied next to the three separate Time Series, in order to smooth the estimates given by experts and to get more precise forecast of service needs in 2015 values.

Application of the Holt-Winter Method to the Lower End

In order to analyze the trend pattern can be properly set to the Lower End and the existence of seasonality, we proceeded to its graphical representation (Fig. 2).

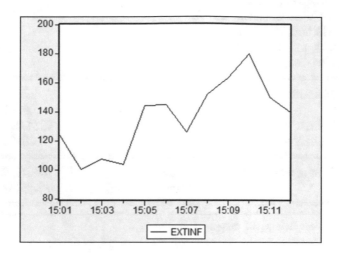

Fig. 2. Lower Extreme. **Source:** Eviews 3.1 output

As can be seen in the X-axis are represented different periods over which the experts issued an opinion, that is, the 12 months of 2015; while the Y-axis shows the lower values each corresponding to said months TFNs.

Moreover, it may be noted that the series analyzed shows a pattern trend, however not observed seasonality, since oscillatory movements aren't manifested at regular intervals. Therefore, it is considered that then on-seasonal Holt-Winter model is best suited to adjust the series and have more accurate forecasts.

Next the corresponding adjustment is made using the value of α and β the program allows, which are those values that minimize the sum of square residual. Corresponding Eviews Output is shown below:

Table 11 shows that the Alpha value provided by the program is 0.75. It is important to clarify that if the number changes quickly, α must be large, because it assigns greater weight to more recent observations. This recommendation is based on both is higher the value of α is less smoothing achieved in the series; however, a small α smoothes over the series, but does not exactly mean that offers the best prognosis. Meanwhile, it appears that the value of β is 0.00, indicating that the series has a rigid trend, growing most of the time throughout the historical period.

Similarly, Beta is reflected the Sum of Square Residual being this of 4531.042, and the Root Mean Squared Error reaching a value of 19.43159. The latter criterion, which is used to compare the results of both Smoothed processes showed that this model was the most suitable to fit the data, as it presented a lower value.

From this Eviews output we can build the model equation for non-seasonal Holt Winter, being as follows:

Table 11. Smoothing of the Lower Extreme series

Date: 04/29/15 Time: 11:25	
Sample: 2015:01 2015:12	
Included observations: 12	
Method: Holt-Winters No Seasonal	
Original Series: EXTINF	
Forecast Series: EXTINFSM	
Parameters: Alpha	0.7500
Beta	0.0000
Sum of Squared Residuals	4531.042
Root Mean Squared Error	19.43159
End of Period Levels: Mean	144.0596
Trend	0.348667

Source: Eviews 3.1 output

$$\hat{Y}_{t+T} = (144.0596 + 0.348667T).$$

Then we proceed to plot the original series and the set one to check how feasible was the setting.

As shown in Fig. 3, the adjustment was good, since the adjusted series how's a similar behavior of the original series. Furthermore, as clearly shown the new series is smoother.

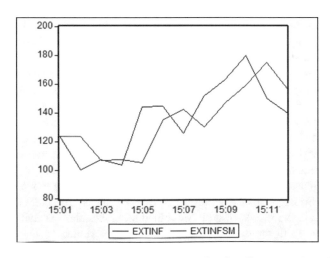

Fig. 3. Behavior of Lower End series and adjusted series. **Source:** Eviews 3.1 output

The values of the original series and forecasts for each period studied are shown in Eviews output as follows (Table 12):

Table 12. Forecasts for Lower Extreme series

obs	EXTINF	EXTINF SM
2015:01	123.4280	123.4280
2015:02	100.5100	123.7767
2015:03	107.1000	106.6750
2015:04	103.6840	107.3424
2015:05	144.2000	104.9472
2015:06	144.4800	134.7360
2015:07	125.5200	142.3928
2015:08	151.5520	130.0867
2015:09	162.9320	146.5346
2015:10	179.7800	159.1815
2015:11	150.0880	174.9793
2015:12	139.8600	156.6592

Source: Eviews 3.1 output

Herein after, will be made the same procedure for the two remaining Time Series, in order to obtain more accurate predictions in each of the values that are made by TFNs.

Application of the Holt-Winter Method to Maximum Presumption
The analysis of the series corresponding to the Maximum Presumption is similar to the one performed for the Lower End, so at first the original series will be plotted (Fig. 4).

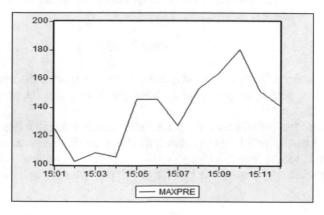

Fig. 4. Behavior of the Maximum Presumption series **Source:** Eviews 3.1 output Machine

As shown, the Maximum Presumption presents a similar behavior to the Lower End, as you can see a pattern trend, but not seasonality. Therefore, it is considered that the Holt-Winter without seasonal pattern is more suitable to fit the data.

Then the corresponding adjustment is made using the α and β value the program facilitates, with the following Eviews output (Table 13):

Table 13. Smoothing of Maximum Presumption Series

Date: 04/29/15 Time: 15:09	
Sample: 2015:01 2015:12	
Included observations: 12	
Method: Holt-Winters No Seasonal	
Original Series: MAXPRE	
Forecast Series: MAXPRESM	
Parameters: Alpha	0.7600
Beta	0.0000
Sum of Squared Residuals	4500.439
Root Mean Squared Error	19.36586
End of Period Levels: Mean	145.0466
Trend	0.333333

Source: Eviews 3.1 output

As noted, the value of α provided by the program is 0.76, while the value of β is 0.00, so it will be assigned more weight to recent observations, and series will present a rigid trend.

For its part, the summation of Squared Residuals acquired a value of 4500.439, while the Root Mean Square Error was 19.36586, obtaining a lower value than the single smoothing, as in previous series. Finally, the non-seasonal Holt-Winter model was composed as follows:

$$\hat{Y}_{t+T} = (145.0466 + 0.33337T).$$

Once plotted the original and the set series, it was found that the adjustment was successful, as the new series maintained a behavior close to the set, as shown in the following chart:

The results of the period under study forecasts that are presented below:

From Fig. 5 and Table 14 we can confirm that the series is a little smoother, because month to month values have not sharp variations, unlike the original values. Next, it proceeds to perform the same procedure for the Upper End corresponding series.

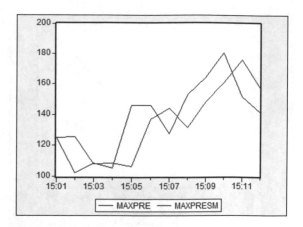

Fig. 5. Behavior of the Maximum Presumption series and the adjusted series. **Source:** Eviews 3.1 output

Table 14. Forecasts for the Maximum Presumption Series

obs	MAXPRE	MAXPRESM
2015:01	125.0800	125.0800
2015:02	101.8400	125.4133
2015:03	108.3240	107.8306
2015:04	104.9580	108.5389
2015:05	145.8800	106.1507
2015:06	145.8800	136.6789
2015:07	127.0800	144.0052
2015:08	153.0320	131.4751
2015:09	164.2200	148.1920
2015:10	180.3140	160.7069
2015:11	151.4020	175.9419
2015:12	141.0750	157.6245

Source: Eviews 3.1 output

Application of the Holt-Winter method to Upper End

As we proceeded with the previous two series, it will begin plotting the Upper End, to analyze its pattern (Fig. 6).

As can be seen, it shows a clear pattern of trend, but not seasonality, which matches earlier analysis in which the non-seasonal Holt-Winter method is best suited to fit the data and provide more accurate forecasts.

It then proceeds to adjust accordingly, using the value of α and β provided by the program, resulting in the following Eviews output (Table 15):

As shown, the α and β values match those of the Presumption Maximum associated series, these being 0.76 and 0.00 respectively. To have a high value of α, can be said to be assigned more weight to recent observations, and the series will present a rigid trend.

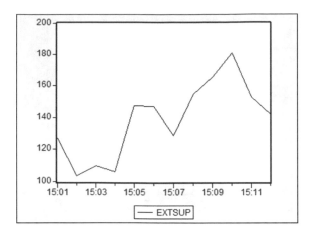

Fig. 6. Behavior of the Upper End series. **Source:** Eviews 3.1 output

Table 15. Smoothing of the Upper End series

Date: 04/29/15 Time: 18:55	
Sample: 2015:01 2015:12	
Included observations: 12	
Method: Holt-Winters No Seasonal	
Original Series: EXTSUP	
Forecast Series: EXTSUPSM	
Parameters: Alpha	0.7600
Beta	0.0000
Sum of Squared Residuals	4472.436
Root Mean Squared Error	19.30552
End of Period Levels: Mean	146.2467
Trend	0.318000

Source: Eviews 3.1 output

For its part, the Sum of Residual Square will take a value of 4472.436, while the Root Mean Square Error will be 19.30552. In this case, this criterion offered a higher error value of the Simple smoothing which was obtained 19.30352. Not being a significant difference and in order to achieve uniformity in the methods used in this research, is chosen to fit the Holt-Winter method without seasonality. Therefore, the equation was structured as follows:

$$\hat{Y}_{t+T} = (146.2467 + 0.3180T).$$

Below the chart shows both series to test the feasibility of adjustment, being as follows (Fig. 7):

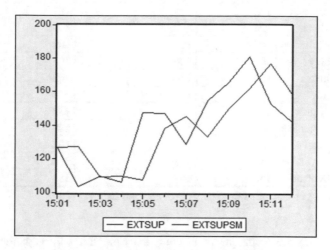

Fig. 7. Behavior Upper End series and adjusted series. **Source:** Eviews 3.1 output

It is noted that the adjustment was good as the original series closely follows the set. Forecasts values for each analyzed period are shown in the following Eviews output (Table 16):

Table 16. Forecasts for Upper End series

obs	EXTSUP	EXTSUPSM
2015:01	126.7320	126.7320
2015:02	103.1700	127.0500
2015:03	109.5480	109.2188
2015:04	106.2320	109.7370
2015:05	147.5600	107.4031
2015:06	147.2800	138.2410
2015:07	128.6400	145.4288
2015:08	154.5120	132.9871
2015:09	165.5080	149.6643
2015:10	180.8480	162.0238
2015:11	152.7160	176.6485
2015:12	142.2900	158.7774

Source: Eviews 3.1 output

Table 17. Forecast service needs for 2015

Months	Lower extreme	Maximum presumption	Upper extreme
January	123	125	126
February	123	125	127
March	106	107	109
April	107	108	109
May	104	106	107
June	134	136	138
July	142	144	145
August	130	131	132
September	146	148	149
October	159	160	162
November	174	175	176
December	156	157	158

Source: Compiled from data obtained from Eviews

After applying the technique of Exponential Smoothing through the simple and non-seasonal Holt-Winter model, it was found that the three Time Series show a pattern trend, but not seasonal, so this latest model is the more suitable to fit. However, through the selection criteria of the Root Mean Square Error could corroborate the above approach, because this model has less errors than simply smoothing in most cases.

Thus, a more accurate prognosis for the company service needs for 2015 was obtained. The results were rounded reference being the integral part thereof, being finally as follows (Table 17):

The needs were expressed through TFNs, resulting in that there is a range of possibilities that should be between the extreme values, thus avoiding subjective nuances and providing greater accuracy.

After obtaining the activity levels forecast for future periods, the company can determine the amount of human resources required to meet those needs. Thus, it is estimated properly and designates the number of employees required in each job in line with the Company's objectives and strategies.

A correct Staff Planning will help SERVICEX to prevent the unnecessarily increases of the personnel, reducing wage costs in the company. Thus decreases the risk of failure, as it ensures labor and economic stability to its Human Resources.

The result reached after applying the process of Staff Planning based on the Theory of Fuzzy Subsets in terms of service needs is not far from reality in SERVICEX as forecasts obtained coincided with the levels of activities demanded by MINCEX in the first two months of 2015.

According to information provided by the company during the course of this investigation, the services demanded by the Department in the months of January and February of 2015 were 126 and 125 respectively, which are within the given range at the ends of the first two TFNs obtained, which demonstrates the effectiveness of the Theory of Fuzzy Subsets.

5   Final Considerations

Staff Planning's main purpose is to anticipate work force needs for conducting future organizational work, which must be assumed by companies as a key process in the sub-heading Human Resources.

The application of tools associated with the Theory of Fuzzy Subsets allows more accurate and precise forecasts that are the basis for decisions regarding staffing needs in the company.

Using Exponential Smoothing technique with TFNs allows smooth expert estimates and avoid too sudden variations produced from one period to another, thus avoiding subjective nuances and giving greater certainty and precision of the results.

The proposed method consists of three stages, ranging from the selection of the Expert Committee to participate in the investigation and collection of their opinions, to the projection of personnel requirements in the company using Exponential Smoothing technique.

References

1. Burack, E.H.: Planificación y aplicacióncreativa de RecursosHumanos. Unaorientaciónestratégica. EdicionesDíaz de Santos, España (1990)
2. Espallargas, I.D., SolísCorvo, M.: Econometría y Series Temporales. Aplicaciones. La Habana (2009)
3. García Rondón, I.: Procedimiento para la selección de los mercadosinternacionales de los servicios de gestiónmedioambientalcubanos. Tesis Doctoral, Universidad de la Habana, Cuba (2010)
4. Kaufmann, A., GilAluja, J.: Introducción de la teoría de los subconjuntosborrosos a la gestión de lasempresas. Editorial Milladoiro, España (1986)
5. Leal Millán, A., RománOnsalo, M., de Prado Sagrega, A., Rodriguez Félix, L.: El factor humano en lasrelacioneslaborales. Manual de Dirección y Gestión. Pirámides, Madrid (2004)
6. Mesa, R.: Apuntessobre la gestión de los RecursosHumanos en la empresa. Revista digital Contribuciones a la Economía, octubre (2009)
7. Miguel Guzmán, M., Campdesuñer, R.P., Hernández, M.N.: Qué es la Planeación de RecursosHumanos Centro de Información y GestiónTecnológica, Holguín (2010)
8. Normalización, O.: Norma Cubana 3002:2007. Sistema de Gestión Integradadel Capital Humano. Implementación. La Habana (2007)
9. Castillo, S., Aguilar, M., Pastor, E.: Dirección de recursoshumanos. Unenfoqueestratégico. McGraw-Hill Interamericana de España, España (2003)
10. Sikula, A., McKenna, T.: Administración de RecursosHumanos. Conceptosprácticos. Editorial Limusa, México (1989)
11. Souto Anido, L.: Aplicación de herramientas de la teoría de los subconjuntosborrosos a los procesos de seleccióndel personal. Tesis de Maestría, Cuba (2013)
12. Werther, W., Davis, K.: Administración de Personal y RecursosHumanos. McGraw Hill, terceraedición, México (1991)

Quantitative Investment Analysis
by Type-2 Fuzzy Random Support Vector
Regression

Yicheng Wei and Junzo Watada[✉]

Graduate School of Information, Production and Systems, Waseda University,
2-7 Hibikino, Wakamatsu-ku, Kitakyushu 808-0135, Fukuoka, Japan
weiyicheng0000@gmail.com, junzow@osb.att.ne.jp

Abstract. Financial markets are connected well these days. One class assets' price performance is usually affected by movements of other classes of assets. The liquidity conduction mechanism usually is: if capital surface of money market is tight, investors may dump short-term treasury bonds to exchange additional liquidities. It may affects the performance of treasury bonds' yield of maturity. Credit bonds' return rate thus will level up hysteretic. Financing cost of companies accordingly raise, then throws effects on their stock prices. However, situation changes along with increase in complexity of markets' behaviors these days. In order to model movements of assets' price performance, analysis of linkage between different markets is thus becoming more and more important. Nothing like stock market, money market or bond market is an over-the-counter market, where assets' prices are often presented in the form of classes of discrete quotations with trader's subjective judgements, thus are hard to analyze. Given concern to this, we define the Type 2 fuzzy random variable (T2 fuzzy random variable) to quantify those bid/offer behaviours in this paper. Moreover, we build a T2 fuzzy random support vector regression scheme to study relationships between these markets. T2 fuzzy random support vector regression is developed from traditional support vector regression and is able to cope with fuzzy data, which has less computation complexity and better generalization performance than linear algorithms.

Keywords: Type-2 fuzzy random variable
Support vector regression model · Creditability theory
Type-reduction · Quantitative investment · Financial market

1 Introduction

For real business transactions, peoples' opinions and judgements sometimes matter much than objective analysis. When a transaction starts, bilateral dealers assess and judge the target on different respects, then offer their prices and prepare to negotiate. Often, human-beings use a linguistic word or a fuzzy price

© Springer International Publishing AG, part of Springer Nature 2018
A. M. Gil-Lafuente et al. (Eds.): FIM 2015, AISC 730, pp. 218–244, 2018.
https://doi.org/10.1007/978-3-319-75792-6_17

range to evaluate the target from various features and characteristics. Assume human-beings' judgement-process is a program with some specific algorithm, the inputs for the program shall be variables with both fuzziness and randomness; The output is then a fuzzy valuation range. In order to model the circumstance mentioned above, there is a need to find a way to quantify those inputs and outputs with fuzziness and randomness, especially in financial market including frequent transactions.

There are two types of prices in financial markets. One is called transaction price and the other is offered price. The former is always continuous and objective based on real transactions, while the latter is discrete and subjective based on quotations. For instance, the open/close price for stock index is a transaction price, which means the price establishes only when real transaction behaviors happened. While the price for Libor (London Inter-Bank Offered Rate) is an offered price. The majority investment banks offered quotations for borrowing/lending rate of money in inter-bank market according to their estimation of the future trend of financial markets. The market regulator collects all those quotations, adds some weight for each bank, then calculates an average result as the current Libor. The Libor contains all majority banks' opinions of the market and will be the basis for valuations of all other financial assets. Moreover, regulators will judge the quotation performance of each bank by some standards (called evaluation system) and give the evaluation to each quoter from time to time. The evaluation will restrict behaviors of quoters meanwhile as a standard to adjust their weight percentage to final result. Usually, offered/quoted price is a leading indicator in financial markets. For two reasons, one is that the offered price is quoted by the most powerful players in the market and issued by the market regulator, thus is quite authoritative. The other is that the quoted price is published before markets start trading transactions. Traders in the market usually use them to understand sentiment of the market and formulate their own trading strategy accordingly. Despite of those merits, there are only a few models applying offered prices as inputs to make estimations or predictions because the price is hard to quantify.

Lotfi A. Zadeh invented the type-2 fuzzy sets (T2 fuzzy sets) in 1975 [1]. A T2 fuzzy set enables our implementation of fuzziness about the membership function into fuzzy set theory. In the case we discussed above, we could use Type 1 fuzzy set (T1 fuzzy set) to take all quotations into consideration and model the regulator's evaluation at T2 fuzzy dimension. We add feature of randomness to T2 fuzzy set using creditability theory and define T2 fuzzy random variable to quantify the quoted prices. The quantification process is based on three proved assumptions: first, the quoted price is offered by traders subjectively and indicated their vague judgements to the market randomly. Second, transaction prices of one period can be described by a candlestick index. Candlestick used in capital market to model prices of financial assets could be viewed as a fuzzy set to some extend. From where open price and close price could construct the lower boundary while highest and lowest price construct upper boundary of the footprint of T2F set. Third, continuous prices with one financial asset of successful transactions in one given time interval contains both fuzziness and randomness naturally.

Many algorithms was invented to deal with T2 fuzzy set. Wei and Watada developed qualitative regression model [2,3] and qualitative classifier based on fuzzy random support vector machine (FSVM) with FER model preset [4]. Mizumoto and Tanaka [5] discussed the kinds of algebraic structures the different grades of T2 fuzzy sets form under negation, join, and meet; Dubois and Prade [6] explored the operations in the logic fuzzy-valued. It is not until recent days that T2 fuzzy sets gradually have been applied successfully to T2 fuzzy logic systems for the sake of handling linguistic and numerical uncertainties [7–12]. In this paper, we develop a T2 fuzzy random support vector regression (T2-FSVR) scheme from traditional support vector regression (SVR) model to study relationships of prices' movements between money market, bond market and stock market in China. Support vector machine (SVM) [13] was first invented by Vapnik and is gaining popularity due to many attractive features. While the traditional NN implements the empirical risk minimization (ERM) principle, SVM implements the structural risk minimization (SRM) principle which seeks to minimize an upper bound on the Vapnik-Chervonenkis (VC) dimension (generalization error). SVM was originally developed for classification problems (SVC) and then extended to support vector regression (SVR) [14]. Based on SRM principle, SVR achieves a balance between the training error and generalization error, leading to better forecasting performance. Traditional SVR is only able to deal with crisp inputs and outputs. Given our type-reduced T2 fuzzy prices' inputs and outputs are fuzzy random interval data, a regression modeling algorithm for interval data called T2 fuzzy support vector regression is then developed based on traditional SVR. The T2-FSVR method is generalized from real number domain to interval number domain, in order to simultaneously keep both fitting errors and prediction errors small and has less computation complexity, better generalization performance.

The remainder of this paper is organized as follows. In Sect. 2, we define T2 fuzzy random variable, meanwhile discussing the expected value and variance of it by using creditability theory with some defuzzification strategy. In Sect. 3, we illustrate the process of quantification of quoted prices and transaction prices by T2 fuzzy random variable. We then build a T2-FSVR scheme by explore the function of SVR to deal with interval data to study the relationships between these markets. We also compare it's performance with Confidence-interval-based fuzzy random regression model and T2 fuzzy expected regression model. Finally, a conclusion is given.

2   T2 Fuzzy Random Variable

Now, the definitions of T2 fuzzy random variable, its expected value and variance operators are presented.

2.1 Definition of T2 Fuzzy Random Variable

Definition 1. *A T2 fuzzy set, denoted by $\widetilde{A}$, is characterized by a T2 member-ship function $\mu_{\widetilde{A}}(x, \mu_A(x))$, where $x \in X$ and $\mu_A(x) \in J_x \subseteq [0,1]$. The elements of the domain of $\mu_A(x)$ are called primary memberships of x in A and the memberships of the primary memberships in $\mu_{\widetilde{A}}(x)$ are called secondary memberships of x in $\widetilde{A}$.*

$$A = \{x, \mu_A(x) | x \in X\} \tag{1}$$

$$\widetilde{A} = \{(x, \mu_A(x)), \mu_{\widetilde{A}}(x, \mu_A(x)) | \\ x \in X \ \mu_A(x) \in J_x \subseteq [0,1]\} \tag{2}$$

in which $\mu_A(x) \in J_x \subseteq [0,1]$ and $\mu_{\widetilde{A}} \subseteq [0,1]$. Figure 1 shows a T2 fuzzy set with Gaussian primary membership function and triangular secondary membership function.

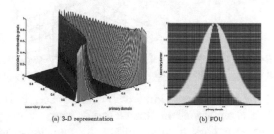

(a) 3-D representation   (b) FOU

Fig. 1. T2 fuzzy set: Gaussian primary membership function and triangular secondary membership function

Definition 2. *Let V be a random variable defined on probability space (Ω, Σ, Pr). Assume there exists $\widetilde{V(\omega)} = \mu_{\widetilde{V(\omega)}} / \mu_{V(\omega)} / V(\omega)$ where $\mu_V(\omega)$ is the primary membership function and $\mu_{\widetilde{V(\omega)}}$ is the secondary membership function.*

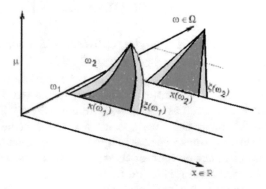

Fig. 2. T2 fuzzy random set

Ω is the probability. We call $\widetilde{V(\omega)}_{random} = (\widetilde{V(\omega)}, \omega)$ is a T2 fuzzy random variable. Figure 2 shows two interval T2 fuzzy random set.

According to the above definition, T2 fuzzy random variable has four levels of uncertainties, which are crisp value, primary membership grade, secondary membership grade and probability. It cannot be modelled or analyzed easily. Let us defined expected value and variance of T2 fuzzy random variable below step by step.

2.2   Expected Value and Variance of T2 Fuzzy Random Variable

Defuzzification

Definition 3 (Embedded Set). Let $\tilde{A}$ be a T2 fuzzy set in X. For discrete universes of discourse X and U, an embedded T2 fuzzy set $\tilde{A}_e$ of $\tilde{A}$ is defined as the following T2 fuzzy set

$$\tilde{A}_e = \Big\{(x_i, (u_i, \mu_{\tilde{A}}(x_i)(u_i)))$$
$$| \ \forall i \in \{1, ..., N\} : x_i \in X, u_i \in J_{x_i} \subseteq U \Big\}. \tag{3}$$

$\tilde{A}_e$ contains exactly one element from $J_{x_1}, J_{x_2}, ..., J_{x_N}$, namely $u_1, u_2, ..., u_N$, each with its associated secondary grade, namely $\mu_{\tilde{A}}(x_1)(u_1), \mu_{\tilde{A}}(x_2)(u_2), \mu_{\tilde{A}}(x_N)(u_N)$.

An embedded set is a special kind of T2 fuzzy set. It relates to the T2 fuzzy set in which it is embedded in this way: For every primary domain value, x, there is a unique secondary domain value, u, plus the associated secondary membership grade, $\mu_{\tilde{A}}(x)(u)$ that is determined by the primary and secondary domain values.

Mendel and John have shown that a T2 fuzzy set can be represented as the union of its T2 embedded sets. This powerful result is known as the T2 fuzzy set Representation Theorem or Wavy-Slice Representation Theorem. There is an example shown in Fig. 3.

Let $\tilde{A}_e^j$ denote the jth T2 embedded set for T2 fuzzy set $\tilde{A}$, i.e.,

$$\tilde{A}_e^j \equiv \Big\{(u_i^j, \mu_{\tilde{A}}(x_i)(u_i^j)), i = 1, ..., N \Big\}. \tag{4}$$

where $\{u_1^j, ..., u_N^j\} \in J_{x_i}$. Then $\tilde{A}$ could be represented as union of the T2 embedded sets itself, i.e., $\tilde{A} = \sum_{j=1}^n \tilde{A}_e^j$, where $n \equiv \prod_{j=1}^N M_i$.

Broadly speaking, defuzzification [15–18] or type-reduction works [19–21] as follows. The Mendel-John representation theorem formalises the notion that a T2 fuzzy set can be represented as a collection of embedded T2 fuzzy sets. Each of these embedded T2 fuzzy sets has a centroid that can be calculated by a number of ways (centroid, centre of set, height) [22–26]. Each of these centroid values provides a point in the domain of the type-reduced set as shown in Fig. 4. The membership grade of a point is found by taking a t-norm of all the secondary grades of the embedded set that produced that point. More formally:

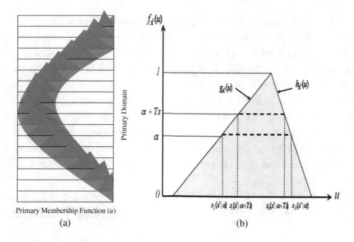

Fig. 3. 2.5D FOU representation of the generalised T2 fuzzy set $\tilde{A}$ with triangular secondary membership function for two different α-levels

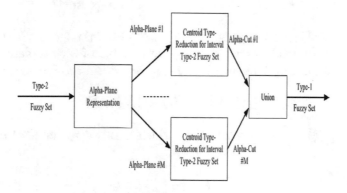

Fig. 4. Process of defuzzification

The step of T2 defuzzification is to get the Type-Reduced Set (T-R-S). Assuming that the primary domain X has been discretised, the T-R-S of a T2 fuzzy set may be defined through the application of Zadeh's Extension Principle. Alternatively the T-R-S may be defined via the Representation Theorem.

Definition 4. *The T-R-S associated with a T2 fuzzy set $\tilde{A}$ with primary domain X discretised into N points is*

$$
C_{\tilde{A}} = \left\{ \left(\frac{\sum_{i=1}^{N} x_i \cdot u_i}{\sum_{i=1}^{N} x_i}, \mu_{\tilde{A}}(x_1)(u_1) * \cdots * \mu_{\tilde{A}}(x_N)(u_N) \right) \right.
$$
$$
\left. | \, \forall i \in \{1, \cdots, N\} : x_i \in X, u_i \in J_{x_i} \subseteq U \right\}.
$$

(5)

The type reduction stage requires the application of a t-norm (*) which Zadah introduced from first to the secondary membership grades. Owing to the product t-norm would not produce very meaningful results for T2 fuzzy in the case of common secondary membership functions. So that it is better to not make use of it. The minimum t-norm is preferred by us.

In order to let this definition of the T-R-S to be meaningful, if the domain X be numeric in nature, the result would be more precisely. The T-R-S is a T1 fuzzy set in U and the computation about it in practical requires us involving the secondary domain U to be as same extend discretised as well. Many exhaustive type-reduction algorithm can be used to compute the T-R-S of a T2 fuzzy set, which we will discuss later.

Centroid of $\tilde{A}$, i.e., $C_{\tilde{A}}$, is in fact the union of the centroid of its entire Interval T2 embedded sets. Its membership function, i.e., $C_{\tilde{A}}(\xi)(\forall X \subseteq R)$, is defined as follows:

$$C_{\tilde{A}}(\xi) = \bigcup \left\{ \min f_x(u)\xi = \frac{\int_{-\infty}^{\infty} xu_{A_e}(x)dx}{\int_{-\infty}^{\infty} u_{A_e}(x)dx} \right\} \tag{6}$$

where $(x, u) \in A_e$. According to Liu [27], centroid of the Generalised T2 Fuzzy Set $\tilde{A}$ can be obtained through aggregating centroid of each α-plane, i.e., $C_{\tilde{A}}$. So the centroid of the Generalised T2 Fuzzy Set $\tilde{A}$ can be expressed as follows:

$$C_{\tilde{A}}(\xi) = \bigcup \alpha / C_{\tilde{A}_\alpha}(\xi)(\forall \xi \in X) \tag{7}$$

where $\alpha / C_{\tilde{A}_\alpha}(\xi)$ is the centroid of $\tilde{A}_\alpha$ which is raised to the level α. And we also get:

$$C_{\tilde{A}}(\xi) = \forall \tilde{A}_e(\alpha) \bigcup \left[\xi = \frac{\int_{-\infty}^{\infty} xu_{A_e(\alpha)}(x)dx}{\int_{-\infty}^{\infty} u_{A_e(\alpha)}(x)dx} \right] = [c_l(\tilde{A}_\alpha), c_r(\tilde{A}_\alpha)] \tag{8}$$

where $c_l(\tilde{A}_\alpha), c_r(\tilde{A}_\alpha)$ can be calculated via Karnik-Mendel algorithm [22]. It should be highlighted that for the sake of easily expression, i.e., $f_x(u)$ will be considered as the combination of two monotonically increasing and decreasing functions called $g_x(u)$ and $h_x(u)$:

$$f_x(u) = \begin{cases} g_x(u) & u \in [S_L(x \mid 0), S_L(x \mid 1)] \\ h_x(u) & u \in [S_R(x \mid 1), S_L(x \mid 0)] \\ l & u \in [S_L(x \mid 1), S_R(x \mid 1)] \\ 0 & Otherwise \end{cases} \tag{9}$$

According to equation above, suppose that the upper (lower) membership values of the Generalised T2 fuzzy set $\tilde{A}$ with the secondary membership value of α be defined as $S_R(x \mid \alpha), S_L(x \mid \alpha)$, respectively. Now, if the slope of the secondary membership function at a for both $g_x(u)$ and $h_x(u)$ are $g'_x(u)$ and

$h'_x(u)$, then connection equations for the upper and the lower membership values at the plane '$\alpha + Ts$' would be defined as follows:

$$S_R(x \mid \alpha + Ts) = S_R(x \mid \alpha) + \frac{Ts}{h'_x(u)} \tag{10}$$

$$S_L(x \mid \alpha + Ts) = S_L(x \mid \alpha) + \frac{Ts}{g'_x(u)} \tag{11}$$

An α-plane of a Generalised T2 fuzzy set $\tilde{A}$ is the union of the entire primary memberships of $\tilde{A}$ whose secondary grades are greater than or equal to $\alpha (0 \leq \alpha \leq 1)$. The α-plane of $\tilde{A}$ is denoted by $\tilde{A}_\alpha$:

$$\tilde{A}_\alpha = \int_{\forall x \in X} \int_{\forall u \in J_x} (x, u) \mid f_x(u) \geq \alpha \tag{12}$$

We can define an α-cut of the secondary membership function is represented by the $S_{\tilde{A}}(x \mid \alpha)$,

$$S_{\tilde{A}}(x \mid \alpha) = [S_L(x \mid \alpha) + S_R(x \mid \alpha)] \tag{13}$$

Hence the composition of all the α-cuts of the whole secondary membership functions is $\tilde{A}_\alpha$,

$$\tilde{A}_\alpha = \int_{\forall x \in X} S_{\tilde{A}}(x \mid \alpha)/x = \int_{\forall x \in X} \left(\int_{\forall u \in [S_L(x \mid \alpha) + S_R(x \mid \alpha)]} \right) x \tag{14}$$

Making use of our theory, the basic concept of the Footprint of Uncertainty (FOU) corresponds to the lowest α-plane $\tilde{A}_0$. According to the process of defuzzification, T2 fuzzy variable $\tilde{A}$ can be type-reduced into a T1 fuzzy variable say $A = [c_l(\tilde{A}), c_r(\tilde{A})]$.

Expected Value of T1 Fuzzy Variable Calculated by Creditability Theory. Given some universe Γ, let Pos be a possibility measure that is defined on the power set $P(\Gamma)$ of Γ. Let R be the set of real numbers. A function Y : $\Gamma \rightarrow R$ is said to be a fuzzy variable defined on Γ. The possibility distribution μ_Y of Y is defined by $\mu_Y(t) = PosY = t, t \in R$, which is the possibility of event $Y = t$. For fuzzy variable Y, with possibility distribution μ_Y, the possibility, necessity, and credibility of event $Y \leq r$ are given as follows:

$$Pos\{Y \leq r\} = sup_{t \leq r}\mu_Y(t)$$

$$Nec\{Y \leq r\} = 1 - sup_{t > r}\mu_Y(t)$$

$$Cr\{Y \leq r\} = \frac{1}{2}(1 + sup_{t \leq r}\mu_Y(t) - sup_{t > r}\mu_Y(t)). \tag{15}$$

where *sup* is the supremum of the function. From equation above, we note that the credibility measure [27] is an average of the possibility and the necessity measure, i.e., $Cr < \cdot > = (Pos < \cdot > + Nec < \cdot >)/2$, and it is a self-dual set function, i.e., $CrA = 1 - CrA^c$ for any A in $P(\Gamma)$. The motivation behind the introduction of the credibility measure is to develop a certain measure, which is a sound aggregate of the two extreme cases, such as the possibility (which expresses a level of overlap and is highly optimistic in this sense) and necessity (which articulates a degree of inclusion and is pessimistic in its nature). Based on credibility measure, the expected value [28] of a T1 fuzzy variable is presented as follows.

Definition 5. *Let Y be a T1 fuzzy variable. The expected value of Y is defined as*

$$E[Y] = \int_0^\infty Cr\{Y \geq r\}dr - \int_{-\infty}^0 Cr\{Y < r\}dr \tag{16}$$

E[Y] is the expected value calculated by creditability theory.

Expected Value and Variance of T2 Fuzzy Random Variable Defined by Creditability Theory. Wang and Watada has defined T1 fuzzy random variable [30]. We proceed their conceptions. Suppose $\tilde{X}$ is a T2 fuzzy variable and $X = [c_l(\tilde{X}), c_r(\tilde{X})]$ is the type-reduced T2 fuzzy variable of $\tilde{X}$ according to defuzzification process mentioned above. Assume there is a probability space ω, for each $\omega \in \Omega$, $\tilde{X}_R = (\tilde{X}, \Omega)$ is a T2 fuzzy random variable and $X_R = ([c_l(\tilde{X}), c_r(\tilde{X}], \omega)$ is the type-reduced T2 fuzzy random variable.

Definition 6. *Assume ξ is a T2 fuzzy random variable. The expected value of the T2 fuzzy random variable (or type-reduced T2 fuzzy random variable) is defined as follows.*

$$E[\xi] = \int_\Omega \left[\int_0^\infty Cr\{\xi(\omega) \geq r\}dr - \int_{-\infty}^0 Cr\{\xi(\omega) \leq r\}dr \right] Pr(d\omega) \tag{17}$$

Notice that the definition above may meet two requirements. (1) When the T2 fuzzy random variable ξ degenerate into a random variable and lose its fuzzy feature. Then for any $\omega \in \Omega$, $\xi(\omega)$ become a crisp random variable, the definition above will degenerate into equation as follows:

$$E[\xi] = \int_0^\infty Pr\{\xi \geq r\}dr - \int_{-\infty}^0 Pr\{\xi \leq r\}dr \tag{18}$$

where Pr is the measure of probability.
(2) When the T2 fuzzy random variable ξ degenerate into a fuzzy variable and lose its random feature. Then the definition above will degenerate into equation as follows:

$$E[\xi] = \int_0^\infty Cr\{\xi \geq r\}dr - \int_{-\infty}^0 Cr\{\xi \leq r\}dr \tag{19}$$

which is the form of Definition 5.

Then we attempt to define the **Variance** of $\tilde{X}_R$. Consider the triangular fuzzy random variable $\tilde{X}_R$. Let us calculate the variance of type-reduced T2 triangular fuzzy random variable X_R. From the distribution of random variable V, we assume that the fuzzy random variable X takes fuzzy variables $X(V_1) = (5,4,6)_T$ with probability 0.25, $X(V_2) = (7,6,8)_T$ with probability 0.4 and $X(V_3) = (8,7,9)_T$ with probability 0.4. From Definition 6, we have $E(X) = 6.9. Then, Var(X) = E[(X(V_1) - 6.9)^2] \cdot 0.25 + E[(X(V_2) - 6.9)^2] \cdot 0.35 + E[(X(V_3) - 6.9)^2] \cdot 0.4$. To obtain Var[X], we need to calculate $E[(X(V_1) - 6.9)^2]$, $E[(X(V_2) - 6.9)^2]$ and $E[(X(V_3) - 6.9)^2]$, where $X(V_1) - 6.9 = (-1.9, -2.9, -0.9)_T$, $X(V_2) - 6.9 = (0.1, -0.9, 1.1)_T$ and $X(V_3) - 6.9 = (1.1, 0.1, 2.1)_T$. Denoting $Y_1 = X(V_1) - 6.9 = (-1.9, -2.9, -0.9)_T$, first, we will calculate $\mu_{Y_1^2}$ and $Cr\{Y_1^2 \geq r\}$. Since

$$\mu_{Y_1^2} = Pos\{Y_1^2 = t\}$$
$$= max\{Pos\{Y_1 = \sqrt{t}\}, Pos\{Y_1 = -\sqrt{t}\}\}$$

where $t \geq 0$, we obtain

$$\mu_{Y_1^2}(t) = \begin{cases} (-0.1 + \sqrt{t}) & 0.1^2 \leq t \leq 1.1^2 \\ (2.1 - \sqrt{t}) & 1.1^2 \leq t \leq 2.1^2 \\ 0 & otherwise. \end{cases}$$

Furthermore, we calculate

$$Cr\{Y_1^2 \geq r\} = \begin{cases} (2 - \mu_{Y_1^2}(r))/2 & 0 \leq r \leq 2.4^2 \\ (\mu_{Y_1^2}(r))/2 & 2.4^2 \leq r \leq 6.4^2 \\ 0 & otherwise. \end{cases}$$

Therefore, from Definition 16, we found $E[(X(V_1) - 6.9)^2] = E[Y_1^2]$ as follows:

$$\begin{aligned} E[Y_1^2] &= \int_0^\infty Cr\{Y_1^2 \geq r\}dr \\ &= \int_{0.1^2}^{1.1^2} \frac{1}{2}\left(2 - (-0.1 + \sqrt{t})4\right)dr \\ &\quad + \int_{1.1^2}^{2.1^2} \frac{1}{2}\left((2.1 - \sqrt{t})\right)dr \\ &= 3.07 \end{aligned} \tag{20}$$

Similarly, we obtain $E[(X(V_2) - 6.9)^2] = E[Y_2^2] = 0.44$ and $E[(X(V_3) - 6.9)^2] = E[Y_3^2] = 2.19$. Thus, $Var(X) = 0.25 \cdot E(X(V_1) - 6.9)^2 + 0.35 \cdot E(X(V_2) - 6.9)^2 + 0.4 \cdot E(X(V_3) - 6.9)^2 = 0.25 \times 3.07 + 0.35 \times 0.44 + 0.4 \times 2.19 = 5.71$.

Through the definition of expected value and variance of $\tilde{X}_R$, the T2 fuzzy random variable can be transferred into the following form:

$$\tilde{X}_R = [E(X_R) - Var[X_R], E(X_R) + Var[X_R]] \tag{21}$$

As we can see, it is much easy to model and analyze.

3   Applying a T2 Fuzzy Random Support Vector Regression Scheme in the Analysis of Linkage Between Different Financial Markets

3.1   Modeling Background

Chinese financial industry has seen a prompt rise in the last decade. There are two main financial markets in China: the inter-bank market and the exchange markets including Shanghai Stock Exchange and Shenzhen Stock Exchange. Basically speaking, listing companies' stocks are traded in Exchange while bonds and money are mainly traded in inter-bank market. Figure 5 shows the regulators and quotation procedure for the inter-bank market. Considering that connections between financial markets become more and more obvious, there emerges study on linked movements between performance of assets' price traded on both markets recently. On the other hand, T2 fuzzy random sets become useful when deployed in a T2 fuzzy random logic system as depicted in Fig. 6. In the deductive reasoning process, T2 fuzzy random sets are usually used to model the to be concerned concepts. Here, we intend to select several index to express assets' price movements in three markets, transfer the price into T2 fuzzy random data and build a T2 fuzzy random support vector regression model as the reasoning logic to study inside connection between money market, bond market and stock market.

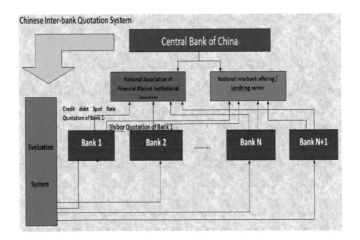

Fig. 5. Chinese inter-bank quotation system

3.2   Selection of Index for the Asset's Price

The SSE Composite Index is a stock market index of all stocks that are traded at the Shanghai Stock Exchange. It is an index reflecting price movement of Chinese stock market.

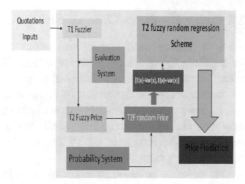

Fig. 6. T2 fuzzy random logic system

The Shanghai Interbank Offered Rate (or Shibor) is a daily reference rate based on the interest rates at which banks offer to lend unsecured funds to other banks in the Shanghai wholesale (or "interbank") money market. There are eight Shibor rates, with maturities ranging from overnight to a year. They are quoted by 20 banks. According to the practical experience, 3MShibor is the best to express the price performance for money market.

Chinese bond includes credit bond and treasury bond. Credit bond market is mainly composite by debt financing instruments of non-financial enterprises (CPs, MTNs, SMECN). 20 investment banks offered quotations for credit bonds' issuing rate by terms and ratings in inter-bank market according to their estimation of the future trend. National Association of Financial Market Institutional Investors (called Nafmii) collects all those quotations (called credit bond's price valuation or CBPV) and judge the quotation performance of each bank by some standards (called evaluation system) and give the evaluation to each quoter from time to time. The evaluation will restrict the behavior of quoters meanwhile adjust their weight to final result according to it. We use CBPV to express price movement of credit bond market in China.

According to the reasoning above, we select SSE Composite Index, 3 months' Shibor, 1 year AAA ranking credit bond's spot rate and 10 year AAA ranking credit bond's spot rate to represent price movements for different financial markets respectively. Those prices are described in the form of T2 fuzzy set naturally. Specifically speaking, for stock price, the pair of open/close prices construct the lower boundary of the FOU and the pair of highest/lowest prices of the day construct the higher boundary of the FOU of T2 fuzzy set. For those quotation prices, because the regulator exerts their effect on the final price, we should consider the evaluation system in our modeling process as well. In this paper, original offer and bid prices range of banks construct the primary membership grades, while evaluations of the regulator for each bank constructs secondary membership grades.

3.3   Collecting T2 Fuzzy Random Data of Stock Index

Exchange market is a floor market with buyers and sellers consummating transactions at continuous bidding. Transaction prices of one period can be described by a candlestick index. Candlestick used in capital market to model prices of financial assets could be viewed as a fuzzy set to some extend, for it contains open price, close price, highest price and lowest price. From where open price and close price could construct the lower boundary while highest and lowest price construct upper boundary of the footprint of T2 fuzzy set. To make it simple, we assume the T2 membership grade is interval.

We assume that the upper membership function of the footprint of stock index for one day follows gaussian distribution and its range on x-axis is the lowest price to highest price of the day. The lower membership function of the footprint is also following gaussian distribution and the range on x-axis is open price to close price. μ is the central of the range and boundary equals 5 times of δ. The assumption is made because highest and lowest price includes the extreme price of the day while open and close price is most important.

Table 1. 2015-3-28's quotation

Rank	Shibor 3M quoted price	1-year AAA quoted price	10-year AAA quoted price
Bank 1	2.83%	3.53%	4.72%
Bank 2	2.85%	3.90%	5.15%
Bank 3	2.86%	3.83%	5.24%
Bank 4	2.85%	3.82%	5.00%
Bank 5	2.91%	3.93%	5.11%
Bank 6	2.90%	3.92%	5.12%
Bank 7	2.93%	3.90%	5.00%
Bank 8	2.89%	4.00%	5.37%
Bank 9	2.93%	3.94%	4.89%
Bank 10	2.90%	4.00%	5.18%
Bank 11	2.90%	3.95%	5.11%
Bank 12	2.89%	3.98%	5.12%
Bank 13	2.92%	3.97%	5.15%
Bank 14	2.87%	4.00%	5.14%
Bank 15	2.86%	3.95%	4.91%
Bank 16	2.90%	3.97%	5.17%
Bank 17	2.92%	4.08%	5.09%
Bank 18	2.91%	4.00%	5.14%
Bank 19	2.95%	3.99%	5.13%
Bank 20	2.80%	4.05%	5.13%

Table 2. Evaluation system of quotation

Evaluation of bond spot rate quotation	Evaluation of Shibor quotation
Quotation promptness (B1)	Deviation of borrowing rate quotation (M1)
Quotation effectiveness (B2)	Deviation of lending rate quotation (M2)
Deviation of transaction rate to quoted rate (B3)	Measurement of quotation error (M3)
Measurement of quotation error (B4)	Stability of bilateral spread (M4)
Investors' assessment (B5)	Trading volume on quoted rate (M5)
Market supervisors' assessment (B6)	Quotation promptness (M6)
Issuers' assessment (B7)	Quotation effectiveness (M7)

Table 3. 2015-6-1's quotation

Rank	Shibor 3M quoted price	1-year AAA e quoted price	10-year AAA quoted price	Evaluation	According fuzziness
Bank 1	2.83%	3.53%	4.72%	Always	0.9
Bank 2	2.85%	3.90%	5.15%	Often	0.5
Bank 3	2.86%	3.83%	5.24%	Sometimes	0.3
Bank 4	2.85%	3.82%	5.00%	Frequently	0.7
Bank 5	2.91%	3.93%	5.11%	Seldom	0.1
Bank 6	2.90%	3.92%	5.12%	Often	0.5
Bank 7	2.93%	3.90%	5.00%	Always	0.9
Bank 8	2.89%	4.00%	5.37%	Frequently	0.7
Bank 9	2.93%	3.94%	4.89%	Often	0.5
Bank 10	2.90%	4.00%	5.18%	Sometimes	0.3
Bank 11	2.90%	3.95%	5.11%	Often	0.5
Bank 12	2.89%	3.98%	5.12%	Often	0.5
Bank 13	2.92%	3.97%	5.15%	Sometimes	0.3
Bank 14	2.87%	4.00%	5.14%	Frequently	0.7
Bank 15	2.86%	3.95%	4.91%	Frequently	0.7
Bank 16	2.90%	3.97%	5.17%	Always	0.9
Bank 17	2.92%	4.08%	5.09%	Always	0.9
Bank 18	2.91%	4.00%	5.14%	Alway	0.9
Bank 19	2.95%	3.99%	5.13%	Often	0.5
Bank 20	2.80%	4.05%	5.13%	Often	0.5

Table 4. Input-output data with confidence interval

Sample	Output	Inputs		
i	$I[e_Y, \sigma_Y]$	$I[e_{X_1}, \sigma_{X_1}]$	$\cdots$	$I[e_{X_K}, \sigma_{X_K}]$
1	$I[e_{Y_1}, \sigma_{Y_1}]$	$I[e_{X_{11}}, \sigma_{X_{11}}]$	$\cdots$	$I[e_{X_{1K}}, \sigma_{X_{1K}}]$
2	$I[e_{Y_2}, \sigma_{Y_2}]$	$I[e_{X_{21}}, \sigma_{X_{21}}]$	$\cdots$	$I[e_{X_{2K}}, \sigma_{X_{2K}}]$
$\vdots$	$\vdots$	$\vdots$	$\vdots$	$\vdots$
i	$I[e_{Y_i}, \sigma_{Y_i}]$	$I[e_{X_{i1}}, \sigma_{X_{i1}}]$	$\cdots$	$I[e_{X_{iK}}, \sigma_{X_{iK}}]$
$\vdots$	$\vdots$	$\vdots$	$\vdots$	$\vdots$
N	$I[e_{Y_N}, \sigma_{Y_N}]$	$I[e_{X_{N1}}, \sigma_{X_{N1}}]$	$\cdots$	$I[e_{X_{NK}}, \sigma_{X_{NK}}]$

Table 5. Sample of transformed data

Sample	Output	Inputs(shibor,1 AAA YTM,10 AAA YTM		
i	$I[e_Y, \sigma_Y]$	$I[e_{X_1}, \sigma_{X_1}]$	$I[e_{X_2}, \sigma_{X_2}]$	$I[e_{X_3}, \sigma_{X_3}]$
week1	$[2003.41, 2078.43]$	$[5.59\%, 4.14\%]$	$[6.42\%, 2.54\%]$	$[6.46\%, 4.42\%]$
week2	$[1985.66, 2025.01]$	$[5.67\%, 2.35\%]$	$[6.17\%, 2.49\%]$	$[6.42\%, 4.44\%]$
$\vdots$	$\vdots$	$\vdots$	$\vdots$	$\vdots$
$week_j$	$[1995.88, 2055.49]$	$[4.67\%, 2.16\%]$	$[5.14\%, 2.29\%]$	$[6.09\%, 2.47\%]$
$\vdots$	$\vdots$	$\vdots$	$\vdots$	$\vdots$
$week_N$	$[3022.39, 3212.56]$	$[4.45\%, 2.05\%]$	$[4.57\%, 3.65\%]$	$[5.34\%, 2.34\%]$

Example 1. Take 2015-3-28 performance of stock into consideration. The open price is 4533.1 and the close price is 4828.74. The lower function's mapping on x-axis thus is [4533.1, 4828.74]. The lowest price is 4615.23 and the highest price is 4829.5. The upper function's mapping on x-axis is [4615.23, 4829.5]. According to our assumption, the lower function and upper function is as follows:

$$\underline{\mu}_{\widetilde{A}} = \frac{1}{21.9 * \sqrt{2 * \pi}} * \exp\left\{-\frac{(x - 4672.4)^2}{2 * 21.9^2}\right\} \tag{22}$$

$$\overline{\mu}_{\widetilde{A}} = \frac{1}{27.9 * \sqrt{2 * \pi}} * \exp\left\{-\frac{(x - 4672.4)^2}{2 * 27.9^2}\right\} \tag{23}$$

On the other hand, many study assume the probability of stock price also follows logarithmic normal distribution. And we collected 10 years stock price in China and formed the probability function. The fuzzy random price of a day is like $\widetilde{P}_{stock} = [4633.1, 4828.74], 0.5$.

3.4   Collecting T2 Fuzzy Random Data of Shibor and Bond Rate

As for Shibor and CBPV, the price for one day is offered by 20 banks. The price range is a T1 fuzzy set naturally. According to the evaluation of regulator, we can decide the T2 fuzziness of these quotation. Thus the prices are modeled into T2 fuzzy variables. Shibor, 1 year AAA ranking credit bond's spot rate and 10 year AAA ranking credit bond's spot rate are quoted by 20 banks as shown in Table 1. They are collected by association of national market of institutional investors.

Regulators judge the quotation performance of each bank by evaluation system and give the evaluation to each quoter from time to time. As we can see from Table 2, Nafmii evaluates the performance of each bank on credit debt spot rate quotation from respects such as quotation promptness, quotation effectiveness, deviation of transaction rate to quoted rate, measurement of quotation error, investors' assessment, market supervisors' assessment and issuers' assessment. China foreign exchange trade center (called CFETS) evaluate the performance of each bank on Shibor quotation from respects deviation of borrowing quotation, deviation of lending quotation, measurement of quotation error, stability of bilateral spread, trading volume on Quoted rate, quotation promptness and quotation effectiveness.

After considering respects mentioned above, the regulator's expertise group will grade each option then give a final assessment for quality of quotation such as always, often, sometimes, frequently, seldom to measure the performance of each bank. The evaluation will restrict the behavior of quoters. We give the fuzziness of the evaluation as 'always, 0.9', 'often, 0.5', 'sometimes, 0.3', 'frequently, 0.7' and 'seldom, 0.1'. And this is the T2 fuzziness of the price of quotation at our assumption.

Example 2. As shown in Table 3, we use the original quoted price as the real number. The evaluation of regulator as primary membership grade. A quoted price not only appear once and its appearance times as its T2 fuzziness. Thus the T2 fuzzy variable for $Shibor_{3M}$ quoted price shall be $0.1/0.7/4.43 + (0.1/0.1 + 0.1/0.7 + 0.1/0.9)/4.6 + 0.1/0.7/4.64 + (0.1/0.5 + 0.1/0.7)/4.65 + (0.1/0.9 + 0.2/0.3)/4.66 + (0.6/0.5 + 0.1/0.3)/4.67 + 0.2/0.9/4.7 + 0.1/0.9/4.71$ and the 1-year AAA rating credit quoted price shall be $0.1/0.9/5.03 + 0.1/0.1/5.19 + 0.1/0.7/5.2 + (0.1/0.5 + 0.1/0.3 + 0.1/0.9 + 0.1/0.7)/5.22 + (0.2/0.5 + 0.1/0.3)/5.23 + (0.1/0.5 + 0.1/0.7)/5.24 + (0.2/0.5 + 0.1/0.3 + 0.1/0.7 + 0.2/0.9)/5.25 + 0.1/0.9/5.26 + 0.1/0.5/5.28$, the 10-year AAA rating credit quoted price shall be $0.1/0.7/5.38 + 0.1/0.5/5.64 + 0.1/0.3/5.65 + (0.1/0.3 + 0.1/0.5)/5.68 + (0.1/0.5 + 0.1/0.9)/5.69 + (0.2/0.5 + 0.3/0.7 + 0.2/0.9)/5.7 + (0.1/0.3 + 0.2/0.5 + 0.1/0.9)/5.71 + 0.1/0.1/5.75 + 0.1/0.9/6.07$.

Still we assume the probability of price distributes at Gaussian way. We take ten years' data to produce and calculate the probability of the current price.

By that, for example, the T2 fuzzy random quotation of one day's $Shibor_{3months}$ could be $[0.1/0.7/4.43 + (0.1/0.1 + 0.1/0.7 + 0.1/0.9)/4.6 + 0.1/0.7/4.64 + (0.1/0.5 + 0.1/0.7)/4.65 + (0.1/0.9 + 0.2/0.3)/4.66 + (0.6/0.5 + 0.1/0.3)/4.67 + 0.2/0.9/4.7 + 0.1/0.9/4.71, 0.3]$

3.5   Defuzzification of the Original T2 Fuzzy Data

Transformation of Continuous Transaction Price of Stock Index.
According to Kavnik-Mendel algorithm (KM algorithm), give an interval T2
fuzzy set A which is defined on an universal set $x \subseteq \Re$, with membership function $\mu_{\tilde{A}}(X)$, its centroid (if it exists) $c(\tilde{A})$ is a closed interval $[c_l, c_r]$ in the
classical sense of mathematics, i.e.,

$$c(\tilde{A}) = [c_l, c_r] \tag{24}$$

where c_l and c_r are respectively the minimum and maximum of all centroid of
the embedded type-1 fuzzy sets in the footprint of uncertainty (FOU) of $\mu_{\tilde{A}}$.
Mendel et al. define continuous version for c_l and c_r of an IT2 fuzzy set $\mu_{\tilde{A}}$:

$$c_l = Min \quad centroid(A_e(l)) \tag{25}$$

$$c_r = Max \quad centroid(A_e(r)) \tag{26}$$

where

$$
\begin{aligned}
centroid(A_e(r)) &= \frac{\int_{-\infty}^{-\infty} x\mu_{A_e(r)}(x)dx}{\int_{-\infty}^{-\infty} \mu_{A_e(r)}(x)dx} \\
&= \frac{\int_{-\infty}^{r} x\underline{\mu}_{\tilde{A}}(x)dx + \int_{r}^{+\infty} x\overline{\mu}_{\tilde{A}}(x)dx}{\int_{-\infty}^{r} \underline{\mu}_{\tilde{A}}(x)dx + \int_{r}^{+\infty} \overline{\mu}_{\tilde{A}}(x)dx}
\end{aligned}
\tag{27}
$$

$$
\begin{aligned}
centroid(A_e(l)) &= \frac{\int_{-\infty}^{-\infty} x\mu_{A_e(l)}(x)dx}{\int_{-\infty}^{-\infty} \mu_{A_e(l)}(x)dx} \\
&= \frac{\int_{-\infty}^{l} x\overline{\mu}_{\tilde{A}}(x)dx + \int_{l}^{+\infty} x\underline{\mu}_{\tilde{A}}(x)dx}{\int_{-\infty}^{l} \overline{\mu}_{\tilde{A}}(x)dx + \int_{l}^{+\infty} \underline{\mu}_{\tilde{A}}(x)dx}
\end{aligned}
\tag{28}
$$

and where $A_e(l)$ and $A_e(r)$ denote embedded type-1 fuzzy sets for which:

$$[\mu_{A_e(l)}] = \begin{cases} (\overline{\mu}_{\tilde{A}}(x)), & x \leq l \\ (\underline{\mu}_{\tilde{A}}(x)), & x \geq l \end{cases} \tag{29}$$

$$\mu_{A_e(l)} = \begin{cases} (\overline{\mu}_{\tilde{A}}(x)), & x \leq l \\ (\underline{\mu}_{\tilde{A}}(x)), & x \geq l \end{cases} \tag{30}$$

According to Mendel, l and r are switch points. Usually we calculate

$$c_0 = \frac{\int_{-\infty}^{-\infty} x(\overline{\mu}_{\widetilde{A}}(x) + \underline{\mu}_{\widetilde{A}}(x))dx}{(\overline{\mu}_{\widetilde{A}}(x) + \underline{\mu}_{\widetilde{A}}(x))dx} \tag{31}$$

We assume c_0 as l or r at first and use irritation algorithm to substitute new $centroid(A_e(l))$ or $centroid(A_e(r))$ until convergence has occurred.

Example 3. The lower function and upper function for 2015-3-28 performance of stock is as follows:

$$\underline{\mu}_{\widetilde{A}} = \frac{1}{21.9 * \sqrt{2 * \pi}} * \exp\left\{-\frac{(x - 4672.4)^2}{2 * 21.9^2}\right\} \tag{32}$$

$$\overline{\mu}_{\widetilde{A}} = \frac{1}{27.9 * \sqrt{2 * \pi}} * \exp\left\{-\frac{(x - 4672.4)^2}{2 * 27.9^2}\right\} \tag{33}$$

and the type-reduced set of it using algorithms above is $[4667.61, 4677.19]$

Thus the T2 fuzzy random price of stock can be re-expressed as $\widetilde{P}_{stock} = \{[4667.61, 4677.19], 0.5\}$, where $[4667.61, 4677.19]$ is the centroid of original T2 fuzzy price and 0.5 is the randomness of the price. According to our Definition 6 of the expectation and variance to type-reduced fuzzy random set, we take five days as a set, we can transfer the data into confidence interval one in the form in Table 4, where $I[e_x, \sigma_x]$ means $I[e_x - \sigma_Y, e_x + \sigma_Y].e_x$ is the expected value and σ_Y is the variance calculated by creditability theory.

Transformation of Discrete Quoted Price of Shibor and CBPV. Take 3–28's quotation into consideration, the T2 fuzzy variable for $Shibor_3M$ quoted price is $\widetilde{A}_1 = 0.1/0.7/4.43 + (0.1/0.1 + 0.1/0.7 + 0.1/0.9)/4.6 + 0.1/0.7/4.64 + (0.1/0.5 + 0.1/0.7)/4.65 + (0.1/0.9 + 0.2/0.3)/4.66 + (0.6/0.5 + 0.1/0.3)/4.67 + 0.2/0.9/4.7 + 0.1/0.9/4.71$ and the 1-year AAA rating credit quoted price is $\widetilde{A}_2 = 0.1/0.9/5.03 + 0.1/0.1/5.19 + 0.1/0.7/5.2 + (0.1/0.5 + 0.1/0.3 + 0.1/0.9 + 0.1/0.7)/5.22 + (0.2/0.5 + 0.1/0.3)/5.23 + (0.1/0.5 + 0.1/0.7)/5.24 + (0.2/0.5 + 0.1/0.3 + 0.1/0.7 + 0.2/0.9)/5.25 + 0.1/0.9/5.26 + 0.1/0.5/5.28$; the 10-year AAA rating credit quoted price is $\widetilde{A}_3 = 0.1/0.7/5.38 + 0.1/0.5/5.64 + 0.1/0.3/5.65 + (0.1/0.3 + 0.1/0.5)/5.68 + (0.1/0.5 + 0.1/0.9)/5.69 + (0.2/0.5 + 0.3/0.7 + 0.2/0.9)/5.7 + (0.1/0.3 + 0.2/0.5 + 0.1/0.9)/5.71 + 0.1/0.1/5.75 + 0.1/0.9/6.07$. They are all discretised T2 fuzzy set. As mentioned in the former passage, there are several methods to defuzzify the discretised T2 fuzzy set. We compare exhaustive type-reduction (centroid type-reduction) [17]. The sampling method [18]. The α-plane representation [26] and vertical slice centroid type-reduction (called VSCTR) [29] here.

After the statistical analysis of our the test results for accuracy. We conclude as follows:

1. For sample sizes up to and including 5000 the exhaustive type-reduction, sampling and VSCTR methods do not produce significantly different defuzzified values. For sample sizes of 10,000 and above there is evidence to support the exhaustive type-reduction and elite sampling method being more accurate than VSCTR.
2. The α-planes method is the least accurate of the techniques assessed apart from when compared with the sampling method at low sample sizes. In these instances, there is no evidence to support the sampling method being more accurate than the α-planes method.
3. For high sample sizes, the elite sampling method is the most accurate of the techniques compared. However, with such high sample sizes, it may be argued that elite sampling is barely distinguishable from the exhaustive method, especially when the set to be defuzzified has a low number of embedded sets.
4. For samples of moderate size, exhaustive type-reduction, VSCTR, sampling, and elite sampling are of equivalent accuracy and efficiency.

We have taken 2014 whole year's data for training. The number of samples are under 1000. Given the benefit and cost analysis above and to show the accurate feature of our new model, we will use exhaustive method (called centroid type-reduction or CTR) to defuzzify the T2 fuzzy set in our case. CTR is obtained by means of a centroid calculation on the T2 FS, which leads to a type-1 set called type-reduced set, i.e., the centroid of a T2 fuzzy set is a T1 fuzzy set.

Using the method, for $\widetilde{A_1}$ mentioned above which has 8 vertical slices defined at $\{4.43, 4.6, 4.64, 4.65, 4.66, 4.67, 4.7, 4.71\}$. According to the exhaustive method, there are 24 type-1 embedded set sets. According to CTR, the minimum point of type-reduced set is 4.36 and the maximum is 4.64. Here, given that the quotation is from top 20 investment banks, we consider there are many other banks' quotation will make the price more continuous. Based on that, after we get the minimum and maximum of the centroid, we assume the range of it is continuous, which is [4.36, 4.64]. The calculation process should be same to $\widetilde{A_2}$ and $\widetilde{A_3}$.

Thus the T2 fuzzy random quotation can be re-expressed as [[4.36, 4.64], 0.4], where [4.36, 4.64] is the centroid and 0.4 is the randomness. According to our Definition 6 of the expectation value and variance to type-reduced fuzzy random set, we take five days as a set, we can transfer the data into confidence interval ones in the form in Table 5.

3.6 Building a T2 Fuzzy Support Vector Regression for the Data with Confidence Interval

Support vector machine (SVM) was developed by Vapnik and his co-workers and is gaining popularity due to many attractive features. While the traditional NN implements the empirical risk minimization (ERM) principle, SVM implements the structural risk minimization (SRM) principle which seeks to minimize an upper bound on the Vapnik-Chervonenkis (VC) dimension (generalization error). SVM was originally developed for classification problems (SVC) and then extended to support vector regression (SVR). Based on SRM principle, SVR achieves a balance between the training error and generalization error, leading to better forecasting performance. Traditional SVR is only able to deal with crisp inputs and outputs. Given our type-reduced price inputs and outputs are fuzzy random interval data, a regression modeling algorithm for interval data called type-2 fuzzy support vector regression is developed based on traditional SVR. The T2-FSVR method is generalized from real number domain to interval number domain, in order to simultaneously keep both fitting errors and prediction errors small and has less computation complexity, better generalization performance.

First we consider the linear situation of traditional SVR. In AA-SV regression, the goal is to find a function f(x) that has at most $\mathring{A}$ deviation from the actually obtained targets y_i, $y_i \in R$, for all the training data x_i, and at the same time is as flat as possible. In other words, we do not care about errors as long as they are less than $\mathring{A}$, but will not accept any deviation larger than this. For pedagogical reasons, we begin by describing the case of linear functions f, taking the form

$$y' f(X) = < W, X > +B, w \in R^d, B \in R \tag{34}$$

where $<, >$ denotes the dot product. Given all the definitions of the coefficients, inputs and outputs, the equation above can be reformed as follows:

$$y' f(X) = [e(B) - \delta(B)+ < W, E(x) > + < |W|, \Delta >, \\ (B) + \delta(B)+ < W, E(x) > + < |W|, \Delta >] \tag{35}$$

Flatness in the case of the equation above means that one seeks a small W. One way to ensure this is to minimize the norm, i.e. $|W| = < W, W >$. We can write this problem as a convex optimization problem:

$$\begin{aligned} & \text{minimize} & & \frac{1}{2} \|w\|^2 \\ & \text{subject to :} \\ & & & y_i - \langle w, x \rangle - b \leq \varepsilon, \\ & & & \langle w, x \rangle + b - y_i \leq \varepsilon, \\ & & & i = 1, \cdots, l \end{aligned} \tag{36}$$

The tacit assumption in the equation above was that such a function f actually exists that approximates all pairs (x_i, y_i) with ε, or in other words, that the convex optimization problem is feasible. Sometimes, however, this may not be the case, or we also may want to allow some errors. Analogously to the "soft margin" loss function which was adapted to SV machines by Cortes and Vapnik. One can introduce slack variables ξ_i, ξ_i^* to cope with otherwise infeasible constraints of the above optimization problem. Hence we arrive at the formulation stated as follows:

$$\text{minimize} \quad \frac{1}{2}\|w\|^2 + C\sum_{i=1}^{l}(\xi_i + \xi_i^*)$$

subject to :

$$y_i - \langle w, x \rangle - b \le \varepsilon + \xi_i,$$
$$\langle w, x \rangle + b - y_i \le \varepsilon + \xi_i^*,$$
$$\xi_i, \xi_i^* \ge 0 \qquad i = 1, \cdots, l$$

(37)

The constant $C \ge 0$ determines the trade-off between the flatness of f and the amount up to which deviations larger than ε are tolerated. This corresponds to dealing with a so called ε-insensitive loss function $\|\xi\|_\varepsilon$ described by

$$\|\xi\|_\varepsilon := \begin{cases} 0, & \|\xi\| \ge \varepsilon \\ |\xi| - \varepsilon, \text{ otherwise} \end{cases}$$

(38)

Then we attempt to develop T2-FSVR. Assume there are n data samples composed by interval fuzzy inputs and outputs (X_i, Y_i), $i = 1, \cdots, n$, where X_i is a d dimensional vector, $X_i = [x_{i1}, x_{i2}, \cdots, x_{id}]^T$, x_{id} is a fuzzy interval inputs, $x_{id} \in I(R)$. Its centroid is its expected value $e(x_{id})$ and its radium is its variance $\delta(x_{id})$, x_{id} can be expressed as $[e(x_{id}) - \delta(x_{id}), e(x_{id}) + \delta(x_{id})]$ or $[e(x_{id}), \delta(x_{id})]$. And $X_i = [e(x_{i1}), \delta(x_i)], [e(x_{i2}), \delta(x_{i2})], \cdots, [e(x_{id}), \delta(x_{id})]$. Where $j = 1, \cdots, d$. We use $E(x_i) - [e(x_{i1}), e(x_{i2}), \cdots, e(x_{id})]$ and $\Delta(x_i) = [\delta(x_{i1}), \delta(x_{i2}), \cdots, \delta(x_{id})]$. The same is to Y_i which is a one dimension interval output. $y_{id} \in I(R)$.

According to our definition of interval inputs and outputs, we can reconstruct the linear relationship of SVR:

$$y' f(X) = W^T X + B, w \in R^d, B \in I(R)$$

(39)

where $W \in R$ and is a d dimensional vector; $B = [e(B) - \delta(B), e(B) + \delta(B)] \in I(R)$ is a interval coefficient; $X \in I(R^d)$ is a d dimensional interval inputs vector. Assume $y' = f(X), \left| y_{iL} - y'_{iL} \right| \le \varepsilon, \left| y_{iR} - y'_{iR} \right| \le \varepsilon$. The Eq. (37) can be rewritten as:

$$\text{minimize} \quad \frac{1}{2}\|w\|^2 + C\sum_{i=1}^{l}(\xi_i + \xi_i^*)$$

subject to :

$$y'_{iL} - y_{iL} \le \varepsilon + \xi_{1i}, y_{iL} - y'_{iL} \le \varepsilon + \xi_{1i}^*,$$
$$y'_{iR} - y_{iR} \le \varepsilon + \xi_{2i}, y_{iR} - y'_{iR} \le \varepsilon + \xi_{2i}^*,$$
$$y'_{iL} = e(B) + \langle w, e(x_i) \rangle - (\delta(B)$$
$$+ \langle |w|, \delta(x_i) \rangle),$$
$$y'_{iR} = e(B) + \langle w, e(x_i) \rangle + (\delta(B)$$
$$+ \langle |w|, \delta(x_i) \rangle),$$
$$i = 1, \cdots, l$$

(40)

where $\xi_i, \xi_i^* \geq 0, k = 1, 2$ and $C = \frac{1}{\tau}$. The coefficient τ is used to adjust the complexity of the regression model. The solution of the problem above will make sure that the model will has the least error of fitting for given data and has the least generalization error for the whole sample space. Till now, we have developed a T2 fuzzy support regression from the traditional SVR. Since T2 FSVR is able to deal with interval fuzzy data. The solution to it is a little bit different than SVR, which we will discuss as follows.

4 Solution to T2-FSVR

The key idea of the solution to T2-SVR is to construct a Lagrange function from the objective function and the corresponding constraints, by introducing a dual set of variables. It can be shown that this function has a saddle point with respect to the primal and dual variables at the solution. We proceed as follows:

$$\text{minimize} \quad \frac{1}{2} \|w\|^2 + C \sum_{i=1}^{l} (\xi_i + \xi_i^*) \tag{41}$$

subject to :

$$y_i - \langle w, x \rangle - b \leq \varepsilon + \xi_i,$$
$$\langle w, x \rangle + b - y_i \leq \varepsilon + \xi_i^*,$$
$$\xi_i, \xi_i^* \geq 0 \quad i = 1, \cdots, l$$

To solve this optimization problem we construct the Lagrangian

$$
\begin{aligned}
L : \\
&= \frac{1}{2} \|w\|^2 + C \sum_{i=1}^{l} (\xi_i + \xi_i^*) \\
&\quad - \sum_{i=1}^{l} (\eta_i \xi_i + \eta_i^* \xi_i^* \\
&\quad - \sum_{i=1}^{l} \alpha_i (\varepsilon + \xi_i - y_i + \rangle w, x_i \rangle + b) \\
&\quad - \sum_{i=1}^{l} \alpha_i^* (\varepsilon + \xi_i^* - y_i + \rangle w, x_i \rangle - b)
\end{aligned}
\tag{42}
$$

and find the saddle point of $L(\mathbf{W}, b, \xi, \alpha, \beta)$. The parameters must satisfy the following conditions:

$$\frac{\partial L}{\partial w} = w - \sum_{i=1}^{l} (\alpha_i - \alpha_i^*) x_i = \mathbf{0}$$

$$\frac{\partial L}{\partial b} = - \sum_{i=1}^{l} (\alpha_i^* - \alpha_i) = 0 \tag{43}$$

$$\frac{\partial L}{\partial \xi_i} = C - \alpha_i - \eta_i = 0$$

$$\frac{\partial L}{\partial \xi_i^*} = C - \alpha_i^* - \eta_i^* = 0$$

By applying these conditions to the Lagrangian, the optimal hyperplane problem can be transformed into

$$
\begin{aligned}
\text{maximize} \quad & -\frac{1}{2}\sum_{i,j=1}^{l}(\alpha_i - \alpha_i^*)(\alpha_j - \alpha_j^*)\langle x_i, x_j\rangle \\
& -\varepsilon\sum_{i=1}^{l}(\alpha_i + \alpha_i^*) + \sum_{i=1}^{l} y_i(\alpha_i - \alpha_i^*) \\
\text{subject to} \quad & \sum_{i=1}^{l}(\alpha_i - \alpha_i^*) = 0 \\
& 0 \le \alpha_i, \alpha_i^* \le C \\
& i = 1, \cdots, l
\end{aligned}
\tag{44}
$$

The Kuhn-Tucker conditions are defined as

$$
\begin{aligned}
\overline{\alpha}_i(y_i(\overline{\mathbf{W}} \cdot \mathbf{z}_i + \overline{b}) - 1 + \overline{\xi}_i) &= 0 \\
i &= 1, \cdots, l \\
(s_i C - \overline{\alpha}_i)\overline{\xi}_i &= 0, \\
i &= 1, \cdots, l
\end{aligned}
\tag{45}
$$

For our problem, according to the aim function and constraints we reconstruct the Lagrange function as well, and make its partial derivative to $e(B); \delta(B); W; \xi_i; \xi_i^*$ equals 0, and we have transformed the optimization problem as follows:

$$
\begin{aligned}
\text{maximize} \quad & -\frac{1}{2}\delta^T H\delta + B^T\delta \\
\text{subject to} \quad & \begin{cases} [-e^T, -e^T]\begin{bmatrix} \delta_1 & \delta_2 \\ \delta_1^* & \delta_2^* \end{bmatrix} \\ \delta_k^*, \delta_k \in [0, C], k = 1, 2 \end{cases}
\end{aligned}
\tag{46}
$$

where $\delta^T = [\delta_1^T, \delta_1^{*T}, \delta_2^T, \delta_2^{*T}]; \delta_k^T = [\delta_{k1}, \cdots, \delta_{kn}]^T; \delta_k^{*T} = [\delta_{k1}^*, \cdots, \delta_{kn}^*]^T, k = 1, 2$ which are the Lagrange multiple vectors. $e = [1, 1, \cdots, 1]_{1 \times n}$.

Suppose $y_L = [y_{1L}, y_{2L}, \cdots, y_{nL}]; y_R = [y_{1R}, y_{2R}, \cdots, y_{nR}];$

$B^T = [y_L^T - \varepsilon e^T, -y_L^T - \varepsilon e^T, -y_R^T - \varepsilon e^T, y_R^T - \varepsilon e^T];$

$$
\begin{bmatrix}
H_1 & -H_1 & -H_3 & H_3 \\
-H_1 & H_1 & H_3 & -H_3 \\
-H_3 & H_3 & H_2 & -H_2 \\
H_3 & -H_3 & -H_2 & H_2
\end{bmatrix}
$$

$H_1 = [< e(x_i) - sgn(w) \odot \delta(x_i), e(x_j) - sgn(w) \odot \delta(x_j)]_{i,j=1,\cdots,n}$

$H_2 = [< e(x_i) + sgn(w) \odot \delta(x_i), e(x_j) + sgn(w) \odot \delta(x_j)]_{i,j=1,\cdots,n}$

$H_2 = [< e(x_i) - sgn(w) \odot \delta(x_i), e(x_j) + sgn(w) \odot \delta(x_j)]_{i,j=1,\cdots,n}$

$$
\begin{aligned}
w = & \sum_{i=1}^{n}(\delta_{1i} - \delta_{1i}^*)(e(x_i) - sgn(w) \odot \delta(x_i) \\
& + \sum_{i=1}^{n}(\delta_{2i} - \delta_{2i}^*)(e(x_i) + sgn(w) \odot \delta(x_i)
\end{aligned}
\tag{47}
$$

$\odot$ means "multiple term by term". $sgn(w)$ is the sign symbol of w. It can be calculated in advance. And we use KKT condition to calculate $B = [B_L, B_R]$

$$\begin{cases} \delta_{1i}(\varepsilon + \xi_{1i} - \langle |w|, e(x_i)\rangle - B_L \\ \qquad + \langle |w|, \delta(x_i)\rangle + y_{iL}) = 0 \\ \delta_{1i}^*(\varepsilon + \xi_{1i} + \langle \|w\|, e(x_i)\rangle + B_L \\ \qquad - \langle \|w\|, \delta(x_i)\rangle - y_{iL}) = 0 \\ (C - \delta_{1i})\xi_{1i} = 0, (C - \delta_{1i}^*)\xi_{1i}^* = 0 \end{cases} \qquad (48)$$

$$B_L = \begin{cases} y_L - \langle w, e(x_i)\rangle - \langle |w|, \delta(x_i)\rangle \\ \qquad + \varepsilon \delta_{1i} \in (0, C), \delta_{1i}^* = 0 \\ y_L - \langle w, e(x_i)\rangle - \langle |w|, \delta(x_i)\rangle \\ \qquad - \varepsilon \delta_{1i}^* \in (0, C), \delta_{1i} = 0 \\ y_L - \langle w, e(x_i)\rangle - \langle |w|, \delta(x_i)\rangle \delta_{1i} \\ \qquad \in (0, C), \delta_{1i}^* \in (0, C) \end{cases} \qquad (49)$$

$$B_R = \begin{cases} y_R - \langle w, e(x_i)\rangle - \langle |w|, \delta(x_i)\rangle \\ \qquad + \varepsilon \delta_{2i} \in (0, C), \delta_{2i}^* = 0 \\ y_R - \langle w, e(x_i)\rangle - \langle |w|, \delta(x_i)\rangle \\ \qquad - \varepsilon \delta_{2i}^* \in (0, C), \delta_{2i} = 0 \\ y_R - \langle w, e(x_i)\rangle - \langle |w|, \delta(x_i)\rangle \delta_{2i} \\ \qquad \in (0, C), \delta_{2i}^* \in (0, C) \end{cases} \qquad (50)$$

The equation above is then the solution of T2-FSVR.

4.1   Comparison with Other Methods

After we build the T2-FSVR to the training data, we find out that the money surface is positively relative to the stock price as well as for the 10 year's credit bond's rate. However, the 1 year's bond's rate is negatively relative to the stock price in most time. We try to explain as follows when money surface is loosen, investors will improve their risk appetite and put much money in high return asset as stock; According to macro economy theory, long term bond's rate indicates the situation of economy. Rise of it means economy is seeing a recover, which stimulates the price of equity assets. Moreover, short-term debt is to some extent an averse indicator to price of equity because when people are attracted by short-term debt, there will be a signal for recession in short-term economy cycle.

Moreover, in order to show the merits of the new model, we model compared it's performance with T2 fuzzy expected value regression model and Confidence-interval-based fuzzy random regression model (CI-FRRM). Confidence-interval-based fuzzy random regression model (CI-FRRM) was developed by Watada et al. [30], which only models T1 fuzziness and randomness of variables without T2 fuzzy uncertainty into consideration. T2 fuzzy expected value regression model was developed by Wei and Watada [31], which use creditability theory to model the T2 fuzzy variable and not consider the randomness.

Before our comparisons, we make some assumptions as follows: (1) we measure the accuracy by overlap between the model's output and the original data, if the overlap is above 60%, then we say it is accurate. (2) as mentioned before, we assume that the probability distribution of a market price follows gaussian distribution and we collect ten years' data to formulate the probability distribution. (3) the T1 fuzzy set of stock price is constructed by pair of open/close prices, not with highest and lowest prices. The T1 fuzzy set of quoted prices of money and bond market are with original bids and asks without regulator's evaluation into consideration, which is for CI-FRRM. (4) The T2 fuzzy set gives a glimpse on highest and lowest price of stock price and evaluation of quotations, which contains more information and is more reasonable. (5) We choose 2014 one year's prices as training samples. We collect each day's fuzzy random price, transform and calculate one week's price range as a unit to train models.

We use matlab platform to implement T2-FSVR, T2-FER and CI-FRRM. Considering parameters of FSVR should be adjusted to reach the best state, we use the cross validation method LOOV to test the model, fix the parameters and evaluate the quality of the T2-FSVR. LOOV method leave one data for test for the prediction accuracy of the model and all the rest samples for training. The model will be built for n times if we have n samples theoretically. Thus we can find the most proper parameters for T2 FSVR. We trained three models to discover the underlying logic of prices between different markets and the result is shown in Table 6. As we can see, T2-FSVR is able to cope with inaccurate fuzzy interval data, which has better generalization performance given that it takes both randomness and T2 fuzzy uncertainties into consideration.

Table 6. Comparisons of performance between T2-FSVR, T2-FER and CI-FRRM

	Accuracy
T2-FSVR	98%
CI-FRRM	81%
T2-FER	70%

5   Concluding Remarks

In this paper, we build a T2 fuzzy support vector regression model to study the relationship between different financial markets. We find out that the money surface is positively relative to the stock price as well as for the 10 year's credit bond's rate. However, the 1 year's bond's rate is negatively relative to the stock price in most time. The innovation of this paper stands on several stakes as follows: (1) we defined the expected value for T2 fuzzy random variable based on creditability theory. (2) we discuss the different strategies for defuzzification and conclude their features (3) we formulated a 2 fuzzy support vector regression model capable of dealing with T2 fuzzy inputs and outputs (4) We have used the new model to study the linkage between financial markets and get good results. This paper generalized our previous work [31,32]. Our further work will try to explore the traditional SVR to a non-linear T2 fuzzy SVR using kernels which projects the variable to higher dimensions to get more accurate result.

References

1. Zadeh, L.A.: Fuzzy sets. Inf. Control **8**(3), 338–353 (1965)
2. Wei, Y., Watada, J.: Building a type-2 fuzzy qualitative regression model. JACII **16**(4) (2012)
3. Wei, Y., Watada, J.: Building a qualitative regression model. In: Proceedings of KES-IDT 2012, 22–25 May (2012)
4. Wei, Y., Watada, J., Witold, P.: A fuzzy support vector machine with qualitative regression preset. In: Proceedings of IEEE-ICGEC 2012, 25–29 August 2012
5. Mizumoto, M., Tanaka, K.: Some properties of fuzzy sets of type-2. Inf. Control **31**(4), 312–340 (1976)
6. Dubois, D., Prade, H.: Operations in a fuzzy-valued logic. Inf. Control **43**(2), 224–240 (1979)
7. John, R.I., Czarnecki, C.: A type 2 adaptive fuzzy inference system. In: Proceedings of IEEE Conference on Systems, Man, Cybernetics, vol. 2, pp. 2068–2073 (1998)
8. Liang, Q., Mendel, J.M.: Interval type-2 fuzzy logic systems: theory and design. IEEE Trans. Fuzzy Syst. **8**(5), 535–549 (2000)
9. Liang, Q., Mendel, J.M.: MPEG VBR video traffic modeling and classification using fuzzy techniques. IEEE Trans. Fuzzy Syst. **9**(1), 183–193 (2001)
10. Karnik, N.N., Mendel, J.M.: Applications of type-2 fuzzy logic systems to forecasting of timeseries. Inf. Sci. **120**(1–4), 89–111 (1999)
11. John, R.I., Czarnecki, C.: An adaptive type-2 fuzzy system for learning linguistic membership grades. In: Proceedings of 8th International Conference on Fuzzy Systems, vol. 3, pp. 1552–1556 (1999)
12. John, R.I., Innocent, P.R., Barnes, M.R.: Type-2 fuzzy sets and neuro-fuzzy clustering or radio-graphic tibia images. In: Proceedings of 6th International Conference on Fuzzy Systems, Barcelona, Spain, pp. 1375–1380 (1997)
13. Vapnik, V.N.: The Nature of Statistical Learning Theory. Springer, New York (1995)
14. Vladimir, N.V.: Statistical Learning Theory. Wiley, New York (1998)
15. Mendel, J.M., John, R.I.B.: Type-2 fuzzy sets made simple. IEEE Trans. Fuzzy Syst. **10**(2), 117–127 (2002)
16. Torshizi, A.D., Zarandi, M.H.F.: Hierarchical collapsing method for direct defuzzification of general type-2 fuzzy sets. Inf. Sci. **277**, 842–861 (2014)
17. Greenfield, S., Chiclana, F., John, I.: Type-reduction of the discretised interval type-2 fuzzy set. In: Proceedings of FUZZ-IEEE, Jeju Island, Korea, pp. 738–743 (2009)
18. Greenfield, S., Chiclana, F., John, R.I., Coupland, S.: The sampling method of defuzzification for type-2 fuzzy sets: experimental evaluation. Inf. Sci. **189**, 77–92 (2012)
19. Liu, F.: An efficient centroid type-reduction strategy for general type-2 fuzzy logic system. Inf. Sci. **178**(9), 2224–2236 (2008)

20. Wu, D., Nie, M.: Comparison and practical implementation of type-reduction algorithms for type-2 fuzzy sets and systems. In: Proceedings of FUZZ-IEEE, Taiwan, pp. 2131–2138 (2011)
21. Greenfield, S., Chiclana, F.: Defuzzification of the discretised generalised type-2 fuzzy set: experimental evaluation. Inf. Sci. **244**, 1–25 (2013)
22. Karnik, N.N., Mendel, J.M.: Centroid of a type-2 fuzzy set. Inf. Sci. **132**(1–4), 195–220 (2001)
23. Salazar, O., Serrano, H., Soriano, J.: Centroid of an interval type-2 fuzzy set: continuous vs discrete. Ingenieria **16**(2), 67–78 (2011)
24. Salazar, O., Serrano, H., Soriano, J.: Centriod of an interval type-2 fuzzy set. Appl. Math. Sci. **6**(122), 6081–6086 (2012)
25. Liu, F.: An efficient centroid type reduction strategy for general type-2 fuzzy logic system. Inf. Sci. **178**(9), 2224–2236 (2008)
26. Zhai, D., Mendel, J.M.: Computing the centroid of a general type-2 fuzzy set by means of the centroid-flow algorithm. IEEE Trans. Fuzzy Syst. **19**(3), 401–422 (2011)
27. Liu, B.: Theory and Practice of Uncertain Programming. Physica-Verlag, Heideberg (2002)
28. Liu, B., Liu, Y.K.: Expected value of fuzzy variable and fuzzy expected value models. IEEE Trans. Fuzzy Syst. **10**(4), 445–450 (2002)
29. Lucas, L.A., Centeno, T.M., Delgado, M.R.: General type-2 fuzzy infernce systems: analysis, design and computational aspects. In: Proceedings of FUZZ-IEEE 2007, pp. 1743–1747 (2007)
30. Watada, J., Wang, S., Pedrycz, W.: Building confidence-interval-based fuzzy random regression model. IEEE Trans. Fuzzy Syst. **11**(6), 1273–1283 (2009)
31. Wei, Y., Watada, J.: Building a type-2 fuzzy regression model based on credibility theory and its application on arbitrage pricing theory. In: IEEE World Congress on Computational Intelligence, Beijing, 6–11 July 2013
32. Wei, Y., Watada, J., Pedrycz, W.: Design of a qualitative classification model through fuzzy support vector machine with type-2 fuzzy expected regression classifier preset. IEEJ Trans. Electr. Electron. Eng. **11**(3), 348–356 (2016)

Modeling and Simulation Techniques

A New Randomized Procedure to Solve the Location Routing Problem

Carlos L. Quintero-Araujo[1,2]([envelope]), Angel A. Juan[1],
Juan P. Caballero-Villalobos[3], Jairo R. Montoya-Torres[4], and Javier Faulin[5]

[1] IN3 - Department of Computer Science, Universitat Oberta de Catalunya,
Av. Carl Friedrich Gauss 5, 08860 Castelldefels, Spain
cquinteroa@uoc.edu
[2] International School of Economics and Administrative Sciences,
Universidad de La Sabana, Autopista Norte Km 7, Chía, Colombia
[3] Industrial Engineering Department, Pontificia Universidad Javeriana,
Carrera 7 No. 40–62, Bogotá, Colombia
[4] Facultad de Ingenieria, Universidad de La Sabana,
Km 7 autopista norte de Bogotá, Chía, Colombia
[5] Statistics and Operations Research Department, Institute of Smart Cities,
Campus Arrosadia, 31006 Pamplona, Spain

Abstract. The Location Routing Problem (LRP) is one of the most important challenging problems in supply chain design since it includes all decision levels in operations management. Due to its complexity, heuristics approaches seem to be the right choice to solve it. In this paper we introduce a simple but powerful approach based on biased randomization techniques to tackle the capacitated version of the LRP. Preliminary tests show that near-optimal or near-BKS can be found in a very short time.

Keywords: Location routing problem · Heuristics
Biased randomization

1 Introduction

The LRP comprises the Facility Location Problem (FLP) and the Vehicle Routing Problem (VRP), which are both known to be NP-hard [1]. In fact, the classical approaches used to tackle the LRP were based on solving a FLP as a first step and then solving the associated VRP. Nowadays, due to the development of computers and optimization techniques, there is an increasing interest in solving the LRP in an integrated way. The LRP has a wide range of applications including food and drink distribution, waste collection, bill deliveries, among others [1]. The benefits obtained by including vehicle routing decisions while locating depots were quantified for the first time by [2]. These authors showed that the classical strategy consisting in solving a location problem and a routing problem separately does not necessarily guarantee optimal solutions.

© Springer International Publishing AG, part of Springer Nature 2018
A. M. Gil-Lafuente et al. (Eds.): FIM 2015, AISC 730, pp. 247–254, 2018.
https://doi.org/10.1007/978-3-319-75792-6_18

There are several variants for the LRP depending on the characteristics of the depots (capacitated or not), vehicles (capacitated or not, homogeneous or heterogeneous fleet), costs (symmetric or asymmetric) and uncertainty or not of other parameters. In this paper, we study the Capacitated version of the LRP (CLRP) i.e. depots and vehicles have both capacity constrains that must be respected when assigning customers to depots and routes.

The remaining of this paper is organized as follows: Sect. 2 gives a description of the CLRP; in Sect. 3 we introduce the concept of biased randomization; Sect. 4 presents the basis of our approach; in Sect. 5 a description of the computational experiments as well as some preliminary results are presented; finally, Sect. 6 provides some conclusions and draws some opportunities for future research.

2   Problem Statement

The Capacitated Location Routing Problem (CLRP) can be defined on a complete, weighted and undirected graph with a homogeneous fleet of vehicles with limited capacities, using the following parameters:

- F_k opening cost of depot k
- W_k capacity of depot k
- d_j demand of customer j
- S number of available vehicles (fleet size)
- Q capacity of each vehicle
- V_f fixed cost per vehicle used
- c_{ij} traveling cost (distance) from i to j

It is assumed that vehicles are shared by all depots (i.e. no depot has a specific fleet) and that c_{ij} satisfy the triangle inequality. The fleet size S is assumed to be a decision variable in our case. Customer demands are deterministic and known in advance. The capacity of each depot and its opening costs are known. Depots might have equal or different capacities between them. Each customer must be serviced from the depot to which it has been allocated by a single vehicle.

A solution for the CLRP consists in determining which depots must be opened, the customer allocation to open depots and the design of vehicle routes for serving customers from their corresponding depot. The following constraints must be satisfied: (i) the total demand of customers assigned to one depot must not exceed its capacity; (ii) each route begins and ends at the same depot; (iii) each vehicle performs at most one trip; (iv) each customer is served by one single vehicle (split deliveries are not allowed); and (v) the total demand of customers visited by one vehicle fits its capacity.

Figure 1 provides an example of a complete CLRP solution where facilities are represented by squares and customers by diamonds. The red squares correspond to closed facilities while the green, light blue and purple squares are the open ones. For each open depot a set of routes starting and finishing at the corresponding depot location is designed to serve all customer demands.

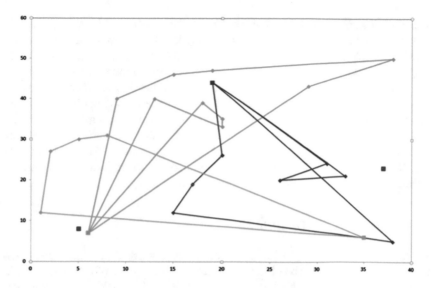

Fig. 1. Graphical representation of a CLRP solution

The objective function is defined as the minimization of the total cost that includes three terms: the cost of opening depots, the costs of vehicles used and the routing cost (usually assumed to be linear to distances).

Very few works in the literature have been dedicated to the study of the CLRP. The first work that include capacity constrains in both facilities and vehicles was [3] when it introduced a new set of instances considering the capacity limit for each depot in order to test their Simulated Annealing algorithm. In [4] a lower bound based on a cutting plane method was proposed for the CLRP and introduced the Barreto set of instances. Exact methods for the CLRP include column generation [5,6] and branch-and-cut [7,8]. Different metaheuristics have been developed in recent years, e.g. the GRASP algorithms proposed in [9,10], the memetic algorithm with population management by [11], the Lagrangean Relaxation Granular Tabu Search by [12]. Hybrid methods combining meta-heuristics with integer linear programming have beaten all previous methods in terms of gap but their principal disadvantage is the great amount of time consumed, especially for medium and large size instances [13].

3   Biased Randomization

A deterministic heuristic can be easily transformed into a probabilistic one by randomizing some of its steps. When a randomized heuristic is executed several times, different results (solutions) can be obtained. The Greedy Randomized Adaptive Search Procedure (GRASP) is one of the well-known meta-heuristics mostly employed in combinatorial optimization that includes randomization [14]. GRASP considers a restricted list of candidates solutions consisting of a subset of most promising solutions ranked according to their potentiality to improve the current solution. Then randomization is employed using a uniform distribution in the order the elements of that restricted list are selected. In this case, most of the logic or common sense behind the original heuristic is still respected.

Biased randomization goes beyond. Indeed, asymmetric probability distributions, such as triangular or geometric distributions, are preferred since the most promising movements -according to a certain criterion- have a higher probability to be chosen [15] and, as a consequence, all movements are potentially eligible at each step of the procedure without losing the logic behind the deterministic heuristic. This idea has been applied in combinatorial optimization problems in their either deterministic or stochastic versions [16–18].

4   Our Solution Approach

Our aim is to develop a simple (no substantial implementation efforts needed), fast and efficient algorithm to solve the CLRP. A key aspect of our approach is to involve all decisions at each phase of the algorithm, i.e. tackle depot location bringing into consideration allocation costs as well as routing costs (or at least one approximation of them), then, when deciding customer allocation to depots we also consider routing costs. And finally, when deciding routing cost, these decisions have already included location and assignment costs.

Our approach consists of two phases, the first one corresponds to the: generation of feasible solutions and the latter is the improvement of these solutions. To generate feasible solutions we decompose the original problem into a set of successive and less complex problems by using simple and fast procedures. In this case, the CLRP can be easily transformed into the Multiple Depot Vehicle Routing Problem (MDVRP) once we have decided the open depots. Once the customers are allocated to the depots, the MDVRP is transformed into a m-CVRP. Each CVRP can be solved by a fast and high quality heuristic, e.g. Clarke & Wright Savings Heuristic. The second phase is the improvement of routing and customer allocation of the top solutions found in the first phase. This is done by means of a perturbation process in which we re-allocate a certain percentage of customers from one open depot to another open depot and then we apply an intensive routing heuristic. In Fig. 2, the flow diagram of our algorithm is presented.

5   Test and Results

The proposed algorithm was coded as a Java application. Some preliminary tests were carried out using six benchmark instances from the Prodhons set and were run in a 2.4 GHz Core i5 with 8 Gb RAM, executed on MAC OsX Yosemite. We have grouped the instances by size into small instances (2o customers and 5 potential facilities), medium instances (50 or 100 customers and 5 or 10 potential facilities) and large instances (200 customers and 10 potential facilities). Two instances were selected from each group.

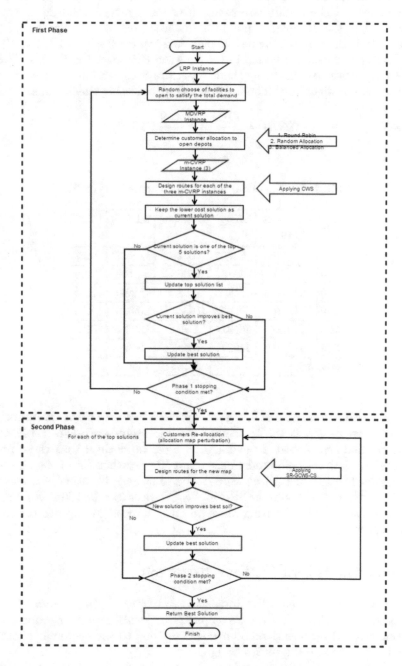

Fig. 2. Flow diagram of our solving approach

Each instance was run 5 times using a single random seed. In Table 1 a brief summary of the results is presented. Column c represents de number of customers, f is the number of potential facilities, W is the capacity of each vehicle, BKS is the value of the best known solution for the instance, GAP(%) is the difference between our best solution and the BKS over the BKS, t(BKS) is the time in seconds consumed by the top performer algorithm and t is the time consumed by our algorithm. The GAP is calculated using the formula:

$$\%GAP = \frac{OBS - BKS}{BKS} * 100\% \tag{1}$$

Table 1. Results of our approach compared to the BKS

Instance specs.			BKS	GAP(%)	t(BKS)	t
c	f	W				
20	5	70	54793	0.41	1.7	3.7
20	5	140	39104	0.00	2.6	4.9
50	5	140	61830	0.28	22.9	13.2
100	10	70	287892	10.24	2621.8	27.4
200	10	140	362817	1 49	1326 4	180.8
200	10	70	449291	0.48	4943	192.1
Average				2.15	1486.4	70.4

The algorithm is capable of find good solutions in a very short computational time, the average gap is 2.15% while the time required to execute is about one minute and ten seconds in average. It must be noticed that these results were obtained with a single random seed and 50 perturbations of the allocation maps obtained in the first phase were executed at each iteration of the second phase. For each perturbation, the SR-GCWS-CS was executed 100 times. Further analysis on the parameter setting must be done in order to improve the results already achieved.

6   Conclusions and Further Research

We have proposed a simple (no intensive parameter setting is required) and fast procedure to efficiently solve the capacitated location routing problem. Our approach is based on biased randomization applied to the customer allocation to depots as well as to the routes design.

The incorporation of stochastic parameters like travel times, demands or capacities as well as the heterogeneous fleet or different objective functions that include environmental aspects could be considered as future research lines.

Acknowledgements. This work has been partially supported by the Spanish Ministry of Economy and Competitiveness (Grant TRA2013-48180-C3-3-P) and FEDER. The work of Jairo R. Montoya-Torres was carried out within the framework of the ANNONA research project funded by the French National Research Agency (ANR) and Universidad de La Sabana (grant number EICEA-86-2014). This paper is also part of the CYTED-SmartLogistics@IB Research Network supported by the Ibero-American Programme of Science, Technology and Innovation (CYTED 515RT0489).

References

1. Nagy, G., Salhi, S.: Location-routing: issues, models and methods. Eur. J. Oper. Res. **177**(2), 649–672 (2007). https://doi.org/10.1016/j.ejor.2006.04.004. ISSN 0377-2217
2. Salhi, S., Rand, G.K.: The effect of ignoring routes when locating depots. Eur. J. Oper. Res. **39**(2), 150–156 (1989)
3. Wu, T.-H., Low, C., Bai, J.-W.: Heuristic solutions to multi-depot location-routing problems. Comput. Oper. Res. **29**(10), 1393–1415 (2002)
4. Barreto, S.: Analysis and modelling of location-routing problems. Ph.D. thesis, University of Aveiro, Portugal (2004)
5. Baldacci, R., Mingozzi, A., Calvo, R.W.: An exact method for the capacitated location-routing problem. Oper. Res. **59**(5), 1284–1296 (2011)
6. Contardo, C., Cordeau, J.-F., Gendron, B.: An exact algorithm based on cut-and-column generation for the capacitated location-routing problem. INFORMS J. Comput. **26**(1), 88–102 (2014)
7. Contardo, C., Cordeau, J.-F., Gendron, B.: A branch-and-cut-and-price algorithm for the capacitated location-routing problem. CIRRELT (2011)
8. Belenguer, J.-M., Benavent, E., Prins, C., Prodhon, C., Calvo, R.W.: A branch-and-cut method for the capacitated location-routing problem. Comput. Oper. Res. **38**(6), 931–941 (2011)
9. Prins, C., Prodhon, C., Calvo, R.W.: Solving the capacitated location-routing problem by a GRASP complemented by a learning process and a path relinking. 4OR **4**(3), 221–238 (2006)
10. Duhamel, C., Lacomme, P., Prins, C., Prodhon, C.: A GRASPxELS approach for the capacitated location-routing problem. Comput. Oper. Res. **37**(11), 1912–1923 (2010)
11. Prins, C., Prodhon, C., Calvo, R.W.: A Memetic Algorithm with Population Management (MA—PM) for the capacitated location-routing problem, pp. 183–194. Springer, Heidelberg (2006)
12. Prins, C., Prodhon, C., Ruiz, A., Soriano, P., Calvo, R.W.: Solving the capacitated location-routing problem by a cooperative lagrangean relaxation-granular tabu search heuristic. Transp. Sci. **41**(4), 470–483 (2007)
13. Contardo, C., Cordeau, J.-F., Gendron, B.: A GRASP + ILP-based metaheuristic for the capacitated location-routing problem. J. Heuristics **20**(1), 1–38 (2014)
14. Feo, Thomas A., Resende, M.G.C.: Greedy randomized adaptive search procedures. J. Global Optim. **6**(2), 109–133 (1995)
15. Juan, A.A., Faulin, J., Grasman, S.E., Rabe, M., Figueira, G.: A review of simheuristics: extending metaheuristics to deal with stochastic combinatorial optimization problems. Oper. Res. Perspect. **2**, 62–72 (2015)

16. Juan, A.A., Faulin, J., Ruiz, R., Barrios, B., Gilibert, M., Vilajosana, X.: Using oriented random search to provide a set of alternative solutions to the capacitated vehicle routing problem, pp. 331–345. Springer, Boston (2009)

17. Montoya Torres, J.R., Franco, E.G., Alfonso-Lizarazo, E.H., Halabi, A.X.: Using randomization to solve the deterministic single and multiple vehicle routing problem with service time constraints. In: Proceedings of the 2009 Winter Simulation Conference (WSC), pp. 2989–2994, December 2009. https://doi.org/10.1109/WSC.2009.5429227

18. Juan, A.A., Pascual, I., Guimarans, D., Barrios, B.: Combining biased randomization with iterated local search for solving the multidepot vehicle routing problem. Int. Trans. Oper. Res. **22**(4), 647–667 (2015). https://doi.org/10.1111/itor.12101. ISSN 1475–3995

A Biased-Randomized Heuristic for the Waste Collection Problem in Smart Cities

Aljoscha Gruler[1(✉)], Angel A. Juan[1], Carlos Contreras-Bolton[2], and Gustavo Gatica[2,3]

[1] Universitat Oberta de Catalunya,
Rambla Poblenou 156, 08018 Barcelona, Spain
{agruler, ajuanp}@uoc.edu
[2] Universidad de Santiago de Chile,
Av. Libertador Bernardo O'Higgins 3363, Santiago, Chile
{carlos.contrerasb, gustavo.gatica}@usach.cl
[3] Universidad Andres Bello, Sazié, Región Metropolitana, Santiago, Chile

Abstract. This paper describes an efficient heuristic to solve the Waste Collection Problem (WCP), which is formulated as a special instance of the well-known Vehicle Routing Problem (VRP). Our approach makes use of a biased-randomized version of a savings-based heuristic. The proposed procedure is tested against a set of benchmark instances, obtaining competitive results.

Keywords: Vehicle routing problem · Biased-Randomized heuristics
Waste collection problem · Smart Cities

1 Introduction

About half of the world's population is currently living in urban areas, a percentage that is expected to increase even more over the next decades, making cities ever more complex systems. Many municipalities struggle to cope with increasing population numbers, as existing urban infrastructure and organization do not meet the necessary standards concerning economic, social and environmental changes in urban areas [6, 20]. In order to make better use of existing resources, many cities across the world are developing the so called Smart City initiatives, helping municipalities to make better use of their resources. A typical Smart City application field is waste management, as inefficient garbage collection can lead to several negative externalities, including traffic congestions, unnecessary pollution, and high noise levels. These externalities can lead to a decreased urban standard of living. Furthermore, waste management–and especially the process of garbage collection– is a high municipal cost factor, calling for an optimization of waste collection activities [12, 13, 17, 19].

The Waste Collection Problem (WCP) can be formulated as a special instance of the well-known Vehicle Routing Problem (VRP), which is described in detail in [5, 18]. Overall, the aim of the VRP is to optimize the costs of serving the demands of a range of customers with a capacitated vehicle fleet, whereby: (*i*) every node is served

© Springer International Publishing AG, part of Springer Nature 2018
A. M. Gil-Lafuente et al. (Eds.): FIM 2015, AISC 730, pp. 255–263, 2018.
https://doi.org/10.1007/978-3-319-75792-6_19

by only one vehicle; *(ii)* all routes start and end at central depot; *(iii)* no customer is served twice; and *(iv)* the vehicle capacity cannot be exceeded.

This paper is structured as follows: Sect. 2 describes the problem setting in more detail. Section 3 discusses the main ideas behind the biased-randomization technique, which is included in our heuristic-based approach to solve the WCP presented in Sect. 4. Numerical results of the presented heuristic in comparison to a benchmark problem set are presented Sects. 5 before a short conclusion is given in Sect. 6.

2   The WCP in the Context of VRP Research

The WCP can be seen as a reversed VRP. The aim is not to deliver, but rather to collect the demand (i.e. waste) from a number of customers (i.e. waste containers). The problem setting can be outlined on a graph $G = (N, E)$. Thereby, the set $N = \{0, 1, ..., k, k + 1, k + m\}$ defines the vertices representing the central depot (node 0), the k waste containers $\{1, 2, ..., k\}$ with a waste level w_i, and m landfills (waste collection sites) $\{k + 1, ..., k + m\}$. The waste levels are picked up by a fleet of collection vehicles stationed at the central depot, who have a (homogeneous) maximum capacity C. An additional constraint in the WCP is that collection vehicles need to leave and return to the central depot empty. As illustrated in Fig. 1, a vehicle collects waste from a number of customers, and then has to empty itself at a waste disposal site before returning to the central depot. Note that multiple trips to one (or different) landfills are possible. A vehicle can thus collect waste until its capacity is reached, empty itself during the route, and then continue the same route with the new available capacity, as long as no further route constraints (e.g. time windows of the central depot) are violated.

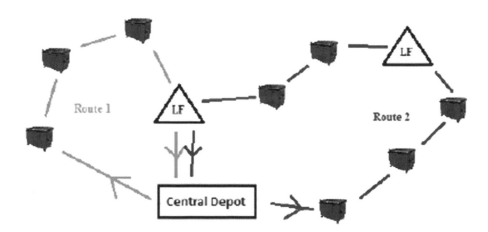

Fig. 1.   A simple example of a WCP with two Routes.

As many other VRPs, the WCP has been the subject of an increasing interest by the scientific community. An extensive literature review on this problem can be found in [1]. The authors of this literature review count a total of 78 publications on the WCP–as a particular case of VRP– since 1971, They also appreciate a clear increase in publications on the topic after 1995, which is in line with the development of the first Smart City initiatives and the rising importance of the WCP as a real-life VRP application.

Work on the WCP can be characterized into many differences, for example concerning the type of waste to be collected (e.g. organic-, non-organic- or hazardous waste) or the business line of the waste collection service provider (e.g. residential-, commercial- or industrial waste) [21]. This paper will focus on residential waste collection with a homogeneous fleet of collection vehicles, multiple landfills, time windows and driver lunch breaks. The same problem setting has been solved in [11], whose authors proposed a Simulated Annealing extended by a version of Solomon's insertion algorithm. They test their approach on a set of publicly available benchmark instances ranging from 102 to 2,100 nodes. These instances are adapted from a real-life case study. Their results are improved in [2] using two metaheuristic algorithms based on Tabu Search and Large Neighborhood Search, respectively. Further improved results are provided in [3], whose authors use an Adaptive Large Neighborhood Search to solve the benchmark instances.

3 Biased Randomization of Heuristics

Combinatorial optimization problems (COPs) like the VRP are often NP-hard in nature, meaning that the solution space grows exponentially as the problem size increases. In order to solve real-life COPs in a reasonable amount of time, heuristic and meta-heuristic methods are usually applied in practice. On the one hand, greedy heuristics build solutions by choosing the most promising movement at each iteration of the solution-construction process, which often leads the algorithm into a local minima. On the other hand, randomized algorithms are characterized by their choice of pseudo-random numbers during their search-space exploration, which is the case in most modern metaheuristics and stochastic local-search processes. Since the solution construction is not deterministic, a randomized algorithm will find different outputs every time it is run. Randomized algorithms include, for example, Genetic Algorithms [15], Simulated Annealing [14], or Variable Neighborhood Search [8].

One of the best-known randomized search algorithms is the so called Greedy Randomized Adaptive Search Procedure (GRASP), which uses an iterative process and uniformly distributed random numbers to choose the next movement in the solution construction out of a restricted candidate list [16]. While most randomized heuristics choose a (non-biased) uniform distribution to assign an equal probability to each possible candidate, biased randomization techniques apply skewed probability distributions, allowing for a more oriented search of the solution space. The heuristic approach to solve the WCP presented in this paper is based on a biased randomization approach as suggested in [10], presented as a Multi-start biased Randomization of classical Heuristics with Adaptive local search (MIRHA). The basic functioning of MIRHA is listed in Fig. 2.

The MIRHA general process is similar to the GRASP algorithm, including two major changes. Firstly, all feasible movements are potentially eligible, unlike the restricted candidate list of the GRASP metaheuristic. Secondly, the probability

```
procedure MIRHA(inputs)
01heuristic ←DefineHeuristic (inputs) % different for each COP
02inititalSolution ←GenerateSolution (heuristic, inputs)
03bestSolution ←ApplyAdaptiveLocalSearch(initialSolution)%for each COP
04 probaDist ←DefineProbabilityDistribution(COP)%different for each COP
05 while stopping criteria is not satisfied do
06        solution ←GenerateRandomSolution (heuristic, probaDist, inputs)
07        solution ←ApplyAdaptiveLocalSearch (solution)
08        bestSolution ←UpdateBestSolution (solution)
09        endwhile
10        return bestSolution
end procedure
```

Fig. 2. Pseudo-code of the MIRHA main procedure.

distribution is not uniformly distributed, but skewed, meaning that more promising candidates have a higher probability of being selected in the solution construction than others. Both differences are illustrated in Fig. 3.

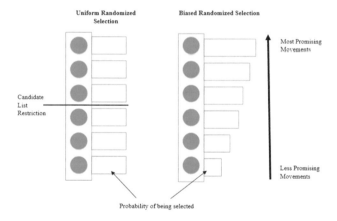

Fig. 3. Comparison between the uniform- and biased randomized selection.

4   A Saving-Based Heuristic

In order to solve the WCP, we proposed a randomized version of the savings-based heuristic proposed in [4]. This well-known heuristic for the VRP starts off by constructing a dummy VRP solution in which each customer is served by a single vehicle. The algorithm then constructs new edges by connecting customers in order to serve them on a single route. In the following, a savings list is constructed in which a savings-value is assigned to each edge, defined by the difference in solution quality by serving each customer with a single vehicle and using a specific edge in the solution construction. An example can be seen in Fig. 4. Instead of serving customers i and k on a single route for each customer, both clients are connected through edge ik. This edge is associated to a savings value $s_{ik} = s_{ki} = c_{0i} + c_{k0} - c_{ki}$. The CWS is an iterative greedy heuristic, as it then goes on to construct a feasible VRP solution by adding edges to construct feasible routes until all customer demand is served, while choosing the available edge with the highest potential savings value at each step.

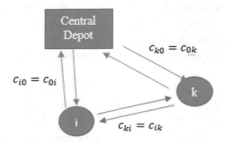

Fig. 4. Edge construction in the CWS.

Our heuristic approach is based on this framework, which we adapt to the WCP and improve through randomization and an algorithm learning mechanism, as proposed in [9]. Also, as proposed in [7], we apply the Untidy-Cities savings based heuristic (USH) to construct an initial dummy solution for the WCP. As a collection vehicle has to visit a landfill before returning to the central depot, the costs of visiting a customer from-, and returning to the customer will be asymmetric (as can be seen in Fig. 5). One possibility would be to double the savings list and including two different savings values for each edge. Rather, we use the average of both savings, i.e. the expected savings associated with an edge.

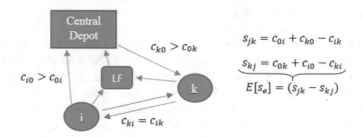

Fig. 5. The USH procedure.

In order to avoid the greedy behavior of the original CWS, a VRP solution is constructed in our approach by iteratively choosing an edge to include in the solution construction according to a biased randomization function. That is, instead of choosing the edge with the highest possible savings potential at each step, every edge on the savings list is potentially eligible, according to a certain probability. The probabilities are thereby biased towards the edges with the highest savings potential. In the implementation of our algorithm, a Geometric distribution with parameter α is employed. Figure 6 shows the distribution of probabilities when choosing $\alpha = 0.25$. Note that a higher value of parameter α will lead the algorithm to be more biased towards edges with high savings potentials and vice versa. Furthermore, we improve found solutions through the use of a simple cache memory technique. Using a hash table, the cache memory stores information about the costs of travelling between a subset of nodes. If a new solution is created, it is checked if identical subsets of customers are served after each other, just in a different order. The hash table is updated if a more efficient way to travel between these nodes is found, or the previously constructed solution is improved using cache memory information.

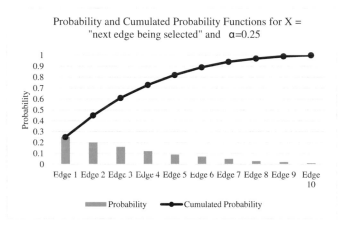

Fig. 6. Probability distribution during the solution construction.

5   Numerical Experiments

To test our approach, we use a set of WCP instances provided in [11], which provide 10 problem settings ranging from 102 to 2,100 nodes. These instances consider multiple landfills, a homogeneous vehicle fleet, time windows for customers and the central depot, as well as driver lunch breaks. Note that the results in [2, 3] are directly comparable to ours as they use the same objective function, while in [1] a multi-objective approach is used instead–this multi-objective approach includes workload balancing as well as the minimization of the number of vehicles used.

The algorithm is implemented in Java. All tests were run on a personal computer (Intel Core i5, 2.3 GHz on Ubuntu 14.04) with a stopping criterion of 300 s for each problem instance, as we consider 5 min to be a reasonable waiting time for a routing solution in practice, whereas the algorithm reaches its best solution after less than 3 min on average. The parameter α defining the probability of the edge with the highest savings potential being selected at each algorithm iteration is set to 0.35. To run the tests, 10 different seeds for the random number generator which is used for the selection of edges during the solution construction are used. Both the average over the solutions using the different seeds and the best solution found are presented. In Table 1

Table 1. Numerical results and comparison to benchmarks.

Instance	Kim et al. (1.1)	Vehicles (1.2)	CT in sec (1.3)	Benja-min & Beasley (2.1)	Vehicles (2.2)	CT in sec (2.3)	Buhrkal et al. (3.1)	Vehicles (3.2)	CT in sec (3.3)	Our Results average (4.1)	Out Best Solution (4.2)	Vehicles (4.3)	Average Gap to Best Bench-mark
102	205.1	3	3	183.5	3	3	**174.5**	3	2	157.7	158	4	−9.1
277	527.3	3	10	**464.5**	3	8	447.6	3	8	463.4	463.3	3	3.6
335	205.0	6	11	204.5	6	11	**182.1**	6	10	168.8	193.1	6	6.2
444	87.0	11	16	89.1	11	18	**78.3**	11	18	85.3	85.3	11	8.9
804	769.5	5	92	725.6	6	92	**604.1**	5	72	624.6	621.4	6	2.9
1051	2370.4	18	329	**2250.2**	17	194	2325.7	17	116	2090.4	2238.9	16	−0.3
1351	1039.7	7	95	915.1	8	105	**871.9**	8	105	876.5	970.6	7	11.3
1599	1459.2	13	212	1364.7	14	231	**1337.5**	13	252	1222.2	1225.2	15	−8
1932	1395.3	17	424	1262.8	16	335	**1162.5**	16	285	1185.2	1178.5	23	1.4
2100	1833.8	16	408	**1749.0**	16	266	1818.9	16	266	1669.1	1701.4	17	−2.7
Average			*160*			*126*			*113*				*−1.2*

below, our results are compared to the results obtained by [11] (columns 1.1–1.3), [2] (column 2.1–2.3), and [3] (columns 3.1–3.3) in regards to overall solution costs, the number of vehicles used and computation times. Our results are illustrated in columns 4.1–4.3. Thereby, 4.1 shows the average over all results obtained by changing the random number generator seed 10 times, and 4.2 shows the best found solution using our approach. The final column shows the percentage difference over our best found solution (column 4.2) and the best solution presented by any other work using the same benchmarks (i.e. the best solution from column 1.1, 2.1 and 3.1).

6 Conclusions

In this work we have presented an efficient heuristic to solve the WCP, which we formulated as a real-life VRP instance. Using a CWS enriched with biased random-ization, we are able to obtain competitive results to a set of WCP benchmarks.

The WCP is a very current issue in many municipalities. As a future work, a case study to implement the proposed heuristic on a realistic problem setting in a city neat Barcelona is planned. Furthermore, the WCP is a typical rich VRP setting, charac-terized by the uncertainty and stochastic nature of its input variables, e.g. travel times or waste levels. In order to take these stochastic inputs into account, it is planned to combine the results of this competitive

Acknowledgments. This work has been partially supported by the Spanish Ministry of Economy and Competitiveness (grant TRA2013-48180-C3-P) and FEDER. Likewise we want to acknowledge the support received by the Department of Universities, Research & Information Society of the Catalan Government (Grant 2014-CTP-00001).

References

1. Beliën, J., de Boeck, L., van Ackere, J.: Municipal solid waste collection and management problems: a literature review. Trans. Sci. **48**, 78–102 (2014)
2. Benjamin, A., Beasley, J.: Metaheuristics for the waste collection vehicle routing problem with time windows, driver rest period and multiple disposal facilities. Comput. Oper. Res. **37**, 2270–2280 (2010)
3. Buhrkal, K., Larsen, A., Ropke, S.: The waste collection vehicle routing problem with time windows in a city logistics context. Procedia-Soc. Behav. Sci. **39**, 241–254 (2012)
4. Clarke, G., Wright, J.: Scheduling of vehicles from a central depot to a number of delivery points. Oper. Res. **12**, 568–581 (1964)
5. Daneshzand, F.: The Vehicle Routing Problem. Logistics Operations and Management, 1st edn. Elsevier, Philadelphia (2011)
6. European Commission: Cities of tomorrow-Challenges, visions, ways forward. European Union, Brussels (2011)
7. Gruler, A., Fikar, C., Juan, A., Hirsch, P.: Un algoritmo basado en aleatorización para la recolección eficiente de residuos en ciudades inteligentes. In: Proceedings of the X Spanish Conference on Metaheuristics, Evolutive and Bioinspired Algorithms (2015)
8. Hansen, P., Mladenovic, N., Brimberg, J., Moreno-Pérez, J.: Variable neighborhood search. In: Handbook of Metaheuristics, Kluwer Academic, Dordrecht (2010)

9. Juan, A., Faulin, J., Jorba, J., Riera, D., Masip, D., Barrios, B.: On the use of Monte Carlo simulation, cache and splitting techniques to improve the Clarke-and-Wright savings heuristic. J. Oper. Res. Soc. **62**, 1085–1097 (2011)

10. Juan, A., Faulin, J., Ferrer, A., Lourenco, H., Barrios, B.: MIRHA: multi-start biased randomization of heuristics with adaptive local search for solving non-smooth routing problems. TOP **18**, 109–132 (2013)

11. Kim, B., Kim, S., Sahoo, S.: Waste collection vehicle routing problem with time windows. Comput. Oper. Res. **33**, 362–3642 (2006)

12. Mattoni, B., Gugliermetti, F., Bisegna, F.: A multilevel method to assess and design the renovation and integration of Smart Cities. Sustain. Cities Soc. **15**, 105–119 (2015)

13. Neirotti, P., De Marco, A., Cagliano, A., Mangano, G., Scorrano, F.: Current trends in Smart City initiatives: some stylized facts. Cities **38**, 25–36 (2014)

14. Nikolaev, A., Jacobson, S.: Simulated annealing. In: Gendreau, M., Potvin, J. (eds.) Handbook of Metaheuristics. International series in operations research management. Kluwer Academic, Dordrecht (2010)

15. Reeves, C.: Genetic algorithms. In: Gendreau, M., Potvin, J. (eds.) Handbook of Metaheuristics. Kluwer Academic, Dordrecht (2010)

16. Resende, M., Ribeiro, C.: Greedy randomized adaptive search procedures: advances, hybridizations and applications. In: Handbook of Metaheuristics, Kluwer Academic, Dordrecht (2010)

17. The World Bank: What a Waste - A Global Review of Solid Waste Management. The World Bank – Urban Development & Local Government Unit (2012)

18. Toth, P., Vigo, D.: An overview of vehicle routing problems. In: Toth, P., Vigo, D. (eds.) The Vehicle Routing Problem. SIAM, Philadelphia (2002)

19. Tranos, E., Gertner, D.: Smart networked cities. Innov. Eur. J. Soc. Sci. Res. **25**, 175–190 (2012)

20. UN Department of Economic and Social Affairs. http://www.un.org/esa/population/publications/wup2007/2007wup.htm

21. Wy, J., Kim, B.I., Kim, S.: The rollon-rolloff waste collection vehicle routing problem with time windows. Eur. J. Oper. Res. **224**, 466–476 (2013)

Innovation Capabilities Using Fuzzy Logic Systems

Víctor G. Alfaro-García[1]($\boxtimes$), Anna M. Gil-Lafuente[1],
and Gerardo G. Alfaro Calderón[2]

[1] Facultad de Economía y Empresa, Universitat de Barcelona,
Av. Diagonal 690, 08034 Barcelona, Spain
valfaro06@gmail.com, amgil@ub.edu
[2] Facultad de Contaduría y Ciencias Administrativas,
Universidad Michoacana de San Nicolás de Hidalgo, Morelia, Mexico
ggalfaroc@gmail.com

Abstract. In recent decades, innovation has been recognized as one of the main
sources of competitive advantage for organizations at an international level.
Recently, numerous studies have been developed, mainly focused on the analyses of innovation processes in large companies, however, little has been done
in the identification of innovative capabilities that allow small and medium-sized
companies to compete in highly uncertain environments. This study presents the
interpreted results of an innovation diagnosis utilizing the Experton method and
the application of the generalized ordered weighted average operator (GOWA)
for the treatment of information. The focus relies on seven organizational areas
of small and medium-sized manufacturing companies. The applied questionnaire presents 32 exploration and 5 control questions applied to a total of 217
small and medium-sized companies in the Latin American city of Morelia,
Mexico. The present work seeks to shed light in the diagnosis of innovation
capabilities in manufacturing companies with the aim of fostering synergies and
effectively allocate resources for the promotion of innovation management as a
source of competitiveness and economic growth.

Keywords: Expertons · Ordered weighted average operator
Innovation diagnose · Small and medium-sized firms

1 Introduction

Innovation management research has evolved over the last decades [10, 22, 32].
Nowadays, no manager or decision maker could affirm that innovation does not carry
competitiveness, is in some way a given fact [1]. Porter [27] states, companies gain an
advantage against the best competitors in the world due to the innovations they generate. The results of innovative activities in companies and organizations can range
from effects on sales and market share to an increased productivity and operational
efficiency.

The concept of innovation includes ideas such as economic progress, business
success, problem-solving, etc. Therefore, knowing how innovative activities affect the

© Springer International Publishing AG, part of Springer Nature 2018
A. M. Gil-Lafuente et al. (Eds.): FIM 2015, AISC 730, pp. 264–276, 2018.
https://doi.org/10.1007/978-3-319-75792-6_20

performance of companies is relevant since it opens the way to attend an important challenge: systematically determine whether the innovative efforts within companies are justified, objectives they are being achieved and ultimately define the adequate incentives for their promotion [8]. The formula to address this issue initiates by gathering the information collected in large studies on success and failure around business innovation, generate a list with the most relevant characteristics and apply a scoring system to the best practices [33]. However, this method faces complicated challenges such as the heterogeneous terminology, dissimilar points of view and dissimilar definitions of the concept of innovation [2]. A correct evaluation, quantification, and comparison of the innovative competences of organizations is difficult since there is no ideal model or technique to evaluate the measurement of innovation [13]. It is in this case that the use of a generalized measurement framework can provide a useful basis for decision-makers so that innovation efforts can be monitored controlled and incentivized [5].

The objective of this research is to diagnose innovation capabilities using methodologies for the treatment of uncertainty [21, 23] in key areas of small and medium-sized businesses (SMEs), following the holistic measurement framework for innovation designed by Adams et al. [1]. The goal is to distinguish the strengths and opportunities that the manufacturing companies of the Latin American city of Morelia, Mexico present in terms of innovative characteristics.

The remainder of the article is as follows, the second section presents a brief revision of the literature mainly focused on the design of the applied questionnaire. The third section focuses on the methodological aspects of the study and the results obtained from the empirical analysis. Finally, the conclusions of the article are presented.

2 Literature Review

It is generally accepted that innovation processes in organizations provide sources of opportunity for the temporary establishment of monopolies, moreover, the continuous introduction of innovative activities are key for the development of long-term business success [29]. The introduction of products, services, innovative processes or business models designed specifically for attractive niches results in opportunities for SMEs [26]. The sustainable introduction of innovations by SMEs creates high entry barriers, which prevent the appearance of new competitors, the position of the industry is strengthened and innovation activities result in favorable and persistent benefits above the market average [26].

Due to its important role and positive effects for SMEs, research on innovation management has attracted a lot of attention in the literature [14, 22, 30] however, the academia has not yet found a definitive model for its administration [16, 24, 28, 31, 34]. Some of the challenges that generate inconsistencies for the unification of criteria and a definitive modeling of the innovative process result from the diverse definitions of innovation, the different perspectives of analysis, the evolution of the understanding of the concept, as well as the intrinsic uncertainty of the phenomenon [2].

This research follows the proposal developed by Adams et al. [1] whose work is based on the revision of six models of innovation measurement [4, 6, 7, 9, 15, 35]. From the seven areas described by the authors, an innovation measurement framework has been adapted to take into account recurrent and relevant factors when quantifying the organizational innovative capacities of the companies analyzed [3].

3   Methodology

3.1   Measurement Approach

A quantitative empirical approach was chosen to obtain information regarding the intensity of the innovative capabilities of the companies within the city of Morelia, México. The observation unit is small and medium-sized firms, taking in count that the first comprehend a minimum of 11 and a maximum of 50 employees, and the second a minimum of 51 and a maximum of 250 employees [19]. A firm is defined as the economic and legal unit that, under a single owner or controlling entity, is mainly engaged in industrial activities for sales to the public; with an operative structure subdivided into branches or with a single physical location [20].

The data was retrieved from a questionnaire on innovation capabilities that includes 32 exploration questions [1, 6, 17], and 5 control questions [20]. The survey was applied personally to a total of 217 SMEs in the city within the period of January to May 2015. This exploration aims to be responded by the managers of the surveyed SMEs that were at least registered in the National Statistical Directory of Economic Units [18].

From the total of 217 questionnaires applied 91 valid responses were gathered i.e. a response rate of 41.9%.

The survey addresses seven specific areas of innovation measurement (Adams et al. [1]), the approach is organizational innovation capacities (Armbruster et al. [3]) and the potential impact that SMEs exercise in the regional economic development of the territories [25].

3.2   Expertons Method

The subjective conditions of the data collected require tools that provide useful, representative and flexible treatment of information for the exploitation of the results and future decision making.

Given the above, in the present work we propose the treatment of the information utilizing Fuzzy Logic tools, specifically, we use the Experton method [21] for the aggregation of the opinions of the administrators for each of the 32 questions asked in the questionnaire. The main advantage of the Experton method is the aggregation of diverse opinions into a unique representative value and on the other hand the treatment of fuzzy numbers.

It is generally known that the properties of the expertons give a monoid and a distribution network with the norms $\wedge$ (min) and $\vee$ (max); then every process that is used with fuzzy sets remains available to the expertons with a monotony restriction.

The expertons are based on probabilistic series when considering probability intervals. On the other hand, any experton does not contain any vertical monotonic growth, except level 0, which always takes the value of 1. Mathematically we have:

$$\forall \alpha \in [0, 1] : a_1(\alpha) \le a_2(\alpha) \ in \ [a_1(\alpha), a_2(\alpha)] \tag{1}$$

$$\forall \alpha, \alpha' \in [0, 1] : (\alpha' > \alpha) \Rightarrow (a_1(\alpha)) \le (a_1(\alpha'), a_2(\alpha) \le a_2(\alpha')) \tag{2}$$

$$(\alpha = 0) \Rightarrow (a_1(\alpha) = 1, a_2(\alpha) = 1) \tag{3}$$

If we consider the metric of each expert that expresses his opinion and we show it on an 11-value scale between 0 and 1 (including both), such opinions could be shown in the following way:

From this point, the next operation is to begin by transforming the subjective opinions into a representative metric, according to the relationships that are mentioned in Table 1. In the first place, it is necessary to obtain statistical information concerning the number of times that the opinions of the experts have repeated. Then, the calculation of the cumulative frequency is obtained and this is divided by the total number of expert's opinions. The result when doing this operation is the expertons. The Experton is itself the aggregate representative view of all the subjective opinions that can be considered in the sample. If a simplified representation of an Experton is required, it would be possible to obtain the mathematical expectation.

Table 1. Scale value

1.0	True
0.9	Practically true
0.8	Almost true
0.7	Fairly true
0.6	More true than false
0.5	Neither false or true
0.4	More false tan true
0.3	Fairly false
0.2	Almost false
0.1	Practically false
0.0	False

The Expertons method facilitates decision-making by providing qualitative data from the dialogue that is maintained with various interest groups surrounding the phenomenon in question. In general, the approach allows a useful tool in the process of aggregation of information, which allows unifying different points of view or expectations of groups with divergent interests. Moreover, the model allows recognizing the levels of distribution in the aggregated values of the characteristic membership function.

3.3   GOWA Operator

Along with the useful information obtained in the previous step, it is also interesting to generate an ordered grouping of the expertons for each of the seven areas of innovation included in the study. Among the large number of operators that exist in the literature [23], the use of the generalized ordered weighted average operator GOWA is proposed.

The GOWA operator is a mathematically generalized decision model that can be used in a wide range of situations. It consists of using the generalized means developed by [11, 12] in the OWA operators [37].

Definition: a $GOWA : R^n \rightarrow R$ function is an *GOWA* operator of n dimension with an associated weighting vector W such that:

(1) $w_j \in [0, 1]$
(2) $\sum_{j=1}^{n} w_j = 1$

y

$$GOWA(a_1, a_2, \ldots, a_n) = \left(\sum_{j=1}^{n} w_j b_j^\lambda \right)^{1/\lambda} \tag{4}$$

where b_j is the *jth* largest of the a_1, and λ is a parameter such that $\lambda \in (-\infty, \infty)$.

The *GOWA* operator is a mean operator that displays the next properties: commutative, idempotency, boundedness and monotonicity [23].

In order to obtain the greatest representativeness of the group of experts, we propose the use of a value $\lambda = -1$ [36], in addition to a weighting vector w_j, suitable for each innovation area.

4   Results

The Fuzzy analysis in our study is proposed in two steps, on the one hand, the use of the Expertons method [21] to generate a representative value of each of the 91 valid opinions obtained from the 32 exploration questions of the questionnaire and in the background, the application of the operator GOWA [23] to these representative values, for the weighted aggregation of the intensity of innovation in the 7 different areas of innovation to be treated.

4.1   Expertons Analysis

Following the guidelines previously defined for the calculation of the expertons, the 91 valid responses were added for each of the 32 reagents formulated in the questionnaire. These reagents comprehend the 7 different areas of measurement proposed by Adams et al. [1]. Below we present the main results obtained.

Table 2 shows the results of the first diagnostic area: innovation strategy. The results show that in general, management places a high value on the professional development of its collaborators, promoting and stimulating learning and teamwork.

Table 2. Innovation strategy

Experton	Question 1.1	Question 1.2	Question 1.3	Question 1.4	Question 1.5
0.0	1.00	1.00	1.00	1.00	1.00
0.1	1.00	1.00	1.00	1.00	1.00
0.2	1.00	1.00	1.00	1.00	1.00
0.3	0.95	0.95	0.96	0.96	0.92
0.4	0.95	0.95	0.96	0.96	0.92
0.5	0.82	0.84	0.70	0.86	0.74
0.6	0.82	0.84	0.70	0.86	0.74
0.7	0.52	0.53	0.46	0.56	0.43
0.8	0.52	0.53	0.46	0.56	0.43
0.9	0.22	0.26	0.22	0.22	0.14
1.0	0.22	0.26	0.22	0.22	0.14
Expected value	0.70	0.71	0.67	0.72	0.65

The firms also identify as a key value the idea that innovation is a process, which must be managed in a formal manner, not improvised, in such a way that it plays an important role in the long-term business. In contrast, there is a high aversion to risk by the administration, it relies mostly on experience without adequately accepting the errors generated in the process. There is also an area of opportunity for effective communication of innovation efforts to workers, customers, and suppliers.

The results of the second area of diagnosis: knowledge management, are shown in Table 3. The surveyed companies find value in systematically analyzing the products of competitors and keep abreast of changes or new legislation. They strive to identify sources of external knowledge and use this information to incorporate new technologies into their products. There is also a high idea of managing the knowledge of its employees in an appropriate manner and have a corporate knowledge hub that is useful and easily accessible to employees.

Table 3. Knowledge management

Experton	Question 2.1	Question 2.2	Question 2.3	Question 2.4
0.0	1.00	1.00	1.00	1.00
0.1	1.00	1.00	1.00	1.00
0.2	1.00	1.00	1.00	1.00
0.3	0.95	0.98	0.91	0.88
0.4	0.95	0.98	0.91	0.88
0.5	0.77	0.84	0.69	0.64
0.6	0.77	0.84	0.69	0.64
0.7	0.49	0.64	0.44	0.36
0.8	0.49	0.64	0.44	0.36
0.9	0.26	0.35	0.10	0.12
1.0	0.26	0.35	0.10	0.12
Expected value	0.69	0.76	0.63	0.60

The summary of the results of the third diagnostic unit: project management is shown in Table 4. The surveyed companies highly value to the continuous and valuable use of advanced tools for the development of their products. The companies are active in the renewal of said instruments, in the development of their own instruments and in the application of the best practices for their use. The design of the products also plays a relevant role in their innovation capabilities, it seeks to improve the functionality of the product, simplify its components or make it aesthetically more attractive.

Table 4. Project management

Experton	Question 3.1	Question 3.2	Question 3.3	Question 3.4	Question 3.5
0.0	1.00	1.00	1.00	1.00	1.00
0.1	1.00	0.99	1.00	1.00	1.00
0.2	1.00	0.99	1.00	1.00	1.00
0.3	0.86	0.87	0.90	0.97	0.80
0.4	0.86	0.87	0.90	0.97	0.80
0.5	0.67	0.73	0.70	0.87	0.60
0.6	0.67	0.73	0.70	0.87	0.60
0.7	0.44	0.40	0.53	0.52	0.32
0.8	0.44	0.40	0.53	0.52	0.32
0.9	0.22	0.16	0.21	0.16	0.14
1.0	0.22	0.16	0.21	0.16	0.14
Expected value	0.64	0.63	0.67	0.70	0.57

Table 5 adds the results of the fourth diagnostic unit: portfolio management. In this unit, it is observed that companies seek structured and systematic processes to identify the degree of affectation implied by the changes and uncertainty of the external environment when making decisions for the selection and development of new

Table 5. Portfolio management

Experton	Question 4.1	Question 4.2	Question 4.3	Question 4.4
0.0	1.00	1.00	1.00	1.00
0.1	1.00	1.00	1.00	1.00
0.2	1.00	1.00	1.00	1.00
0.3	0.86	0.81	0.89	0.82
0.4	0.86	0.81	0.89	0.82
0.5	0.65	0.55	0.74	0.71
0.6	0.65	0.55	0.74	0.71
0.7	0.35	0.33	0.31	0.47
0.8	0.35	0.33	0.31	0.47
0.9	0.12	0.14	0.10	0.18
1.0	0.12	0.14	0.10	0.18
Expected value	0.60	0.57	0.61	0.64

projects. The results obtained from this unit also show areas of opportunity in terms of formal procedures for the selection of concepts to be generated. The development of new products is mainly driven by the management's criteria, not on an analysis of profitability, viability or market. There is little involvement of employees or diverse areas of the company to systematically limit the challenges posed by the impulse of the new concepts.

The results on internal drivers are shown in Table 6. According to the answers obtained, companies make an appreciable use of the most appropriate tools for the definition and control of production processes. It also highlights the use of techno-logical surveillance mechanisms over competitors and a continuous scrutiny of changes in the legislative environment. A high value is given to the selection and hiring of personnel with skills of ideation, creativity, abstraction, and problem-solving. As well as an allocation of specific resources for the development of new production processes. The area of opportunity is the allocation of Human Resources for specific innovation projects, facilities for the optimal performance of a team aimed at generating new projects, with clear objectives and specific goals.

Table 6. Internal drivers

Experton	Question 5.1	Question 5.2	Question 5.3	Question 5.4	Question 5.5
0.0	1.00	1.00	1.00	1.00	1.00
0.1	1.00	1.00	1.00	0.99	0.99
0.2	1.00	1.00	1.00	0.99	0.99
0.3	0.69	0.81	0.93	0.93	0.79
0.4	0.69	0.81	0.93	0.93	0.79
0.5	0.51	0.63	0.69	0.76	0.65
0.6	0.51	0.63	0.69	0.76	0.65
0.7	0.25	0.37	0.46	0.41	0.40
0.8	0.25	0.37	0.46	0.41	0.40
0.9	0.12	0.19	0.18	0.19	0.12
1.0	0.12	0.19	0.18	0.19	0.12
Expected value	0.51	0.60	0.65	0.65	0.59

Table 7 includes the results in the area: organization and structure. It is observable that companies highly value the organizational climate of the company, efforts are generated aimed at integration activities and good staff relationship. Teamwork is also an important asset to the company, it supports structured programs for the integration of teams oriented towards the improvement of processes. However, there is the pos-sibility of acquiring or improving organizational management tools organizational manuals with the delimitation of tasks and responsibilities. In addition to this, there is an important gap in terms of the lack of specific mechanisms to stimulate the creative capacity of workers, the contribution of new ideas and the innovative spirit. It can be said that there are rigid control systems that limit the contribution of ideas or suggestions.

Table 7. Organization and structure

Experton	Question 6.1	Question 6.2	Question 6.3	Question 6.4
0.0	1.00	1.00	1.00	1.00
0.1	0.99	0.99	0.99	0.99
0.2	0.99	0.99	0.99	0.99
0.3	0.87	0.79	0.90	0.91
0.4	0.87	0.79	0.90	0.91
0.5	0.62	0.68	0.76	0.79
0.6	0.62	0.68	0.76	0.79
0.7	0.38	0.49	0.56	0.57
0.8	0.38	0.49	0.56	0.57
0.9	0.19	0.31	0.32	0.22
1.0	0.19	0.31	0.32	0.22
Expected value	0.61	0.65	0.71	0.70

The results of the last diagnostic area: external drivers are shown in Table 8. Based on the responses of the companies, it is observed that the relationship with customers is of the utmost importance. Efforts are generated to maintain a close post-sale contact, moreover, it is highly valued to have direct attention lines and there is a structured complaint management system that allows identifying improvements. It is also observed that there is flexibility in the processes of commercialization of the developed products, the price, communication, and distribution are not pre-set factors. The possibility of using new information technologies to position products in new environments is accepted. However, there is an area of opportunity regarding the development of mechanisms to know and measure the best business management practices of competing companies.

Table 8. External drivers

Experton	Question 7.1	Question 7.2	Question 7.3	Question 7.4	Question 7.5
0.0	1.00	1.00	1.00	1.00	1.00
0.1	0.99	0.99	0.99	0.99	0.99
0.2	0.99	0.99	0.99	0.99	0.99
0.3	0.92	0.88	0.88	0.92	0.90
0.4	0.92	0.88	0.88	0.92	0.90
0.5	0.76	0.74	0.73	0.79	0.71
0.6	0.76	0.74	0.73	0.79	0.71
0.7	0.42	0.49	0.46	0.63	0.49
0.8	0.42	0.49	0.46	0.63	0.49
0.9	0.11	0.19	0.22	0.34	0.19
1.0	0.11	0.19	0.22	0.34	0.19
Expected value	0.64	0.66	0.65	0.73	0.66

4.2 Análisis GOWA

The mathematical tools designed for the aggregation and grouping of information have grown substantially in recent years [38]. Applications of these have been observed in various environments e.g. finance, human resources management, decision making in companies, selection of athletes, see [23].

For the specific case of the city of Morelia, Mexico, the operator GOWA [23] is used to establish a hierarchical order of each innovation measurement area [1]. This is done in line with the previously defined methodology in Sect. 3, establishing vectors Wj of specific weights for each representative single value obtained using Expertons [21]. Table 9 shows the vectors for each of the innovation measurement areas.

Table 9. Weighting W_j vectors by experton and innovation measurement area

Innovation area	w_1	w_2	w_3	w_4	w_5
1. Innovation strategy	0.30	0.10	0.40	0.15	0.05
2. Knowledge management	0.40	0.30	0.20	0.10	
3. Project management	0.15	0.30	0.40	0.10	0.05
4. Porfolio management	0.30	0.10	0.20	0.40	
5. Internal drivers	0.05	0.4	0.1	0.15	0.3
6. Organization and structure	0.4	0.3	0.2	0.1	
7. External drivers	0.30	0.15	0.10	0.40	0.05

Fuente: elaborado a partir de opiniones de expertos.

These vectors Wj are subject to the preferences of 8 expert appraisers in public policy areas, thus offering specific results for the case of the city. The use of vectors Wj seeks a closer approach to the conditions of the specific environment of the city, thus generating flexible results, supporting future decision making by managers or institutions.

The results generated by the GOWA operator are shown in Fig. 1. From the treatment, ordering, and aggregation of the information, the diagnosis in innovation capacities of the manufacturing companies of the city of Morelia can be observed globally.

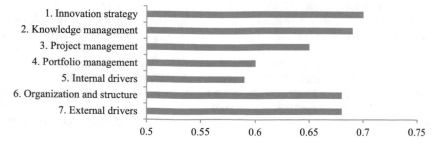

Fig. 1. GOWA results for innovation management areas

5 Conclusion

The purpose of this study is to quantify, based on a holistic model of innovation measurement, the intensity of innovative efforts by small and medium-sized manufacturing enterprises within the city of Morelia, Mexico, the objective is to generate a diagnosis making use of Fuzzy Logic tools appropriate to the environment of the studied companies. The analysis allows us to observe the strengths and opportunities of organizational capacities of the selected firms.

An empirical study is proposed based on an innovation questionnaire comprising 32 exploration questions and 5 control questions addressing 7 key areas of the company: Innovation strategy, knowledge management, project management, portfolio management, internal drivers, organization and structure and external drivers. The questionnaire was applied personally to 217 companies, of which 91 valid answers were obtained.

It is proposed to treat the information obtained from Fuzzy tools, specifically the Experton method and applying the generalized ordered weighted averages (GOWA) operator, providing key information for the understanding of the variables that promote high performance in areas such as innovation strategy, knowledge management and organization and structure, also describing the global variables that imply poor performance such as: portfolio management and internal drivers.

The present study seeks to shed light on the understanding of the characteristics that describe the innovation process on the part of the Morelian companies, thus enabling a comparative evaluation of a company's innovation capabilities based on the results.

Acknowledgements. The first author expresses his gratitude to the Mexican Council of Science and Technology (CONACYT) for the financial support given to this research project with the scholarship no. 381436 to graduate studies. Research supported by "Red Iberoamericana para la Competitividad, Innovación y Desarrollo" (REDCID) project number 616RT0515 in "Programa Iberoamericano de Ciencia y Tecnología para el Desarrollo" (CYTED).

References

1. Adams, R., Bessant, J., Phelps, R.: Innovation management measurement: a review. Int. J. Manag. Rev. **8**(1), 21–47 (2006)
2. Alfaro García, V.G., Gil-Lafuente, A.M., Alfaro Calderon, G.G.: A fuzzy logic approach towards innovation measurement. Glob. J. Bus. Res. **9**(3), 53–71 (2016)
3. Armbruster, H., Bikfalvi, A., Kinkel, S., Lay, G.: Organizational innovation: the challenge of measuring non-technical innovation in large-scale surveys. Technovation **28**(10), 644–657 (2008)
4. Burgelman, R.A., Christensen, C.M., Wheelwright, S.C.: Strategic Management of Technology and Innovation, 4th edn. McGraw Hill/Irwin, New York (2004)
5. Cebon, P., Newton, P.: Innovation in firms: towards a framework for indicator development. Melbourne Business School, no. 99-9 (1999)
6. Chiesa, V., Coughlan, P., Voss, C.A.: Development of a technical innovation audit. J. Prod. Innov. Manag. **13**(2), 105–136 (1996)

7. Cooper, R.G., Kleinschmidt, E.J.: Benchmarking the firm's critical success factors in new product development. J. Prod. Innov. Manag. **12**(5), 374–391 (1995)
8. Cordero, R.: The measurement of innovation performance in the firm: an overview. Res. Policy **19**, 185–192 (1990)
9. Cormican, K., O'Sullivan, D.: Auditing best practice for effective product innovation management. Technovation **24**(10), 819–829 (2004)
10. Drejer, A.: Situations for innovation management: towards a contingency model. Eur. J. Innov. Manag. **5**(1), 4–17 (2002)
11. Dujmovic, J.J.: Weighted conjunctive and disjunctive means and their application in system evaluation. Univ. Beograd. Publ. Elektrotechn. Fak **483**, 147–158 (1974)
12. Dyckhoff, H., Pedrycz, W.: Generalized means as model of compensative connectives. Fuzzy Sets Syst. **14**(2), 143–154 (1984)
13. Frenkel, A., Maital, S., Grupp, H.: Measuring dynamic technical change: a technometric approach. Int. J. Technol. Manag. **20**, 429–441 (2000)
14. Gimbert, X., Bisbe, J., Mendoza, X.: The role of performance measurement systems in strategy formulation processes. Long Range Plan. **43**(4), 477–497 (2010)
15. Goffin, K., Pfeiffer, R.: Innovation Management in UK and German Manufacturing Companies. Anglo-German Foundation for the Study of Industrial Society, London (1999)
16. Hobday, M.: Firm-level innovation models: perspectives on research in developed and developing countries. Technol. Anal. Strateg. Manag. **17**, 121–146 (2005)
17. i Ohme, E.T.: Guide for Managing Innovation: Part 1, Diagnosis. Generalitat de Catalunya, CIDEM (2002)
18. Instituto Nacional de Estadística y Geografía. Directorio Estadístico Nacional de Unidades Económicas. Recuperado de (2015). http://www3.inegi.org.mx/sistemas/mapa/denue/default.aspx. Último acceso: Mayo 2015
19. Instituto Nacional de Estadística y Geografía. Las empresas en los Estados Unidos Mexicanos: Censos Económicos 2009. INEGI, México, c2012 (2009)
20. Instituto Nacional de Estadística y Geografía. Resultados de los módulos de innovación tecnológica: MIT 2008, 2006 y 2001. INEGI, México, c2010 (2010)
21. Kaufmann, A., Gil-Aluja, J.: Técnicas especiales para la gestión de expertos. Milladoro, España (1993)
22. Keupp, M.M., Palmié, M., Gassmann, O.: The strategic management of innovation: a systematic review and paths for future research. Int. J. Manag. Rev. **14**(4), 367–390 (2012)
23. Merigó, J.: Nuevas Extensiones a los Operadores OWA y su Aplicación en los Métodos de Decisión. Doctorando, Universitat de Barcelona, España (2008)
24. Nilsson, S., Wallin, J., Benaim, A., Annosi, M.C., Berntsson, R., Ritzen, S., Magnusson, M.: Re-thinking innovation measurement to manage innovation-related dichotomies in practice. In: Proceedings of the Continuous Innovation Network Conference, CINet 2012, Rome, Italy (2012)
25. Oke, A., Burke, G., Myers, A.: Innovation types and performance in growing UK SMEs. Int. J. Oper. Prod. Manag. **27**(7), 735–753 (2007)
26. Porter, M.E.: Competitive Strategy: Techniques for Analyzing Industries and Competition. Free Press, New York (1980)
27. Porter, M.E.: The competitive advantage of nations. Harvard Bus. Rev. **68**, 73–93 (1990)
28. Rothwell, R.: Successful industrial innovation: critical factors for the 1990s. R&D Management **22**, 221–239 (1992)
29. Schumpeter, J.A.: The Theory of Economic Development: An Inquiry into Profits, Capital, Credit, Interest, and the Business Cycle, 55th edn. Transaction Publishers, London (1934)
30. Simons, R.: The role of management control systems in creating competitive advantage: new perspectives. Acc. Organ. Soc. **15**(1–2), 127–143 (1990)

31. Tidd, J.: A Review of Innovation Models. Imperial College London, London (2006)
32. Tidd, J.: Innovation management in context: environment, organization and performance. Int. J. Manag. Rev. **3**(3), 169–183 (2001)
33. Tidd, J., Bessant, J.: Managing Innovation: Integrating Technological, Market and Organizational Change, 5th edn. Wiley, Hoboken (2013)
34. Velasco Balmaseda, E., Zamanillo Elguezabal, I.: Evolución de las propuestas sobre el proceso de innovación: ¿qué se puede concluir de su estudio? Investigaciones Europeas de dirección y economía de la empresa **14**(2), 127–138 (2008)
35. Verhaeghe, A., Kfir, R.: Managing innovation in a knowledge-intensive technology organization (KITO). R&D Manag. **32**, 409–417 (2002)
36. Yager, R.R.: Generalized OWA aggregation operators. Fuzzy Optim. Decis. Making **3**(1), 93–107 (2004)
37. Yager, R.R.: On ordered weighted averaging aggregation operators in multicriteria decision making. IEEE Trans. Syst. Man Cybernet. **18**(1), 183–190 (1988)
38. Yager, R.R., Kacprzyk, J.: The Ordered Weighted Averaging Operators: Theory and Applications, 1st edn. Springer Science & Business Media, New York (1997)

Association Rule-Based Modal Analysis for Various Data Sets with Uncertainty

Hiroshi Sakai[✉] and Chenxi Liu

Graduate School of Engineering, Kyushu Institute of Technology, Tobata,
Kitakyushu 804-8550, Japan
sakai@mns.kyutech.ac.jp, o100414r@mail.kyutech.jp

Abstract. We coped with association rule-based modal data analysis, and generated rules by using the *getRNIA* software tool. This tool is powered by the *NIS-Apriori* algorithm, and handles tables with non-deterministic information. Moreover, the *NIS-Apriori* algorithm is logically complete for the defined rules. In this paper, we consider the *NIS-Apriori* algorithm with the granules defined by the descriptors. By using these granules, we can uniformly apply *NIS-Apriori* algorithm to various types of uncertain data sets. The problem is reduced to the definition of the descriptors and the granules. We show the property of this algorithm and some examples.

Keywords: Association rules · Rule generation · Apriori algorithm
Granularity · Descriptor · Uncertainty

1 Introduction

Rough set theory proposed by Pawlak is employed as the mathematical framework for analyzing table data, and this theory is applied to several research areas [8,9,15]. We are also interested in rough set-based rule generation, and proposed some algorithms for handling the tables with non-deterministic information as well as the tables with deterministic information [11]. In this research, we adjusted the well-known *Apriori* algorithm [1–3] for deterministic information to the *NIS-Apriori* algorithm for non-deterministic information. The computational complexity of the *NIS-Apriori* algorithm will be about twice the complexity of the *Apriori* algorithm [12,14]. Based on this algorithm, we have implemented the software tool *getRNIA* [17].

In this paper, we extend this framework for handling table data sets with non-deterministic information to the framework for handling various types of data sets with uncertainty.

© Springer International Publishing AG, part of Springer Nature 2018
A. M. Gil-Lafuente et al. (Eds.): FIM 2015, AISC 730, pp. 277–286, 2018.
https://doi.org/10.1007/978-3-319-75792-6_21

Table 1. A table Φ_1 with non-deterministic information and intervals.

OB	sex	height	weight
x_1	$\{female\}$	$[158.0, 162.0]$	$[63.0, 68.0]$
x_2	$\{male, female\}$	$[160.0, 164.0]$	$[64.2, 64.8]$
x_3	$\{male, female\}$	$[162.0, 166.0]$	$[66.0, 70.0]$
x_4	$\{male\}$	$[168.0, 176.0]$	$[71.0, 75.0]$
x_5	$\{male\}$	$[150.0, 160.0]$	$[67.0, 70.0]$
x_6	$\{male, female\}$	$[160.0, 168.0]$	$[56.0, 58.0]$

Table 2. A possible table ψ_1 for the attribute sex from Table 1.

OB	sex
x_1	$female$
x_2	$male$
x_3	$female$
x_4	$male$
x_5	$male$
x_6	$female$

2   Preliminary of Rule Generation

This section briefly surveys the previous work.

2.1   Rule Generation in Standard Tables

Let us consider Table 1. There are six persons and three attributes, sex, $height$, and $weight$. For handling information incompleteness, the set and the interval expressions are employed for sex, $height$ and $weight$, respectively. We interpret the set $\{male, female\}$ in x_2, x_3, and x_6 as that the value of sex is either $male$ or $female$. We call such information *non-deterministic information*. Lipski and Orłowska handled information incompleteness in the tables by using non-deterministic information [4–7].

For an attribute A_i and an attribute value val_i, a pair $[A_i, val_i]$ is called a *descriptor*, and the formula $\tau : \wedge_i[A_i, val_i] \Rightarrow [Dec, val]$ is called an *implication*. Here, Dec is called a *decision attribute*. In a standard table, we call an implication τ satisfying the constraint below (a candidate of) a *rule* [8, 15].

> For two threshold values $0 < \alpha, \beta \leq 1.0$,
> $support(\tau)(= Num(\tau)/Num(OB)) \geq \alpha$,
> $accuracy(\tau)(= Num(\tau)/Num(\wedge_{A \in CON}[A, val_A])) \geq \beta$,   (1)
> Here, $Num(*)$ means the number of objects satisfying
> the formula $*$, OB means a set of all objects.

In the subsequent sections, we extend the definition of the rule in the standard tables to the tables with non-deterministic information.

2.2   Properties in Tables with Non-deterministic Data

Let Φ be a table with non-deterministic information. We technically employ a mapping $g : OB \times \{attribute\} \rightarrow P(\{attribute\ values\})$ (power set), for example $g(x_1, sex) = \{female\}$ and $g(x_2, sex) = \{male, female\}$. In Φ, we pick up an attribute value in every non-deterministic information, and we obtain a standard table. In Table 1, there are eight cases for the attribute sex, and Table 2 is a possible case. We see there is an actual case in these eight cases. Let $DD(\Phi)$ be a set of all possible cases from now on.

In Φ, we say an *object x supports $\tau : \wedge_i[A_i, val_i] \Rightarrow [Dec, val]$*, if $val_i \in g(x, A_i)$ for every A_i and $val \in g(x, Dec)$. In order to clarify this object x, we employ the notation τ^x instead of τ. We call τ^x is an *instance* of τ in Φ.

In each $\psi \in DD(\Phi)$, we can calculate $support(\tau^x)$ and $accuracy(\tau^x)$, however two values depend upon ψ. We face with the problem that which ψ we should employ for evaluating τ^x. For this problem, we introduced the following [12,14].

(1) $minsupp(\tau^x) = min_{\psi \in DD(\Phi)}\{support(\tau^x)$ in $\psi\}$,
(2) $minacc(\tau^x) = min_{\psi \in DD(\Phi)}\{accuracy(\tau^x)$ in $\psi\}$,
(3) $maxsupp(\tau^x) = max_{\psi \in DD(\Phi)}\{support(\tau^x)$ in $\psi\}$, $\qquad$ (2)
(4) $maxacc(\tau^x) = max_{\psi \in DD(\Phi)}\{accuracy(\tau^x)$ in $\psi\}$,
If τ^x does not occur in ψ, we deine $support(\tau^x) = accuracy(\tau^x) = 0$ in ψ.

The computational complexity of these values depend upon $|DD(\Phi)|$, which increases exponentially. For example, in the *mammographic* data set and the *hepatitis* data set in UCI machine learning repository [16], $|DD(\Phi)|$ is more than 10 power 100. However, we can solve this problem.

We define the following two granules of objects, i.e., *inf* and *sup*, for each descriptor $[A, val]$ [12].

(1) $inf([A, val]) = \{x \in OB \mid g(x, A) = \{val\}\}$,
(2) $inf(\wedge_i[A_i, val_i]) = \cap_i\ inf([A_i, val_i])$,
(3) $sup([A, val]) = \{x \in OB \mid val \in g(x, A)\}$, $\qquad$ (3)
(4) $sup(\wedge_i[A_i, val_i]) = \cap_i\ sup([A_i, val_i])$.

In a standard table, $inf([A, val]) = sup([A, val])$ holds, however $inf([A, val]) \subset sup([A, val])$ holds in Φ. By using inf and sup, we divide the set OB of objects to the nine granules, ①, ②, $\cdots$, ⑨ in Table 3, for example,

① $= \{x \in OB \mid x \in inf(\zeta) \cap inf(\eta)\}$,
④ $= \{x \in OB \mid x \in (sup(\zeta) \setminus inf(\zeta)) \cap inf(\eta)\}$. $\qquad$ (4)

Each granule also defines the set of the obtainable implications. For example, we have an implication $\zeta \Rightarrow \eta$ from ①, and we have four implications $\zeta \Rightarrow \eta$, $\zeta \Rightarrow \eta'$, $\zeta' \Rightarrow \eta$, and $\zeta' \Rightarrow \eta'$ from ⑤. The selection of an implication from a granule implicitly causes to fix the attribute value. Thus, we have Table 4 for causing $minsupp(\tau^x)$ and $minacc(\tau^x)$. Similarly, we have Table 5 for causing $maxsupp(\tau^x)$ and $maxacc(\tau^x)$.

Based on Tables 4 and 5, there is at least one $\psi_{min} \in DD(\Phi)$ causing $support(\tau^x)$ and $accuracy(\tau^x)$ the minimum, and there is at least one $\psi_{max} \in DD(\Phi)$ causing the $support(\tau^x)$ and $accuracy(\tau^x)$ the maximum. Furthermore, we can easily calculate the following for $x \in ① \neq \emptyset$ [13,14].

$$
\begin{aligned}
&(1) \ minsupp(\tau^x) = |①|/|OB|, \\
&(2) \ minacc(\tau^x) = |①|/(|①| + |②| + |③| + |⑤| + |⑥|), \\
&(3) \ maxsupp(\tau^x) = (|①| + |②| + |④| + |⑤|)/|OB|, \\
&(4) \ maxacc(\tau^x) = (|①| + |②| + |④| + |⑤|)/(|①| + |②| + |③| + |④| + |⑤|).
\end{aligned}
\tag{5}
$$

Table 3. Nine granules defined by $\tau : \wedge_i[A_i, val_i] \Rightarrow [Dec, val]$. For simplicity, ζ denotes $\wedge_i[A_i, val_i]$ and η denotes $[Dec, val]$. Furthermore, ζ' denotes $\wedge_i[A_i, val_i']$ where $val_i \neq val_i'$ at least one i and η' does $[Dec, val']$ where $val \neq val'$.

Granule	ζ	η	Obtainable implications
①	$inf(\zeta)$	$inf(\eta)$	$\zeta \Rightarrow \eta$
②	$inf(\zeta)$	$sup(\eta) \setminus inf(\eta)$	$\zeta \Rightarrow \eta$, $\zeta \Rightarrow \eta'$
③	$inf(\zeta)$	$OB \setminus sup(\eta)$	$\zeta \Rightarrow \eta'$
④	$sup(\zeta) \setminus inf(\zeta)$	$inf(\eta)$	$\zeta \Rightarrow \eta$, $\zeta' \Rightarrow \eta$
⑤	$sup(\zeta) \setminus inf(\zeta)$	$sup(\eta) \setminus inf(\eta)$	$\zeta \Rightarrow \eta$, $\zeta \Rightarrow \eta'$, $\zeta' \Rightarrow \eta$, $\zeta' \Rightarrow \eta'$
⑥	$sup(\zeta) \setminus inf(\zeta)$	$OB \setminus sup(\eta)$	$\zeta \Rightarrow \eta'$, $\zeta' \Rightarrow \eta'$
⑦	$OB \setminus sup(\zeta)$	$inf(\eta)$	$\zeta' \Rightarrow \eta$
⑧	$OB \setminus sup(\zeta)$	$sup(\eta) \setminus inf(\eta)$	$\zeta' \Rightarrow \eta$, $\zeta' \Rightarrow \eta'$
⑨	$OB \setminus sup(\zeta)$	$OB \setminus sup(\eta)$	$\zeta' \Rightarrow \eta'$

Table 4. The assignment causing both $accuracy(\tau^x)$ and $support(\tau^x)$ the minimum.

Granule	Selected implications
①	$\zeta \Rightarrow \eta$
②	$\zeta \Rightarrow \eta'$
③	$\zeta \Rightarrow \eta'$
④	$\zeta' \Rightarrow \eta$
⑤	$\zeta \Rightarrow \eta'$
⑥	$\zeta \Rightarrow \eta'$
⑦	$\zeta' \Rightarrow \eta$
⑧	$\zeta' \Rightarrow \eta$
⑨	$\zeta' \Rightarrow \eta'$

Table 5. The assignment causing both $accuracy(\tau^x)$ and $support(\tau^x)$ the maximum.

Granule	Selected implications
①	$\zeta \Rightarrow \eta$
②	$\zeta \Rightarrow \eta$
③	$\zeta \Rightarrow \eta'$
④	$\zeta \Rightarrow \eta$
⑤	$\zeta \Rightarrow \eta$
⑥	$\zeta' \Rightarrow \eta'$
⑦	$\zeta' \Rightarrow \eta$
⑧	$\zeta' \Rightarrow \eta$
⑨	$\zeta' \Rightarrow \eta'$

After all, these results are summarized in Fig. 1. We do not have to enumerate each $\psi \in DD(\Phi)$, and we only consider two possible cases ψ_{min} and ψ_{max}.

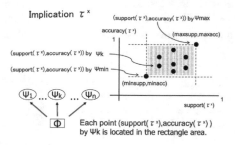

Fig. 1. A distribution of pairs $(support(\tau^x), accuracy(\tau^x))$ by ψ ($\psi \in DD(\Phi)$).

2.3   Rule Generation in Tables with Non-deterministic Data

In Φ, we defined two types of rules with modalities based on $DD(\Phi)$ below:

(Certain rule). An implication τ is a *certain rule*, if there is an instance τ^x satisfying $support(\tau^x) \geq \alpha$ and $accuracy(\tau^x) \geq \beta$ in each $\psi \in DD(\Phi)$.

(Possible rule). An implication τ is a *possible rule*, if there is an instance τ^x satisfying $support(\tau^x) \geq \alpha$ and $accuracy(\tau^x) \geq \beta$ in at least one $\psi \in DD(\Phi)$.

In this definition, we say τ^x satisfying the constraint the *evidence* of rule τ. If $DD(\Phi)$ is a singleton set, namely Φ is a standard table, the certain rule and the possible rule define the same implication τ. Therefore, this definition completely include the definition of the rule in a standard table. Even though the computational complexity in the above definition depends upon $|DD(\Phi)|$, we can solve this problem by using Fig. 1. We have the following:

(Certain rule). An implication τ is a *certain rule*, if and only if there is an instance τ^x satisfying $minsupp(\tau^x) \geq \alpha$ and $minacc(\tau^x) \geq \beta$.

(Possible rule). An implication τ is a *possible rule*, if and only if there is an instance τ^x satisfying $maxsupp(\tau^x) \geq \alpha$ and $maxacc(\tau^x) \geq \beta$.

By using the above property, we adjusted the well-known *Apriori* algorithm [1–3] to the *NIS-Apriori* algorithm. The computational complexity of *NIS-Apriori* will be about the twice of the original *Apriori*. Moreover, *NIS-Apriori* is logically complete for the rules. Namely, *NIS-Apriori* generates only defined rules, and each defined rule (except the redundant rules) is obtainable by using *NIS-Apriori* [14].

3   Rule Generation from Various Types of Uncertain Data

This section considers rule generation from various types of uncertain data. Here, we propose rule generation in Figs. 2 and 3. If we specify the descriptors and the granules *inf* and *sup* in Fig. 2, we can similarly consider *Apriori*-based rule generation. Furthermore, we may cope with any types of data sets except table data in Fig. 3.

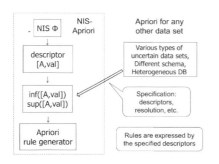

Fig. 2. Apriori-based rule generation based on the granules inf and sup.

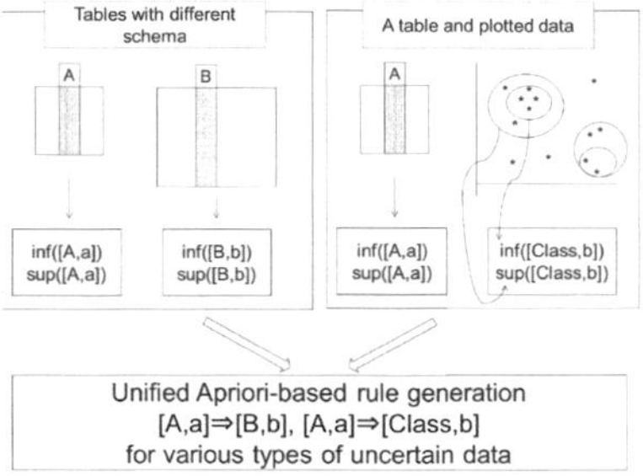

Fig. 3. Unified Apriori-based rule generation for various types of uncertain data.

3.1   Possible Cases, Granules Inf and Sup for Intervals

In Table 1, we see intervals for *height* and *weight*. As for interval data expressing information incompleteness, we similarly define possible cases by using the *resolution*, which specifies the length of the interval [14]. If we fix the resolution of *height* to 3.0, we see that $[158.0, 161.0]$, $[158.2, 161.2]$, $[158.234, 161.234]$, $[159, 162]$ are all possible cases in x_1. There are theoretically infinite number of intervals. For a table Φ with non-deterministic data and intervals, let $DD(\Phi)$ denote a set of all possible tables.

As for intervals, we see $g(x, A)$ means a subinterval $[low_{x,A}, up_{x,A}]$ $(min_A \leq low_{x,A} \leq up_{x,A} \leq max_A)$ for the total interval $[min_A, max_A]$. For a specified descriptor $[A, [low, up]]$, we define $inf([A, [low, up]])$ and $sup([A, [low, up]])$ in the following.

(1) $inf([A, [low, up]]) = \{x \in OB \mid g(x, A) \subset [low, up]\}$,
(2) $inf(\wedge_i[A_i, [low_i, up_i]]) = \cap_i inf([A_i, [low_i, up_i]])$,
(3) $sup([A, [low, up]]) = \{x \in OB \mid |g(x, A) \cap [low, up]| \geq resolution\}$,   (6)
(4) $sup(\wedge_i[A_i, [low_i, up_i]]) = \cap_i sup([A_i, [low_i, up_i]])$.
Here, $|g(x, A) \cap [low, up]|$ means the length of the interval.

3.2   An Example

In Table 1, we simply have two descriptors, $[sex, male]$ and $[sex, female]$ for the attribute *sex*, and the system generates the following lists.

```
infsup(1,1,[sex,male],[4,5],[2,3,4,5,6]).
infsup(1,2,[sex,female],[1],[1,2,3,6]).
```

The first formula shows $inf([sex, male]) = \{x_4, x_5\}$ and $sup([sex, male]) = \{x_2, x_3, x_4, x_5, x_6\}$.

Specification 1. For two attributes *height* and *weight*, we specify the following 6 and 7 attribute values, respectively.

```
attrib_values(2,height,6,[[150.0,155.0],[155.0,160.0],[160.0,
    165.0],[165.0,170.0],[170.0,175.0],[175.0,180.0]]).
resolution(2,interval,5.0).
attrib_values(3,weight,7,[[45.0,50.0],[50.0,55.0],[55.0,60.0],
    [60.0,65.0],[65.0,70.0],[70.0,75.0],[75.0,80.0]]).
resolution(3,interval,5.0).
```

By using this specification, the translation program generates *inf* and *sup* for each specified descriptors. For two descriptors [*height*, [160.0, 165.0]] and [*weight*, [60.0, 65.0]], we have the following.

```
infsup(2,3,[height,[160.0,165.0]],[2],[2,6]).
infsup(3,4,[weight,[60.0,65.0]],[2],[2]).
```

The resolution of the interval controls *inf* and *sup*. Since [160.0, 164.0] $\subset$ [160.0, 165.0] holds in x_2, $x_2 \in inf(height, [160.0, 165.0])$. For x_6, [160.0, 168.0] $\cap$ [160.0, 165.0]=[160.0, 165.0] and the length of the interval is greater than 5.0, therefore $x_6 \in sup(height, [160.0, 165.0])$. The system generates rules by using the above *inf* and *sup*.

```
File=[male7|rs] Support=0.1,Accuracy=0.5
------ 1st STEP ------------------------------------------------
------ Lower System --------------------------------------------
  [1] [height,[160.0,165.0]]==>[weight,[60.0,65.0]](0.167,0.5)
      Objects:[2]
  The Rest Candidates:[[[1,1],[3,5]],[[1,1],[3,6]]]
  (Next Candidates are Remained)
------ Upper System --------------------------------------------
  [1] [sex,male]==>[weight,[65.0,70.0]](0.333,0.667)
      :      :      :
  [8] [height,[160.0,165.0]]==>[weight,[60.0,65.0]](0.167,1.0)
      Objects:[2]
  [9] [height,[170.0,175.0]]==>[weight,[70.0,75.0]](0.167,1.0)
      Objects:[4] LOWER:Indefinite GC [4]
  The Rest Candidates: [[[1,1],[3,3]],[[1,1],[3,4]]]
  (Next Candidates are Remained)
EXEC_TIME=0.0(sec)
```

Based on the property of the *NIS-Apriori* algorithm, the obtained implication $\tau : [height, [160.0, 165.0]] \Rightarrow [weight, [60.0, 65.0]]$ satisfies $support(\tau^2) \geq 0.1$ and $accuracy(\tau^2) \geq 0.5$ in any possible case. This τ is not affected by information incompleteness. So, we will see τ a certain rule for the specified descriptors.

Specification 2. If we employ $resolution = 2.0$ for $height$ and $resolution = 2.0$ for $weight$, we have the following inf and sup.

```
infsup(2,3,[height,[160.0,165.0]],[2],[1,2,3,6]).
infsup(3,4,[weight,[60.0,65.0]],[2],[1,2]).
```

In this case, the above rule $\tau : [height, [160.0, 165.0]] \Rightarrow [weight, [60.0, 65.0]]$ in the lower system is no longer a rule, because $minacc(\tau^2) = |\{x_2\}|/|\{x_1, x_2, x_3, x_6\}| = 0.25 < 0.5$.

Specification 3. For descriptors $[height, [100.0, 200.0]]$ and $[weight, [45.0, 80.0]]$, we have the following.

```
infsup(2,1,[height,[100.0,200.0]],[1,2,3,4,5,6],[1,2,3,4,5,6]).
infsup(3,1,[weight,[45.0,80.0]],[1,2,3,4,5,6],[1,2,3,4,5,6]).
```

In this case, $\tau : [height, [100.0, 200.0]] \Rightarrow [weight, [45.0, 80.0]]$ is obtained as a certain rule, because $minsupp(\tau^1) = 1.0$ and $minacc(\tau^1) = 1.0$. However, this implication is trivial and meaningless.

3.3   Granules Inf and Sup for Various Types of Uncertain Data

In Fig. 2, we previously considered the left side rectangle area, namely only for non-deterministic data. We naturally employed descriptors $[A, val]$, where val appears as an attribute value. However, as we described in the interval data, we can similarly handle uncertain data, if the set of descriptors and granules inf, sup are given. After defining them, we can consider Table 3 for each implication $\tau : \wedge_i[A_i, val_i] \Rightarrow [Dec, val]$, and *NIS-Apriori* generates rules by using $minsupp(\tau^x)$, $minacc(\tau^x)$, $maxsupp(\tau^x)$, and $maxacc(\tau^x)$.

Like this, we can similarly obtain the certain rules and the possible rules by the specified descriptors. We propose the following procedure.

Apriori-based rule generation for various types of uncertain data
(Part 1). We specify the descriptors in each attribute.
(Part 2). We define $inf([A, val])$ and $sup([A, val])$ for each descriptor $[A, val]$. A rule is expressed by an implication of the specified descriptors.
(Part 3). The rule generator handles two types of implications below:
(Part 3-1). An implication τ satisfying $support(\tau^x) \geq \alpha$ and $accuracy(\tau^x) \geq \beta$ in each $\psi \in DD(\Phi)$ is obtained as a certain rule.
(Part 3-2). An implication τ satisfying $support(\tau^x) \geq \alpha$ and $accuracy(\tau^x) \geq \beta$ in at least one $\psi \in DD(\Phi)$ is obtained as a possible rule.

4   Concluding Remarks

In this paper, we considered the possibility of *Apriori*-based rule generation to various types of uncertain data. Since we can investigate (Part 1) and (Part 2) independently, it will be more convenient to develop *Apriori*-based rule generation. In order to present our framework, we are now coping with an experimental system. Some actual examples will be more comprehensive for us. Moreover, in (Part 1), the selection of the descriptors probably depends upon user's intension. It will be necessary for us to consider what are the proper descriptors for user's intension, and what are the proper granules inf and sup for user's intension. We need to customize them in each data set. In (Part 2), it is necessary to realize the environment for generating inf and sup from the specified descriptors. For handling big data, it also will be necessary to realize the rule generation environment in SQL [10].

Acknowledgment. This work is supported by JSPS (Japan Society for the Promotion of Science) KAKENHI Grant Number 26330277.

References

1. Agrawal, R., Srikant, R.: Fast algorithms for mining association rules. In: Proceedings of VLDB, pp. 487–499 (1994)
2. Agrawal, R., Mannila, H., Srikant, R., Toivonen, H., Verkamo, A.: Fast discovery of association rules. In: Advances in Knowledge Discovery and Data Mining, pp. 307–328. AAAI/MIT Press (1996)
3. Ceglar, A., Roddick, J.F.: Association mining. ACM Comput. Surv. **38**(2), 5 (2006)
4. Lipski, W.: On semantic issues connected with incomplete information data base. ACM Trans. DBS. **4**, 269–296 (1979)
5. Lipski, W.: On databases with incomplete information. J. ACM **28**, 41–70 (1981)
6. Nakata, M., Sakai, H.: Twofold rough approximations under incomplete information. Int. J. Gen. Syst. **42**(6), 546–571 (2013)
7. Orłowska, E., Pawlak, Z.: Representation of non-deterministic information. Theoret. Comput. Sci. **29**, 27–39 (1984)
8. Pawlak, Z.: Rough Sets. Kluwer Academic Publishers, Dordrecht (1991)
9. Pedrycz, W., Skowron, A., Kreinovich, V. (eds.): Handbook of Granular Computing. Wiley, Chichester (2008)
10. RNIA software logs. http://www.mns.kyutech.ac.jp/~sakai/RNIA
11. Sakai, H., Okuma, A.: Basic algorithms and tools for rough non-deterministic information analysis. Trans. Rough Sets **1**, 209–231 (2004)
12. Sakai, H., Ishibashi, R., Nakata, M.: On rules and apriori algorithm in non-deterministic information systems. Trans. Rough Sets **9**, 328–350 (2008)
13. Sakai, H., Wu, M., Nakata, M.: Division charts as granules and their merging algorithm for rule generation in nondeterministic data. Int. J. Intell. Syst. **28**(9), 865–882 (2013)
14. Sakai, H., Wu, M., Nakata, M.: Apriori-based rule generation in incomplete information databases and non-deterministic information systems. Fundam. Informaticae **130**(3), 343–376 (2014)

15. Skowron, A., Rauszer, C.: The discernibility matrices and functions in information systems. In: Intelligent Decision Support - Handbook of Advances and Applications of the Rough Set Theory, pp. 331–362. Kluwer Academic Publishers (1992)
16. UCI Machine Learning Repository. http://mlearn.ics.uci.edu/MLRepository.html
17. Wu, M., Nakata, M., Sakai, H.: An overview of the getRNIA system for non-deterministic data. Procedia Comput. Sci. **22**, 615–622 (2013)

A Biased-Randomized Algorithm for the Uncapacitated Facility Location Problem

Jesica de Armas[(✉)], Angel A. Juan, and Joan Manuel Marquès

Computer Science Department, IN3 - Open University of Catalonia, Barcelona, Spain
jde_armasa@uoc.edu

Abstract. Facility Location Problems (FLPs) have been widely studied in the fields of Operations Research and Computer Science. This is due to the fact that FLPs have numerous practical applications in different areas, from logistics (*e.g.*, placement of distribution or retailing centers) to Internet computing (*e.g.*, placement of cloud-service servers on a distributed network). In this paper we propose a biased iterated local search algorithm for solving the uncapacitated version of the FLP. Biased randomization of heuristics has been successfully applied in the past to solve other combinatorial optimization problems in logistics, transportation, and production -*e.g.*, different vehicle and arc routing problems as well as scheduling problems. Our approach integrates a biased randomization within an Iterated Local Search framework. Several standard benchmarks from the literature have been used to prove the quality and efficiency of the proposed algorithm.

Keywords: Facility location problem
Biased-randomized algorithms · Iterated local search · Metaheuristics

1 Introduction

The Facility Location Problem (FLP), originally introduced by Stollsteimer [22] and Balinski [2], involves locating an undetermined number of facilities (decision variable) to minimize the sum of the setup costs of these facilities and the costs of serving the customers from these facilities. Most versions of the problem assume that the alternative sites where the facilities can be located are predetermined, and also that the demand associated with each customer is known in advance. Moreover, the uncapacitated version of the FLP assumes that there is no limit on the demand (*i.e.*, number of customers) that can be served from each single facility. Usually, decisions on facility location are difficult to reverse due to the fixed costs associated with opening a facility. The FLP is useful to model allocation problems in very diverse application fields, from logistics and inventory planning -*e.g.*, where to allocate distribution or retailing centers in a supply chain- to telecommunication and computing networks -*e.g.*, where to allocate cloud-service servers in a distributed network or cabinets in optical fiber networks.

© Springer International Publishing AG, part of Springer Nature 2018
A. M. Gil-Lafuente et al. (Eds.): FIM 2015, AISC 730, pp. 287–298, 2018.
https://doi.org/10.1007/978-3-319-75792-6_22

According to Verter (2011), the uncapacitated FLP, where both the alternative facility locations and the customer positions are discrete points in the graph, is considered to be the simplest version of FLP. Even without the constraint on the facility capacity, the FLP is proved to be NP-hard [4]. An example of a FLP problem instance can be found in Fig. 1. This assumes that the alternative sites have been predetermined and the demand in each customer zone is concentrated at the customer point representing that region. FLP focuses on the production and distribution of a single commodity over a single time period, during which the demand is (usually) assumed to be known with certainty. The distinguishing feature of this basic discrete location problem, however, is the decision maker's ability to determine the size of each facility without any restriction. More formally, the uncapacitated FLP is defined over an undirected graph $G = (F, C, E)$, where F is a non-empty subset of facilities (each of them with unlimited capacity), C is a non-empty set of customers to be served, and E is a set of edges connecting each customer $j \in C$ with some of the facilities in F (Fig. 1). Delivering a customer $j \in C$ throughout a facility $i \in F$ has a service cost, $c_{ij} \geq 0$. Also, each facility $i \in F$ has a fixed opening cost, $f_i \geq 0$. Let X denote the set of open facilities (decision variable), with $\emptyset \leq X \subseteq F$. Let $\sigma : C \to X$ be a function assigning to each customer $j \in C$ a facility $\sigma(j) \in X$ satisfying that $c_{\sigma(j)j} = \min_{i \in X} \{c_{ij}\}$. Under these circumstances, the uncapacitated FLP consist in minimizing the total cost of providing service to all customers, *i.e.*:

$$\min \sum_{i \in X} f_i + \sum_{j \in C} c_{\sigma(j)j}, \text{ subject to } \emptyset \leq X \subseteq F \tag{1}$$

In the literature, the uncapacitated FLP is also called the simple facility location problem, the simple warehouse location problem or the simple plant location problem. Many successful application of the problem model have been reported [16, 18, 24]. Also, practical problems without facilities to locate, such as cluster analysis, machine scheduling, economic lot sizing, portfolio management and computer network design, can be successfully modeled as uncapacitated FLP problems [23]. Having this wide range of application makes the uncapacitated FLP very interesting to be studied. Solutions methods have been proposed since more than 40 years ago [1, 7, 14, 15, 20].

For an extensive literature review on the FLP, the reader is addressed to Drezner [5], Snyder [21], and Fotakis [6]. However, the use of fast biased-randomized heuristics has not been explored for this problem in the past. In this paper we propose a new algorithm which combines biased randomization of a heuristic with a local search framework for solving the uncapacitated FLP.

The remainder of this paper is structured as follows: Sect. 2 offers an overview of the proposed algorithm; some numerical experiments that allow to test the efficiency of our approach are discussed in Sect. 3; finally, Sect. 4 highlights the main conclusions of this paper and proposes some open research lines.

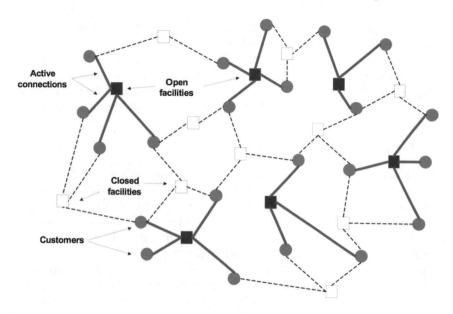

Fig. 1. Illustrating example of the uncapacitated FLP

2 Our Approach

In this paper we propose a hybrid approach that combines an Iterated Local Search (ILS) metaheuristic [17] with biased-randomization techniques [9,11] to deal with the FLP.

First, we define some basic ideas (which we could call as a basic heuristic) for solving the problem with the main property of being of fast execution while obtaining results not so far from the optima. Then, by randomizing some steps of the deterministic heuristic, it is transformed into a randomized procedure. Therefore, it can be run multiple times in order to obtain different solutions for the problem. These executions, being independent one from another, can be executed either in sequential or parallel mode.

Notice that it is possible to perform this randomization process without losing the logic behind the deterministic heuristic by using a biased probability distribution in the procedure. To achieve this, the process can give higher weight to the candidates that are of best choices according to the deterministic heuristic.

With this randomized process, alternative feasible solutions of a similar quality are easily generated. Being an algorithm with few input parameters, our approach represents an interesting and competitive alternative to other state-of-the-art metaheuristics, which usually require more cumbersome, time-costly and instance-dependent fine-tuning processes. In practice, these approaches are usually harder to implement and often not reproducible. Furthermore, Kant et al. [12] suggest that most efficient heuristics and metaheuristics are not used in practice because of the difficulties that arise when dealing with real-life scenarios.

To the best of our knowledge, this is the first proposal that uses biased-randomization techniques to efficiently solve the FLP.

The proposed method, similarly to GRASP [20], relies on a biased constructive procedure to generate new solutions. This class of methods has been less studied in the literature and can offer competitive results when compared to more sophisticated approaches, which are difficult to implement in practice.

2.1   Overview of Our Approach

The proposed algorithm is detailed on Fig. 2 as a flow chart and works as follows. First, it loads the problem instance. Then, it generates an initial random solution, which is the starting point for the ILS procedure. To generate this initial solution, the algorithm chooses a random number p_i between $|F|/2$ and $|F|$, and picks randomly p_i facilities to open. After that, it computes the total cost of the generated solution, which is selected as starting point.

The basis for selecting always more than the half of facilities to open in the initial solution is not casual. Regarding computational costs, the fact of closing a facility on the solution is cheaper than opening a solution. This is due to the way that the costs are calculated. When a facility is selected to be closed, for updating the solution costs we only have to relocate the customers that were assigned to the closed facilities, but the rest of customers assigned to the other facilities remain unmodified. However, when opening a facility in the solution, every single client assigned to a facility must be evaluated in order to determine if the newly open facility is the one with lesser assignment cost for this customer among all the open facilities in the current solution. For this reason, if starting with a solution with a higher number of open facilities, latter there will be more probabilities of improving the solution by performing closing movements which are less costly. Therefore, the whole process will be cheaper.

After this initial solution is generated, a local search procedure is applied to refine it. This procedure combines independent closing and opening movements, and then swap movements. Therefore, it is divided in two parts. Firstly, it starts by trying to close an open facility in the current solution, and regardless of whether it improves the solution (i.e. a solution with lower total cost is obtained) or not, then an opening movements is also tried. This process is iterated until a certain number of consecutively non-improving iterations is reached. On the second stage, it performs a swap movement, which consists of replacing an open facility with a closed facility, until a certain number of consecutively non-improving iterations is reached. Further details on the way the facilities are selected to be open or closed are specified in Sect. 2.2.

The solution generated will finally be the starting point for the ILS framework and considered as base solution. The ILS is mainly an iterative process which, at each iteration, generates a new feasible solution with chances to outperform the base solution. Our ILS consists of three steps:

1. Destruction/construction of the solution (perturbation)
2. Refinement of the solution (local search)
3. Acceptance criteria of the solution

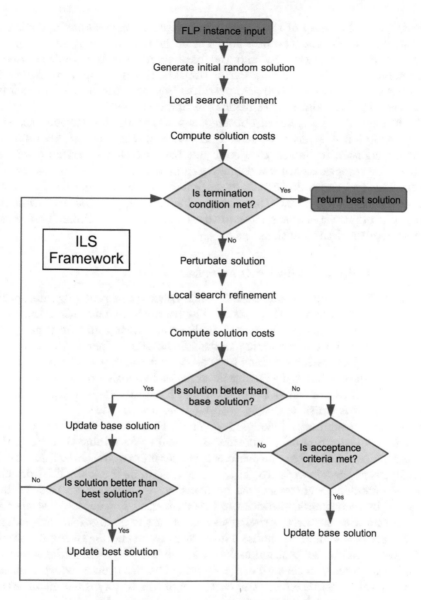

Fig. 2. Flow diagram of the proposed approach

On the first step, a perturbation operator is applied to the solution. This operator basically destructs some part of the solution by removing open facilities, and then reconstructs it by opening new facilities. This operator always opens more facilities than the amount of closed ones, so it benefits from the fact that the closing movement is less computationally expensive, as happened with the generation of the initial solution. This solution is then refined by the same local search operator used on the initial solution.

Finally, the last step of the ILS is the acceptance criteria for updating the best and base solutions. The best solution is the one that will be returned as the result of our methodology and the base solution is the solution used as initial solution for the next iteration of the ILS. If the solution obtained from the perturbation and local search procedures improves the best solution found so far, then the best and base solutions are updated.

Additionally, if the solution obtained is worse than the current base solution, an acceptance criterion is defined. This acceptance criterion allows a non-improving solution to be accepted as a new base solution if certain conditions are met. The acceptance criteria define a gap in which the solution is allowed to worsen. This gap varies during the execution of the algorithm, allowing greater gaps at the beginning of the execution, and smaller gap as the execution time passes. With that, we enable a method to escape from local minima and explore different regions of the solutions space.

2.2   Friendship and Enmity Concepts

In general, if two facilities are close to each other, then probably the optimal solution will not contain both facilities. The rationale behind this is that these facilities are competing in the same small area, which could only needs one facility to serve the corresponding nodes of the zone. Therefore, taking into account this aspect, we have defined two concepts: friendship and enmity between facilities. On the one hand, if a facility is very close to another one, it is considered that both facilities are enemies. On the other hand, if two facilities are far one another, then it is considered that both facilities are friends.

Since the benchmarks belonging to the UFLP do not contain distances between facilities or any other information useful to determine them (*e.g.* their coordinates), we have taken one customer as reference to sort all facilities by distance. It means that, in order to get a way to determine if two facilities are close or far one another, a customer has been taking as reference, so that facilities are sorted by increasing distance from this customer. The chosen customer has been the one with smallest variance according to the distance to each existing facility, since it is similar to choose a customer at one end. Selecting a customer at leftmost or rightmost position makes that the majority of facilities are at one side of the customer, so that we are preventing the situation in which two facilities at the same distance from the customer but in totally different directions are said to be enemies but they are not.

This way we always have all facilities increasingly sorted by distance to the reference customer in a list (hereinafter Sorted List), and if it is necessary to open a new facility, this facility will be a friend of one open facility, *i.e.*, a close facility to its left or its right in the Sorted List. On the contrary, if it is necessary to close an open facility, this facility will be an enemy of one open facility, *i.e.*, an open facility far from its left or its right in the Sorted List.

The selection of the reference open facility with friends or enemies is randomly made. In order to avoid the selection of the same facility as reference when it does not involve an improvement of the solution, we have applied a number of

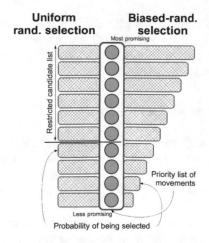

Fig. 3. Uniform randomization vs. biased randomization

iterations in which the same facility cannot be selected. This number of iterations makes a work similar to a tabu list.

2.3 Randomization Issues

The approach presented in this paper is based on applying biased-randomization techniques at some stages. Roughly speaking, biased random sampling can be defined as a set of techniques that make use of random numbers and asymmetric probability distributions to solve certain stochastic and deterministic problems. When properly combined with heuristics, random sampling has proven to be extremely useful for solving stochastic problems like the Arc Routing Problem with Stochastic Demands [8]. Similarly, it can also solve efficiently deterministic routing problem such as the Capacitated Vehicle Routing Problem [11].

One of the key processes on our algorithm is the randomization on the selection of facilities in different stages of the procedure. This randomization will allow us to widely explore the space solutions. For example, the local search procedures use randomization when exploring facilities, sorting them randomly. Finally, this randomization process has to be done without introducing too many parameters in the algorithm. Otherwise, a non-trivial, time-consuming and often instance-dependent fine-tuning process should be performed. Hence, biased probability functions, such as the single-parameter geometric distribution or the parameter-free decreasing triangular distribution, may be exploited.

We named this general methodology as Multi-start biased Randomization of classical Heuristics with Adaptive local search (MIRHA). When compared to GRASP, our approach is not restricting the list of candidates to be evaluated, so we are leaving out one parameter. Still, the randomization is biased, since all the candidate facilities are not considered with the same probability (Fig. 3). One of the probability distribution functions which have demonstrated to obtain competitive results when combined with the MIRHA is the single-parameter geometric distribution. The proposed algorithm will use the geometric distribution, which will add the single parameter for the algorithm: the beta parameter for the probability distribution.

3   Computational Experiments

In order to assess the performance of the proposed algorithm, several computational experiments were performed. The proposed algorithm was implemented as a Java 7 SE application. We performed all the tests on a commodity desktop computer with an Intel(R) Xeon(R) E5603 at 1.60 GHz and 8 GB RAM running Windows 7.

Even though Java is a programming language executed in a virtual machine (JVM) and we are aware it may show poorer performance than others like C or C++, the vast amount of tools available in the standard API (such as advanced structures or garbage collection) and its object-orientation eased the development process. In addition, the execution on the JVM offers better replicability and repeatability than other languages.

With the aim of testing the efficiency of the proposed algorithm, three different classes of problem instances obtained from Hoefer (2014) [10] were used. The selected datasets were chosen with the criteria of testing the algorithm against instances of small, medium and medium-large size (in terms of facilities and customers included in the graph). We briefly describe next the used sets, but the reader is referred to Hoefer [10] for further details of each class of instances.

- BK: a small-sized class introduced by Bilde and Krarup [3]. Includes 220 instances in total divided in 22 subsets, with the number of facilities varying from 30 to 50 and the number of customers from 80 to 100. These instances were artificially generated by the authors, selecting the assignment costs randomly on the range [0, 1000], and opening costs being always greater than 1000.
- GAP: a medium-sized class also introduced by Kochetov and Ivanenko (2005) [13]. Consists of three subsets, each with 30 instances: GAPA, GAPB and GAPC, being GAPC the hardest. These instances are considered to be especially hard for dual-based methods.
- FPP: a medium-sized class introduced by Kochetov and Ivanenko (2005) [13]. Consists of two subsets, each with 40 instances: FPP11 and FPP 17. Although optimal solutions in this class can be found in polynomial times, the instances are hard for algorithms based on flip and swap local search, since each instance has a large number of strong local optima.

The experiments were run by setting a stopping criteria based on a time limit for every problem instance. Every instance is solved by generating feasible solution until the termination criteria of time has been met (see Fig. 2). These termination criteria will be different depending on the size and the class of instances being solved. So, the greater the instances are, the more time it is given to the algorithm (see Table 1). We have set these times the great enough to guarantee that the algorithm reaches a reasonably good solution. We record the time when the best solution was found for every instance, so we can know how fast the convergence to this best solution was.

Table 1. Termination criteria for every class of instances

Class	No. of facilities (n)	Time (secs.)
BK	80–100	10
GAPA	100	20
GAPB	100	20
GAPC	100	20
FPP11	133	100
FPP17	307	200

Next, we show the results obtained when executing our algorithm in the earlier explained benchmarks. Each instance has been run 10 times with different seeds, and the average has been calculated. We compare our results with the ones declared by Resende and Werneck [19] using a GRASP algorithm, since to best of our knowledge they reached the closest results to the best-known solution for a wide range of different instance sets. We would like to remark that our experiments were performed on a low-end commodity desktop computer, as opposed to the supercomputer used by Resende and Werneck [19]; also, we used a standard Java SE application, instead of the specifically compiled application utilized by Resende and Werneck [19].

We started evaluating the methodology in the set of tests with simplest and smallest problem instances, the Bilde and Krarup [3] benchmark. Note that the last column of each table shows the gap between the best known solutions and the best solution obtained from our 10 runs with different seeds (BEST10), while the other gaps are calculated taking into account the average results after 10 runs (AVG10).

For this test, our algorithm outperforms the GRASP proposal. We find the optimal solution for practically every instance in the benchmark in much shorter execution times (Table 2). Then, we performed the next evaluation over the GAP benchmark, known as a hard test for dual-based oriented methods. Neither our algorithm nor GRASP were thought specifically for dual-based problems, so finding good quality results in these instances is a challenging task (Table 3).

For this set of tests, our algorithm obtains slightly better quality solutions. Although the running times employed by our proposal are slightly higher than the ones reported in Resende's experiments, we have run all our experiments on a desktop computer. Note that for this set of test the gap between the best known solutions and our best solutions among the 10 runs is much smaller. This means that running our proposal a few times it is possible to obtain a

Table 2. Results obtained for the 22 BK subsets of instances

Instance	# customers	# facilities	GRASP AVG10	t (s)	ILS-UFLP AVG10	t (s)	BEST10
B	100	50	0.000	0.310	0.000	< 0.010	0.000
C			0.016	0.450	0.000	< 0.010	0.000
D01	80	30	0.000	0.223	0.001	< 0.010	0.001
D02			0.000	0.211	0.000	< 0.010	0.000
D03			0.000	0.199	0.000	< 0.010	0.000
D04			0.000	0.170	0.000	< 0.010	0.000
D05			0.000	0.162	0.000	< 0.010	0.000
D06			0.000	0.186	0.000	< 0.010	0.000
D07			0.000	0.174	0.000	< 0.010	0.000
D08			0.000	0.166	0.000	< 0.010	0.000
D09			0.000	0.175	0.000	< 0.010	0.000
D10			0.000	0.166	0.000	< 0.010	0.000
E01	100	50	0.000	0.476	0.000	< 0.010	0.000
E02			0.000	0.588	0.000	< 0.010	0.000
E03			0.019	0.512	0.013	< 0.010	0.000
E04			0.000	0.464	0.000	< 0.010	0.000
E05			0.000	0.376	0.000	< 0.010	0.000
E06			0.000	0.408	0.000	< 0.010	0.000
E07			0.000	0.416	0.000	< 0.010	0.000
E08			0.000	0.418	0.000	< 0.010	0.000
E09			0.000	0.352	0.000	< 0.010	0.000
E10			0.000	0.353	0.000	< 0.010	0.000
Average			0.002	0.316	0.001	< 0.010	0.001

Table 3. Results obtained for GAP subsets of instances

Instance	# customers	# facilities	GRASP AVG10	t (s)	ILS-UFLP AVG10	t (s)	BEST10
GAP A	100	100	5.140	1.41	4.660	3.403	0.55
GAP B			5.980	1.81	5.882	3.493	1.88
GAP C			6.740	1.89	6.560	4.217	1.68
Average			5.953	1.700	5.701	3.704	1.369

high quality solution very close to the optimal. The next test was performed over the FPP benchmark, known to include very hard instances for swapping algorithms. Our algorithm is not purely in this family, since we also close and open an independent number of facilities at each overcoming iteration.

In this experimentation, our methodology clearly outperforms the quality of the results obtained by the Resende's proposal. Additionally, our approach is able to find the optimal solution for many of these individual tests in some runs, so that in the last column we can see really small averages of the best gaps for FPP11 and FPP17 groups of instances (Table 4).

With all the shown benchmarks, we can conclude our methodology is quite competitive at least in small and medium-scale scenarios, especially on those with short lists of open facilities. Thus, our algorithm could be a valuable tool for reduced or clustered scenarios, on which a large amount of clients could be served by a small amount of facilities.

Table 4. Results obtained for FPP subsets of instances

Instance	# customers	# facilities	GRASP AVG10	t (s)	ILS-UFLP AVG10	t (s)	BEST10
FPP 11	133	133	8.480	2.58	1.200	4.290	0.020
FPP 17	307	307	58.270	25.18	40.192	46.886	0.060
Average			33.375	13.88	20.696	25.597	0.041

4 Conclusions

In this paper we have presented a new algorithm for solving the uncapacitated FLP. This algorithm makes use of biased-randomized techniques inside an Iterated Local Search metaheuristic framework. Having fast execution times, it does not require any time-consuming fine-tuning process, as it only requires of few parameters. These parameters correspond to the characterization of the geometric distribution applied in the biased-randomization processes used during the algorithm execution, and to the percentages of construction/destruction of the solutions in the perturbation phase. The algorithm has proven to obtain competitive results for the problem.

Acknowledgments. This work has been partially supported by the Spanish Ministry of Economy and Competitiveness (TRA2013-48180-C3-P) and FEDER. Likewise we want to acknowledge the support received by the Department of Universities, Research & Information Society of the Catalan Government (2014-CTP-00001).

References

1. Alves, M., Almeida, M.: Simulated annealing algorithm for the simple plant location problem: a computational study. Rev. Invest. Operacional **12**, 1–31 (1992)
2. Balinsky, M.L.: On Finding Integer Solutions to Linear Programs (1964)
3. Bilde, O., Krarup, J.: Sharp lower bounds and efficient algorithms for the simple plant location problem. In: Hammer, P.L., Johnson, E.L., Korte, B.H., Nemhauser, G.L. (eds.) Studies in Integer Programming, Annals of Discrete Mathematics, vol. 1, pp. 79–97. Elsevier (1977)
4. Cornuejols, G., Nemhauser, G., Wolsey, L.: The uncapacitated facility location problem. In: Mirchandani, P., Francis, R. (eds.) Discrete Location Theory, pp. 119–171. Springer (1990)
5. Drezner, Z.: Facility Location: A Survey of Applications and Methods. Springer, New York (1995)
6. Fotakis, D.: Online and incremental algorithms for facility location. SIGACT News **42**(1), 97–131 (2011)
7. Ghosh, D.: Neighborhood search heuristics for the uncapacitated facility location problem. Eur. J. Oper. Res. **150**(1), 150–162 (2003). o.R. Applied to Health Services
8. Gonzalez, S., Riera, D., Juan, A., Elizondo, M., Fonseca, P.: Sim-RandSHARP: a hybrid algorithm for solving the arc routing problem with stochastic demands. In: Proceedings of the 2012 Winter Simulation Conference (WSC), pp. 1–11, December 2012

9. González-Martín, S., Juan, A.A., Riera, D., Castellà, Q., Muñoz, R., Pérez, A.: Development and assessment of the SHARP and RandSHARP algorithms for the arc routing problem. AI Commun. **25**(2), 173–189 (2012)

10. Hoefer, M.: (2014). http://www.mpi-inf.mpg.de/departments/d1/projects/benchmarks/UflLib/

11. Juan, A., Faulin, J., Jorba, J., Riera, D., Masip, D., Barrios, B.: On the use of monte carlo simulation, cache and splitting techniques to improve the clarke and wright savings heuristics. J. Oper. Res. Soc. **62**, 1085–1097 (2011)

12. Kant, G., Jacks, M., Aantjes, C.: Coca-Cola enterprises optimizes vehicle routes for efficient product delivery. Interfaces **38**(1), 40–50 (2008)

13. Kochetov, Y., Ivanenko, D.: Computationally difficult instances for the uncapacitated facility location problem. In: Ibaraki, T., Nonobe, K., Yagiura, M. (eds.) Metaheuristics: Progress as Real Problem Solvers, Operations Research/Computer Science Interfaces Series, vol. 32, pp. 351–367. Springer, US (2005)

14. Kuehn, A.A., Hamburger, M.J.: A heuristic program for locating warehouses. Manag. Sci. **9**(4), 643–666 (1963)

15. Lai, M.C., suk Sohn, H., Tseng, T.L.B., Chiang, C.: A hybrid algorithm for capacitated plant location problem. Expert Syst. Appl. **37**(12), 8599–8605 (2010)

16. Lee, G., Murray, A.T.: Maximal covering with network survivability requirements in wireless mesh networks. Comput. Environ. Urban Syst. **34**(1), 49–57 (2010)

17. Lourenço, H., Martin, O., Stuetzle, T.: Iterated local search: framework and applications. In: Gendreau, M., Potvin, J.Y. (eds.) Handbook of Metaheuristics, International Series in Operations Research and Management Science, vol. 146, pp. 363–397. Springer, US (2010)

18. Marić, M., Stanimirović, Z., Božović, S.: Hybrid metaheuristic method for determining locations for long-term health care facilities. Ann. Oper. Res. **227**(1), 3–23 (2013)

19. Resende, M.G., Werneck, R.F.: A hybrid multistart heuristic for the uncapacitated facility location problem. Eur. J. Oper. Res. **174**(1), 54–68 (2006)

20. Resende, M., Ribeiro, C.: Greedy randomized adaptive search procedures: advances, hybridizations, and applications. In: Gendreau, M., Potvin, J.Y. (eds.) Handbook of Metaheuristics, International Series in Operations Research and Management Science, vol. 146, pp. 283–319. Springer, US (2010)

21. Snyder, L.V.: Facility location under uncertainty: a review. IIE Trans. **38**(7), 547–564 (2006)

22. Stollsteimer, J.: The Effect of Technical Change and Output Expansion on the Optimum Number, Size, and Location of Pear Marketing Facilities in a California Pear Producing Region. University of California, Berkeley (1961)

23. Sun, M.: Solving the uncapacitated facility location problem using tabu search. Comput. Oper. Res. **33**(9), 2563–2589 (2006)

24. Thouin, F., Coates, M.: Equipment allocation in video-on-demand network deployments. ACM Trans. Multimedia Comput. Commun. Appl. **5**(1), 5:1–5:22 (2008)

A Methodology for the Valuation of Woman's Work Culture in a Fuzzy Environment

Anna M. Gil-Lafuente[1] and Beatrice Leustean[2(✉)]

[1] University of Barcelona, Barcelona, Spain
amgil@ub.edu
[2] Politehnica University of Bucharest, Bucharest, Romania
leusteanbeatrice@yahoo.com

Abstract. Academic research in gender payment gaps and woman's economic behavior has been developing over the last decades. The European Project itself edifies on the Lisbon Treaty's values of respect for human dignity, freedom, equality, etc. Therefore important contributions have been published in the main journals of the field of economic and social research. This paper analyzes the results of the research concerning woman's work culture in all economic fields by using bibliometric indicators. The main results are summarized in two fundamental issues. First, the citation structure, in economics and social sciences, is presented. Next, the paper studies the influence of the journals by using a wide range of indicators including publications, citations and the h-index. There is found the results are in accordance with the expectations, that the research on the woman's economic role is very narrow and needs to be enriched with scholar further researches.

Keywords: Bibliometric evaluation · Web of science · Journal rankings
Woman's work culture

JEL: C80 · I24 · Z130

1 Introduction

This paper utilize a bibliometric evaluation [1–3, 8] to comprise the field of research and to quantitatively describe the scientific contributions to the culture of work, in general, and to the specific topic of women's work culture [4]. The authors generally consider that the culture of origins is an important field of research in locating "the evil" in the gender payment gaps [4], disparities of incomes among regions, un-synchronicities and the lack or the delays in the convergence to the medium levels of the EU. There are empirical evidences that the culture of women's work has many similarities across Europe. A coherent tool is presumed to be found, from this study and further. This will be applied for finding relevant results in what regards the role of the culture of women's work on the socio-economical behavior.

This approach is preferred to the qualitative state of the art because of the need to measure the results of the scientific works on the general and on the specific topic, it's impact and it's diffusion. The bibliometric study's main goal is to regain the strength for

© Springer International Publishing AG, part of Springer Nature 2018
A. M. Gil-Lafuente et al. (Eds.): FIM 2015, AISC 730, pp. 299–306, 2018.
https://doi.org/10.1007/978-3-319-75792-6_23

continuing the search of finding the most appropriate indicators or transfer papers and also evaluating the efficiency of scientific results in a world of change and competition [6].

2 Woman's Work Culture in the World of Science

Questioning World of Science databases as considered with the most influence on the scientific community after the key words: "women's work culture" there were found 1904 articles published in more than 100 journals from 1950 until 2015. Refining the search to better comprehend the field of research towards social sciences and business economics or sociology, there could be found 527 articles from all databases, 2296 times cited with an average citation per item of 4.36 and an h index of 26. The most cited 20 articles from our area of research are the following ones:

1. Cain, M.; Khanam, S.R.; Nahar, S., *Class, Patriarchy, and Women's Work in Bangladesh,* POPULATION AND DEVELOPMENT REVIEW, Volume: 5, Issue: 3 Pages: 405–438, Published: 1979, https://doi.org/10.2307/1972079, Times cited: 132
2. Post, RC; Siegel, RB; *Legislative constitutionalism and section five power: Policentric interpretation of the family and medical leave act,* YALE LAW JOURNAL Volume: 112, Issue: 8, Pages:1943-+, Published: JUN 2003, https://doi.org/10.2307/3657473, Times cited: 110.
3. Fernandez, R.; Fogli, A.; *Culture: An Empirical Investigation of Beliefs, Work, and Fertility,* AMERICAN ECONOMIC JOURNAL-MACROECONOMICS, Volume: 1, Issue: 1, Pages: 146–177, Published: JAN 2009, https://doi.org/10.1257/mac.1.1.146, Times cited: 92
4. Lerner, M.; Brush, C.; Hisrich, R.; *Israeli women entrepreneurs: An examination of factors affecting performance,* JOURNAL OF BUSINESS VENTURING, Volume: 12, Issue: 4, Pages: 315–339, Published: JUL 1997, https://doi.org/10.1016/s0883-9026(96)00061-4, Times cited: 67
5. Lehrer, E.; Nerlove, M.;*Female Labor-Force Behavior and Fertility in the United-States,* ANNUAL REVIEW OF SOCIOLOGY, Volume: 12, Pages: 181–204, Published: 1986, Times cited: 58
6. Greene, K.; Brinn, L.S., *Messages influencing college women's tanning bed use: Statistical versus narrative evidence format and a self-assessment to increase perceived susceptibility,* Conference: Annual Conference of the National-Communication-Association Location: New Orleans, Louisiana, Date: Nov, 2002, JOURNAL OF HEALTH COMMUNICATION, Volume: 8, Issue: 5, Pages: 443–461, Published: SEP-OCT 2003, https://doi.org/10.1080/10810730390233271, Times cited: 58.
7. Shortall, S.; *Gendered agricultural and rural restructuring: A case study of Northern Ireland,* SOCIOLOGIA RURALIS, Volume: 42, Issue: 2, Pages: 160-+, Published: APR 2002, https://doi.org/10.1111/1467-9523.00208, Times cited: 52.
8. Fernandez, Raquel; Alfred Marshall Lecture - Women, work, and, culture, Conference: 21th Annual Congress of the European-Economic-Association Location: Vienna, AUSTRIA Date: Aug 24–28, 2006, JOURNAL OF THE EUROPEAN ECONOMIC ASSOCIATION, Volume: 5, Issue: 2–3, Pages: 305–332, Published: Apr-May 2007, https://doi.org/10.1162/jeea.2007.5.2-3.305, Times cited: 50.

9. Boyd, M.; *At a Disadvantage - The Occupational Attainments of Foreign Born Women In Canada*, INTERNATIONAL MIGRATION REVIEW, Volume: 18, Issue: 4, Pages: 1091–1119, Published: 1984, https://doi.org/10.2307/2546074, Times cited: 49

10. Sunder, M; *Piercing the veil*, YALE LAW JOURNAL, Volume: 112, Issue: 6, Pages: 1399-+, Published: APR 2003, https://doi.org/10.2307/3657449, Times cited: 46

11. Deem, R; *Gender, organizational cultures and the practices of manager-academics in UK universities*, Conference: Symposium on the Management of Universities Location: London, England, Date: Jun 20, 2001, GENDER WORK AND ORGANIZATION, Volume: 10, Issue: 2, Pages: 239–259, Published: Mar 2003, https://doi.org/10.1111/1468-0432.t01-1-00013, Times cited: 46.

12. Buttel, F.H.; Gillespie, G.W., The Sexual Division of Farm Household Labor – *An Exploratory Study of the Structure of On-Farm and Off-Farm Labor Allocation Among Men and Women*, RURAL SOCIOLOGY Volume: 49 Issue: 2 Pages: 183–209 Published: 1984, Times cited: 46

13. Van Vianen, AEM; Fischer, AH, *Illuminating the glass ceiling: The role of organizational culture preferences*, JOURNAL OF OCCUPATIONAL AND ORGANIZATIONAL PSYCHOLOGY, Volume: 75, Pages: 315–337, Part: 3 Published: SEP 2002, Times cited: 45

14. Bolton, SC, *Women's work, dirty work: The gynaecology nurse as "other"*, GENDER WORK AND ORGANIZATION, Volume: 12, Issue: 2, Pages: 169–186, Published: MAR 2005, https://doi.org/10.1111/j.1468-0432.2005.00268.x, Times cited: 44

15. Huisman, K A; *Wife battering in Asian American communities. Identifying the service needs of an overlooked segment of the U.S. population*, VIOLENCE AGAINST WOMEN, Volume: 2, Issue: 3, Pages: 260–83, Published: 1996-Sep, https://doi.org/10.1177/1077801296002003003, Times cited 39

16. Saugeres, L; *Of tractors and men: Masculinity, technology and power in a French farming community*, SOCIOLOGIA RURALIS, Volume: 42, Issue: 2, Pages: 143-+, Published: APR 2002, https://doi.org/10.1111/1467-9523.00207, Times cited: 36

17. Enchautegui, M.E.; *Welfare payments and other economic determinants of female migration*, JOURNAL OF LABOR ECONOMICS, Volume: 15, Issue: 3, Pages: 529–554, Part: 1, Published: Jul 1997, https://doi.org/10.1086/209871, Times cited: 34.

18. Powell, A.; Bagilhole, B.; Dainty, A., *How Women Engineers Do and Undo Gender: Consequences for Gender Equality*, GENDER WORK AND ORGANIZATION Volume: 16, Issue: 4, Pages: 411–428, Published: JUL 2009, https://doi.org/10.1111/j.1468-0432.2008.00406.x, Times cited: 31

19. Fratkin, E.; Smith, K.; *Women's Changing Economic Roles with Pastoral Sedentarization – Varying Strategies in Alternate Rendille Communities*, HUMAN ECOLOGY, Volume: 23, Issue: 4, Pages: 433–454, Published: Dec 1995, https://doi.org/10.1007/bf01190131, Times cited: 30

20. Bokemeier, J.L.; Tickamyer, A.R., *Labor-Force Experiences of Nonmetropolitan Women*, RURAL SOCIOLOGY, Volume: 50, Issue: 1, Pages: 51–73, Published: 1985, Times cited: 31

Continuing to present the distribution of the published articles concerning women's work culture in the least 40 years, between 1974 and 2014, we find out the following distribution. Figure 1 shows that there is a low interest in scientific records of women's work culture comparing with the 9607 results found searching after the key words: "working culture". These articles were 93327 cited, with 11.11 average cites per item and an h index of 119 (Table 1).

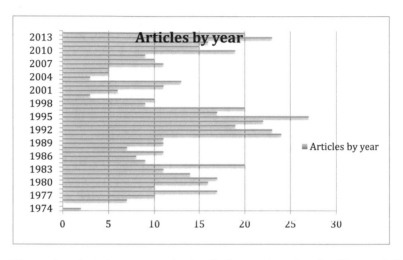

Fig. 1. The number of articles per year returning the "women's work culture" keywords Source: Author's elaboration based on the Web of Science data

Table 1. Articles per year in Web of Science that are containing "women's work culture" Data Source: Web of Science

1974	2	1988	7	2002	11
1975	0	1989	11	2003	13
1976	7	1990	11	2004	3
1977	10	1991	24	2005	5
1978	17	1992	23	2006	5
1979	10	1993	19	2007	11
1980	16	1994	22	2008	10
1981	17	1995	27	2009	9
1982	14	1996	17	2010	19
1983	11	1997	20	2011	15
1984	20	1998	9	2012	16
1985	9	1999	10	2013	23
1986	8	2000	3	2014	20
1987	11	2001	6		

We notice that there is somehow an increase between 1992 and 1997 and after 2010 but the maximum level of publications on this topic was only 27, in 1995 and there were 23 in 2013.

3   The Ranking of the Reviews that Return "Woman's Work Culture" in the Social Sciences

Following the methodology found in Merigo and Gil-Lafuente [5–7], in this section we describe the 2013th ranking of the reviews that returned the most cited searched results, classified after the h-index (Table 2).

Table 2. The h-index review classification, in 2013, after "women's work culture" search in the Web of Science

Rank	Title	ISSN	SJR	H index	Country
289	Social science and medicine	ISSN 02779536	1.789	149	United Kingdom
747	Harvard business review	ISSN 00178012	0.424	104	United States
942	Annual review of sociology	ISSN 15452115	5.801	94	United States
1047	Journal of business venturing	ISSN 08839026	4.357	90	United States
1047	Journal of business venturing	ISSN 08839026	4.357	90	United States
1540	Journal of business ethics	ISSN 15730697	0.962	75	Netherlands
1662	Human relations	ISSN 1741282X	1.864	72	United Kingdom
2053	Journal of labor economics	ISSN 15375307	5.263	65	United States
2166	Journal of occupational and organizational psychology	ISSN 09631798	2.44	63	United States
2265	BMC public health	ISSN 14712458	1.233	61	United Kingdom
2532	Psychology and marketing	ISSN 15206793	1.016	58	United States
2580	International journal of human resource management	ISSN 14664399	0.734	57	United Kingdom
2610	Work and Stress	ISSN 14645335	1.772	57	United Kingdom
2619	British journal of management	ISSN 14678551	1.597	56	United Kingdom
2888	International migration review	ISSN 01979183	1.458	53	United Kingdom

(continued)

Table 2. (*continued*)

Rank	Title	ISSN	SJR	H index	Country
3023	Violence against women	ISSN 15528448	0.782	52	United States
3221	Sociologia ruralis	ISSN 14679523	1.131	50	United Kingdom
3489	Journal of family issues	ISSN 0192513X	0.798	47	United States
3492	Journal of health communication	ISSN 10870415	1.192	47	United States
3926	Journal of the european economic association	ISSN 15424774	7.726	43	United States
4219	Yale Law Journal	ISSN 00440094	2.354	41	United States
4269	Gender, work and organization	ISSN 14680432	0.922	40	United Kingdom
4273	Human ecology	ISSN 00468169	0.481	40	United States
4531	Health and social care in the community	ISSN 09660410	0.801	38	United Kingdom
4749	Rural sociology	ISSN 15490831	0.82	37	United States
5233	Academy of management learning and education	ISSN 1537260X	1.959	33	United States
6387	Comparative studies in society and history	ISSN 14752999	0.716	27	United Kingdom
6396	East African medical journal	ISSN 0012835X	0.152	27	Kenya
8088	Economic and industrial democracy	ISSN 0143831X	0.504	20	United Kingdom
9967	Sahara J	ISSN 17290376	0.354	15	South Africa
13412	American economic journal: macroeconomics	ISSN 19457715	6.979	8	United States

Source: Author's elaboration based on data available at: http://www.scimagojr.com/journalrank.php [7]

Figure 2 shows the papers published by countries, revealing that the most part of the articles about the women's work culture were published in the United States of America and in European reviews.

However, the questioning after "fuzzy logic" and "women's work culture" in the same database returned no results.

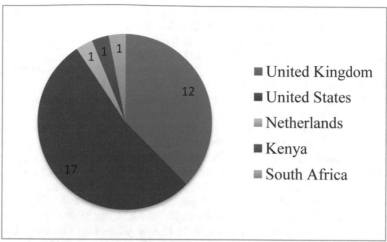

Source: Author's elaboration based on the Web of Science data

Fig. 2. The most cited articles containing "woman's work culture" after WOS by countries
Source: Author's elaboration based on the Web of Science data

4   Conclusions

As seen above, the topic of this research has a limited scientific background so the author considers that there is a real need to enhance knowledge in this area, discussing about women's work culture for two main reasons: the European growing ageing population problem and the pressure of the migration of the workforce from other regions. European behavior is considered to be a personal conduct that overlays European values according to a stated common goal (that of sustaining the European Project by convergence and synchronicity) and is determined by exterior factors as by personal regulators. A European economic behavior might be the woman's decision to have a job (in order to increase the national income, to contribute to social security and pension system etc.) but also to have children (in order to support the demography pyramid). This topic cannot be dissociated from the general agenda of equality of chances promoted by the Lisbon Treaty and European Commission and must also be treated starting from the national or regional system of values and rules regarding women's work.

References

1. Alonso, S., Cabrerizo, F.J., Herrera-Viedma, E., Herrera, F.: H-index: a review focused on its variants, computation and standarization for different scientific fields. J. Inf. **3**(4), 273–289 (2009)
2. Bar-Ilan, J.: Informetrics at the beginning of the 21st century-a review. J. Inf. **2**(1), 1–52 (2008)

3. Chan, K.C., Chang, C.H., Chang, Y.: Ranking of finance journals: some google scholar citation perspectives. J. Empir. Fin. **21**, 241–250 (2013)
4. Currie, R.R., Pandher, G.S.: Finance journal rankings and tiers: an active scholar assessment methodology. J. Bank. Fin. **35**, 7–20 (2011)
5. Garcia, A.: Gender earnings gap among young European higher education graduates higher education. Int. J. High. Educ. Educ. Planning **53**(4), 431–455 (2007)
6. Gil-Lafuente, A.M.: Fuzzy Logic in Financial Analysis, p. 450. GmbH & Co., Springer-Verlag, Berlin (2010)
7. Merigó, J.M., Yang, J.B., Xu, D.L.: A bibliometric overview of financial studies. In: Gil-Aluja, J., Terceño-Gómez, A., Ferrer-Comalat, J., Merigó-Lindahl, J., Linares-Mustarós, S. (eds.) Scientific Methods for the Treatment of Uncertainty in Social Sciences. Advances in Intelligent Systems and Computing, vol. 377, pp. 245–254. Springer, Cham (2015). https://doi.org/10.1007/978-3-319-19704-3_20
8. Merigó, J.M., Gil-Lafuente, A.M., Yager, R.R.: An overview of fuzzy research with bibliometric indicators. Appl. Soft Comput. **27**, 420–433 (2015)
9. Olheten, E., Theoharakis, V., Travlos, N.G.: Faculty perceptions and readership patterns of finance journals: a global view. J. Fin. Quant. Anal. **40**, 223–239 (2011)

Asian Academic Research in Tourism with an International Impact: A Bibliometric Analysis of the Main Academic Contributions

Onofre Martorell Cunill[1]([⊠]), Anna M. Gil-Lafuente[2],
José M. Merigó[3], and Luis Otero González[4]

[1] Department of Business Economics, University of the Balearic Islands,
c/de Valldemosa Km 7.5, 07122 Palma de Majorca, Balearic Islands, Spain
onofre.martorell@uib.es
[2] Department of Economics and Business Organization,
University of Barcelona,
Gran Via de les Corts Catalanes 585, 08007 Barcelona, Spain
amgil@ub.es
[3] Department of Management Control and Information Systems,
University of Chile, Avenida Diagonal Paraguay 257, 8330015 Santiago, Chile
jmerigo@fen.uchile.cl
[4] Department of Finance and Accounting,
University of Santiago de Compostela,
avda Burgo das Nacions, 15782 Santiago de Compostela, Spain
luis.otero@usc.es

Abstract. Asian academic research in tourism is a very recent field of research, which has significantly developed over the last decade due to the strong expansion of the tourism industry worldwide, and also owing to the strong evolution of search engines via the Internet. This article analyses the main contributions to Asian academic research in tourism over recent years using bibliometric indicators. The results obtained are based on the information contained in the Web of Science database. These results focus on explaining three fundamental questions. Firstly, we study the publication structure of Asian articles in tourism over recent decades, as well as the citations these articles have received. Secondly, we present a ranking of the most important tourism journals in Asia through the use of a series of indicators such as the number of publications in said journals, the number of citations, and the h-index. Finally, we present a list of the 50 most cited Asian articles in tourism (and hence the ones that can be considered the most influential) of all times. The results show how, in Asian terms, the most influential journals in this field are Tourism Management (TM), the Annals of Tourism Research (ATR) and the International Journal of Hospitality Management (IJHM).

Keywords: Bibliometrics · Tourism · Asia · Web of science · h-Index

© Springer International Publishing AG, part of Springer Nature 2018
A. M. Gil-Lafuente et al. (Eds.): FIM 2015, AISC 730, pp. 307–325, 2018.
https://doi.org/10.1007/978-3-319-75792-6_24

1 Introduction

During recent decades, due to globalisation and the development of technology, among other things, the travel market has increased considerably, as has the flow of tourists travelling all over the world. Another factor that may explain this spread of tourism is the economic development of many countries which previously found themselves in more precarious conditions. Thus, the study of tourism and of the different techniques that enable people to make the most of this tourism boom has recently gained more and more importance. Hence, lately many studies and important contributions have begun to be published which have made it possible to improve knowledge on tourism, and which have enriched research in this field.

One way of analysing the state of research in tourism is, amongst other alternatives, by carrying out a bibliometric study. Bibliometrics is an approach which through the use of certain techniques and tools makes it possible to study an area of research by analysing, amongst other indicators, the number of publications of articles in this area and the number of citations these articles receive. In tourism, during the last decade, bibliometric studies analysing this area in question have sprung up. However, although there is one bibliometric study that analyses China tourism [22], there is as yet none that refers to Asian tourism in general.

The aim of this article is to present a bibliometric analysis of Asian academic research in tourism, by analysing the main academic contributions in this area. In order to conduct this study, information was gathered from the Web of Science (WoS), as it is considered the most influential database for scientific research. With this information a series of rankings was drawn up, which are commented on below, including a set of variables (number of citations, number of publications, h-index, etc.), since the ranking order varies depending on the variable used. Thereby, by including more variables, we are able to carry out a broader analysis.

First of all, the number of publications of Asian articles in tourism is analysed over the last three decades. Thereby, it is possible to analyse the evolution of the volume of research that has been conducted recently in this area. If we observe this evolution, a great increase in the number of Asian publications in tourism can be appreciated, mainly due to the improvement in information technologies over the last decade and the development of the Asian economy, which has enabled the existence of a greater number of Asian universities, as well as a greater number of researchers.

Subsequently, we also conducted an analysis of the number of citations the articles published in this field receive, distinguishing between articles according to their year of publication. It is possible to observe how the level of citations is rather low in comparison with other areas of research, such as economics and even accountancy or finance. The main reason for this is that, despite the fact that there are few journals that collect tourism publications, most of these have been indexed in the Web of Science (WoS) for only a short time, wherefore there are not many publications in this field (and even less so if only Asian ones are considered); besides, most of them were conducted very recently. As a result, not enough time has gone by for them to have received a considerable number of citations.

Thirdly, a ranking of the most relevant journals in tourism is made, by analysing the journals strongly oriented towards tourism research that are indexed in the WoS. It must be taken into account that the WoS only includes the journals that meet high standards of quality, hence the journals included in this study (there are only 20, since, as we pointed out above, there are not many journals specialising in tourism) are the most relevant ones in terms of tourism research. In order to analyse the influence and relevance of these journals, several indicators are used, namely the number of articles published in said journals, the number of citations these articles receive, the h-index, and the Impact Factor provided by Journal Citation Reports (JCR). The results reveal that, as far as Asia is concerned, the most influential journals in this field are Tourism Management (TM), the Annals of Tourism Research (ATR) and the International Journal of Hospitality Management (IJHM).

Finally, a ranking of the most influential Asian articles in tourism research is presented, in accordance with the results found in the WoS. In order to draw up this ranking, the variable that is taken into consideration is the number of citations received by each article. Thereby, the 50 most cited Asian articles in tourism journals are analysed, with the aim of identifying the most influential contributions in this field.

The rest of the article is arranged as follows. In the next section, a brief analysis is made of the literature review of the bibliometric articles in general, and particularly of those related to tourism. Section 3 refers to the methodology used to carry out this study. In Sect. 4, we analyse the main results, which refer to the publication and citation structure over recent decades; to the most influential tourism journals in Asia according to number of publications; and to the Asian publications in this field that have received the greatest number of citations. Finally, in Sect. 5, the main results are summarised and the conclusions of the article are written.

2 Literature Review

Recently, it has become very popular to use bibliometric tools and techniques in order to quantitatively assess the bibliographic material of a scientific discipline [5]. Therefore, bibliometric studies are now very common, since due to the strong development of computers and of the Internet, information is readily available to any scientific institution or researcher. Specifically a bibliometric study enables you to obtain a general overview as to the state of an area of research. Recently, various studies have attempted to expand the definition that Broadus made of bibliometrics in 1987, by integrating this methodology into a much wider discipline that also covers scientometrics and informetrics [4].

In recent decades in the literature, many studies have been presented that provide a very broad vision of an area of research, by finding out the most productive and influential publications, journals, authors, institutions and countries in the area in question. Among the main bibliometric contributions in different areas stands out the article published by Podsakoff et al. [19], which provides a full picture of the field of management, through an analysis of the most influential authors and institutions in this field. In the field of operation management, Pilkington and Meredith [18] analysed the most influential papers using the citation analysis approach. Meanwhile, in the discipline of business initiative stands out the publication of Landström et al. [13], which recently provided a full picture of this discipline.

In economics, many authors have published bibliometric articles, one of the most influential of which is the paper by Stern [20], which analyses the most relevant journals in economic research. In specific issues in economics stand out the publications of [11] who focused on ecological economics, Baltagi [3] who studied the most productive authors, institutions and countries in econometrics, and Wagstaff and Culyer [23] who developed a bibliometric analysis in the field of health economics.

In financial research, many bibliometric articles have also been published. The most representative is the one published by Alexander and Mabry [2], which presents a very broad view of this field of research, providing rankings of the most influential journals, authors and institutions, and the most cited articles to date. Meanwhile, in accounting, some of the main bibliometric contributions are those by Brown and Gardner [6] and Brown [7], who analysed the most influential articles, institutions and authors in this field, through the use of a citation analysis.

As regards tourism, the area of research that we focus on in this paper, a series of bibliometric studies has also been published. Some of these focused on analysing the main authors in this area of research, such as [15], who identified the authors who were engaged in publishing in the most cited tourism journals over two periods of time (1970–2007 and 1998–2007). Meanwhile, Zhao and Brent Ritchie [24] in their paper identified the 57 most influential authors that had made publications in tourism journals in the period between 1985 and 2004. Other studies, such as Jogaratnam et al. [12] and Law et al. [14], sought to find out the most influential institutions and countries in the tourism research community.

Meanwhile, other authors focused on developing a bibliometric analysis based on the main tourism journals, for instance, Svensson et al. [21] in their paper described the empirical characteristics of journals specialising in tourism and the hospitality industry, while McKercher et al. [17] and Hall [9] focused on drawing up rankings of the most influential journals in tourism. Apart from this contribution, McKercher [16] also proposed a measure to assess the relative influence of tourism journals in their field of research, which consisted of calculating the percentage of citations a journal receives by taking into account the total number of citations received by all the journals in this field. Cheng et al. [8], meanwhile, after reviewing a set of journals related to tourism and the hospitality industry identified the main trends therein.

Finally, as in other disciplines, in tourism some bibliometric studies have been conducted where a specific country or region is analysed. As pointed out above, Tsang and Hsu [22] carried out a regional analysis of tourism in China, and Albacete-Sáez et al. [1] did the same for Spain.

3 Methodology

As already mentioned above, all the information used to carry out the analysis performed in the present study is gathered from the Web of Science (WoS) database. Owing to the development of information science and the Internet, this is becoming a very practical tool in the search for research information, as the information available therein is accessible to any researcher. Even though there are other important databases that also enable information of this type to be gathered, such as SCOPUS, Google Scholar and Econ Lit, we chose to work with the WoS due to the fact that it is generally considered the most influential database in scientific research, as all the publications included therein are considered high quality. Currently, the WoS includes over 15,000 journals and 50 million papers. Research is classified in 251 categories and 151 areas. In order to carry out this work, we selected the category "Hospitality, Leisure, Sport and Tourism", and in this way 20 journals (which can be observed in Table 1) were selected which focus broadly speaking on research in tourism and related areas.

Table 1. List of journals included in the analysis

Acronym	Journal title
ATR	Annals of Tourism Research
APJTR	Asia Pacific Journal of Tourism Research
CHQ	Cornell Hospitality Quarterly
CIT	Current Issues in Tourism
IJCHM	Int. J. Contemporary Hospitality Management
IJHM	Int. J. Hospitality Management
IJTR	Int. J. Tourism Research
JHLST	J. Hospitality, Leisure, Sport & Tourism Education
JHTR	J. Hospitality & Tourism Research
JLR	J. Leisure Research
JST	J. Sustainable Tourism
JTCC	J. Tourism and Cultural Change
JTR	J. Travel Research
JTTM	J. Travel & Tourism Marketing
LS	Leisure Sciences
LSt	Leisure Studies
SJHT	Scandinavian J. Hospitality and Tourism
TE	Tourism Economics
TG	Tourism Geographies
TM	Tourism Management

Some of the journals clearly focus on research in tourism, while others also cover other research topics but equally have a strong connection with tourism. The latter type of journals mainly refers to leisure and to issues related to travel, so they are related to tourism research. Since the WoS does not have many journals indexed on tourism issues, both types of journals were included in the study.

Based on the articles published in the aforementioned journals, we applied another filter, in order to obtain exclusively Asian articles, which is what the whole bibliometric analysis of the paper is based on.

There is no single widely-accepted mechanism for conducting a bibliometric study. In fact, many bibliometric studies analyse information through totally dissimilar variables. In order to perform this analysis and to avail ourselves of the most complete possible view of a set of publications, we selected the main indicators, which are the number of publications, and the number of citations, and we also used the h-index [10]. The number of publications is an indicator of the productivity of the journal, whereas the number of citations is an indicator of the influence of this journal. Meanwhile, the h-index is an indicator that enables the value of a set of articles to be measured, whilst taking into account the number of publications and the number of citations. For instance, if a journal has an h-index of 35, this means that of all the articles included in this journal, there are 35 with at least 35 citations each.

Nevertheless, the three indicators reveal a series of drawbacks and limitations. On the one hand, basing a bibliometric study on the number of publications is criticised as it is not the same if an author publishes an article in the most influential tourism journal as in another lower ranking journal, and the WoS does not distinguish between high quality and lower quality journals when it comes to conducting a recount of the number of publications. Moreover, the number of citations is an indicator that is criticised due to the fact that certain topics receive a higher number of citations than others, since they are published in journals of greater prestige or because of the nature of the field in question, which could be more attractive than other fields where articles of equal or even higher quality are published, but which receive a lower number of citations. As far as the h-index is concerned, its main limitation is that at times it can distort the perception of information, in such a way that if in a journal there are over 200 articles published and 5 of these receive over 1,000 citations, but the rest receive fewer than 5 citations, the h-index of this journal will be 5, despite the fact that the value of this journal is much higher.

In an attempt to mitigate these limitations, and especially the fact that the WoS when conducting a recount of the publications does not distinguish between the "levels of quality" of journals, an alternative would be to assign a value to each journal. Therefore, if a top quality journal is assigned a score of three and a medium quality journal is assigned a score of one, a publication in the higher quality journal would be three times more important than a publication in the less good journal. However, it is extremely difficult to assign a score to each journal, as these scores would have to be determined based on many criteria. Currently, the alternative approach that comes nearest to solving this problem is to take into consideration the Impact Factor of each journal.

The Impact Factor is a measure that is widely known throughout the scientific community and is used as one of the main indicators to identify the value of a journal. The Impact Factor is provided by the WoS through JCR, and is calculated in the following way:

$$IF = \frac{citations_{n-1} + citations_{n-2}}{papers_{n-1} + papers_{n-2}}. \tag{1}$$

By performing this equation, the number of citations received by the articles published in year $n-1$ and $n-2$ in this journal during a given year is calculated with respect to the number of articles that this journal published in year $n-1$ and $n-2$. This is the 2-year Impact Factor (IF). The 5-year Impact Factor (IF5), is exactly the same, but instead of considering only 2 years, the last 5 years are included. Lately, there has been much criticism of the 2-year Impact Factor, as this can be easily manipulated through the use of self-citation and other related techniques; and, as a result, occasionally it does not give a true picture of the influence of journals. Hence, more and more importance is being placed on the 5-year Impact Factor, since as it considers a longer time period, it is more difficult to manipulate.

4 Results and Analysis of the Results

In this section we present the main results we have obtained from the analysis of the information compiled in the WoS. These results are divided into three parts. First of all, reference is made to the structure of the number of publications and citations of Asian articles in tourism. Secondly, a list of the most influential journals in tourism in Asia is drawn up, on the basis of a series of indicators. Finally, a ranking of the most cited Asian articles in tourism research is drawn up.

4.1 Structure of the Number of Publications and Citations of Asian Articles in Tourism

Tourism is a field of research that currently does not hold a significant position in the Web of Science (WoS), as there are barely 20 indexed journals related to tourism. However, over the last 5 years, the number of Asian articles in tourism research indexed in the WoS has undergone a considerable rise. One of the main reasons for this is the regional expansion carried out by the WoS over recent years, which has enabled the entry of many new or already existing journals.

To be precise, in August 2014, in the 20 tourism journals chosen to perform the analysis, the WoS included 2,079 Asian articles in tourism research. Nevertheless, it is worth mentioning that many articles on issues related to tourism are not published in journals that strictly speaking concern tourism research, and therefore are not included in this analysis. These 2,079 articles are the ones that were considered as a basis for performing the bibliometric analysis.

Figure 1 shows the number of Asian articles published in tourism every year, since 1982.

As can be appreciated in the above Figure, until the mid-90s, practically no Asian article had been published in tourism, due to the fact that there were hardly any tourism journals indexed in the WoS. Since 1995, there has been a slight yet steady rise in the volume of publications of Asian articles, until in 2006 and 2007 when around 50 articles were published per year. However, since 2008 this increment in volume of publications has been exponential. In fact, in 2013 twice as many Asian articles in tourism were published as in 2009, with the figure reaching close to 370 articles per year; and it seems as if the number of publications per year is going to continue to rise in the future.

As for the number of citations received by these 2,079 Asian articles in tourism research, it is worth pointing out that in comparison with other disciplines, such as

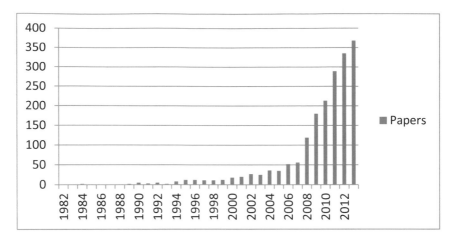

Fig. 1. Asian papers in tourism during the last 31 years (1982–2013)

finance or economics, the number of journals and of articles published in this field is relatively low, as is the number of citations received.

For the purpose of analysing the citation structure of the Asian articles in tourism, minimum citation thresholds were used, which enable the number of articles (in a certain year) which have received a number of citations above this threshold to be detected. Thereby, it is possible to detect in which period the articles receiving a greater number of citations are to be found. Table 2 illustrates the citation structure of Asian articles in tourism, considering various citation thresholds; and conducts an annual analysis from 1995 onwards.

From the results reflected in the above Table, it can be concluded that in the field of tourism research in Asia, a highly cited article is one receiving at least 50 citations, as only 2.68% of all Asian articles in tourism exceeded this citation threshold. In reference to the 100 citation threshold, this was only exceeded by 0.48% (specifically 9 articles) of all the articles in this field. Most of the Asian articles in tourism (79% of them), have currently received fewer than 10 citations.

The majority of the most cited articles in this field were published between 2002 and 2009, while the articles published in the last five years have not received a very high number of citations. The main reason why recently published articles have received a lower number of citations is because in order to receive a more or less consolidated number of citations (in this field we could consider this as between 30 and 50 citations), an article needs several years of maturity before it is known by the scientific community. For this reason, despite the fact that it is in recent years when a greater volume of Asian articles in tourism has been published, these articles have not yet received a considerable number of citations because they lack a couple of years to reach maturity. Nevertheless, most of these more recent articles are expected to receive many citations, as a large percentage of them have already received between one and five citations.

Table 2. General citation structure in Asia in tourism research according to WoS

	≥ 100	≥ 50	≥ 20	≥ 10	≥ 5	≥ 1	Total
Pre 1995	2	3	7	10	14	18	31
1995	0	0	1	3	6	7	12
1996	0	1	7	7	9	11	12
1997	0	1	4	7	8	11	11
1998	0	1	3	7	9	10	11
1999	1	6	6	8	9	10	14
2000	0	3	9	9	13	16	18
2001	0	0	4	6	10	17	20
2002	1	3	14	18	19	23	27
2003	1	3	10	17	22	23	25
2004	0	4	16	27	29	35	36
2005	1	5	17	25	30	33	35
2006	0	6	19	37	42	48	52
2007	1	7	20	36	43	50	56
2008	2	4	21	49	83	110	119
2009	0	1	23	45	94	159	180
2010	0	1	18	50	90	187	213
2011	0	1	4	25	97	231	289
2012	0	0	0	5	35	224	335
2013	0	0	0	0	3	102	368
Total	9	50	203	391	665	1325	1864
Percent.	0.48%	2.68%	10.89%	20.98%	35.68%	71.08%	100%

Abbreviations: ≥ 100, ≥ 50, ≥ 20, ≥ 10, ≥ 5, ≥ 1 = number of papers with more than 100, 50, 20, 10, 5 and 1 citations; Percent. = Percentage of papers.

In fact, it is interesting to note that a highly representative percentage of all Asian articles in tourism ever published have received at least one citation, specifically we are talking about 71% of all of them.

On the other hand, it is also worth mentioning that the oldest articles are sometimes not easily available on the Internet and, because of this, neither do they receive a very high number of citations.

4.2 Analysis of the Most Influential Journals in Tourism Research in Asia

In this section a ranking of the most influential journals in tourism research in Asia is conducted, according to the data available in the WoS. In order to draw up the ranking, we used the 20 journals that appear in Table 1, and the indicator taken into account was the number of publications. Thereby, the top journal in the ranking is the one that published the greatest number of articles. However, with the aim of obtaining a full picture of the quality and influence of the journal, many other indicators were taken into consideration. These other indicators are:

- The number of citations
- The h-index
- The ratio between the number of citations and the number of publications (C/P), which represents the mean number of citations received by each article published in the journal.
- The 2-year and 5-year Impact Factor. (There are some journals for which there are not enough data available to be able to calculate the IF5, as they have been included in the WoS for less than 6 years).

Apart from these indicators, for each journal we calculated the %TP and %TC, where %TP represents a percentage of Asian articles in tourism published by the journal in relation to the total number of articles published by the journal worldwide. While %TC represents the number of citations received by the Asian articles in tourism published in the journal with respect to the total number of citations received by all the articles published in the journal worldwide.

Please note that if two journals are tied in number of publications, the journal with the greatest number of citations is chosen in first place. In Table 3 the ranking of the 20 journals included in the analysis is presented.

As can be observed, depending on the number of publications, undoubtedly, the most productive journals in this field are TM, the IJHM and the ATR, as these are the ones that encompass the greatest number of publications of Asian articles in tourism. However, by analysing the ranking according to the TC/TP variable, it can be observed that the most influential journals in this field are, by far, TM and the ATR, while the IJHM appears quite a long way down. The main reason for this is that the Asians publish a lot in the IJHM (42.99% of the articles published in this journal are Asian), and owing to this, in the ranking based on the number of publications it appears above the ATR. One of the reasons why TM and the ATR so clearly dominate this ranking is due to the fact that these are journals that are already highly consolidated, whereas many of the other journals have only been indexed in the Web of Science (WoS) for a short time.

By analysing the rest of the most influential journals in Asian tourism according to the TC/TP ratio, we observe how, in a second block, two other journals with strong connections with tourism research stand out, namely the JLR and the LS, while the JTR is also rather influential.

By analysing the h-index, we reach the same conclusions as we have just mentioned, observing that the journals with the best score are TM and the ATR, by quite a long way, followed by the JTR. Meanwhile, among the journals most related to leisure, the best ranked ones are the JLR and the LS.

As far as the Impact Factor is concerned, as commented at the beginning of the article, the 5-year Impact Factor is considered more representative than the 2-year Impact Factor, as it avoids being manipulated so easily. Hence, in order to conduct the analysis, the IF5 was taken into consideration. The results obtained are in consonance with the rest of the indicators, where the journals obtaining the highest scores are TM and the ATR, closely followed by the IJHM, the JTR and the JST.

One interesting aspect is to observe which journals are the ones in which Asians have the greatest weight. By analysing the %TP column, we can observe how, in tourism, the presence of the Asian market is very important; for instance, in TM (which is the most influential journal in tourism), around 20% of the articles published there in are Asian. However, the journals in which the Asians have most weight are the APJTR (it must be born in mind that the APJTR is an Asian journal), the IJHM, and the JTTM. As stated above, this is the main reason why these three journals are well placed in the Asian ranking, since in the world ranking they would appear in much lower positions. On the other hand, we see how the participation of Asians in the ATR, the JLR and the LS is quite scarce.

With the aim of examining the most influential documents published in these journals, we included three citation threshold columns, which indicate the number of Asian articles in tourism with over 100, 50 and 20 citations published in these journals. Returning to the same conclusions that we have just mentioned, we see how TM and the ATR are the only journals in which Asian articles in tourism research with over 100 citations have been published, whereas, apart from these two journals, only the IJHM has published articles with over 50 citations (in this case, two). As for the articles that have exceeded the threshold of 20 citations, practically all of these are included in these three journals.

Finally, as can be observed in the last column of Table 3, in the ranking of the 50 most influential Asian articles in tourism appearing in Table 4, all of these, except for two, were published by TM, the ATR or the IJHM, with TM, having published 35, by far the one that has published the majority of these 50 most influential articles.

Table 3. Most influential tourism journals in Asia according to WoS

R	Name	TP	TC	H	TC/TP	%TP	%TC	>100	>50	>20	IF	IF5	T50
1	TM	495	7648	43	15,45	19.27%	25.75%	6	36	130	2.377	3.382	35
2	IJHM	319	1370	18	4,29	42.99%	42.13%	0	3	13	1.837	2.466	3
3	ATR	202	2450	26	12,13	7.20%	6.86%	2	10	36	2.795	3.216	10
4	APJTR	161	128	4	0,80	67.65%	59.81%	0	0	0	0.566	–	0
5	JTTM	152	328	9	2,16	41.19%	37.19%	0	0	0	0.695	0.966	0
6	TE	104	113	5	1,09	23.01%	13.36%	0	0	0	0.573	0.901	0
7	IJTR	92	211	7	2,29	29.30%	21.12%	0	0	0	1.024	1.498	0
8	IJCHM	90	264	9	2,93	25.35%	30.99%	0	0	1	1.623	–	0
9	JST	63	258	9	4,10	14.62%	11.12%	0	0	3	2.392	3.134	0
10	JTR	57	395	12	6,93	17.54%	21.76%	1	1	5	1.884	2.487	2
11	CHQ	57	236	8	4,14	15.88%	21.75%	0	0	2	1.165	1.694	0
12	JHTR	51	196	8	3,84	32.08%	35.25%	0	0	1	1.125	1.602	0
13	CIT	41	45	4	1,10	13.90%	6.68%	0	0	0	0.958	1.241	0
14	JHLST	41	44	3	1,07	21.47%	15.83%	0	0	0	0.062	0.325	0
15	TG	28	74	4	2,64	11.86%	12.11%	0	0	1	1.327	1.302	0
16	LS	26	172	9	6,62	3.66%	2.13%	0	0	2	1.109	1.862	0

(*continued*)

Table 3. (*continued*)

R	Name	TP	TC	H	TC/TP	%TP	%TC	>100	>50	>20	IF	IF5	T50
17	JLR	20	213	9	10,65	1.23%	1.11%	0	0	4	0.592	1.382	0
18	JTCC	16	10	2	0,63	10.39%	11.49%	0	0	0	0.238	–	0
19	LSt	7	5	2	0,71	2.41%	1.05%	0	0	0	1.096	1.237	0
20	SJHT	5	38	1	7,60	2.59%	6.34%	0	0	1	0.882	1.087	0

Abbreviations: R = Rank; H = *h*-index; TC and TP = Total citations and papers; %TP = Percentage of Asian papers to total papers published in the journal; %TC= Percentage of citations to Asian papers to total citations to papers published in the journal; >100, >50, >20 = number of papers with more than 100, 50 and 20 citations; IF = Impact Factor 2013; IF5 = 5-Year Impact Factor 2013; T50 = Number of papers in the Top 50 list shown in Table 4; Journal abbreviations: Are available in Table 1

4.3   The Most Influential Articles in Tourism Research in Asia

In this section, the most influential articles are identified through the bibliometric analysis of the 50 most cited Asian articles of all times, published in the tourism journals specified in Table 1, according to the WoS. The number of citations received by an article enables us to identify the most influential articles, which are normally those that have provided the most significant contribution to tourism research.

However, this ranking may present various limitations, as the most cited article is not always the most relevant one, but rather there may be some exceptions. One instance of this is that a very important article which is not very attractive to the scientific community, as it is very specific, may receive a much lower number of citations than another article that is less important yet much more attractive to the scientific community.

One noteworthy aspect is that through the automatic search of the WoS, it is not possible to find information directly from articles published in a journal which was not indexed at that time. In an attempt to tackle this problem, we carried out a manual search process, using the tool "Cited Referenced Search", so as to identify any articles that were not indexed in the WoS, but which received a number of citations that enables them to enter the ranking of the 50 most cited Asian articles of all times, published in tourism journals. This ranking is given in Table 4.

Among the top 10, most of these were published by TM, except for two which were published by the ATR, and one by the JTR.

By analysing the C/Y column, we can identify the articles that receive most citations per year. We observe that the article by Law and Buhal is, published in 2008, which appears in the fifth position in the ranking, is the one that receives most citations per year, 31 to be precise. Another article also published in 2008 which has received a high number of citations per year, is the one that appears in sixth position of the ranking, published by Song and Li, which receives 28 citations per year. In third place, the article published by Yoon and Uysalin 2005, which appears in the second position of the ranking, receives approximately 26 citations per year.

The most recent articles that appear in the ranking were published in 2009, 2010 and 2011, receiving 11, 14 and 18 citations per year, respectively. No article published

Table 4. The 50 most cited Asian papers in tourism according to WoS

R	J	TC	Title	Author/s	Year	C/Y
1	ATR	292	Rethinking authenticity in tourism experience	N Wang	1999	19
2	TM	234	An examination of the effects of motivation and satisfaction on destination loyalty: a structural model	Y Yoon, M Uysal	2005	26
3	TM	227	The evaluation of airline service quality by fuzzy MCDM	SH Tsaur, TY Chang, CH Yen	2002	19
4	ATR	199	Attitude determinants in tourism destination choice	S Um, JL Crompton	1990	8
5	TM	184	Progress in information technology and tourism management: 20 years on and 10 years after the internet – the state of eTourism research	D Buhails, R Law	2008	31
6	TM	165	Tourism demand modelling and forecasting – A review of recent research	H song, Gang Li	2008	28
7	TM	130	How destination image and evaluative factors affect behavioural intentions?	CF Chen, DC Tsai	2007	19
8	TM	130	Using data envelopment analysis to measure hotel managerial efficiency change in Taiwan	SN Hwang, TY Chang	2003	12
9	JTR	119	Measuring destination attractiveness: a contextual approach	YZ Hu, JRB Ritchie	1999	8
10	TM	99	Tourism destination competitiveness: a quantitative approach	MJ Enright, J Newton	2004	10
11	TM	92	Tourism expansion and economic development: The case of Taiwan	HJ Kim, MH Chen, SS Jang	2006	12
12	ATR	84	Korea's destination image formed by the 2002 World Cup	CK Lee, YK Lee, BK Lee	2005	9
13	TM	83	Predicting behavioral intention of choosing a travel destination	T Lam, CHC Hsu	2006	10
14	TM	82	A neural network model to forecast Japanese demand for travel to Hong Kong	R Law, N Au	1999	5
15	ATR	82	Evaluating tourist risks from fuzzy perspectives	SH Tsaur, GH Txeng, KC Wang	1997	5

(*continued*)

Table 4. (*continued*)

R	J	TC	Title	Author/s	Year	C/Y
16	TM	81	Segmentation of festival motivation by nationality and satisfaction	CK Lee, YK Lee, BE Wicks	2004	8
17	TM	80	Application of importance-performance model in tour guide's performance: evidence from mainland Chinese outbound visitors in Hong Kong	HQ Zhang, I Chow	2004	8
18	TM	75	Back-propagation learning in improving the accuracy of neutral network-based tourism demand forecasting	R Law	2000	5
19	TM	74	A structural equation model of residents' attitudes for tourism development	DW Ko, WP Steward	2002	6
20	ATR	71	A DEA evaluation of Taipei hotels	WE Chiang	2004	7
21	TM	70	Tourism development and economic growth: A closer look at panels	CC Lee, CP Chang	2008	12
22	ATR	70	Measuring novelty seeking in tourism	TH Lee, J Crompton	1992	3
23	TM	69	An importance-performance analysis of hotel selection factors in the Hong Kong hotel industry: a comparison of business and leisure travellers	RKS Chu, T Choi	2000	5
24	TM	68	Modelling and forecasting tourism demand for arrivals with stochastic nonstationary seasonality and intervention	C Goh, R Law	2002	6
25	TM	66	Rating tourism and hospitality journals	B McKercher, R Law, T Lam	2006	8
26	ATR	66	Heritage and postmodern tourism	W Nuryanti	1996	4
27	TM	63	A chaos approach to tourism	B McKercher	1999	4
28	IJHM	63	Service quality, customer satisfaction, and customer value: a holistic perspective	Oh HaeMoon	1999	4
29	TM	63	Tourism destination image modification process: marketing implications	KS Chon	1991	3

(*continued*)

Table 4. *(continued)*

R	J	TC	Title	Author/s	Year	C/Y
30	TM	62	Support vector regression with genetic algorithms in forecasting tourism demand	KY Chen, CH Wang	2007	9
31	ATR	61	Tourism development and cultural policies in China	THB Sofield, FMS Li	1998	4
32	TM	57	Investigating the relationships among perceived value, satisfaction, and recommendations: the case of the Korean DMZ	CK Lee, YS Yoon, SK Lee	2007	8
33	TM	56	Experience quality, perceived value, satisfaction and behavioral intentions for heritage tourists	CF Chen, FS Chen	2010	14
34	ATR	56	Touristic quest for existential authenticity	H Kim, T Jamal	2007	8
35	TM	56	Critical reflections on the economic impact assessment of a mega-event: the case of 2002 FIFA World Cup	CK Lee, T Taylor	2005	6
36	TM	56	A comparison of three different approaches to tourist arrival forecasting	V Cho	2003	5
37	TM	55	Applying the stochastic frontier approach to measure hotel managerial efficiency in Taiwan	CF Chen	2007	8
38	IJHM	54	The impact of online user reviews on hotel room sales	Q Ye, R Law, B Gu	2009	11
39	TM	54	Motivations for ISO 14001 in the hotel industry	ESW Chan, SCK Wong	2006	7
40	TM	54	The relationship between brand equity and firms' performance in luxury hotels and chain restaurants	HM Kim, WG Kim	2005	6
41	TM	53	The establishment of a rapid natural disaster risk assessment model for the tourism industry	CH Tsai, CW Chen	2011	18
42	IJHM	53	The impact of the website quality on customer satisfaction and purchase intentions: Evidence from Chinese online visitors	B Bai, R Law, I Wen	2008	9
43	JTR	53	International student's imagine of rural Pennsylvania as a travel destination	Po Ju Chen, DL Kerstetter	1999	4
44	TM	52	The development of an e-travel service quality scale	CI Ho, YL Lee	2007	7

(continued)

Table 4. (*continued*)

R	J	TC	Title	Author/s	Year	C/Y
45	TM	52	The impact of the 2002 World Cup on South Korea: comparisons of pre- and post-games	HJ Kim, D Gursoy, SB Lee	2006	7
46	TM	51	Tourists' satisfaction, recommendation and revisiting Singapore	TK Hui, D Wan, A Ho	2007	7
47	ATR	51	Antecedents of revisit intention	S UM, K Chon, YH Ro	2006	6
48	TM	51	The relationship among tourist's involvement, place attachment and interpretation satisfaction in Taiwan's national parks	SN Hwang, C Lee, HJ Chen	2005	6
49	TM	51	Critical service features in group package tour: An exploratory research	KC Wang, AT Hsiesh, TC Huan	2000	4
50	TM	50	Passenger expectations and airline services: A Hong Kong based study	D Gilbert, RKC Wong	2003	5

Abbreviations are available in Tables 1, 2 and 3 except for: J = Journal name; C/Y = Citations per year.

after 2011 appears, as it must be born in mind that they have not been published for long enough to be highly cited and to appear in the top 50 most cited Asian articles of all time in tourism journals.

5   Conclusions

In this paper we present a general view of academic research in tourism in Asia over recent decades, through the use of different bibliometric indicators. In particular, we analyse the fundamental results with respect to the publication and citation structure of Asian articles in tourism research; which tourism journals in Asia are the most productive; and the most influential Asian articles in tourism research. All the information was compiled through the Web of Science (WoS), which is considered the main database in academic research.

As far as the study of the publication structure in this discipline is concerned, it is concluded that in Asia, academic research in tourism is a relatively small area in comparison with other fields, such as economics or finance. In the 80s and 90s, the publication of Asian articles in this area was practically non-existent, until at the end of the 90s a few more began to be published, reaching a mean of around 50 articles per year in 2007. However, in the last five years the number of publications in this area has risen a lot, to reach nearly 370 articles published in 2013. Even so, these figures are totally irrelevant compared to other areas, such as economics.

It is worth highlighting that the number of tourism journals indexed in the WoS is very low in comparison with other areas; therefore, only 20 journals were included in this study. Besides, most of these journals have been indexed in the WoS during the last six years through the regional expansion carried out by the WoS. This is another of the reasons explaining the low number of Asian publications in this field, as it also affects the number of citations received by these articles, since as they were included recently, these articles are not yet mature enough to be known throughout the scientific community.

Regarding the citation structure, it is worth noting that only 9 articles (0.48% of the total Asian articles published in tourism journals) have received over 100 citations, while only 2.68% of them have received over 50 citations. Thus, it can be clearly seen that as happens with the number of publications, the level of citation of said articles, is very low compared to that of other fields of research. However, it is worth mentioning that over 70% of all Asian articles in tourism published throughout history have received at least one citation.

In reference to the analysis of the most productive tourism journals indexed in the WoS, there is no doubt that the most influential journal in this field, where the majority of the most cited Asian articles are published, is TM. In Asian terms, in second place is the IJHM and in third place the ATR.

Given the fact that it is difficult to establish a ranking of journals based on a single indicator (in this case the number of publications), other indicators were included in order to attain a more global vision as to the quality of each journal. Amongst other indicators, we included the total number of citations, the mean citations per article published in the journal (TC/TP), the h-index, the Impact factor, and certain citation thresholds. Based on some of these other indicators, it is clear that although the IJHM occupies a very high position in the ranking, TM and the ATR are by far the most influential journals, as they are the ones that receive the greatest number of citations per article published. In a second block, we find the JLR and the LS, while the JTR is also quite influential. The 5-year Impact Factor merely confirms what has been said thus far, reflecting the leadership of TM and the ATR, closely followed by the IJHM and the JTR.

Another aspect to highlight as far as the analysis of the journals is concerned, is the fact that the journals where most articles are published in Asia are the APJTR, the IJHM and the JTTM, in which out of all the articles published in these journals, 67.65%, 42.99% and 41.19% are, respectively, Asian.

Concerning the analysis of the most cited Asian articles in tourism research, the ranking of the 50most cited Asian articles in tourism enabled us to identify the most important and popular main contributions in this field. The most cited article is entitled "Rethinking authenticity in tourism experience", which was published in the ATR, in 1999 by N. Wang and currently has 292 citations. Two other articles have received over 200 citations, and eight articles have received over 100 citations.

It is worth emphasising the fact that four journals have monopolised the publication of the 50 most influential Asian articles in tourism, two of which stand out far above the others. Of these 50 most influential articles, 35 were published by TM, ten by the ATR, three by the IJHM and two by the JTR. Besides, it is worth noting that if any author has rather a lot of influence in this ranking, it is R. Law, who has published seven of the 50 most influential Asian articles in tourism.

Although this ranking is classified in accordance with the number of citations, the mean number of citations received by each article per year was included as a variable. This indicator enables us to analyse the importance of the article by taking into account the time elapsed since it was published. Based on this indicator, the article by Law and Buhalis, published in 2008, which appears in fifth position in the ranking and receives 31 citations per year, stands out. In second place, receiving 28 citations per year, appears the article published by Song and Li, which occupies the sixth position in the ranking.

The main conclusions to be drawn from this study are useful to reflect a general overview of the state of tourism research in Asia, according to the bibliometric information included in the WoS. However, this paper presents a series of limitations to be taken into account. First of all, the results obtained refer to the information gathered from the WoS and, as has been mentioned above, this database and especially some of its indicators, have a series of limitations that must be born in mind. Secondly, it must also be taken into account that many journals or articles that occupy low positions in the rankings presented, when using a different indicator from the one used to draw up the ranking, could occupy a much higher position therein, or vice versa. For this reason, this paper does not aim to provide official rankings, but rather it aims to be merely informative, through the provision of a wide range of indicators. The reason why the information is presented through rankings is because this is a way in which the information can be analysed very easily.

Despite these limitations, on the whole, this study provides a general overview which can be extremely useful in order to understand the main trends in academic research in tourism, with respect to the most influential journals y articles in Asia.

References

1. Albacete-Saéz, C.A., Fuentes-Fuentes, M.M., Haro-Domínguez, M.C.: Spanish research into tourism with an international impact (1997–2011). A perspective from the economy and company management. Cuadernos de Economía y Dirección de Empresa **16**, 17–28 (2013)
2. Alexander Jr., J.C., Mabry, R.H.: Relative significance of journals, authors, and articles cited in financial research. J. Finan. **49**, 697–712 (1994)
3. Baltagi, B.H.: Worldwide econometrics rankings: 1989–2005. Econometric Theor. **23**, 952–1012 (2007)
4. Bar-Ilan, J.: Informetrics at the beginning of the 21st century – a review. J. Inf. **2**, 1–52 (2008)
5. Broadus, R.N.: Toward a definition of "Bibliometrics". Scientometrics **12**, 373–379 (1987)
6. Brown, L.D.: Influential accounting articles, individuals, PhD granting institutions and faculties: a citational analysis. Acc. Organ. Soc. **21**, 723–754 (1996)
7. Brown, L.D., Gardner, J.C.: Applying citation analysis to evaluate the research contributions of accounting faculty and doctoral programs. Account. Rev. **60**, 262–277 (1985)
8. Cheng, C.K., Li, X., Petrick, J.F., O'Leary, J.T.: An examination of tourism journal development. Tour. Manag. **32**, 53–61 (2011)

9. Hall, C.M.: Publish and perish? Bibliometric analysis, journal ranking and the assessment of research quality in tourism. Tour. Manag. **32**, 16–27 (2011)
10. Hirsch, J.E.: An index to quantify an individual's scientific research output. Proc. Natl. Acad. Sci. U.S.A. **102**, 16569–16572 (2005)
11. Hoepner, A.G.F., Kant, B., Scholtens, B., Yu, P.S.: Environmental and ecological economics in the 21st century: an age adjusted citation analysis of the influential articles, journals, authors and institutions. Ecol. Econ. **77**, 193–206 (2012)
12. Jogaratnam, G., Chon, K., McCleary, K., Mena, M., Yoo, J.: An analysis of institutional contributors to three major academic tourism journals: 1992–2001. Tour. Manag. **26**, 641–648 (2005)
13. Landström, H., Harirchi, G., Aström, F.: Entrepreneurship: exploring the knowledge base. Res. Policy **41**, 1154–1181 (2012)
14. Law, R., Leung, R., Buhalis, D.: An analysis of academic leadership in hospitality and tourism journals. J. Hospitality Tourism Res. **34**, 455–477 (2010)
15. McKercher, B.: A citation analysis of tourism scholars. Tour. Manag. **29**, 1226–1232 (2008)
16. McKercher, B.: Influence ratio: an alternate means to assess the relative influence of hospitality and tourism journals on research. Int. J. Hospitality Manag. **31**, 962–971 (2012)
17. McKercher, B., Law, R., Lam, T.: Rating tourism and hospitality journals. Tour. Manag. **27**, 1235–1252 (2006)
18. Pilkington, A., Meredith, J.: The evolution of the intellectual structure of operations management – 1980–2006: a citation/co-citation analysis. J. Oper. Manag. **27**, 185–202 (2009)
19. Podsakoff, P.M., MacKenzie, S.B., Podsakoff, N.P., Bachrach, D.G.: Scholarly influence in the field of management: a bibliometric analysis of the determinants of university and author impact in the management literature in the past quarter century. J. Manag. **34**, 641–720 (2008)
20. Stern, D.I.: Uncertainty measures for economics journal impact factors. J. Econ. Lit. **51**, 173–189 (2013)
21. Svensson, G., Svaeri, S., Einarsen, K.: Empirical characteristics of scholarly journals in hospitality and tourism research: an assessment. Int. J. Hospitality Manag. **28**, 479–483 (2009)
22. Tsang, N.K.F., Hsu, C.H.C.: Thirty years of research on tourism and hospitality management in China: a review and analysis of journal publications. Int. J. Hospitality Manag. **30**, 886–896 (2011)
23. Wagstaff, A., Culyer, A.J.: Four decades of health economics through a bibliometric lens. J. Health Econ. **31**, 406–439 (2012)
24. Zhao, W., Brent Ritchie, J.R.: Test article sample title placed here. Tour. Manag. **28**, 476–490 (2007)

Check for
updates

Academic Contributions in Asian Tourism Research: A Bibliometric Analysis

Onofre Martorell Cunill[1]([✉]), Anna M. Gil-Lafuente[2],
José M. Merigó[3], and Luis Otero González[4]

[1] Department of Business Economics, University of the Balearic Islands,
Palma de Majorca, c/de Valldemosa Km 7.5, 07122 Balearic Islands, Spain
onofre.martorell@uib.es
[2] Department of Economics and Business Organization,
University of Barcelona, Gran Via de les Corts Catalanes,
585, 08007 Barcelona, Spain
amgil@ub.es
[3] Department of Management Control and Information Systems,
University of Chile, Av. Diagonal Paraguay 257, 8330015 Santiago, Chile
jmerigo@fen.uchile.cl
[4] Department of Finance and Accounting,
University of Santiago de Compostela, avda Burgo das Nacions,
15782 Santiago de Compostela, Spain
luis.otero@usc.es

Abstract. Bibliometrics is a fundamental field of information science that helps to draw quantitative conclusions about bibliographic material. During the last decade, the use of techniques and bibliometric studies has experienced a significant increase due to the improvement of information technology and its usefulness to organize knowledge in a scientific discipline. This paper presents an overview of the most productive and influential Asian universities and countries in academic tourism research through the use of bibliometric indicators, according to information found in the database Web of Science (WoS). This database is considered one of the main tools for the analysis of scientific information. In order to analyze the information obtained, several rankings of universities and countries have been carried out, both global and individual, based on a series of bibliometric indicators, such as the number of publications, the number of citations and h-index. Analyzing the results, we observe that within tourism research in Asia, the most influential countries are China, Taiwan and South Korea, and that the leading university is Hong Kong Polytechnic University.

Keywords: Bibliometrics · Tourism · Asia · Web of science · H-Index

1 Introduction

Historically, tourism research has been a field in which there have been very few scientific publications, especially compared to other areas, such as finance or economics. However, especially during the last decade, thanks to the evolution of information technology, the economic expansion in many countries, globalization and,

© Springer International Publishing AG, part of Springer Nature 2018
A. M. Gil-Lafuente et al. (Eds.): FIM 2015, AISC 730, pp. 326–342, 2018.
https://doi.org/10.1007/978-3-319-75792-6_25

mainly, the expansion of tourism, the number of publications in this field has increased very considerably. Thereby, in recent years there have been many papers that have helped to improve knowledge on this field, because how to face tourism demand has become a major concern in many countries.

To carry out a study on academic research in a particular field, there are different methodologies. Nonetheless, the most common is the one that we used in this study, which is called bibliometrics. Bibliometrics is defined as a discipline that quantitatively analyzes the bibliographical material [8]. This methodology is becoming very popular within the scientific community, particularly thanks to the development of the Internet, which facilitates access to bibliometric databases such as the Web of Science (WoS). More recently, the scope of this methodology has expanded and been integrated into a broader discipline that also includes scientometrics and informetrics [4].

Many studies have used bibliometric techniques to provide a complete overview of a field of research. There are many areas in which these kinds of studies have been developed. The most important bibliometric contributions in different fields are reflected below.

In the field of management, Podsakoff et al. [37] provided a comprehensive overview of the same, identifying through an analysis of the structure of publication and citations the most influential institutions and authors on the basis of the 20 most influential journals in management. In the field of production management and operations, Pilkington and Meredith [36] analyzed the most influential works using the citation analysis approach and Hsieh and Chang [25] presented an overview of the discipline, identifying the most influential authors, institutions and countries.

In the field of entrepreneurship, bibliometric studies have also been published; being one of the most important and recent works that of Landström et al. [28], which has provided a comprehensive overview of this discipline.

In economics there are a lot of studies examining the state of the art through a bibliometric approach. For example, Laband and Piette [27] studied the influence of economic journals for the period 1970–1990. A more modern study of the most influential journals in this field is that of Stern [38]. Other works, such as Dusansky and Vernon [19] and García-Castrillo et al. [20], have focused their study in order to determine the most productive and influential institutions in this field. In the meantime, other authors have devoted to make a bibliometric analysis of a particular country or region, such as Europe [16, 30], China [18] or Germany [39], among others.

Being economics a very wide field of research, there are many bibliometric studies that have focused on a specific topic. Wagstaff and Culyer [42] developed a bibliometric analysis in health economics that provides a complete overview of this research area during the last forty years, analyzing the authors and the most influential institutions as well as the most cited papers in this area. Moreover, Baltagi [3] and Hall [21] studied the authors, institutions and most productive countries in econometrics. In ecological economics, Hoepner et al. [24] published a bibliometric study providing an overview of the field.

In finance, many bibliometric studies have also been published. Among these there is Alexander and Mabry [2], which provided an overview of financial research, presenting some rankings with respect to the most influential authors, institutions and journals in this field, as well as a list of the 50 most cited papers within the financial community. Besides, Borokhovich et al. [6] also presented a ranking of the leading authors and institutions in finance. In the meantime, other publications were based on analyzing the influence of financial magazines, such as studies of Borokhovich et al. [7] and Currie and Pandher [17]. Finally, we can emphasize that a considerable amount of bibliometric studies in finance have focused on analyzing specific regions or countries, such as Chan et al. [11], which analyzes Europe, Chan et al. [12], which focuses on Asia and Chan et al. [13], which examines Canada.

Accounting is also a discipline that has been analyzed by many bibliometric studies. Among others, there are studies that have analyzed the quality of accounting journals [5]. For his part, Brown and Gardner [10] and Brown [9], were devoted to provide rankings using citation analysis of papers, authors and institutions leading the field. As in finance and economics, bibliometric studies examining a specific country or region have also been published, as the study of Chan et al. [14], which presented an overview of research in accounting and finance in Australia and New Zealand from 1991 to 2010.

As for academic tourism research, which is the scientific discipline that will be analyzed in this study, it can be noted that a considerable number of bibliometric studies have also been published. Among these studies, some have been devoted to develop rankings of journals, some to rankings of institutions and countries, and others to rankings of authors. Eventually, regional analyses have also been carried out.

Among the authors who have focused their study analysis in magazines, we can highlight McKercher et al. [33], which presented the results of a study made over 70 journals in tourism and hospitality, through surveys to 314 experts in tourism and 191 experts in hospitality. Subsequently, Svensson et al. [40] studied the empirical characteristics of tourism and hospitality journals. Later, Hall [22] developed a ranking of the most influential magazines in tourism, based on the information obtained in the SCOPUS database. That very same year, Cheng et al. [15] focused on identifying which were the most published topics in major travel magazines. In 2012, McKercher [32] introduced a measure to assess the relative influence of a particular journal, calculating the number of citations received by a tourism magazine, divided by the number of citations received by all the Tourism-Related magazines as a whole.

Other bibliometric studies in tourism have preferred to focus on identifying the authors, institutions and influential countries in this field. Among these is the study of Jogaratnam et al. [26], which analyzed the major tourism institutions and countries, based on the publication of three major tourism research journals: the ATR, the JTR, and TM. Meanwhile, Law et al. [29] analyzed which were the most influential countries and institutions in tourism and hospitality research, concluding that the United States was the main publishing country in this field, while universities specializing in tourism and hospitality were the leaders in this area of research. As for the identification of the most influential authors, Zhao and Brent Ritchie [43] focused on identifying the authors who had received more citations during the period between 1985 and 2004. McKercher [31] did the same, though analyzing the periods 1970–2007 and 1998–2007.

Furthermore there is the study of Palmer et al. [35], which aimed to make a bibliometric analysis of the use of statistical methods in tourism research. Based on 12 tourist magazines published in a period of 5 years, it determined the percentage of papers in which statistical techniques were applied.

Finally, we can emphasize that regional bibliometric analysis have been also conducted in tourism. First, Tsang and Hsu [41] did a regional analysis of tourism in China and later Albacete-Sáez et al. [1] developed a regional analysis of tourism in Spain.

The main objective of this paper is to present a comprehensive and updated overview of the most influential universities and Asian countries in tourism research, by applying a series of techniques and bibliometric indicators. All the information under study is collected from the database Web of Science (WoS), which is considered the most influential database in academic research, since it only indexes magazines with high quality standards. The main advantage of this study is that the information is analyzed taking into account a wide range of indicators, such as the total number of publications, the total number of citations, the h-index and certain citation thresholds, allowing us to have a very broad view of the most important universities and Asian countries in tourism.

This paper is divided into two parts. First, the most influential Asian universities in tourism research are discussed, thus presenting a general and a specific ranking for each magazine included in the analysis, from Asian universities that have published a large number of papers. Other indicators, such as the total number of citations and the h-index, are also included. Moreover, an identical analysis of universities is developed, but this time for the most influential Asian countries in academic tourism research.

The paper is organized as follows. In Sect. 2, the methodology used to conduct this study is described. In Sect. 3, the main results of the study are analyzed, which, on the one hand, refers to the most productive and influential Asian universities according to WoS, and on the other hand refers to the most relevant Asian countries. Section 4 ends the article summarizing the main conclusions and results of the study.

2 Methodology

In this study, given that there is no set methodology to determine the value of a set of papers in a university or country, information concerning tourism research in Asia is analyzed mainly through the combination of three indicators [34]: the total number publications, the total number of citations and the h-index. This way, it is possible to develop a full analysis of this information.

Regarding the total number of publications it is necessary to underline that it is often associated with a measure that determines the productivity of a university or country. However, using the total number of publications as a single indicator to make a bibliometric analysis has a number of limitations. First, not all the papers have the same number of pages, there may be papers of 3 pages and others with more than 20 pages, and WoS assigns a unit to each publication regardless of the length of each paper. Secondly, it is difficult to compare the publications of two different magazines, yet it is not the same publishing paper in the top journal of tourism than in an average quality magazine. The value of publishing a paper in the best magazine of the field should be much higher than publishing in another journal, and WoS does not consider this distinction, because it assigns a unit to each publication regardless of where it has been published.

Given this last limitation, the optimal solution would be to assign a value to each magazine, so that if the best magazine has a value of 5 and another magazine has a value of 1, the act of publishing paper in the best magazine is equivalent to publishing five papers in another journal. Still, this is not an easy task. The closest thing to this solution is to use the Impact Factor. The Impact Factor, provided by the Web of Science (WoS) through the JCR, is a measure used to identify the value of a magazine. This indicator is calculated as follows:

$$IF = \frac{citations_{n-1} + citations_{n-2}}{papers_{n-1} + papers_{n-2}}$$

Specifically, this indicator represents the number of citations during the year n received by the papers published in years n–1 and n–2 in the magazine, with respect to the number of journal papers published in years n–1 and n–2. As we can observe, to calculate this indicator, only the last two years are taken as reference. Because of this, the 2-year Impact Factor has received considerable criticism since it is relatively easy to manipulate its value through certain techniques, such as the use of self-citations. For this reason, the 5-year Impact Factor, which is calculated in exactly the same way as the Impact Factor 2-year, but including five years instead of two, is becoming increasingly important. By including a longer period of time, it is more difficult to manipulate and, therefore, it usually represents a more accurate value of reality, as the most popular magazines are always those that tend to obtain a highest 5-year Impact Factor.

Another important limitation presented by WoS arises when the total number of citations as a single indicator is used to make a bibliometric analysis. The number of citations is used as a measure to identify the influence of a paper, a university or a country, among other things. However, there are a number of limitations. First, much like the number of publications, it is not the same receiving a citation in a paper published in a top journal than in an average quality journal. Given this limitation, we have already mentioned that the Impact Factor serves to somehow reduce it. Yet another limitation is that some topics always receive more citations than others, regardless of the quality of the papers, only because they are published in the best-known magazines or because the nature of that particular topic attracts more researchers. Accordingly, very high quality papers may receive fewer citations than papers of lesser quality.

As regards the h-index [23], it is a measure that allows evaluating the quality of a set of documents, combining the number of publications and the number of citations under the same approach. Thus, if a college has an h-index of 43 it means that of all the works published the university, there are 43, which have received at least 43 citations. Despite being a very useful measure it presents a number of limitations. Its main limitation is that it cannot distinguish between different levels of citation, so that if a university has just published 15 papers and these have received more than 1,000 citations (this is a very extreme example), the h-index is 15. If another university has published 15 papers with 15 citations each, their h-index will also be 15, although it is clear that the first university has a much greater influence.

Concerning all the information that has been used in this work to reach the conclusions that will be discussed below, the database that was used to collect this information is the WoS, currently owned by Thomson and Reuters. Overall it is a very practical and easy-to-use database, where most of the information can be found directly through the use of keywords in the search bar, and once obtained the results, these can be filtered according to different options offered by the database. Despite this, it has the limitation that only material that is indexed in the WoS can be found directly through the automatic scanning process. Therefore, if we are looking for a very old paper or one of the first papers published by a magazine that was not indexed since the very first volume, it will probably not appear automatically. In order to try to find it, the paper should be searched manually using the "Cited Referenced Search", which allows searching for any paper that has received at least one citation.

We should keep in mind that apart from the WoS, there are other databases that could be used to perform this analysis, such as Scopus, Google Scholar and EconLit. However, it seems that the WoS is best suited to the purpose of this analysis, since it provides objective and important information for the most relevant journals and papers. Currently the WoS includes more than 15,000 journals and 50,000 papers, which are classified into 251 thematic categories and 151 areas of more general research. Among these 251 categories, the category "Hospitality, Leisure, Sport & Tourism" is included, and within this category a selection of 20 journals concerning tourism research and related areas, which can be seen in Table 1, has been made.

Table 1. List of journals included in the analysis

Acronym	Journal title	Impact factor	5 year impact factor
ATR	Annals of Tourism Research	2.795	3.216
APJTR	Asia Pacific Journal of Tourism Research	0.566	–
CHQ	Cornell Hospitality Quarterly	1.165	1.694
CIT	Current Issues in Tourism	0.958	1.241
IJCHM	Int. J. Contemporary Hospitality Management	1.623	–
IJHM	Int. J. Hospitality Management	1.837	2.466
IJTR	Int. J. Tourism Research	1.024	1.498
JHLST	J. Hospitality, Leisure, Sport & Tourism Education	0.062	0.325
JHTR	J. Hospitality & Tourism Research	1.125	1.602
JLR	J. Leisure Research	0.592	1.382
JST	J. Sustainable Tourism	2.392	3.134
JTCC	J. Tourism and Cultural Change	0.238	–
JTR	J. Travel Research	1.884	2.487
JTTM	J. Travel & Tourism Marketing	0.695	0.966
LS	Leisure Sciences	1.109	1.862
LSt	Leisure Studies	1.096	1.237
SJHT	Scandinavian J. Hospitality and Tourism	0.882	1.087
TE	Tourism Economics	0.573	0.901
TG	Tourism Geographies	1.327	1.302
TM	Tourism Management	2.377	3.382

3 Analysis of Results

In this section the main results of the analysis of the information collected in WoS are analyzed. Analysis of these results can be divided into two parts. First, a ranking, global and individual, is made of the most influential and productive magazines at the Asian level regarding the publication of research papers in tourism universities. Moreover, an identical ranking of universities is presented, though concerning the most influential and productive Asian countries in terms of tourism research.

3.1 The Most Influential and Productive Asian Universities in Tourism Research

Many Asian universities have published papers contributing to tourism research. Table 2 presents a ranking of the 50 most productive Asian universities, classified by the number of papers published in one of the 20 journals listed in Table 1. In order to obtain a more detailed picture of each university and identify which are the most influential ones, other indicators such as the total of citations received, the h-index and certain citation thresholds are also included.

Table 2. The most productive and influential institutions in Asia in tourism research

R	Institution	Country	TP	TC	H	>100	>50	>20
1	Hong Kong Polytechnic U	CHN	561	4238	31	2	15	61
2	Kyung Hee U	KOR	127	861	15	0	2	12
3	Sejong U	KOR	110	1007	19	0	2	19
4	Sun Yat Sen U	CHN	52	354	5	1	1	1
5	Dong A U	KOR	42	398	13	0	0	6
6	National Chiayi U	TW	37	184	7	0	0	3
7	National Cheng Kung U	TW	35	363	8	1	3	5
8	National Kaoh U Hospitality and Tourism	TW	34	39	3	0	0	0
9	Ming Chuan U	TW	33	689	12	2	4	8
10	National Taiwan Normal U	TW	32	161	6	0	0	2
11	National Chung Cheng U	TW	29	197	6	0	1	3
12	Chinese Culture U	TW	28	591	11	1	2	9
13	Jinwen U Science and Technology	TW	27	118	6	0	0	1
14	National Chi Nan U	TW	26	96	4	0	0	2
15	National U of Singapore	SG	25	350	11	0	1	9
16	Pusan National U	KOR	24	103	5	0	0	1
17	U of Hong Kong	CHN	23	284	8	0	1	4
18	National Taiwan U	TW	23	134	7	0	0	2
19	Fu Jen Catholic U	TW	23	121	6	0	0	2
20	U of Macau	CHN	21	57	5	0	0	0
21	Nanyang Technological U	SG	20	119	7	0	0	1
22	Chinese U Hong Kong	CHN	20	95	6	0	0	0

(continued)

Table 2. (*continued*)

R	Institution	Country	TP	TC	H	>100	>50	>20
23	Kaohsiung Medical U	TW	19	323	12	0	0	6
24	National Chiao Tung U	TW	19	173	7	0	0	3
25	U Malaya	MY	19	64	5	0	0	0
26	Seoul National U	KOR	16	69	4	0	0	1
27	Beijing International Studies U	CHN	15	91	6	0	0	1
28	Peking U	CHN	14	118	5	0	0	3
29	Hong Kong Baptist U	CHN	14	55	5	0	0	0
30	Macau U Science and Technology	CHN	14	46	3	0	0	1
31	Kyonggi U	KOR	13	436	7	1	4	4
32	Tunghai U	TW	13	90	5	0	0	2
33	National Chin-Yi U Technology	TW	13	80	4	0	0	2
34	Hanyang U	KOR	13	50	4	0	0	0
35	U Sains Malaysia	MY	13	32	4	0	0	0
36	National Dong Hwa U	TW	12	86	4	0	0	2
37	Nanjing U	CHN	12	37	3	0	0	0
38	National Central U	TW	11	149	5	0	1	3
39	National Chung Hsing U	TW	11	149	5	0	1	3
40	Kyungnam U	KOR	11	101	6	0	0	1
41	Feng Chia U	TW	11	82	5	0	0	2
42	National Sun Yat Sen U	TW	11	77	4	0	0	2
43	Pai Chai U	KOR	10	127	4	0	1	3
44	South Chinese U Technology	CHN	10	114	6	0	0	2
45	City U Hong Kong	CHN	10	42	4	0	0	0
46	Asia U Taiwan	TW	10	33	2	0	0	0
47	U Science and Technology of China	CHN	10	22	3	0	0	0
48	Chung Hua U	TW	9	146	5	0	0	4
49	Keimyung U	KOR	9	75	6	0	0	0
50	Dongseo U	KOR	9	25	1	0	0	1

Abbreviations: >100, >50, >20 = number of papers with more than 100, 50 and 20 citations; TP, TC and H = Total papers, citations and *h*-index in tourism journals indexed in WoS.

Analyzing the above table, we can observe that Hong Kong Polytechnic University is the most productive Asian university in tourism by far, having published 561 papers, whereas the second-ranked has published only 127. Despite having published many fewer papers than the first, the second and third in the ranking, Kyung Hee University and Sejong University, also stand well above the rest of universities, with 127 and 110 publications, respectively. The fourth most productive university has only about 50 publications.

According to other indicators such as the number of citations and h-index, the composition of the top 3 ranking does not vary The Hong Kong Polytechnic University is the most influential one with an overwhelming advantage over the rest, while the

Kyung Hee University and Sejong University swap positions in the ranking, being a little more influential Sejong University.

In the same way that we have just now analyzed a ranking of the 50 most productive in tourism research Asian universities, the same ranking (including only the top 10 or top 5) is individually made for most journals included in Table 1. Thus, we can see which are the most dominant in each of the most influential magazines in Asian tourism research universities. This ranking can be observed in Table 3.

We observe that the monopoly of Hong Kong Polytechnic University is very clear, occupying the first position in the ranking in each of the 12 most influential journals in tourism. In the journals related to leisure, Hong Kong Polytechnic University does not dominate so clearly, since, for example, neither the LS nor the JLR are present in the top 5. However, in the 20 remaining journals if it does not occupy the first position, it is always within the top 5.

As in the case of the general ranking, Kyung Hee University and Sejong University are always on top of the ranking, normally occupying the second or third position in each of the magazines. Nevertheless, it can be noted, first, that in the top 10 of the ATR, Kyung Hee University does not appear, and second, that in the TE Sejong University is not present. Apart from these three universities, there is none that stands out, since depending on the magazine, they appear in the ranking.

To end this section, it must be highlighted that one of the limitations of applying bibliometric techniques to universities, is that a university can become productive and influential not only by publications of its own researchers, but also by the collaboration of researchers from other universities. Furthermore, it has to be kept in mind that university researchers can get in and out of it at any time.

3.2   The Most Influential Asian Countries in Tourism Research

This section presents a ranking of the 17 Asian countries that have published a paper in one of the 20 journals listed in Table 1. This ranking is sorted according to the number of publications in these 20 most influential tourism journals, so that the first one in the ranking will be the most productive country, and the last one the least productive. As we have done for universities, apart from the number of publications, in order to have a broader view of each country other indicators such as the total of citations, the h-index, certain citation thresholds and, finally, the productivity of each country, have been included. This ranking is reflected in Table 4.

By analyzing the table above, it is very clear that China is the most productive and influential Asian country in this field, since it obtains the best results in all variables (except productivity), with significant difference with respect to the second in the ranking. In the second position there is Taiwan, whose value of indicators such as the total of publications, total of citations and h-index are very similar to those of South Korea, which appears in third position. Fourth is Malaysia. Other Asian countries with a significant number of publications are Japan, Singapore, Thailand, India and Iran.

The other countries in the ranking (which are countries that are developing, such as Bangladesh, Sri Lanka, Nepal, Laos, Vietnam, etc.) have just published 5 papers. Consequently, these countries do not have a strong influence on the field tourism research. Yet, we should stress the case of Malaysia, since despite being a country that

Table 3. Institutions with the highest number of papers in eighteen selected journals

R	TM Institution	TP	IJHM Institution	TP	ATR Institution	TP	APTR Institution	TP	JTTM Institution	TP	TE Institution	TP
1	H. Kong Pol. U	120	H. Kong Pol. U	82	H. Kong Pol. U	87	H. Kong Pol. U	50	H. Kong Pol. U	53	H. Kong Pol. U	14
2	Sejong U	37	Kyung Hee U	11	Nat U of Singapore	27	Sejong U	15	Sejong U	8	Kyung Hee U	10
3	Kyung Hee U	27	Sejong U	8	Sejong U	22	N. Kaoh. U H. Tour	11	Kyung Hee U	6	Nat Chung Cheng U	8
4	Nat Cheng Kun. U	19	Dong A U	5	Kyonggi U	12	Fu Jen Catholic U	6	Dong A U	6	Nat Cheng Kung U	6
5	Nat Taiwan U	14	U of Macau	4	Sun Yat Sen U	12	Jinwen U Sci Tech	6	Hanyang U	5	Nat Chi Nan U	6
6	Ming Chuan U	13	Nat Taiw Norm U	4	Chinese Culture U	10	Kyung Hee U	5	N. Kaoh. U H. Tour	5	Sun Yat Sen U	5
7	Kaohsiung Med U	12	Sun Yat Sen U	4	Ming Chuan U	9	Sun Yat Sen U	5	Pusan Nat U	5	Chinese Culture U	4
8	Nat Chiayi U	11	Hong Kong Bapt U	3	Chin U Hong Kong	8	Dong A U	4	Feng Chia U	4	U Scienc. Tec. Chn	4
9	Chinese Culture U	10	Chin U Hong Kong	3	Peking U	8	I Shou U	4	Jeonju U	4	Chiba U	3
10	Nat Chiao Tung U	9	N. Kaoh. U H. Tour	3	U Keban. Malaysia	7	Pusan Nat U	4	MacauU Sci. Tech.	4	Monash U	3

R	JTR Institution	TP	JST Institution	TP	IJCHM Institution	TP	IJTR Institution	TP	CHQ Institution	TP	JHTR Institution	TP
1	H. Kong Pol. U	11	H. Kong Pol. U	39	H. Kong Pol. U	19	H. Kong Pol. U	19	H. Kong Pol. U	34	H. Kong Pol. U	28
2	Kyung Hee U	6	Kyung Hee U	11	Kyung Hee U	11	Kyung Hee U	11	Kyung Hee U	5	Sejong U	4
3	Sun Yat Sen U	4	Sun Yat Sen U	4	Dong A U	4	Nanyang Tech U	4	Chn U Hong Kong	3	Dong A U	2
4	Mac. U Sci.. Tech.	3	Beijing Int St.. U	4	Pusan Nat U	4	U Malaya	4	Sun Yat Sen U	2	Kyung Hee U	2
5	Nat Chiayi U	2	Dong A U	3	Sejong U	3	Nat Chiayi U	3	Tunghai U	2	Nat Chung Che. U	2
6	Sejong U	2	Fu Jen Catho. U	2	City U Hong Kong	3	N Ka. U H Tour	3	Dong A U	2	Pusan Nat U	2
7	South Chn U Tec.	2	Jinw. U Sci Tech	2	Jinan U	2	Aletheia U	2	Ming Chuan U	2	Cheongju U	1
8	U of Hong Kong	2	Kyungnam U	2	Jinwen U Sci Tech	2	Dong A U	2	Nat U Singapore	2	Chinese Culture U	1
9	Asia U Taiwan	2	Nat. Ec. U Vietnam	2	Kyungnam U	2	Nat Chi. Tung U	2	Pusan Nat U	2	Chn U Hong Kong	1
10	Bangkok U	2	Nat Taiw Norm U	2	N. Kaoh. U H. Tour	2	Nat Taiw Norm U	2	Sejong U	1	City U Hong Kong	1

R	CIT Institution	TP	JHLST Institution	TP	TG Institution	TP	LS Institution	TP	JLR Institution	TP	JTCC Institution	TP
1	U Malaya	6	Fu Jen Catholic U	6	Kyung Hee U	6	Nat Chiayi U	7	Seoul Nat U	3	Chongqi. Norm. U	3
2	H. Kong Pol. U	5	N. Kaoh. U H. Tour	3	H. Kong Pol. U	5	Nat Dong HWA U	3	U of Illi. Urb Champ.	3	Dongseo U	3
3	Sun Yat Sen U	4	Jinwen U Sci Tech	3	Dong A U	4	Sejong U	2	Nat Chiayi U	3	Hebrew U	2
4	Nan Kai U Techn.	4	Nat Taiw Norm U	2	Nanjing U	4	Honam U	2	U of Hong Kong	2	H. Kong Pol. U	2
5	Nat Taiw Ocean U	3	H. Kong Pol. U	2	Sun Yat Sen U	3	Hong Kong Bapt U	2	Chn U Hong Kong	2	MacauU Sci. Tech.	1

Abbreviations are available in Table 1.

Table 4. The most productive countries in Asia in tourism research

R	Name	TP	TC	H	>100	>50	>20	Pop	Prod.
1	China	875	5913	41	3	18	77	1,350,695	0,648
2	Taiwan	519	3973	31	3	13	61	22,814.636	22,749
3	South Korea	427	3698	30	2	11	55	49,540	8,619
4	Malaysia	79	225	7	0	0	1	29,628.392	2,666
5	Japan	54	201	8	0	0	3	126,695.683	0,426
6	Singapore	53	525	13	0	1	11	5,353.494	9,900
7	Thailand	51	141	8	0	0	0	65,493.298	0,779
8	India	36	94	5	0	0	1	1,241,492	0,029
9	Iran	18	64	4	0	0	1	75,853.9	0,237
10	Indonesia	5	85	2	0	1	1	237,556.363	0,021
11	Philippines	4	22	2	0	0	0	99,084	0,040
12	Vietnam	4	0	0	0	0	0	91,519.289	0,044
13	Pakistan	3	1	1	0	0	0	182,565.320	0,016
14	Sri Lanka	2	10	1	0	0	0	20,277.597	0,099
15	Laos	2	0	0	0	0	0	6,677.534	0,300
16	Nepal	2	0	0	0	0	0	30,485.798	0,066
17	Bangladesh	1	10	1	0	0	0	167,671	0,006

Abbreviations: TP and TC = Total papers and citations in all the tourism journals indexed in WoS; >100, >50, >20 = number of papers with more than 100, 50 and 20 citations; H = h-index. Pop = Population in thousands; Prod = Productivity – Number of papers per million of inhabitants.

is not well developed, it ranks fourth in the ranking overcoming Japan, a fact that deserves great credit.

One thing that draws a lot of attention is that although Taiwan is the 4th ranked country with a smaller population, it appears in second position behind China. Whereupon, the productivity of this country is very high, publishing 22 papers per million inhabitants. To be aware of the amount of papers that are per million inhabitants, it is appropriate to compare it with the second ranked country with higher productivity, which is Singapore. With a quarter of the population of Taiwan, it occupies the sixth position in the rankings with an output of nearly 10 papers per million inhabitants. On the opposite side there is be India, which although being the largest Asian country (along with China) appears only in the eighth position of the ranking with a productivity of 0,029 papers per million inhabitants.

Finally it must be noted that only China, Taiwan and South Korea have managed to publish papers in tourist magazines, which have received more than 100 citations, and apart from these three countries, only Singapore and Indonesia have published one with more than 50 citations.

As it has been done with universities, was also prepared an individual ranking for each journal, in which the number of publications in tourism that every Asian country has published in each of the magazines mentioned. This classification can be seen in Table 5.

Table 5. Countries classified by the twenty tourism journals included in the analysis

		TM	IJHM	ATR	APJTR	JTTM	TE	IJTR	IJCHM	JST	JTR	CHQ	JHTR	CIT	JHLST	TG	LS	JLR	JTCC	LSt	SJHT	Total
1	China	184	134	92	80	90	33	36	47	28	41	28	32	14	9	13	5	3	4	1	1	875
2	Taiwan	154	95	30	49	24	39	24	12	12	7	12	8	7	22	3	12	3	2	2	2	519
3	South Korea	104	78	28	25	40	18	19	24	8	10	12	12	5	10	10	11	9	2	1	1	427
4	Malaysia	14	11	6	5	4	7	8	4	6	0	0	0	10	0	2	0	0	1	0	1	79
5	Japan	16	2	8	2	5	6	2	1	4	1	1	0	1	0	0	1	2	1	1	0	54
6	Singapore	15	1	22	2	1	0	5	0	1	0	3	0	0	0	1	0	1	1	0	0	53
7	Thailand	11	2	8	6	3	4	3	0	2	2	0	1	0	1	1	0	0	2	1	0	51
8	India	9	9	7	0	1	2	0	3	0	0	2	0	0	0	0	0	0	2	1	0	36
9	Iran	5	2	1	0	1	0	4	2	0	0	0	0	1	0	0	0	1	0	0	1	18
10	Indonesia	0	1	3	0	0	1	0	0	1	0	0	0	0	0	0	0	0	0	0	0	5
11	Philippines	1	1	1	1	0	0	0	0	0	0	0	0	0	0	0	0	0	0	0	0	4
12	Vietnam	0	0	1	0	0	0	0	0	2	0	0	0	0	0	0	0	0	1	0	0	4
13	Pakistan	0	1	0	1	0	0	0	0	1	0	0	0	0	0	0	0	0	0	0	0	3
14	Laos	0	0	0	2	0	0	0	0	0	0	0	0	0	0	0	0	0	0	0	0	2
15	Nepal	0	0	0	0	0	2	0	0	0	0	0	0	0	0	0	0	0	0	0	0	2
16	Sri Lanka	1	0	0	1	0	0	0	0	0	0	0	0	0	0	0	0	0	0	0	0	2
17	Bangladesh	0	0	0	0	0	0	0	0	0	0	0	0	0	0	0	0	1	0	0	0	1

Abbreviations are available in Table 1.

Table 6. Publication evolution by years and countries

Rank	Country	<94	94	95	96	97	98	99	00	01	02	03	04	05	06	07	08	09	10	11	12	13
1	China	0	3	1	0	2	2	4	9	8	7	10	12	9	16	23	61	91	107	136	118	159
2	Taiwan	0	1	0	2	1	2	0	1	1	6	4	13	9	19	14	30	42	52	58	112	102
3	South Korea	3	1	0	3	2	2	1	4	6	7	9	10	15	14	14	18	40	35	65	69	76
4	Malaysia	3	0	1	0	0	0	0	0	0	0	0	0	0	0	0	2	5	10	15	17	13
5	Japan	2	0	0	1	0	1	0	0	1	1	0	0	2	2	4	5	5	7	5	9	5
6	Singapore	6	1	4	3	2	2	2	3	5	3	3	0	0	1	2	3	3	1	3	3	2
7	Thailand	1	0	3	1	0	1	0	0	0	2	0	1	0	1	1	7	2	6	5	8	8
8	India	3	0	0	1	1	0	0	1	0	1	0	0	0	0	2	0	3	3	8	4	8
9	Iran	0	0	0	0	0	0	0	0	0	0	0	0	0	0	1	0	0	2	3	5	5
10	Indonesia	1	0	1	1	0	0	0	0	0	0	0	0	0	0	0	0	0	0	0	0	1
11	Philippines	0	0	0	0	0	0	0	0	0	0	0	0	0	2	0	0	0	0	0	1	1
12	Vietnam	0	0	0	0	0	0	0	0	0	0	0	0	0	0	0	0	0	1	1	0	2
13	Pakistan	0	0	0	0	0	0	0	0	0	0	0	0	0	0	0	0	0	0	0	0	3
14	Laos	0	0	0	0	0	0	0	0	0	0	0	0	0	0	0	0	0	0	0	0	1
15	Nepal	0	0	0	0	0	0	0	0	0	0	0	0	1	0	0	0	0	0	0	0	1
16	Sri Lanka	0	0	0	0	0	0	0	0	0	0	0	0	0	0	0	0	0	0	0	1	0
17	Bangladesh	0	0	0	0	1	0	0	0	0	0	0	0	0	0	0	0	0	0	0	0	0

Analyzing this table, it is observed that China is the most influential Asian country in all tourism magazines that appear in this study, except in the TE, the JHLST, the LS, the LSt and the SJHT, which dominates Taiwan, and the JLR, which dominates South Korea. Apart from this, it can be mentioned that Singapore, in relation to the number of publications of other countries, and taking into account the position of each one of them in the ranking, appears well positioned in the ATR, while the same applies to Malaysia in the CIT. Finally, let us look into the publication evolution of Asian countries between 1994 and 2013. Table 6 presents the number of articles that each country has published in tourism journals between 1994 and 2013.

To conclude this section, it must be stressed that we must keep in mind that one of the constraints on rankings by country, is that a country takes into account the papers published in the universities within its very same borders, forgetting about the nationality of the researchers who publish such papers. The problem is that many good researchers usually move to the best universities. As a consequence, their publications only consider the country where the university is located (and in which researchers are working) and do not take into account the country of origin. Then, as for the Asian scope, the best universities are to be found in China, Taiwan and South Korea, which are precisely the three nations leading the ranking. Despite the fact that because of this limitation the nationality of researchers is not reflected, it is reasonable to develop this ranking, since the goal of this section is to find those Asian countries where the best papers in tourism research are published, regardless of who the authors are.

4   Conclusions

Through this study, a general bibliometric view of the most productive and influential Asian universities and countries in tourism research has been reflected. This analysis has been developed through the use of a set of techniques and bibliometric indicators applied to information gathered in the WoS, which is a database regarded as the most influential scientific research.

To try to have this complete view of universities and most influential Asian countries in the tourism research, rankings have been drawn on these countries and universities. In all rankings, the classification has been carried out taking into account the "number of publications" variable, an indicator that measures the productivity of the country or university in question. However, to try to get this overview, we have included more indicators such as the total number of citations, the h-index or certain citation thresholds, among others.

With regard to universities, there is no doubt that Hong Kong Polytechnic University is the most productive Asian university in tourism by far, having published 561 papers, while the second of the ranking, which is Kyung Hee University, has published only 127. In the third place there is Sejong University with 110 publications. The rest of Asian universities are much less productive, having published altogether less than 50 papers in tourism. These three universities, besides being the most productive, are also the most influential, receiving more citations, especially the Hong Kong Polytechnic University.

Analyzing the influence of these universities on each of the magazines in particular, we observe that the Hong Kong Polytechnic University is the one that has more publications in almost all journals under study. In addition, Kyung Hee University and Sejong University are always on the top of the ranking, normally occupying the second or third position in each of the magazines.

Referring to the most influential tourism research in Asian countries, China is clearly the most influential Asian country in this field. Second is Taiwan, followed closely by South Korea, which appears in third position.

Logically, the developing countries occupy the lowest positions of the ranking because they devote few resources to research. Yet, Malaysia stands out, since despite being a nation that is not well developed, it ranks fourth.

The case of Taiwan is striking. It occupies the second position in the ranking, being the 4th country in the list with a smaller population. Known for its high productivity, it publishes an average of 22 papers per million inhabitants, far ahead of the second most productive country, that is, Singapore, with about 10 papers per million inhabitants.

Analyzing this same ranking for each particular journal, we observe that China is the most influential Asian country in every magazine except the TE, the JHLST, the LS, the LSt and the SJHT, which dominates Taiwan, and the JLR, which dominates South Korea. The rest of countries have little influence on the magazines, as these three nations cover most of the papers published within their borders.

Once explained the main conclusions of the study, it should be noted that apart from the limitations of the different indicators, which are explained in the "Methodology" section of this work, this study has other limitations. First, it must be noted that rankings have been developed based on a given indicator. Nonetheless, depending on the indicator used, this classification may change its order easily. For this reason, the aim of this work is not to present an official ranking of universities or countries. On the contrary, the objective is to be a merely informative study, providing an overview of key information on tourism in Asia, through of a broad range of indicators. Finally, we must take into account that the overview of this field is presented basing on the information that was indexed in WoS in June 2014. Hence, it is not a dynamic view. Truthfully, it is rather static, since each week the database is updated, continually adding new contributions. Therefore, the results of this study reflect the situation of June 2014, a situation that may vary over time.

References

1. Albacete-Saéz, C.A., Fuentes-Fuentes, M.M., Haro-Domínguez, M.C.: Spanish research into tourism with an international impact (1997–2011). A perspective from the economy and company management. Cuadernos de Economía y Dirección de Empresa, **16**, 17–28 (2013)
2. Alexander Jr., J.C., Mabry, R.H.: Relative significance of journals, authors, and articles cited in financial research. J. Finance **49**, 697–712 (1994)
3. Baltagi, B.H.: Worldwide econometrics rankings: 1989–2005. Econometric Theor. **23**, 952–1012 (2007)
4. Bar-Ilan, J.: Informetrics at the beginning of the 21st century – a review. J. Informetrics **2**, 1–52 (2007)

5. Bonner, S.E., Hesford, J.W., Van der Stede, W.A., Young, S.M.: The most influential journals in academic accounting. Acc. Organ. Soc. **31**, 663–685 (2006)
6. Borokhovich, K.A., Bricker, R.J., Brunarski, K.R., Simkins, B.J.: Finance research productivity and influence. J. Finance **50**, 1691–1717 (1995)
7. Borokhovich, K.A., Bricker, R.J., Simkins, B.J.: An analysis of finance journal impact factors. J. Finance **55**, 1457–1469 (2000)
8. Broadus, R.N.: Toward a definition of "Bibliometrics". Scientometrics **12**, 373–379 (1987)
9. Brown, L.D.: Influential accounting articles, individuals, PhD granting institutions and faculties: a citational analysis. Acc. Organ. Soc. **21**, 723–754 (1996)
10. Brown, L.D., Gardner, J.C.: Applying citation analysis to evaluate the research contributions of accounting faculty and doctoral programs. Acc. Rev. **60**, 262–277 (1985)
11. Chan, K.C., Chang, C.H., Chen, C.R.: Financial research in the European region: a long-term assessment (1990–2008). Eur. Financ. Manag. **17**, 391–411 (2011a)
12. Chan, K.C., Chen, C.R., Lee, T.: A long-term assessment of finance research performance among Asia-Pacific academic institutions (1990–2008). Pac.-Basin Finance J. **19**, 157–171 (2011b)
13. Chan, K.C., Chang, C.H., Chen, Y.: Financial research in Canada: a long-term assessment of journal publications. Can. J. Adm. Sci. **28**, 101–113 (2011c)
14. Chan, K.C., Lai, P., Liano, K.: A threshold citation analysis in marketing research. Eur. J. Mark. **46**, 134–156 (2012)
15. Cheng, C.K., Li, X., Petrick, J.F., O'Leary, J.T.: An examination of tourism journal development. Tour. Manag. **32**, 53–61 (2011)
16. Coupé, T.: Revealed performances: worldwide rankings of economists and economics departments, 1990–2000. J. Eur. Econ. Assoc. **1**, 1309–1345 (2003)
17. Currie, R.R., Pandher, G.S.: Finance journal rankings and tiers: an active scholar assessment methodology. J. Bank. Finance **35**, 7–20 (2011)
18. Du, Y., Teixeira, A.A.C.: A bibliometric account of chinese economics research through the lens of the China economic review. China Econ. Rev. **23**, 743–762 (2012)
19. Dusansky, R., Vernon, C.J.: Rankings of US economics departments. J. Econ. Perspect. **12**, 157–170 (1988)
20. García-Castrillo, P., Montañés, A., Sanz-Gracia, F.: A worldwide assessment of scientific production in economics (1992–1997). Appl. Econ. **34**, 1453–1475 (2002)
21. Hall, A.D.: Worldwide rankings of research activity in econometrics: an update: 1980–1988. Econometric Theor. **6**, 1–16 (1990)
22. Hall, C.M.: Publish and perish? bibliometric analysis, journal ranking and the assessment of research quality in tourism. Tour. Manag. **32**, 16–27 (2011)
23. Hirsch, J.E.: An index to quantify an individual's scientific research output. Proc. Natl. Acad. Sci. U.S.A. **102**, 16569–16572 (2005)
24. Hoepner, A.G.F., Kant, B., Scholtens, B., Yu, P.S.: Environmental and ecological economics in the 21st century: an age adjusted citation analysis of the influential articles, journals, authors and institutions. Ecol. Econ. **77**, 193–206 (2012)
25. Hsieh, P.N., Chang, P.L.: An assessment of world-wide research productivity in production and operations management. Int. J. Prod. Econ. **120**, 540–551 (2009)
26. Jogaratnam, G., Chon, K., McCleary, K., Mena, M., Yoo, J.: An analysis of institutional contributors to three major academic tourism journals: 1992–2001. Tour. Manag. **26**, 641–648 (2005)
27. Laband, D.N., Piette, M.J.: The relative impacts of economics journals: 1970–1990. J. Econ. Lit. **32**, 640–666 (1994)
28. Landström, H., Harirchi, G., Aström, F.: Entrepreneurship: Exploring the knowledge base. Res. Policy **41**, 1154–1181 (2012)

29. Law, R., Leung, R., Buhalis, D.: An analysis of academic leadership in hospitality and tourism journals. J. Hospitality Tourism Res. **34**, 455–477 (2010)
30. Lubrano, M., Kirman, A., Bauwens, L., Protopopescu, C.: Ranking economics departments in Europe: a statistical approach. J. Eur. Econ. Assoc. **1**, 1367–1401 (2003)
31. McKercher, B.: A citation analysis of tourism scholars. Tour. Manag. **29**, 1226–1232 (2008)
32. McKercher, B.: Influence ratio: an alternate means to assess the relative influence of hospitality and tourism journals on research. Int. J. Hospitality Manag. **31**, 962–971 (2012)
33. McKercher, B., Law, R., Lam, T.: Rating tourism and hospitality journals. Tour. Manag. **27**, 1235–1252 (2006)
34. Merigó, J.M., Gil-Lafuente, A.M., Yager, R.R.: An overview of fuzzy research with bibliometric indicators. Appl. Soft Comput. **27**, 420–433 (2015)
35. Palmer, A.L., Sesé, A., Montaño, J.J.: Tourism and statistics: bibliometric study 1998–2002. Ann. Tourism Res. **32**, 167–178 (2005)
36. Pilkington, A., Meredith, J.: The evolution of the intellectual structure of operations management—1980–2006: a citation/co-citation analysis. J. Oper. Manag. **27**, 185–202 (2009)
37. Podsakoff, P.M., MacKenzie, S.B., Podsakoff, N.P., Bachrach, D.G.: Scholarly influence in the field of management: a bibliometric analysis of the determinants of university and author impact in the management literature in the past quarter century. J. Manag. **34**, 641–720 (2008)
38. Stern, D.I.: Uncertainty measures for economics journal impact factors. J. Econ. Lit. **51**, 173–189 (2013)
39. Sternberg, R., Litzenberger, T.: The publication and citation output of German faculties of economics and social sciences–a comparison of faculties and disciplines based upon SSCI data. Scientometrics **65**, 29–53 (2005)
40. Svensson, G., Svaeri, S., Einarsen, K.: Empirical characteristics of scholarly journals in hospitality and tourism research: an assessment. Int. J. Hospitality Manag. **28**, 479–483 (2009)
41. Tsang, N.K.F., Hsu, C.H.C.: Thirty years of research on tourism and hospitality management in China: a review and analysis of journal publications. Int. J. Hospitality Manag. **30**, 886–896 (2011)
42. Wagstaff, A., Culyer, A.J.: Four decades of health economics through a bibliometric lens. J. Health Econ. **31**, 406–439 (2012)
43. Zhao, W., Brent Ritchie, J.R.: An investigation of academic leadership in tourism research: 1985–2004. Tour. Manag. **28**, 476–490 (2007)

The Managerial Culture and the Development of the Knowledge Based Society – A Bibliometric Assessment –

Cristina Chiriţă[✉]

National Institute of Economic Research 'Costin C. Kiritescu' - Romanian Academy, Bucharest, Romania
christichirita@yahoo.com

Abstract. The main purpose of this article is to reflect the extent of the scientific research done about two important concepts as the managerial culture and the knowledge based society, in the last two decades, in the areas of business economics and management, by the method of the bibliometric research. Bibliometry is broadly used today to assess the state of the art of a research subject, as the ease of access to scientific information via internet enabled this.

A sound literature review allows us to delimit the field of the research, for the more precise goal of finding the correlations between the managerial culture and the knowledge society achievement. The papers presents, as main findings, the scientific interest shown in the last two decades for this two concepts, using bibliometric indicators such as number of published items, number of citations received by an article or in a journal, ranking of journals and H-index. Following, we found that there are potential determinations between the two concepts, which are suggested by the results of the bibliometric assessment of the theme.

Keywords: Managerial culture · Knowledge based society
Building the knowledge society · Bibliometry · Bibliometric indicators

1 Introduction

The bibliometric research has used the Web of Science (WoS) as the source of all investigations for the said purpose. The WoS was initially created in Philadelphia at the Institute for Scientific Information, in 1960, and was bought in 1992 by Thomson Scientific, also bought in 2008 by Reuters Agency - therefore having the Thomson-Reuters ownership nowadays (http://thomsonreuters.com/en/). WoS is the largest scientific data bases collection worldwide, being the most valued for purposes of valid and reliable desk researches in many areas of scientific interests, being it for theoretical type or for empirical type published works. The papers published in the WoS are scientific articles, proceedings of congresses, other scientific communications, academic studies and books; some of them are available in full text format, some others only with the abstract.

Bibliometry is usually defined as the discipline that studies the bibliographic material quantitatively (Broadus 1987). The identification and the quantitative evaluations of the

© Springer International Publishing AG, part of Springer Nature 2018
A. M. Gil-Lafuente et al. (Eds.): FIM 2015, AISC 730, pp. 343–360, 2018.
https://doi.org/10.1007/978-3-319-75792-6_26

articles published on the before mentioned topics, using the search engine of WoS (accessed at http://thomsonreuters.com/en/ and via http://crai.ub.edu/), gives us a good perception about the scientific interest (Merigó ct al. 2015) and also about the extent of usefulness, utility that this papers demonstrated, the effects they had on the progress in the pertaining research areas after their publication.

2   Methodology for the Literature Review

The present bibliometric study first describes quantitatively the amount of published work for each of the two concepts during the chosen time span, as follows:

a. Searching for the syntagm "**managerial culture**" (MC), within the range of 1994–2014 period of years, we obtain a total of 2,548 articles containing the syntagm MC as for the general field of scientific interests. Refining the search for the area of Business economics, there are a number of 1.314 articles for managerial culture (*1,314 records matched your query of the 94,288,271 in the data limits you selected*). From all these 1,314 articles, the first 150 generate a sum of 12,382 as Times Cited, with an Average Citations per Item of 82.55 and an H-index of 60 (Figs. 1 and 2).

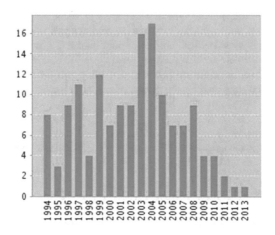

Fig. 1. Number of published articles on MC in each year (graph generated by the WoS)

b. Searching for the syntagm "**knowledge based society**" (KBS), within the same time span of two decades (1994–2014), the search returns a number of 2,726 articles containing it; refining the search for the area of Business economics, there are a number of 579 articles containing the knowledge based society syntagm (*579 records matched your query of the 94,288,271 in the data limits you selected*). From all these 579 articles, the first 150 generate a sum of 916 as Times Cited, with an Average Citations per Item of 6,11 and an H-index of 18. => > *The citation parameters are significantly different in comparison with previous ones – for the "managerial culture" syntagm – they are much lower* (Figs. 3 and 4).

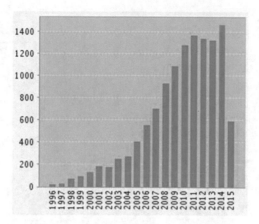

Fig. 2. Number of citations for MC in each year (graph generated by the WoS)

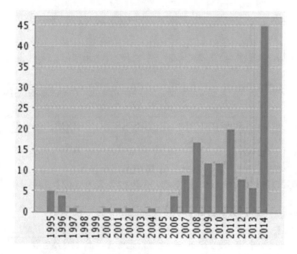

Fig. 3. Number of published articles on KBS in each year (graph generated by the WoS)

c. A search for the topic "*building the knowledge society* "gives us the following results: 2,741 articles (from All Databases), and only 464 articles in the field of Business economics. As for the syntagm of "building the knowledge society", there are only 18 relevant articles, going as a number of citations from 198 to only 1 citation (from All Databases) – see the list below (Table 1):

d. Searching for articles including **both terms, Managerial Culture and Knowledge Based Society,** same time span, there are only 20 results returned by the engine of WoS. They generate a sum of 341 as Times Cited, with an Average Citations per Item of 17.05 and an H-index of 7 (Figs. 5 and 6).

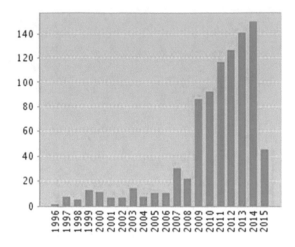

Fig. 4. Number of citations for KBS in each year (graph generated by the WoS)

Table 1. List of relevant authors and articles for the syntagm "building the knowledge society" with TC

Nr:	Author(s)/Editor(s), Title, Source, Author identifiers, Year, ISSN/ISBN	Times cited
1	**CRITICAL SKILLS AND KNOWLEDGE REQUIREMENTS OF IS PROFESSIONALS - A JOINT ACADEMIC-INDUSTRY INVESTIGATION** By: LEE, DMS; TRAUTH, EM; FARWELL, D MIS QUARTERLY Volume: 19 Issue: 3 Pages: 313–340 Published: SEP 1995	198
2	**STRATEGIC ENTREPRENEURSHIP, COLLABORATIVE INNOVATION, AND WEALTH CREATION** By: Ketchen, David J., Jr.; Ireland, R. Duane; Snow, Charles C. STRATEGIC ENTREPRENEURSHIP JOURNAL Volume: 1 Issue: 3–4 Pages: 371–385 Published: DEC 2007	38
3	**UNDERSTANDING OPPORTUNITY DISCOVERY AND SUSTAINABLE ADVANTAGE: THE ROLE OF TRANSACTION COSTS AND PROPERTY RIGHTS** By: Foss, Kirsten; Foss, Nicolai J. STRATEGIC ENTREPRENEURSHIP JOURNAL Volume: 2 Issue: 3 Special Issue: SI Pages: 191–207 Published: SEP 2008	18

(continued)

Table 1. (*continued*)

Nr:	Author(s)/Editor(s), Title, Source, Author identifiers, Year, ISSN/ISBN	Times cited
4	**A new paradigm for knowledge-based competition: Building an industry through knowledge sharing** By: Chen, S Conference: Competing through Knowledge Conference Location: BRISTOL, ENGLAND Date: MAY 06–07 1997 Sponsor(s): Bristol Business Sch TECHNOLOGY ANALYSIS & STRATEGIC MANAGEMENT Volume: 9 Issue: 4 Pages: 437–452 Published: DEC 1997	10
5	**BENCHMARKING THE SYSTEM DYNAMICS COMMUNITY - RESEARCH RESULTS** By: SCHOLL, GJ SYSTEM DYNAMICS REVIEW Volume: 11 Issue: 2 Pages: 139–155 Published: SUM 1995	9
6	**The Paradigm of Knowledge Management in Higher Educational Institutions** By: Sedziuviene, Natalija; Vveinhardt, Jolita INZINERINE EKONOMIKA-ENGINEERING ECONOMICS Issue: 5 Pages: 79–89 Published: 2009	8
7	**Corporate social responsibility and competitiveness** By: Serbanica, Daniel; Militaru, Gheorghe AMFITEATRU ECONOMIC Volume: 10 Issue: 23 Pages: 174–180 Published: FEB 2008	6
8	**Accumulated Knowledge and Technological Progress in Terms of Learning Rates: A Comparative Analysis on the Manufacturing Industry and the Service Industry in Malaysia** By: Asgari, Behrooz; Yen, Lim Wee ASIAN JOURNAL OF TECHNOLOGY INNOVATION Volume: 17 Issue: 2 Pages: 71–99 Published: DEC 2009	2
9	**Principles for Constructing Alternative Socio-economic Organizations: Lessons Learned from Working Outside Institutional Structures** By: Barkin, David REVIEW OF RADICAL POLITICAL ECONOMICS Volume: 41 Issue: 3 Pages: 372–379 Published: SEP 2009	2
10	**AN EXPERT-SYSTEM SOLUTION TO MORTGAGE ARREARS PROBLEMS** By: DOHERTY, N; POND, K SERVICE INDUSTRIES JOURNAL Volume: 15 Issue: 2 Pages: 267–285 Published: APR 1995	2
11	**The evolution of sodoeconomic order in the move to a market economy** By: Hodgson, Geoffrey M. REVIEW OF INTERNATIONAL POLITICAL ECONOMY Volume: 1 Issue: 3 Pages: 387–404 Published: 1994	2

(*continued*)

Table 1. (*continued*)

Nr:	Author(s)/Editor(s), Title, Source, Author identifiers, Year, ISSN/ISBN	Times cited
12	**DEVELOPMENT OF UNIVERSITIES' SOCIAL RESPONSIBILITY THROUGH ACADEMIC SERVICE LEARNING PROGRAMS** By: Peric, Julia Edited by: Tonkovic, AM Conference: 1st International Scientific Symposium Economy of Eastern Croatia - Yesterday, Today, Tomorrow Location: Osijek, CROATIA Date: MAY 17–18 2012 Sponsor(s): Minist Entrepreneurship & Crafts; Univ J J Strossmayer; Fac Econom; Croatian Acad Arts & Sci; Minist Poduzetnistvo & Obrto (MINPO); Hypo Alpe Adria; UniCredit Grp, Zagrebacka Banka; HEMCO; Gradska Banka d d Stecaju; Grafika Osijek; Duro Dakovic Holding d d; Studio hs Internet 1. MEDUNARODNI ZNANSTVENI SIMPOZIJ GOSPODARSTVO ISTOCNE HRVATSKE - JUCER, DANAS, SUTRA Book Series: Medunarodni Znanstveni Simpozij Gospodarstvo Istocne Hrvatske-Jucer Danas Sutra Pages: 365–375 Published: 2012	1
13	**A POSSIBLE MODEL FOR DEVELOPING STUDENTS' SKILLS WITHIN THE KNOWLEDGE-BASED ECONOMY** By: Plumb, Ion; Zamfir, Andreea AMFITEATRU ECONOMIC Volume: 13 Issue: 30 Pages: 482–496 Published: JUN 2011	1
14	**TOWARDS A MODEL OF DESIGNING AN ORGANIZATIONAL STRUCTURE IN A KNOWLEDGE BASED SOCIETY** By: Varzaru, Mihai; Jolivet, Eric AMFITEATRU ECONOMIC Volume: 13 Issue: 30 Pages: 620–631 Published: JUN 2011	1
15	**Modelling Knowledge Sharing Into a Medical Facility Using Human and Virtual Agents (Knowbots)** By: Maracine, Virginia; Iandoli, Luca; Scarlat, Emil; et al. Edited by: Lehner, F; Bredl, K Conference: 12th Annual European Conference on Knowledge Management (ECKM) Location: Univ Passau, Passau, GERMANY Date: SEP 01–02 2011 PROCEEDINGS OF THE 12TH EUROPEAN CONFERENCE ON KNOWLEDGE MANAGEMENT, VOLS 1 AND 2 Pages: 578–589 Published: 2011	1
16	**SERVICE SECTOR AS A STIMULUS OF KNOWLEDGE-BASED ECONOMY DEVELOPMENT** By: Wegrzyn, Grazyna TRANSFORMATIONS IN BUSINESS & ECONOMICS Volume: 9 Issue: 2 Special Issue: SI Supplement: B Pages: 362–381 Published: 2010	1

(*continued*)

Table 1. (*continued*)

Nr:	Author(s)/Editor(s), Title, Source, Author identifiers, Year, ISSN/ISBN	Times cited
17	**Knowledge Economy and Knowledge Society-Role of University Outreach Programmes in India** By: Narasimharao, B. Panduranga SCIENCE TECHNOLOGY AND SOCIETY Volume: 14 Issue: 1 Special Issue: SI Pages: 119–151 Published: JAN–JUN 2009	1
18	**Multilevel cross-linking and offering of organisational knowledge** By: Schumann, Christian-Andreas; Tittmann, Claudia Book Author(s): Martins, B Edited by: Remenyi, D Conference: 8th European Conference on Knowledge Management Location: Consorci Escola Ind, Barcelona, SPAIN Date: SEP 06–07 2007 Sponsor(s): ACL PROCEEDINGS OF THE 8TH EUROPEAN CONFERENCE ON KNOWLEDGE MANAGEMENT, VOL 1 AND 2 Pages: 878–883 Published: 2007	1

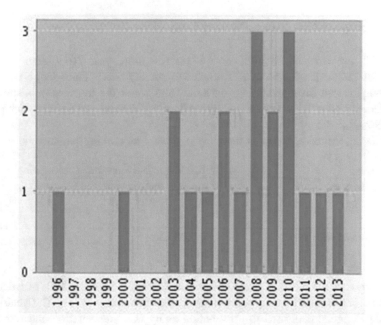

Fig. 5. Number of published articles for both MC and KBS in each year (graph generated by the WoS)

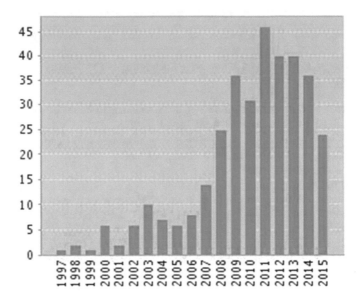

Fig. 6. Number of citations for both MC and KBS in each year (graph generated by the WoS)

3   Ranking of the Journals and Scientific Publications

Next Table presents the H index review classification, year 2013, after the search queries for "Managerial culture", "Knowledge society" and "Knowledge based society" syntagms and their pertaining journals. Data about the journals publishing most cited of identified articles were manually verified in the results of searches achieved on www.scimagojr.com.

The list in the table below is made by choosing as ranking parameters:

Subject Area:	Business, Management and Accounting
Subject Category:	Business and International Management
Region/Country:	All
Order by:	H Index
Display journals with at least:	3 Total Cites
Year:	2013

SCImago is a research group from the Consejo Superior de Investigaciones Científicas (CSIC), University of Granada, Extremadura, Carlos III (Madrid) and Alcalá de Henares, dedicated to information analysis, representation and retrieval by means of visualisation techniques[1].

---

[1] http://www.scimagojr.com/aboutus.php

Criteria we used to assess the influence of the papers are: H-index and SJR – the SCImago Journal Rank indicator[2], being the number of citations received in a journal, related to the quantity of documents published by this journal in same one year - developed by SCImago from the widely known algorithm Google PageRank™. This indicator shows the visibility of the journals contained in the Scopus® database from 1996. The H-index (Hirsh, 2005) is an indicator that describes the scientific productivity and the impact of researchers, by the number of citations they receive for their papers, but can also be used to measure the productivity and impact of scientific journals.

The SCImago Journal & Country Rank is a portal that includes the journals and country scientific indicators developed from the information contained in the Scopus® database (Elsevier B.V.). These indicators can be used to assess and widely analyse scientific domains and authors (Cantos 2009; Merigó et al. 2015).

The ranking of the journals is followed by a visual comparison for the two selected indicators for the influence of the papers, SJR and H-index (Fig. 7).

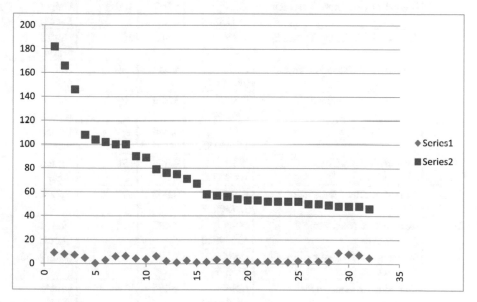

Fig. 7. A comparison for SJR (Series 1) and H-index (Series 2) visualized by a scatter diagram

This diagram of the two indexes shows the impact and the visibility, at same time, for the selected journals. The visibility is more or less constant, but the impact rate is sensitively decreasing, from the top journals to the last journals identified for the scope of the search. As we can see in Table 2, noticeable and maybe even expected, American journals are taking the center stage with 65.625% in the list, the British ones having a 28.125 percentage and two are Danish journals, which means 6.25%.

---

[2] http://www.scimagojr.com/files/SJR2.pdf

Table 2. Ranking of the journals and scientific publications by H index for the items MC, KS, KBS - 2013

Rank	Title	ISSN	SJR	H index	Country
1	Academy of Management Journal	ISSN 00014273	9.067	182	United States
2	Strategic Management Journal	ISSN 10970266	7.909	166	United Kingdom
3	Journal of Marketing	ISSN 15477185	7.284	146	United States
4	Journal of International Business Studies	ISSN 14786990	4.835	108	United Kingdom
5	Harvard Business Review	ISSN 00178012	0.424	104	United States
6	Journal of the Academy of Marketing Science	ISSN 00920703	3.022	102	United States
7	Journal of Consumer Research	ISSN 15375277	6.057	100	United States
8	Journal of Marketing Research	ISSN 15477193	6.465	100	United States
9	Journal of Business Venturing	ISSN 08839026	4.357	90	United States
10	Journal of Management Studies	ISSN 14676486	3.806	89	United Kingdom
11	Marketing Science	ISSN 1526548X	6.248	79	United States
12	Leadership Quarterly	ISSN 10489843	2.069	76	United States
13	Journal of Business Ethics	ISSN 15730697	0.962	75	Netherlands
14	Academy of Management Perspectives	ISSN 15589080	2.482	71	United States
15	International Journal of Project Management	ISSN 02637863	1.092	67	Netherlands
16	MIT Sloan Management Review	ISSN 15329194	1.465	58	United States
17	Entrepreneurship Theory and Practice	ISSN 15406520	3.254	57	United Kingdom
18	R and D Management	ISSN 14679310	1.441	56	United Kingdom
19	International Journal of Electronic Commerce	ISSN 10864415	1.57	54	United States
20	Journal of Advertising	ISSN 00913367	1.415	53	United States
21	Journal of Interactive Marketing	ISSN 15206653	1.301	53	United Kingdom

(*continued*)

Table 2. (*continued*)

Rank	Title	ISSN	SJR	H index	Country
22	Economy and Society	ISSN 14695766	1.58	52	United States
23	Journal of World Business	ISSN 10909516	1.757	52	United States
24	Technological Forecasting and Social Change	ISSN 00401625	1.265	52	United Kingdom
25	Transportation Research, Part E: Logistics and Transportation Review	ISSN 13665545	2.155	52	United States
26	International Journal of Forecasting	ISSN 01692070	1.611	50	United Kingdom
27	Journal of Corporate Finance	ISSN 09291199	1.84	50	United States
28	Journal of International Marketing	ISSN 15477215	1.747	49	United States
29	International Business Review	ISSN 09695931	9.067	48	United States
30	Journal of Consumer Marketing	ISSN 07363761	7.909	48	United States
31	Journal of Productivity Analysis	ISSN 15730441	7.284	48	United States
32	Business Strategy and the Environment	ISSN 10990836	4.835	46	United Kingdom

4   Main Findings of the Articles Containing Both Key Terms

In the next table we extracted some interesting trends in scientific research as they result from the abstracts of published and returned in search-queries items. We found that only 8 articles can be downloaded (are available in full text) via WoS (Table 3).
In total, the analysis shows that:

1. only eight articles were found available for download, in the first twenty articles per search categories: four articles for Managerial Culture syntagm, three articles for Knowledge Based Society and one article for both key terms; one was published in 1996, two were in 2002, one in 2003, one in 2004, one in 2007 and two in 2008;
2. three articles are published in the Subject Area of Business, Management and Accounting Journals, Subject category: Business and International Management and five articles are in different, various categories like: Economics, Econometrics and Finance (ASIA PACIFIC JOURNAL OF MANAGEMENT), Computer Sciences category (JOURNAL OF STRATEGIC INFORMATION SYSTEMS), Social Sciences category (HUMAN RELATIONS) or Social Sciences Area (YALE LAW JOURNAL), even Psychology Area (JOURNAL OF APPLIED PSYCHOLOGY).

Table 3. Findings of the articles containing MC and/or KBS terms (downloaded from WoS)

Author/s	Title of article	Journal	Year	TC	H index	Rank	Key words
Alvesson, M.; Willmott, H.	Identity regulation as organizational control: Producing the appropriate individual	JOURNAL OF MANAGEMENT STUDIES	2002	445	89	10[a]/ 421[b]	*Management Culture*

Main findings from abstract: The regulation of identity (the legal norms) can determine the organizational control, by the "self-images and work orientations that are deemed congruent with managerially defined objectives". This empirical study is focused"on identity extends and themes developed within other analyses of normative control", showing the influence of "managerial intervention (...) more or less intentionally and in/effectively, to employees' self-constructions in terms of coherence, distinctiveness and commitment". The organizational control is "in tension with other intra. and extra-organizational claims upon employees' sense of identity in a way that can open a space for forms of micro-emancipation".(Alvesson and Willmott 2002)

| Tett, R.P.; Burnett, D.D. | A personality trait-based interactionist model of job performance | JOURNAL OF APPLIED PSYCHOLOGY | 2003 | 331 | 157 | 232[c] | *Management Culture* |

Main findings from abstract: The "situational specificity of personality-job performance" effects helps the "understanding of how personality is expressed as valued work behavior". The model proposed "distinguishes among 5 situational features relevant to trait expression (job demands, distracters, constraints, releasers, and facilitators), operating at task, social, and organizational levels. Trait-expressive work behavior" can minimize the impact of personality use in the selection process "if distinguished from (valued) job performance items. The model frames linkages between situational taxonomies (e.g., J. L. Holland's 1985. RIASEC model) and the Big Five and promotes useful discussion of critical issues, including situational specificity, personality-oriented job analysis, team building and work motivation."(Tett and Burnett 2003)

| O'Reilly, C. A.; Tushman, M.L. | The ambidextrous organization | HARVARD BUSINESS REVIEW | 2004 | 325 | 104 | 5[d]/ 9487[e] | *Management Culture* |

Main findings from abstract: Ambidextrous organization are that ones capable to run existing productive capacities that ensure high rates of income and profits, meanwhile also having some "new, exploratory units "(e.g. research and development departments) – but "at the expense of its traditional business"; they possess the ability to have "different processes, structures, and cultures"; a "tight links across units at the senior executive level". Such "ambidextrous organizations," as the authors call them, allow executives to pioneer radical or disruptive innovations while also pursuing incremental gains. Of utmost importance to the ambidextrous organization are ambidextrous managers -executives who have the ability to understand and be sensitive to the needs of very different kinds of businesses. "This mental balancing act is one of the toughest of all managerial challenges- it requires ambidextrous executives", that are"rigorous cost cutters and free-thinking entrepreneurs "and possess "the objectivity required to make difficult trade-offs" - only a senior team management has the will and also the ability to integrate the supposed separation of structures and functions. (O'Reilly and Tushman 2004)

| Benkler, Y. | Coase's penguin, or, Linux and The Nature of the Firm | YALE LAW JOURNAL | 2002 | 299 | 41 | 1020[f] | *Management Culture* |

*Main findings from abstract: The "commons-based peer production" Is, in the author's vision, an excellent productive activity – "a third mode of production" and a new concept of working style, introduced by the free software industry (or open source software). This specific type of work, that distinguishes itself "from the property- and contract-based modes of firms and markets "– "groups of individuals successfully collaborate on large-scale projects following a diverse cluster of motivational drives and social signals rather than market prices or managerial commands" (...) "has systematic advantages over markets and managerial hierarchies in the digitally networked environment when the object of production is information or culture". This peer production has also the explicit advantage of "information opportunity cost" – less information is lost when searching for the right person, for a given job – and the implicit one, coming from a range of strategies used to solve "the collective action problems usually solved in managerial and market-based systems by property, contract, and managerial commands." (Benkler 2002)

(*continued*)

Table 3. (*continued*)

Author/s	Title of article	Journal	Year	TC	H index	Rank	Key words
Baucus, D.A. Baucus, M.S.	Consensus in franchise organizations: A cooperative arrangement among entrepreneurs	JOURNAL OF BUSINESS VENTURING	1996	42	90	343[g]	*Knowledge Based Society*

Main findings from abstract: Consensus among representants of firms being involved in a form of collaboration (franchise systems, network partnerships or constellations of firms) brings more competitive advantage to all partnering firms. Around this main idea, the authors have deployed some realistic arguments and a brief record of evolution in the scientific research of this consensus about the means and the ends - or final scopes - of cooperative arrangements. "Entrepreneurs reaching consensus with their franchise, network, or constellation partners on means and ends may secure the advantages of cooperation and outperform rivals; those lacking consensus likely face recurring conflicts, disappointing performance, and the eventual dissolution of partnerships." Some indicators "to examine perceived consensus" and the relation "between the consensus and performance" in such cooperative arrangements were proposed and analysed in the article, for the specific case of franchise systems – chosen for being the most popular among entrepreneurs who are driven to partnerships. The analyse leads to the conclusion that "dissension arises as franchisees move from an initially knowledge disadvantage relative to franchisors, to a knowledge advantage associated with the accumulation of local experience"(...) "local knowledge about their markets, exercise entrepreneurial initiative, and adopt their own standards for quality and conduct", even improvements in operational standarMs. (Baucus and Baucus 1996)

| von Krogh, Georg; Spaeth, Sebastian | The open source software phenomenon: Characteristics that promote research | JOURNAL OF STRATEGIC INFORMATION SYSTEMS | 2007 | 40 | 50 | 685[h] | *Knowledge Based Society* |

Main findings from abstract: The phenomenon of open source software has been, in the last fifteen years, one of the most interesting for researches in many fields of scientific interest - economy and business, technology and informatics, social sciences and culture. The authors of the article present here the five characteristics of this phenomenon, that make it so attractive and able to undergo "a plethora of research methods" and also a "transdisciplinary research dialog". Here are the five traits: "(1) impact: open source software has an extensive impact on the economy and society; (2) theoretical tension: the phenomenon deviates sharply from the predictions and explanations of existing theory in different fields; (3) transparency: open source software has offered researchers an unprecedented access to data; (4) communal reflexivity: the community of open source software developers frequently engage in a dialog on its functioning (it also has its own research community); (5) proximity: the innovation process in open source software resembles knowledge production in science (in many instances, open source software is an output of research processes)". The open source software can be seen, as well, as a "model of innovation to other areas than software". (von Krogh and Spaeth 2007)

| Lu, Yuan; Tsang, Eric W. K.; Peng, Mike W. | Knowledge management and innovation strategy in the Asia Pacific: Toward an institution-based view | ASIA PACIFIC JOURNAL OF MANAGEMENT | 2008 | 33 | 40 | 40[i]/ 1347[j] | *Knowledge Based Society* |

Main findings from abstract: Knowledge management and innovation strategy are key words in defining businesses, today, and new ways of competing among Asia Pacific companies – and not only there. The institutional environment in Asia Pacific plays an important role in sustaining this perspective, in a "knowledge-intensive society" – by offering "source of knowledge, and allocate incentives and resources for innovation", and this institutional support will help the development, also in the future, of the "knowledge management and innovation strategy in Asia Pacific firms". (Lu et al. 2008)

(*continued*)

Table 3. (*continued*)

Author/s	Title of article	Journal	Year	TC	H index	Rank	Key words
Costea, B.; Crump, N.; Amiridis, K.	Managerialism, the therapeutic habitus and the self in contemporary organizing	HUMAN RELATIONS	2008	28	72	1536[k]	*Both key terms: Managerial Culture and Knowledge Based Society*

Main findings from abstract: This is a niche article, offering some interesting views on the effects of managerialism in the proliferation of the 'soft capitalism', by analysing some dimensions of this like: "'culture', 'performativity', 'knowledge' and 'wellness'." The four listed dimensions may not be among the generally accepted concepts that describe management, but they are being more and more present in research literature, and therefore are suggesting "a more stable cultural tendency of management discourses to capture subjectivity in its general agenda". The author makes here his own "historical-cultural interpretation" of these (yet) managerial concepts, following the evolution of 'soft capitalism' "as a regime of governance". (Costea et al. 2008)

[a]The rank in the previous analyse of Ranking of the Journals and Scientific publications section – see Table 2.
[b]The rank in the general Ranking of the Journals and Scientific publications as on www.scimagojr.com (29,385 listed publications).
[c]Ibidem 4
[d]Ibidem 3
[e]Ibidem
[f]Ibidem 4
[g]Ibidem 4
[h]Ibidem 4
[i]Ibidem 3
[j]Ibidem 4
[k]Ibidem 4

Table 4. Total per year (sum) of the TC for all related articles in the data bases of WoS

Year	Management Culture, MC	Knowledge Based Society, KBS	Building the Knowledge Society	Both key terms, MC and KBS	Total
1994	1,035	64	2	0	**1,101**
1995	211	33	209	0	**453**
1996	866	96	0	141	**1,103**
1997	826	10	10	0	**846**
1998	229	0	0	0	**229**
1999	963	0	0	0	**963**
2000	638	31	0	33	**702**
2001	515	2	0	0	**517**
2002	1,103	1	0	0	**1,104**
2003	1,598	0	0	5	**1,603**
2004	1,426	13	0	43	**1,482**
2005	594	0	0	33	**627**
2006	934	6	0	30	**970**
2007	496	158	39	23	**716**
2008	482	241	24	28	**775**
2009	155	68	13	0	**236**
2010	159	38	1	5	**203**
2011	70	102	3	0	**175**
2012	53	33	1	0	**87**
2013	28	18	0	0	**46**
2014	0	2	0	0	**2**
Total	**12,381**	**916**	**302**	**341**	**13,940**
%					**100%**

To better visualize the amount of figures obtained as total citations for the four variants of syntagms put in the search queries, and to enable a more comprehensive evaluation of the results, we first made a table that is easy to translate into a graph (Table 4):

The graphic shows that for certain periods of time, years 1994–1996 and 2007–2010, there was a common interest for both concepts and their impact in theoretical or empirical research studies. The research interest for the Managerial culture has rapidly decreased after 2008, this fact suggesting, in my opinion, a trend of dramatic changes in the perception of the term and in the perception of the managerial work in its whole. At the same time, last years have brought more interest in the effects of the Knowledge Based Society (Fig. 8).

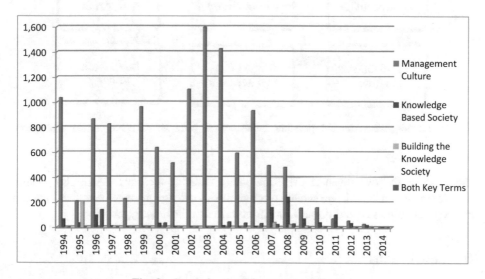

Fig. 8. Graph for the TC amounts evaluation

As a conclusion we can say that the areas and categories of journals are correlated as in real life for the impact of the terms of the study. They can cover widespreaded areas of management, economy and social sciences. Also, by the rank the journals have in the general ranking of www.scimagojr.com made for the specified areas and categories, we can give a second opinion about their presence.

From the personal point of view and interest in Managerial Culture research, all previous findings suggest mutual influences or effects between Managerial Culture, Knowledge Based Society and the managerial concepts displayed in the figure below (Fig. 9).

The substitution into the search engine of "knowledge based society" with "building the knowledge society" syntagm did not generate any supporting hunch for

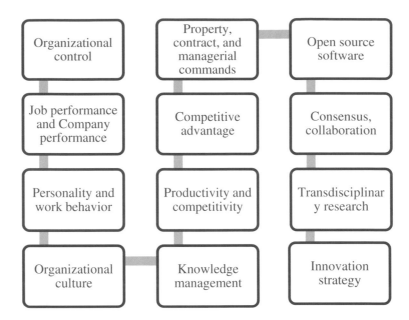

Fig. 9. Findings-based mutual possible influences of MC, KS, KBS on other managerial concepts

the links of the two, even if there are many ideas about the link for the general development of a knowledge society through the knowledge management, that includes managerial culture or/and organizational culture as a firm background.

5   Conclusions

Bibliometric studies can offer reliable information about the visibility of the articles, of the journals, also about the impact of published articles or about the degree of scientific (or public) interest for their topics, being today an useful manner to quantify the scientific productivity - in terms like bibliometric indicators: Total Citations, H index, SJR, country of origin, rankings etc.

They can offer a comprehensive approach about the interest in a given research theme and its state of the art, about main tendencies in future researches, related topics and areas of study, also about biases and possible drifts in the interpretation of available published works' findings. They enable assessment of authors' productive capacity, the research performances and demand for future studies in the given areas – therefore comparissons can be done easily by bibliometric evaluation or, to say it better, they inherently come out.

The use of this bibliometric work for the chosen topics contributes to a better understanding of them: as we can ellaborate from all previously presented results, it brings to the conclusion that it is a complex endeavour to identify trends for the connections between the two syntagms taken into analysis – management culture and knowledge based society. This paper being the first on the proposed topic, I estimate that further investigations will shed more light on the chosen subjects and their possible mutual influences.

The emergence of the phenomenon of 'knowledge based society' offered interesting research subjects, that catched the curiosity of management and business researchers - and they proceeded with their interest within along with that one for the managerial culture, but - as a personal hint - in the last 15 years the managerial culture as 'pairing' concept was replaced by the entrepreneurial culture, more congruent with the inner values of the knowledge society.

Acknowledgement. This paper is supported by the Sectorial Operational Programme Human Resources Development (SOP HRD), financed from the European Social Fund and by the Romanian Government under the contract number SOP HRD/159/1.5/S/136077.

References

Alvesson, M., Willmott, H.: Identity regulation as organizational control: producing the appropriate individual. J. Manage. Stud. **39**(5), 619–644 (2002). https://doi.org/10.1111/1467-6486.00305

Baucus, D.A., Baucus, M.S.: Consensus in franchise organizations: a cooperative arrangement among entrepreneurs. J. Bus. Ventur. **11**(5), 359 (1996). https://doi.org/10.1016/0883-9026(96)00055-9

Benkler, Y.: Coase's penguin, or, Linux and "The Nature of the Firm". Yale Law J. **112**(3), 369 (2002). https://doi.org/10.2307/1562247

Broadus, R.N.: Towards a definition of 'Bibliometrics'. Scientometrics **12**, 373–379 (1987). https://doi.org/10.1007/bf02016680

Cantos, A.M.: Scientific Information Resources and Bibliometric Indicators for the Research Performance Assessment (2009). http://www.researchgate.net/publication/39711248

Costea, B., Crump, N., Amiridis, K.: Managerialism, the therapeutic habitus and the self in contemporary organizing. Hum. Relat. **61**(5), 661–685 (2008). https://doi.org/10.1177/0018726708091763

Lu, Y., Tsang, E.W.K., Peng, M.W.: Knowledge management and innovation strategy in the Asia Pacific: toward an institution-based view. Asia Pac. J. Manage. **25**(3), 361–374 (2008). https://doi.org/10.1007/s10490-008-9100-9

Merigó, J.M., Gil-Lafuente, A.M., Yager, R.R.: An overview of fuzzy research with bibliometric indicators. Appl. Soft Comput. **27**, 420–433 (2015). https://doi.org/10.1016/j.asoc.2014.10.035

Merigó, J.M., Yan, J.-B., Xu, D.-L.: A bibliometric overview of financial studies. In: Scientific Methods for the Treatment of Uncertainty in Social Sciences, vol. 4, pp. 245–254 (2015). https://doi.org/10.1007/978-3-319-19704-3_20

O'Reilly, C.A., Tushman, M.L.: The ambidextrous organisation. Harvard Bus. Rev. **82**(4), 74 (2004). https://doi.org/10.2307/41165852

Tett, R.P., Burnett, D.D.: A personality trait-based interactionist model of job performance. J. Appl. Psychol. **88**(3), 500–517 (2003). https://doi.org/10.1037/0021-9010.88.3.500

von Krogh, G., Spaeth, S.: The open source software phenomenon: characteristics that promote research. J. Strateg. Inf. Syst. **16**(3), 236–253 (2007). https://doi.org/10.1016/j.jsis.2007.06.001

Websites

http://crai.ub.edu/. Accessed June, July 2015
http://thomsonreuters.com/en/. Accessed June, July 2015
http://www.scimagojr.com/. Accessed June, July 2015

Applications in Business and Engineering

A Methodological Approach for Analysing Stakeholder Dynamics in Decision-Making Process: An Application in Family Compensation Funds

Fabio Blanco-Mesa[1,2(✉)] and Anna M. Gil-Lafuente[2]

[1] Universidad Pedagógica y Tecnológica de Colombia,
Av. Central del Norte 39-115, 150001 Tunja, Colombia
fabio.blanco01@uptc.edu.co
[2] Universidad de Barcelona, Av. Diagonal, 690, 08034 Barcelona, Spain
amgil@ub.edu

Abstract. The aim of the paper is to examine the stakeholder dynamics through causality relationship process. The study proposes a methodological perspective of stakeholders, which allows analysing the linking of relationship from the relative intensity and linked relations of entire stakeholders. An application is developed to help decision-making process in uncertainty concerning the ordering according to their importance algorithm and linking of relation method, which are based on notions of relation, gather and order. The case of study is focused on The Family Compensation Funds (FCF) in Colombia. The results show how the ambiguity and fuzziness of the stakeholders and appraisal subjective of decision-maker can be dealt with, helping to make a decision according to its individual estimations. The linked relations between each stakeholders and relative intensity are depicted. As the ties relations of incidence and relative impact on stakeholder behaviours are explained. The main implication of this proposition is to enable to deal with the subjective appraisal of the decision-maker to do a better interpretation of environment and subjectivity factors. Furthermore, it contributes to aid to the strategic planning and decision-making process for operative unit within uncertainty environment in the short-term.

Keywords: Decision making · Strategy · Level of importance
Stakeholder dynamics · Causality incidences · Sport business

1 Introduction

Firms should handle relationships with its business partners, unions, controllers and community in order to meet the strategic aims and social responsibility. They should take into account multiple factors, which have a relative influence and cannot be controlled by them affecting decision-making process and behaviour. In this sense, stakeholder theory has tried to explain the influence of stakeholders on firm's decision-making and behaviour [1, 2]. This theory has focused on the identification of relationship stakeholders groups [2] from point of view descriptive, normative and

© Springer International Publishing AG, part of Springer Nature 2018
A. M. Gil-Lafuente et al. (Eds.): FIM 2015, AISC 730, pp. 363–380, 2018.
https://doi.org/10.1007/978-3-319-75792-6_27

instrumental [3]. In Freeman [4] has shown a synthetic approach of the relations between firms and a set of actors around it. The relation has depicted on a visual scheme, in which internal and external relationships around the firm are intended to be explained in order to be comprehended, simplified and aggregate complex information about it [5]. However, several authors [3, 6, 7] argue that this model is a static representation that does not consider changes over the time and heterogeneity. The stakeholders are always in a dynamic situation with important changes [7], which occur at different levels of influence [8] and different levels in the environment [9].

The relationship between firms and its stakeholders change in a dynamic and uncertain process. *Dynamic process* implies a change in the interaction of the relationship within actors that takes part into the firm's environment. *Uncertainty* implies unexpected facts of environment, fuzzy boundaries and unclear levels of organization. In this sense, a dynamic view is suggested by [6], in which is proposed a new categorization and classification of stakeholders showing a more dynamic perspective of stakeholder management. Stakeholders are divided into three categories: *real stakeholders, stakewatchers, stakekeepers* [6], which are composed by subgroups. On the one hand, from this categorization it is thinkable to understand the dynamics of the relations between them in a descriptive manner. On the other hand, this approaching can aid to know dynamic and change of relations, since it can define better boundaries of the firm and identify and select stakeholders groups in different levels of the environment. Hence, a methodological perspective of the stakeholder relationship can be developed from descriptive perspective using fuzzy techniques for decision making in uncertainty.

Based on the above, the paper aims is to examine the stakeholder dynamics through causality relationship process. A methodological application is proposed from descriptive perspective of the stakeholders, which provides the basis to study the intensity and linked relations of entire stakeholders. In order to explain both proposals, ordering according to their importance algorithm and linking of relations method are used, which are based on notions of relation, gather and order [10]. The main implication of this proposition is to enable to deal with the subjective appraisal of the decisor to do a better interpretation of environment and subjectivity factors. Furthermore, it contributes to aid to the strategic planning and decision-making process for operative unit within uncertainty environment in the short-term. The case of study is considered The Family Compensation Funds (FCF) in Colombia and sport movement. FDF's corporations in Colombia have multi-functionality role fulfilling with policies and programs of social development, which can be managed both private and public manner. Sport is characterized by its transversality with other markets and industries since it is used for doing business, selling and persuasion [11]. Implications of sport go beyond its original boundaries and its effects have been disseminated throughout the world in different ways. Illustrative example is developed in order to see useful of this proposition. The relative intensity and linked relations of entire stakeholders has centred its attention on salience characteristics where administrative body of FDF is seeking to take advantage of new opportunities that can offer the sport movement for its leisure operative unit in the short time. Both concepts are explained as relative impact on stakeholder behaviours and the ties relations of incidence.

The structure of this paper is as follows: Firstly, the theoretical framework is composed by dynamic relationships of the stakeholders and the main contribution of mathematical models in decision making in uncertainty related to the incidence of relations. Secondly, a case of study is presented. Thirdly, methodology is defined by a mathematical model, which refers to comparison notion and causality and incidence of the relation concepts. Fourthly, it is explained the illustrative example and mathematical application model and the main results. Finally, the conclusion and implications are presented.

2 Theoretical Framework

2.1 Dynamic Relationship of the Stakeholders

Stakeholder theory has helped to understand firm's environment and its relations. This theory has tried to explain and predict how organizations should act by taking into consideration the influences of stakeholders [2]. Furthermore, it has provided account of how stakeholders try to act and to influence on the firm's decision-making and behaviour [1]. In fact, stakeholders theory is used as part of the strategic management approach in order to evaluate the environment and to identify the relations between groups that constitute stakeholders; although there is not provision for understanding how to manage change [2]. The analyses of the stakeholder is focused mainly on a descriptive, normative and instrumental point of view, [3, 8, 12], in which several authors have developed different analysis methods that explain the relations of the stakeholders [1, 3, 6, 7, 13–17]. In broad terms, these methods follow a two-parts process: (a) specific stakeholders and their stakes are identified in order to evaluate them; (b) strategy path is defined using the results obtained [18]. Likewise, some of these methods such as: studies of salience concept [13], network theory [14], resources dependence theory [1] and stakeholders dynamics perspective [7, 15] are focused on the influence, power and impact of stakeholders.

The analysis of stakeholder relationships is also depicted in visual schemes, which aid to comprehend, simplify and aggregate complex information [5]. One of the graphical representations most widely accepted is Freeman's stakeholder model. [4, 17, 19] has made an approaching synthetic of the relations between firms and a set of actors around it. The model depicts two ways of relationship, (a) Internal stakeholders: bi-directional arrows towards and from central oval and (b) external stakeholders: without arrows [19]. Likewise, this model tries to express the three-management perception legitimacy, power and urgency from a visual perspective [13]. Notwithstanding, this model has been criticized because it is a static representation that does not consider change over the time and heterogeneity [3, 6]. According to [5] this model does not provide graphical representation of the imperfection and complex reality of the stakeholders, such as: the heterogeneity within stakeholder groups, multiple inclusion, the variability in the dependence among stakeholders and salience, the existence of a central place within the model, the multiple linkages and the network relationships. These considerations suggest that the stakeholders' relationship is imperfect, heterogeneous and dynamic. The stakeholders are always in a dynamic situation with important changes, which can occur simultaneously within its internal relationship structure affecting the organizational results and internal composition of the stakeholders [7].

Under this perspective, network approach is proposed to order to aid to explain interaction of stakeholders. This approach is focused on different characteristics of the relationship, which can be explained by *dyadic relationship*, *Ego-network*, *Multiple Interaction* and *Complete Network* [18]. The network analysis [14] suggests a broader perspective, which goes beyond the dyadic linkages between the firm and each stakeholder. It is focused on how the firm replies to the influence of its stakeholders and considers the multiple and interdependent interactions that simultaneously exist in stakeholder environments [14]. Furthermore, it is shown the way in which stakeholder interactions can impact on the organization within relative web of relationships [18] giving a descriptive approximation on how firm behaves to the influence of indirect stakeholder relationship. Likewise, this method has aid to identify the types of network structures based on the number of existing relationships and relative position in the network [14]. Moreover, it is indicated how and when the stakeholder group attempts to influence the focal organization [14] that can aid to give a predictive response to the firm. However, this approach has not shown how the relation within the network could vary in facing threats and opportunities of changing the environment [3]. The relationships between firms and their stakeholders change in a dynamic [7, 9, 15] and uncertain process [3, 7, 9, 20].

In this sense, from levels of environment [9], salience concept [13, 19] model and managerial and legal approach, [6] has suggested a new categorization and classification of stakeholders. Stakeholders are divided into three categories: real stakeholders, stakewatchers and stakekeepers [6], which are composed by subgroups. Thus, from this descriptive perspective of the relation it is possible to develop a methodological perspective using fuzzy techniques of decision-making in uncertainty.

2.2 Decision Making in Uncertainty: Incidence of the Relations

Decision making in the real world is taken in uncertain environments, wherein the precise consequences are not known accurately. The implications, limitations, social, and economic benefits of the aims and strategies formulated are also uncertain. Based on Fuzzy Set Theory [21], the development of studies on mathematics of uncertainty has led to create a basis of a new theory of decision in uncertainty applied on economy and business management [22–24]. This theory is based on the principle of gradual simultaneousness and it is explained by four main concepts: relation, assignment, association and order. Gil-Aluja [10] has explored widely notion of *"relation"* and it has been defined: *"Relation is any type of grouping that is capable of bringing to light the levels of connection existing between mental or physical objects that are members of one and the same set or between objects of different sets"*. Indeed, a wide range of connection from elements of the same set to elements of one set between other ones (normally finite) is taken into account by this notion [10]. The relations are found either in individuals – subject – or physical or mental objects. These relations have certain characteristics whereby they have linked. The characteristics of each of these links allow any subject or object to be related to other subject or object either directly or through other subjects or objects [10]. Thus, the linking of the relations is becoming denser although certain links are disappeared during the time. The links between them can be strengthened or weakened by the variation of the intensity of relations [25].

This variation and intensity are explained in different means. On the one hand, the intensity will be expressed by certain values $\mu_{ij} \in [0, 1]$, i.e. when μ_{ij} gets closer to 1 the relation between elements a_i and a_j are intensified, but when it gets closer to 0 the relation between them is weakened. On the other hand, the variation occurs either during the course of the time or succession of stages [10] and it is given by composition or convolution max-min. This process generates a great deal of direct and indirect relations that are accumulated. This kind of accumulative relations is created by behaviour of these relations. In fact, primary relations between the elements of one set and another one, and relations between elements of each set with itself are established by the accumulative relations, i.e. the direct and indirect accumulated relations between the elements of two sets are established [10]. The accumulations of these kinds of relations are called "total relations of first and second degree", and they are directly related to the relations of causality and incidences. This type of relation is independent of the time factor and it is focused on accumulated effects obtained of direct and indirect relations. Besides, incidences of the first and second order showing on matrix form or arrow graph are established.

The "*relation*" perspective, it has suggested useful techniques for establishing direct and indirect relationships between physical and mental objects, linking of relation and relation of causality based on incidence concept [26]. In fact, incidence is a difficult subjective concept to measure and rarely properly justified. Thus, these techniques allow leading to a process of analysis of subjective attributes, i.e. it is considered the appraisal of decision-maker according to some notable characteristics. In this process the opinion of decision-maker is more significant than in other methodologies. The decision-maker offers its estimations based on the quality or quantity of data received, i.e. statistics, reports, surveys information is used as a means of guidance by decision-maker. From this standpoint several authors have made several applications of this methodology within business and economic field, such as: finance [27], strategy [28, 29], stakeholders [30] among others, which have shown its usefulness on decision-making in uncertainty. These applications have the advantage that the appraisal of decision-maker can be assessed showing several alternatives, intensities and importance of relations. Hence, this methodology allows reflecting attitudinal character of decision-maker focusing on decision-making business and economic problems.

3 Case of Study

3.1 The Family Compensation Funds

In Colombia, The Family Compensation Funds (FCF) was conceived as non-profit corporations for fulfilling the functions established on law 21/1982. These organizations have been assigned several functions in order to benefit poor sectors of population through providing family allowance, which is intended for low-wage earners and service allowance through health, education, marketing and leisure programs. In fact, it is recognized that FCF's system has multi-functionality role, which is oriented to fulfil with policies and programs of social development for diminishing vulnerability and improving quality life of Colombian citizens. Hence, FCF's is considered as a very special organization that it is organized under private laws for managing activities of general interest [31]. In this sense, FCF's can develop public and private functions simultaneously. Currently, there are 69 FCF's in Colombia with 7 millions affiliates, which provide service as health, education, behalf, culture, leisure, official protection housing and credit [32]. All these activities gather a great dealt of groups with different purposes over entity. These groups are identified as stakeholders, who can be general and specific according to the linking activity. The business activities of FCF's should respond to the principles of sustainably, equity, good governance, transparency and social responsibility. Hence, when FCF's make strategy and development plans should take into account their stakeholder and a possible effect that plans can have over them. In general way, FCF's have already identified their stakeholders in their corporate status, such as: administration body, employees, suppliers, creditors, competitors workers and firms affiliated, customers, partners, unions, controlling body, local, regional and national government and community.

As noted above, some of FCF's services can be developed as privately without losing its social function, such as: leisure. Leisure programs are one of the main activities that let to obtain financial resources without depending on government. Nowadays, FCF's offer six kinds of services with different levels of coverage and categories (see Table 1).

Table 1. Kind and categories of leisure programs

Kind of program	Coverage	Categories	Coverage
Entrance to sport facilities	31,3%	Affiliated type A	32,3%
Special sport service	23,0%	Affiliated type B	10,8%
Sport programs	20,5%	Affiliated type C	10,2%
Recreational programs	20,3%	Not affiliated D	17,1%
Holiday sport and leisure programs	4,50%	Enterprises	29,7%
Tourism	0,05%		

Source: Elaboration own based on data of the Superintendence of Family Subsides (2014)

It is noted that entrance to sport facilities, special sport service, sport programs, recreational programs, are services programs more used and categories A and D and enterprises are niche markets that most use leisure services. Hence, sport facilities, special sport service, sport programs can lead to not affiliated and enterprises to obtain further profitability. In this sense, FCF wants to take advantage of the sport movement and their influence to attract new customers and to keep customers loyalty. Therefore, this activity has gathered a great deal of stakeholders, which are influenced by the management policy and decisions adopted by administration body.

3.2   Sport as Factor of Influence

Sport movement has become one of the global events the most influential and powerful over the world. Mass media and major sport events have facilitated the creation of mass culture that forms a public awareness [33]. This activity has affected social and economic structures in several ways. Furthermore, sport's industrial activity has contributed to the economic regeneration, social development and the change of image of cities [34]. In fact, implications of sport go beyond its original boundaries and its effects have been disseminated throughout the world in different ways. Sport is characterized by its transversely with other markets and industries since it is used for doing business, selling and persuasion [11]. Sport is a dynamic and interactive sector, which is able to influence on markets and industries, whether they are associated directly with sport sector or no. Furthermore, its power is presented on society affecting multiple contexts and going through boundaries in the business world. Hence, sport phenomena can surpass boundaries that others sectors cannot.

4   Methodology

In this Section, we briefly review some basic concepts about comparison notion and fuzzy relation composed by the importance of characteristics algorithm and fuzzy composition represents in a square fuzzy matrix.

4.1   The Square Fuzzy from Comparison Index

In the decision making in uncertainty is used to link relation and establish the relation of incidence or causality through the nuances of their relation levels. The vectors are given by subjective preferences that in turn are parameterized by the importance of characteristics algorithm. Therefore, results obtained by comparison index can be represented on a square fuzzy matrix.

4.1.1   Ordering According to the Importance of Characteristics Algorithm

The importance of the characteristics [10] is a useful technique for establishing the relative importance in a causality relation between two objects considering their characteristics. The importance of the characteristics is composed by dominant eigenvalue and dominant eigenvector.

Definition 1. Matrix reciprocal $[\tilde{R}]$ collects all characteristic compared by the time it has been preferred. For each characteristic C_j is carried out a comparison two by two, $C_i, C_k; i, k = 1, 2, \ldots, n$ using a quotient, which determines the time that it is preferred to the other one, such as:

$$\mu_{ik} = \frac{f_i}{f_k}, \quad i, k = 1, 2, \ldots, n, \tag{1}$$

where C_i represents the times is preferred to C_k.

Notes that matrix is built by the collection of all μ_{ik} and it is reciprocal and coherent/consistent. It is reciprocal because it is complied with $\mu_{ii} = 1$; $\mu_{ik} = 1/\mu_{ki}$ where $\mu_{ik} \in R_o^+$, $i, k = 1, 2, \ldots, n$. It is coherent/consistent because it is complied with $\forall\, i, k, l \in \{1, 2, \ldots, n\}$; $f_i/f_k * f_k/f_l = f_i/f_l$, i.e. $\mu_{ik} * \mu_{kl} = \mu_{il}$.

Therefore, matrix must comply with transpose property, which is given by:

$$\sum_{k=1}^{n} \mu_{ik} * f_k = \sum_{k=1}^{n} \frac{f_i}{f_k} * f_k = n * f_i, \tag{2}$$

and proportionality property, which is given by:

$$\frac{\mu_{ik}}{\mu_{lk}} = \frac{f_i/f_k}{f_l/f_k} = \frac{f_i}{f_l}, \tag{3}$$

also:

$$\frac{\mu_{ik'}}{\mu_{lk'}} = \frac{f_i/f_{k'}}{f_l/f_{k'}} = \frac{f_i}{f_l}, \tag{4}$$

therefore:

$$\frac{\mu_{ik}}{\mu_{lk}} = \frac{\mu_{ik'}}{\mu_{lk'}}. \tag{5}$$

Definition 2. A Dominant Eigenvalue E_{va} of dimension n, is a mapping $E_{va} : [0, 1]^n \times [0, 1]^n \to [0, 1]$ that has an associated limit weighting vector $\lambda_1^{(c)}$, with $w_j \in [0, 1]$ and $\sum_{j=1}^{n} w_j \geq 1$, such as:

$$E_{va}(\langle x_i, y_k \rangle, \ldots, \langle x_n, y_m \rangle) = \sum_{k=1}^{n} maxw_j(\mu_{ik} * y_k), \tag{6}$$

where x_i and y_k represents the jth largest of sets X and Y.

Therefore:

$$\lambda_1^{(c)} = E_{va}max \tag{7}$$

Definition 3. A Dominant eigenvector $V^{(c)}$ has an associated weighting vector $\lambda_1^{(c)}$, with $w_j \in [0, 1]$ and $\sum_{j=1}^{n} w_j \leq 1$, such as:

$$V^c(\langle x_i, y_k \rangle, \ldots, \langle x_n, y_m \rangle) = \sum_{k=1}^{n} \frac{(\mu_{ik} * y_k)}{max(\mu_{ik} * y_k)}, \tag{8}$$

and normalizing:

$$N^{(c)} = \frac{V^{(c)}}{\sum V^{(c)}}. \tag{9}$$

Therefore, Relative Importance is shown within representative of the importance matrix $[\tilde{R}]$ by each characteristic. This matrix is given by:

$$[\tilde{R}]^* = N^{(c)} * [\tilde{R}], \tag{10}$$

where $[\tilde{R}]$ is the ith arguments of the set X.

Hence, following the process above, it is obtained aesulting matrix $[\tilde{R}]^*$, which represents a *square fuzzy matrix*.

4.1.2   The Square Fuzzy Matrix

The square fuzzy matrix [10] is a useful for representing direct and indirect relationship between physical and mental objects. In the decision making in uncertainty is used to link relation and establish the relation of incidence or causality through the nuances of their relation levels. Therefore, the distance relatives can be represented on a square fuzzy matrix:

$$[\tilde{R}]^* = \begin{array}{c|cccc}
\tilde{r} & a_1 & a_2 & \cdots & a_i \\
a_1 & (x_1, y_1) & (x_1, y_2) & \cdots & (x_1, y_i) \\
a_2 & (x_2, y_1) & (x_2, y_2) & \cdots & (x_2, y_i) \\
\vdots & \vdots & \vdots & \cdots & \vdots \\
a_i & (x_i, y_1) & (x_1, y_2) & \cdots & (x_i, y_i)
\end{array}, \tag{11}$$

where $[\tilde{R}]^*$ represents the ith arguments of the sets X and Y.

Note that this matrix can comply with the reflexive, transitive, symmetry and fuzzy anti-symmetry properties. It is reflexive because the relation of elements of the set $x \in E$ with itself that is with $x \in E$ is total and the main diagonal is full of 1. Therefore, it must be accomplished with $\forall a_i \in E$ where $.i = 1, 2, \ldots, n$: $\mu_{ij} = 1, i = j$. and $\mu_{ij} \in [0, 1], i \neq j$ where a_i are the ith arguments of the set E. It is transitive because the indirect relation between three elements of the referential $E(a_i, a_j, a_k)$ can be considered of the same manner, i.e. that the indirect relation between a_i and a_k cannot be greater than the direct relation a_j and a_k. Therefore it must be accomplished with $\forall a_i, a_j, a_k \in E$: $\mu_{aijk} \geq \vee (\mu_{aiaj} \wedge \mu_{ajak})$. It is symmetry because the intensity of the relation from a_i to a_j is considered the same as a_j to a_i. Therefore it must be accomplished with $\forall a_i, a_j \in E, a_i \neq a_j$ and $\mu_{ai} = \mu_{aj}$ where a_i and a_j are the ith arguments of the set E. It is fuzzy anti-symmetry because the intensity of the relation from a_i to a_j is not considered the same as a_j to a_i. Therefore it must be accomplished with $\forall a_i, a_j \in E, a_i \neq a_j$ and $\mu_{ij} \neq \mu_{ji}$ or $\mu_{ij} = \mu_{ji} = 0$ where a_i and a_j are the ith arguments of the set E.

4.1.3   Fuzzy Composition

Fuzzy composition or convolution max-min [10] is useful technique for associating between physical and mental objects. In decision making on uncertainty is used to represent the degree of belonging or the lack of association and interaction or interconnection of fuzzy relation between elements of itself set or two or more fuzzy sets. For elements of itself set or two or fuzzy sets, the convolution max-min can be defined as follows:

Definition 4. A fuzzy composition $R \circ S$ is defined as a fuzzy relation $U \times W$ and it is associated with their characteristic functions $\mu_R(x, y)$ and $\mu_S(y, z)$, which is given by composition max-min, such as:

$$\mu_{R \circ S}(x, z) = \vee_{y \in V}(\mu_R(x, y) \wedge \mu_s(y, z)), \tag{12}$$

where $(x, z) \in (U, W)$.

Therefore, the relative intensity is established by the convolution of fuzzy matrix $[\tilde{R}]$ with itself. The behaviour of relation can be observed through evolution over time or no temporal stage.

Definition 5. The max-min composition of matrix $[\tilde{R}]$ is given by:

$$[\tilde{R}] \circ [\tilde{R}] = [\tilde{R}]^2, [\tilde{R}] \circ [\tilde{R}] \circ [\tilde{R}] = [\tilde{R}]^2 \circ [\tilde{R}] = [\tilde{R}]^3. \tag{13}$$

Therefore:

$$[\tilde{R}] \circ [\tilde{R}] = [\tilde{R}]^n \circ [\tilde{R}] = [\tilde{R}]^{n+1}, \tag{14}$$

when $[\tilde{R}]^n = [\tilde{R}]^{n+1}$ the process is stopped.

5   Application

In this section we present an application of the method proposed above. The main advantage on using the ordering importance of characteristics method and linking of relations is that it can parameterize the importance of the information of each characteristic according to appraisal of the decision-maker. This application is focused on stakeholder analysis of FCF's example in the sport sector.

5.1   Strategic Analysis Approach

The method describes the procedure for the development of application allowing a linking and ordering between each characteristic. To approach its design five steps are followed:

Step 1: We have analysed and determined the attributes – characteristic – [13] for each stakeholder category [6] for FCF's in Colombia. It is assumed that administrative body of a FCF want to analysis the immediate firm environment for carrying out strategic planning and decision-making process by leisure operative unit. The FCF's stakeholders are considered as immediate firm environment (Table 2), which will be analysed through their characteristics. Each characteristic for each specific stakeholder is estimated according to power/influence dominance (P). Likewise, it also has assumed that administrative body wants to consider the level of importance of the environment (L_{IE}), which is given by its stakeholder categories: R_S: real stakeholders; S_W: stakewatchers; S_K: stakekeepers (see Table 3). Each characteristic of the set of stakeholders is considered a property. This first step allows us to make a holistic appraisal of the immediate firm environment, since it is taking into account each category and sub-set around the firm.

Table 2. Immediate firm environment

Classification of stakeholders	Kind of relation	Stakeholders FCF's
Firm	Growth business	Administration body
Employees	Labour laws	Employees
Business	Contracts and agreements	Suppliers, creditors, partners
Customers	Customers	Workers and firms affiliated
Unions and association	Unions and safety groups	Unions
Competitors	Marked, competitors	FCF's competitors
Institutions and auditors	Public interest group	Controlling body
Local organization and government	Legal activities control	Local, regional and national government
Media and others	Diffuser and observer	Media and communication
Civil society	Civil, environmental and human rights	Community

Table 3. Characteristics and categories of each stakeholder

Category		Stakeholder group	Characteristic
Real stakeholders R_S	a	Firm	P
	b	Employees	P
	c	Business	P
	d	Customers	P
Stakewatchers S_W	e	Unions and association	P
	f	Competitors	P
	g	Institutions and auditors	P
Stakekeepers S_K	h	Local organization and government state	P
	i	Media and others	P
	j	Civil society	P

Source: Own elaboration based on Fassin (2008) and Mitchell *et al.* [23].

Table 4. Subjective preference matrix between characteristics

	Power	Legitimacy	Responsibility
Power	1	1 2/7	1 1/2
Legitimacy	7/9	1	1 1/6
Responsibility	2/3	6/7	1

Step 2: It has fixed the level of preference for each of the characteristics in order to form subjective preference matrix – reciprocal matrix. Here, each of the estimates could be composed by quality or quantity of data received, i.e. statistics, reports, surveys information, which is used as a way of guidance by administrative body. It has assumed that administrative body suggests level of preference between power, legitimacy and responsibility (see Table 4) and power estimation for each of the stakeholders in order to form power subjective preference matrices between each stakeholder (see Table 5).

Table 5. Subjective preference matrix of power relation

P	a	b	c	d	e	f	g	h	i	j
a	1	1,40	1,00	1,17	1,40	1,17	1,17	1,00	1,00	1,40
b	0,71	1	0,71	0,83	1,00	0,83	0,83	0,71	0,71	1,00
c	1,00	1,40	1	1,17	1,40	1,17	1,17	1,00	1,00	1,40
d	0,86	1,20	0,86	1	1,20	1,00	1,00	0,86	0,86	1,20
e	0,71	1,00	0,71	0,83	1	0,83	0,83	0,71	0,71	1,00
f	0,86	1,20	0,86	1,00	1,20	1	1,00	0,86	0,86	1,20
g	0,86	1,20	0,86	1,00	1,20	1,00	1	0,86	0,86	1,20
h	1,00	1,40	1,00	1,17	1,40	1,17	1,17	1	1,00	1,40
i	1,00	1,40	1,00	1,17	1,40	1,17	1,17	1,00	1	1,40
j	0,71	1,00	0,71	0,83	1,00	0,83	0,83	0,71	0,71	1

Step 3: It has fixed the levels of importance for each stakeholder in order to form the actual condition of each stakeholder and characteristic (C_A). The levels of importance for each stakeholder are obtained from the average and normalization of subjective preference established above (see Table 6).

Table 6. Level of importance for each stakeholder and characteristic

C_A	a	b	c	d	e	f	g	h	i	j
L_{ISP}	0,4454	0,1224	0,0542	0,0399	0,1215	0,0662	0,1701	0,1350	0,1357	0,0559	0,0991

Note that L_{ISP} (Power), has been Normalized (N) to establish the weight of each stakeholder and characteristic (C_A).

Step 4: It has fixed the levels of importance environment in order to determine the influence of certain sector at a specific moment. The level of importance of the environment is defined by each category according to external information and experience of experts about specific sector. In this case, sport sector is taken into account. It has assumed that administrative body suggests the level of importance of the environment as weighted factor of the sport sector for each stakeholder (see Table 7).

Table 7. Level of importance of the environment

	a	b	c	d	e	f	g	h	i	j
Category	Real stakeholders (R_S)				Stakewatchers (S_W)			Stakekeepers (S_K)		
L_{IE}	0,3				0,4			0,2		

Step 5: For obtaining main fuzzy matrix technical comparison between subjective preference matrix and determine relative level of importance are considered as starting point. In this application, each result obtained is considered as vector of importance forming a fuzzy matrix and the multiplication of L_{ISP} with L_{IE}, which is considered as the relative level of importance for each stakeholder and characteristic (RL_I's).

5.2   Results

The following section shows the main results of the application. The RL_I, dominant eigenvalue (E_{va}), dominant eigenvector ($V^{(c)}$) and fuzzy matrix is obtained, which allow establishing the intensity relative among stakeholders. Thus, the adjustability of these algorithms to assume the decision-makers preference is shown. Finally, the obtained results of stakeholders and characteristics are depicted, linked and grouped. We have aggregated subjective information to obtain representative results for P category proposed within specific and uncertainty environment. The multiplication of L_{ISP} with L_{IE} is considered as the relative level of importance for each stakeholder and characteristic (RL_I's) (see Table 8). From different RL_Is all (E_{va}) and ($V^{(c)}$) are obtained (see Table 9). In this case, P and C_A are gathered to obtain (E_{va}) and ($V^{(c)}$) for T_R. In order to determine fuzzy relative matrix (FR_M) is multiplied each subjective preference matrix with $V^{(c)}$ normalized (see Table 10). This matrix shows how ambiguity and fuzziness of the stakeholders and subjective appraisal of decision-maker can be dealt with. Into each matrix should be considered that 1 defines relationship of each stakeholder with itself is total and decimal zero – 0,00 – defines weakest incidence relation of the order of $<10^{-3}$.

Table 8. Relative level of importance

C_A	a	b	c	d	e	f	g	h	i	j
RL_{ISP}	0,1336	0,0367	0,0163	0,0120	0,0364	0,0265	0,0681	0,0540	0,0271	0,0112	0,0198

For getting intensity relative matrix (IR_M) is processed FR_M through max-min composition (see Table 11). A simulation of relationship evolution in the short time is

Table 9. Dominant eigenvalue and dominant eigenvector

	E_{va}	$V^{(c)}$	N
a	0,327	0,733	0,110
b	0,163	0,366	0,055
c	0,159	0,355	0,053
d	0,446	1,000	0,150
e	0,254	0,568	0,085
f	0,425	0,952	0,142
g	0,400	0,896	0,134
h	0,378	0,848	0,127
i	0,192	0,430	0,064
j	0,239	0,535	0,080

Note that dominant eigenvector has been
Normalized (N) to establish the weight of
each stakeholder and salience.

Table 10. Fuzzy relative matrix of power

	a	b	c	d	e	f	g	h	i	j
a	1	0,219	0,164	0,137	0,164	0,073	0,066	0,073	0,088	0,274
b	0,027	1	0,088	0,068	0,055	0,018	0,011	0,000	0,000	0,011
c	0,035	0,033	1	0,066	0,000	0,027	0,027	0,011	0,000	0,000
d	0,120	0,120	0,120	1	0,150	0,150	0,449	0,449	0,000	0,150
e	0,057	0,085	0,000	0,085	1	0,071	0,085	0,061	0,000	0,085
f	0,214	0,427	0,285	0,142	0,171	1	0,085	0,095	0,142	0,712
g	0,223	0,670	0,268	0,045	0,134	0,223	1	0,134	0,000	0,000
h	0,190	0,000	0,634	0,042	0,178	0,190	0,127	1	0,127	0,127
i	0,080	0,000	0,000	0,000	0,000	0,064	0,000	0,064	1	0,129
j	0,032	0,400	0,000	0,080	0,080	0,032	0,000	0,080	0,040	1

shown. It is noted that relationship apparently non-existent before are discovered, which means that interactive relationship between stakeholders within changing environment can be analysed in the short-time by dynamic process.

Intensity and linked relations for each characteristic are shown in visual schemes in order to understand the dynamics of relationships under specific environmental conditions (see Figs. 1 and 2). Intensity represents the possible impact on relation behaviour that it can have each stakeholder on entire structure and firm. It is established by mid-point, lower and upper threshold of each relative intensity matrix. In intensity analysis is shown the relative impact on stakeholders behaviour according to characteristic analysis, where black colours indicate lowest impact (L_{IM}), dark grey colours indicate medium impact (M_{IM}) and light grey colours indicate highest impact (H_{IM}). Thus, d has H_{IM}, a has M_{IM} and c has L_{IM}. Hence, if the level of impact is high, the pressure exerted is higher and the pressure received is less. If the level of impact is low, the pressure exerted is lower and the pressure received is greater.

Table 11. Relative intensity matrix of power

	a	b	c	d	e	f	g	h	i	j
a	1	0,274	0,164	0,137	0,164	0,137	0,137	0,137	0,137	0,274
b	0,068	1	0,088	0,068	0,068	0,068	0,068	0,068	0,068	0,068
c	0,066	0,066	1	0,066	0,066	0,066	0,066	0,066	0,066	0,066
d	0,223	0,449	0,449	1	0,178	0,223	0,449	0,449	0,137	0,223
e	0,085	0,085	0,085	0,085	1	0,085	0,085	0,085	0,085	0,085
f	0,214	0,427	0,285	0,142	0,171	1	0,142	0,142	0,142	0,712
g	0,223	0,670	0,268	0,142	0,171	0,223	1	0,142	0,142	0,223
h	0,190	0,190	0,634	0,142	0,178	0,190	0,142	1	0,142	0,190
i	0,080	0,129	0,088	0,080	0,080	0,080	0,080	0,080	1	0,129
j	0,080	0,400	0,088	0,080	0,080	0,080	0,080	0,080	0,080	1

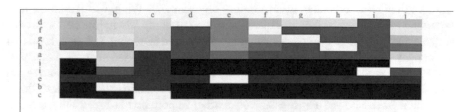

Fig. 1. Graphical representation of relative importance of Power

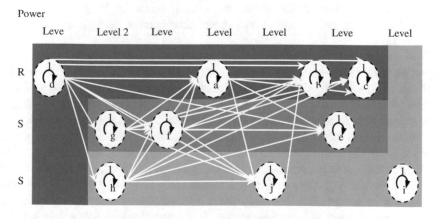

Fig. 2. Linked relation for each stakeholder and category

Linked relation represents the ties relations of incidence – possible influence – of each stakeholder on entire structure and firm. It is also established by mid-point, lower and upper threshold of each intensity relative matrix. In linked relation analysis is shown relative order of each stakeholder within different levels and categories. For establishing order of relation is used order relation and ordering by grouping concepts [10].

Depicts is shown that power has only unidirectional relations. In the power depicts is shown that d – customers – has greatest power of influence over other ones, while a – firm – exercises its power over b – employees, c – business, e – unions – and j – community – and f – competitors, g – auditors – and h – government – exercises its power over a. In conclusion, the analysis of the immediate firm environment shows the relative impact and influence of entire stakeholders and firm according to power of influence, for carrying out strategic planning and decision-making process by leisure operative unit within uncertainty environment.

6 Conclusions

We have analysed the relative impact and influence of entire stakeholders and firm relationship within a dynamic environment. It has studied the intensity and linked relations of entire stakeholders in uncertainty environment. We have focused on new categorization and classification of stakeholders suggested by Fassin [6]. This approach facilitates the identification and selection of stakeholders within different levels of environment with an ambivalent position of pressure groups and regulators. It has used fuzzy techniques to develop a methodological perspective of the new characterization of stakeholders. Application of the importance of characteristics algorithm and fuzzy composition have allowed us to establish relative importance in a causality relation based on incidence concept [26] and lead to a process of analysis of subjective attributes. It has applied these algorithms to obtain the *fuzzy subjective preference* and *relative intensity* between each stakeholder. We have made application of these algorithms considering FDF's in Colombia and sport. Implications of sport go beyond its original boundaries and its effects have been disseminated throughout the world in different ways.

It is shown how the ambiguity and fuzziness of the stakeholders and appraisal subjective of decision-maker can be dealt with. A simulation of relationship evolution in the short time is carried out by fuzzy relative matrix in which is shown different degrees of intensity according to the levels of importance. The results are depicted to show the possible impact on relation behaviour and the ties relations of incidence of each stakeholder on entire structure and firm. It has developed an application of fuzzy logic tools, which can be used in the decision-making field. It has obtained specific results from subjective preference and specific information, which are only valid in the short-term. Even though the model will continue being fully valid, specific results will be evolving in the medium and large term according to functional trend that the dynamic process itself will provide. Hence, it is important that estimations are dealt with fuzzy tools.

References

1. Blanco-Mesa, F.R.: Técnicas para la toma de decisiones en contextos inciertos: identificación de oportunidades socio- económicas en el ámbito deportivo (2015)
2. Donaldson, T., Preston, L.E.: The stakeholder theory of the corporation: concepts, evidence and implications. Acad. Manag. Rev. **20**, 65–91 (1995)
3. Fassin, Y.: Imperfections and shortcomings of the stakeholder model's graphical representation. J. Bus. Ethics **80**, 879–888 (2007)
4. Fassin, Y.: The stakeholder model refined. J. Bus. Ethics **84**, 113–135 (2009)
5. Fassin, Y.: A dynamic perspective in Freeman's stakeholder model. J. Bus. Ethics **96**, 39–49 (2010)
6. Freeman, R.E.: Strategic Management: A Stakeholder Approach. Pitman, Boston (1984)
7. Freeman, R.E.: The stakeholder approach revisited. Zeitschrift Für Wirtschafts- Und Unternehmensethik **5**, 228–241 (2004)
8. Freeman, R.E.: Managing for stakeholders: trade-offs or value creation. J. Bus. Ethics **96**, 7–9 (2011)
9. Friedman, A.L., Miles, S.: Developing stakeholder theory. J. Manag. Stud. **39**, 1–21 (2002)
10. Frooman, J.: Stakeholder influence strategies. Acad. Manag. Rev. **24**, 191–205 (1999)
11. Gil-Aluja, J.: Towards a new paradigm of investment selection in uncertainty. Fuzzy Sets Syst. **84**, 187–197 (1996)
12. Gil-Aluja, J.: Elements for a Theory of Decision in Uncertainty. Kluwer Academic Publishers, Dordrecht (1999)
13. Gil-Aluja, J.: Lances y desventuras del nuevo paradigma de la teoría de la decisión. In: Proceedings del III Congreso SIGEF, Buenos Aires, pp. 11–37 (2000)
14. Gil-Lafuente, A.M., Barcellos dc Paula, L.: Una aplicación de la metodología de los efectos olvidados: los factores que contribuyen al crecimiento sostenible de la empresa. Cuad. del CIMBAGE **12**, 23–34 (2010)
15. Gil-Lafuente, A.M., de Paula, L.B.: Algorithm applied in the identification of stakeholders. Kybernetes **42**, 674–685 (2013)
16. Gil Lafuente, A.M., Blanco, F.R.: Elementos para la toma de decisiones en las entidades deportivas en entornos de incertidumbre. In: Maqueda, F., Barquero, D. (eds.) Economía y Deporte: Gestión de Entidades Deportivas, pp. 155–176. Furtwangen, Barcelona (2014)
17. Gratton, C., Shibli, S., Coleman, R.: Sport and economic regeneration in cities. Urban Stud. **42**, 985–999 (2005)
18. Holzer, B.: Turning stakeseekers into stakeholders: a political coalition perspective on the politics of stakeholder influence. Bus. Soc. **47**, 50–67 (2007)
19. Kaufmann, A., Gil-Aluja, J.: Grafos Neuronales para la Economía y la Gestión de Empresas. Ediciones Pirámide, Madrid (1995)
20. Kaufmann, A., Gil-Aluja, J.: Models per a la Recerca d'Efectes Oblidats (1988)
21. Koppet, L.: Sports Illusion, Sports Reality: A Reporter's View of Sports, Journalism, and Society. University of Illinois Press, Illinois (1994)
22. Martorell-Cunill, O., Gil-Lafuente, A.M., Socias Salvà, A., Mulet Forteza, C.: The growth strategies in the hospitality industry from the perspective of the forgotten effects. Comput. Math. Organ. Theory **20**, 195–210 (2013)
23. Mitchell, R.K., Agle, B.R., Wood, D.J.: Toward a theory of stakeholders identification and salience: defining the principle of who and what really. Acad. Manag. Rev. **22**, 853–886 (1997)
24. Pesqueux, Y., Damak-Ayadi, S.: Stakeholder theory in perspective. Int. J. Bus. Soc. **5**, 5–21 (2005)

25. Post, J.E., Preston, L.E., Sachs, S.: Managing the extended enterprise: the new stakeholder view. Calif. Manag. Rev. **45**, 6–28 (2002)
26. Rodríguez Arévalo, V.: La naturaleza jurídica de las cajas de compensación en colombia: una visión frente a su control. Justicia Juris **8**, 9–21 (2012)
27. Rowley, T.I., Moldoveanu, M.: When will stakeholder groups act? An interest- and identity-based model of stakeholder group mobilization. Acad. Manag. Rev. **28**, 204–219 (2003)
28. Salazar-Garza, R.: The Mexican peso: exchange risk coverage management through the forgotten effects theory. J. Econ. Financ. Adm. Sci. **17**, 53–73 (2012)
29. Sciarelli, M., Tani, M.: Network approach and stakeholder management. Bus. Syst. Rev. **2**, 175–190 (2013)
30. Superintendencia del Subsidio Familiar (SSF): El Subsidio Familiar., Bogotá (2013)
31. Wagner, E., Alves, H., Raposo, M.: Stakeholder theory: issues to resolve. Manag. Decis. **49**, 226–252 (2011)
32. Wagner, E., Alves, H., Raposo, M.: A model for stakeholder classification and stakeholder relationships. Manag. Decis. **50**, 1861–1879 (2012)
33. Windsor, D.: The role of dynamics in stakeholder thinking. J. Bus. Ethics **96**, 79–87 (2011)
34. Zadeh, L.A.: Fuzzy sets. Inf. Control **8**, 338–353 (1965)

Identification of the Exchange Risk Exposure by Applying the Model of Forgotten Effects

Gumaro Alvarez Vizcarra[1]([⊠]), Anna M. Gil-Lafuente[2]([⊠]),
and Ezequiel Avilés Ochoa[3]([⊠])

[1] Tecnológico de Monterrey, Sinaloa, Culiacán, Sinaloa, Mexico
galvarez@itesm.mx
[2] Department of Economics and Business Management,
University of Barcelona, Barcelona, Spain
amgil@ub.edu
[3] Universidad de Occidente, Culiacán, Sinaloa, Mexico
ezequiel.aviles@udo.mx

Abstract. The management of foreign exchange exposure is essential to reduce vulnerability of firms to unexpected changes in the exchange rate, which adversely affect the profit margin, cash flow and value of the business. In the early eighties, Adler and Dumas lead the discipline of managing risk exposure with the help of econometric techniques that are effective in stable conditions. However, the main feature of modern economies is the uncertainty and instability of the environment. Under such circumstances, the traditional models are insufficient and ineffective in making decisions. It is therefore necessary to address decisions under uncertain conditions, from another perspective. The objective of the present work is to identify the determinants of exchange risk exposure, by applying of forgotten effects model. The results show that the *lack of information, the poor financial planning, and the entrepreneur attitude*, are important aspects that have been forgot in the identification of exchange risk exposure. The findings, help companies redefine the action strategies to develop an efficient program of risk management.

Keywords: Exposure · Exchange risk · Forgotten effects · Fuzzy logic

1 Introduction

Global markets increase the interdependence of economies, expressed Barcellos [3], increasing their vulnerability to financial imbalances, thereby generating greater uncertainty. The volatility of financial variables, say Muñoz and Aviles [28], issued by uncertain economic environment, it has adverse effects on the competitiveness of organizations. In an environment of high volatility, says Pascale [29], the world becomes a riskier place. In this context, it is necessary, speak Preciado, España and Lopez [31] and Danhani [7], all companies make efforts to develop a management of foreign exchange risk exposure, to help reduce the effects of the financial collapse. Regardless, tell Martinez [27], Rodriguez [35], Doukas, Hall and Lang [8], the type of the activity or operation level to record in domestic and international markets.

© Springer International Publishing AG, part of Springer Nature 2018
A. M. Gil-Lafuente et al. (Eds.): FIM 2015, AISC 730, pp. 381–399, 2018.
https://doi.org/10.1007/978-3-319-75792-6_28

Manage exchange risk exposure, is highly complex due to the intense flow of information and the accelerated rate of mutability of economic and financial relations. Effective management, demand efforts in three areas, say Popov and Stutzmann [30]: the identification, measurement and coverage. To date, most of the actions have focused on measurement and less protection, somehow avoiding identification. Studies have been conducted from the perspective of deterministic models, tools very useful in stable conditions, nevertheless in uncertain conditions, lose effectiveness, and is evident, mention Gil [13] and Martinez et al. [26], its offset from the requirements of an environment marked by growing uncertainty.

In this way, Gil [17] notes that the mechanisms of certainty are inoperative when the reality deviates from elaborate scheme, as the stochastic techniques difficult to adapt when the data has been impregnate with subjectivity. Adding, for a treatment adequacy of the phenomena under uncertainty, it is necessary to recourse to the theory of fuzzy subsets like the more general of theories capable of describing fuzzy environments. For this, it is perfectly suited to the treatment of the subjective and uncertain, collects phenomena as they occur in real life and makes their treatment without attempting deform to make them accurate and true. In this context, Gil [17] concludes, against a background of rapid and constant change, it is necessary to have instruments whose key features are *flexibility* and *adaptability*. Therefore, it is considering that the implementation of the theory of fuzzy subsets through non-numerical model called Forgotten Effects.

2   Literature Review

It's generally accepted that unexpected movements in the exchange rate contemplated cash flows by altering the value of the revenue in local currency and the cost of inputs in foreign currency. The effects of exposure to exchange rate risk is not exclusive tradable goods firms. Doukas, Hall and Lang [8], Shapiro [37] and Allayannis and Ofek [2], Chaieb and Mazzotta [5], believe that exposure is operational, that relating to exports, imports and source inputs. In this context, firms non-tradable goods face changes in demand; if that is elastic, augmenting costs, increase in working capital requirements and intensifying competition. In this manner, conclude Martinez [27], Rodriguez [35], in a globalized and uncertain environment, the effects of unexpected variations in the exchange parity disturb the companies, regardless of location, size and activity.

2.1   The Management of Exchange Rate Exposure

Definitions of foreign exchange exposure are converging. Adler and Dumas [1], Allayannis and Ofek [2], Flood and Lesard [9], Bartram [4] and Doukas et al. [8], defined as the elasticity of cash flow to unanticipated changes in the exchange rate. From the perspective of Jorion [20], represents the sensitivity of the value of firms to unexpected variations of parity, coinciding with the approach of Adler and Dumas [1]. However, he makes a fundamental difference: specified as endogenous variable the *value of the firms*, their statement is multivariate and the value of the company is not define exclusively by the cash flow.

The pioneering research focuses on determining the sensitivity of cash flow and value of firms to unexpected movements in the exchange rate [1, 9, 32, 37]. (Contributions, says Rodríguez [35], are aim at studying the impact of exchange rate fluctuations on the level of operating cash flow of firms.

Today, companies have persistent links with the economic environment, with aspects independent to their field of choice. This interaction, immersed in systemic volatility of economic and financial variables, turn into in an increased exposure to foreign exchange risk, which, if not neutralized, deteriorating competitive position and reduces the market value of the organizations [1, 20]. The company, to be of tradable goods or not, it is exposed to risk, and therefore, required to design and develop a management strategy aimed to minimize losses arising from unanticipated events in the volatility of foreign exchange market [7, 31].

The management of risk exposure, conceptualized Geissler [10], Redja [33], Wallace [40], is a *systematic and interactive process* designed to detect, and assess their impact and design the most appropriate hedging technique and activities, which allows arriving at goals. Popov and Stutzmann [30] and Rahnema [34], they point out, as a process consisting of three phases: *identifying, measuring and hedging*. In this case, only we address the first phase, which is the scope of this investigation.

In the *identification* phase, efforts have been few. According to Jegher [19], the first step is the identification, understood as the phase profiled to identify the causes of exposure, which allowing design strategies to address it. Popov and Stutzmann [30] note that identifying exposure; it must first be defining and functionally expressed. Some variables that have been discuss, as determinants of exchange risk exposure, are show in the following Table 1.

Table 1. Determinants of exchange rate risk exposure

Determinants	Author
Foreign sales	Jorion (1990)
Firm size	Doukas et al. (1996)
Leveraging	He y Ng (1998)
Hedging through derivatives	Allayannis y Ofek (1997)
Market position	Allayannis y Ihrig (2001)
Shared markets	Williamson (2001)
Industrial concentration	Bartram y Karolgy (2003)
Liquidity of the firm	Bartram (2004)

Source: Authors with Bartram Information (2008).

Gil [13] point out, the sophistication in the use of techniques and tools has not been success factor in making decisions in an environment of increasing uncertainty. Hence the pop instrumental hypothesis, which postulates that: the traditional methods of binary math does not have the expected efficiency in a volatile and globalized environment, opening a space to deal the subjectivity, vagueness and uncertainty from another perspective.

2.2   Fuzzy Logic

In literature there are many models used to explain and solve problems in environments of low uncertainty [6, 36, 38, 39]. However, today, says Gil [14], efforts are focus at structuring a theoretical and methodological framework, for explain with more analytical rigor the problem of decision makers in a complex and diffuse environment. The tools used to complement the classical analysis, inscribed Gil [13], have their origin in the evolution of scientific knowledge and the development of mathematical uncertainty.

Zadeh [41] introduces[1] the theory of fuzzy sets, which objects to provide the basis of approximate reasoning using imprecise premises as a tool for formulating knowledge. With this theory, the logic is crystallized in a new concept that gives consistency to subjective thought; fuzzy logic as an extensive universe, where it converge with the classical Boolean logic; as a complementary alternative to discrete logic, since it uses membership degrees in the range [0, 1], rather than the specific binary values 0 or 1. There are few events strictly considered like entirely true or false; all manifest as a degree of membership. Fuzziness means multivalence, where everything is a matter of degree, including truth and falseness in his belonging to a set [12, 13, 15, 23, 41].

With well define and differentiated from each other propositions, there aren't objection to form a set. However, in real life, situations are not as clear as expected. For better reflection of nature, the notion of fuzzy subsets is introduce [23], so that the membership characteristic function $f_A = (x)$ can assume values in the segment [0, 1].

$$\mu_{\tilde{A}} : E \to [0, 1], \quad x \to \mu_{\tilde{A}}(x).$$

The higher value is assigned when satisfy the proposition or characteristic established. A fuzzy subset is represent as follow:

$$\tilde{E} = \begin{array}{|c|c|c|c|c|c|}
\hline
a & b & C & d & e & F \\
\hline
.2 & .5 & 1 & .9 & 1 & 0 \\
\hline
\end{array}$$

$$\tilde{A} = \left[(x; \mu_{\tilde{A}}(x)) \right] \quad \forall x \in E.$$

Being $\mu_{\tilde{A}}(x)$ the degree of membership of x to $\tilde{A}$ and [0, 1] the membership space.

2.3   Forgotten Effects Theory

In his attempt to examine, explain and solve a particular problem, decision makers often made mistakes or omissions. You can make use of sophisticated technologies to alleviate human failure. Errors or omissions, revealed in an increased exposure, note Kaufmann and Gil [24], are not detectable immediately, because usually are undetected

---

[1] In fact, the origin could be set in 1922, when Lukasiewicz questioned classical Boolean logic (true and false values) and proposed a logic of certain values in the unit interval as a generalization of his trivaluada logic. In the thirties were valued logics proposals for any number of certain values (equal to or greater than 2), identified by rational numbers in the range [0, 1].

and more destructives, under which are effects of an accumulation of causes. Human intelligence need to rest on tools and models capable of creating a technical basis on which can be worked with the information available, contrasting with those obtained from the environment and bring out all the relations of direct and indirect causation, that can be detached [24].

From problems showing impact relationships of a set of expressions on another, or on itself, the *incidence* concept emerge, which can be link to the function concept commonly used, conception highly subjective and difficult to quantify. However, incidence spread concatenated way, causing many steps are obviated; conclusions are forgotten, causing the called second-generation effects, whose relative idea, pointing Kaufmann and Gil [24], is due to Fourastié. However, attend the incidence, can understand everyday life, improve reasoned performance and influence decision making at all levels of management. The effects of second generation, are detected when using forgotten effects model, assert [11], the design or improvement strategies.

The mathematical expression model, according to Kaufmann and Gil [24], is expressed as follows, is the discrete set of causes identified: $C = \{c_1, c_2, \ldots, c_n\}$. That has an impact on another set of effects identified: $E = \{e_1, e_2, \ldots, e_n\}$.

You can say that c_1 has an effect on e_1 if the value of the characteristic membership function is different of zero [18, 24]. The pair (c_1, c_2) and (e_1, e_2) is valued at, $[0, 1]$ that is to say, the degree of incidence each c_i on each e_j is expressed by a function: $\mu : C \to E \in [0, 1]$, where μ represents the values of the Cartesian plane formed by the set C and the set E, determined in the range $[0, 1]$.

So that $\forall \; (c_i, c_j) \in \; C \to E, \mu \; (c_i, e_j) \in [0, 1]$.

Presumably, when the expert is ask to value[2] the incidence of C on E, in binary values as 0 or 1 find some difficulty in trying to express their appreciation, because it does not always locate the very definite relationship. Therefore, the convenience of using fuzzy matrices is demonstrated, using the range $[0, 1]$, values that allow the expert have 11 options to assess the incidence between c_n and e_m. Pairs valued assembly thus form what is call *original incidence matrix*, which is denote as matrix $[CE]$.

The matrix analysis of incidence, point Kaufmann and Gil [22], can be do with a single matrix through a process of self-convolution, the analysis is limited to an incidence of the first order, but to pretend to know the effects of second order, then the analysis must be taken at least three matrices: $[CE]$, $[CC]$, $[EE]$.

To determine the incidence of $[CE]$ about $[EE]$ knowing the incidence of $[CE]$ about $[CC]$, it is necessary that the matrices observe the property of transitivity. To ensure the presence of this property, use *maxmin* model, through two operators:($\vee$) that means maximum, the largest of the numbers, and ($\wedge$) that means least, that is, the smaller. Using this model guarantees the condition of transitivity; the value of secondary pair is less than the value of the primary pair. To define the mathematical formula to construct the matrix of secondary effects, it is important that peer relationships are expressed as follows: if the valuation of the box (c_1, e_1) is = 0 is written $\mu(c_1, e_1) = 0$, where the valuation is μ, that way we proceed with all rows $(c_1, c_2, c_3, \ldots, c_n)$. Once you have the

---

[2] The *valuation* concept is similar to a numerical assignment made subjectively. Subjectivity is the characteristic that distinguishes the notion of measure, eminently objective (Gil et al., 2004).

valuations of each box provided by the expert, the procedure *maxmin* is perform to know the valuation given to each cell in the matrix $[M*] = [CE] \circ [CC] \circ [EE]$ (where $\circ$ denote the application of composition *maxmin*). The equation from which it is possible to estimate the values of $[M*]$ is as follows:

$$\mu(c_1, b_1) = (\mu(c_1, e_1) \wedge \mu(c_1, c_1)) \vee (\mu(c_1, e_2) \wedge \mu(c_1, e_2))$$
$$\vee (\mu(c_1, e_3) \wedge \mu(c_1, c_3)) \vee (\mu(c_1, e_4) \wedge \mu(c_1, c_4))$$
$$= (0 \wedge 0) \vee (1 \wedge 0) \vee (0 \wedge 1) \vee (0 \wedge 1)$$
$$= (0) \vee (0) \vee (0) \vee (0) = 0$$

$\mu(c_1, b_1) = 0$, in this way are the values each cell of the matrix $[M*]$. The general formula for this procedure is define as:

$$\mu(c_n, b_k) = \vee_m (\mu(c_n, e_m) \wedge \mu(e_m, b_k)).$$

The above procedure, you may do so through method called FICO, which is confronting a row with the column of the matrices subject to convolution, proceeding to do the same for all rows and columns to fill the box that form the matrix $[M*]$ call secondary effects. We proceed in the same way, regardless if working with ordinary matrices or fuzzy matrices. The relationships between sets of variables in a fuzzy matrix and a sagittate graph are expresses as follows:

$$\forall (c_n, b_k) \in M : v(c_n, b_k) \in [0, 1].$$

When experts are asked to evaluate relationships between entities of two sets, using values between $[0 \; y \; 1]$, postulate Kaufmann and Gil [24], their consideration is nuanced, allowing locate levels of truth in the notion of incidence. In response, Kaufmann and Gil [24], propose using a semantic edecadaria scale. Which provides two advantages: the first, allowed by the semantic processing, faithfully modify the opinions of respondents in numerical values. Second, because it consists of 11 values [. 0.1, 0.2, 0.3, 0.4, ... 1] admits a better adaptation and treatment of the concepts discussed, because people are accustomed to think and work in decimal form [16].

From the construction of the matrix of secondary effects, it is able to find the forgotten effects, for it is necessary to separate the cumulative effects shown in the matrix $[M*]$, from the direct that appear in the matrix $[CE]$. This is possible, to perform a simple algebraic subtraction $[M * *] = [M*] - [CE]$, then, the resulting matrix $[M * *]$ shows the forgotten effects.

3   Methodological Decisions

The characteristics of scientific work, has evolved and require less orthodox forms. Which, in the research practice combine objectivity with subjectivity. In this sense, new approaches must be oriented primarily on the properties defined in the Theory of fuzzy subsets and its infinite variations. Kaufmann and Gil [25], whose use, nowadays, promotes convergence between objectivity and subjectivity, where objective data, arising from past events, impregnated with "certainty" that does not correspond to the growing status of a dimly uncertain contemporary world [13], supplemented by subjective information, but accepted as reasonable.

Transforming traditional models of numeric character in the field of uncertainty, based on the replacement of the precise numbers for numbers uncertain, enhances its effectiveness in the treatment of reality. However, the complex reality of the present, demand more than this transformation, says Gil [13], requires models developed from non-numerical mathematics, they are able to fill the void, deepening in the study of phenomena that escape, not only the measurement but also to the valuation. The flexibility of non-numerical models, to bind together subjectivity as a quasi-truth, provide freedom to find the *relationships* in existence. Under these circumstances, for the realization of this, was use the Forgotten Effects model, as an expression of non-numerical mathematics.

3.1   Integration of the Expert Group and Collection of Information

Implementation of the model of the forgotten effects demand working with groups of experts; therefore, a non-probability sampling was use, from type of key informant, which is to select people whose skills, experience, knowledge and participation in decision-making is essential for research. The group was integrate for directors, managers and financial executives, small and medium manufacturing enterprises and financial institutions. The selection was made of conventional deterministic manner. Inclusion criteria were: *(a)* Established in Sinaloa; *(b)* With antiquity of operation at least five years; *(c)* With participation in national and international markets.

The process of data collection was conducted through an interview and a questionnaire, through the following steps; first a semi-structured interview is conducted, where the pledge of confidentiality is established, then one questionnaire, from which the variables which according to experts are the determinants of exchange risk exposure are identified. Finally, incidence matrices are create and experts are ask the valuated.

4   Results

Exposure to exchange risk, from the perspective of the experts consulted, is a phenomenon that affect directly in the operation of small and medium enterprises.

During the interviews, point out the existence of phenomena that affect and complicate the operation of firms. Among others, the following: *(1)* the high volatility in the value of the Mexican peso. *(2)* The economic and financial globalization, whose effects, are differently according to the type and condition of the company. *(3)* The bias of entrepreneurs facing the exchange risk exposure by addressing other problems such as logistics operations, relationship between business and family, succession processes, organizational climate, credit, collections, among others. *(4)* The absence of financial training and updated information on macroeconomic indicators inhibits them understand and manage exchange risk exposure.

Corroborates that efforts about the administration of foreign exchange risk exposure in companies, is extremely fragile. Only 14.28% of the companies interviewed has or has had a program for foreign exchange risk management, while the remaining 85.72% has not recorded a single experience. Decisions in this respect, they taken by the owner or by the counter. There is no specialized risk management department.

4.1   Definition of Variables

Kaufmann and Gil [22] argue that the starting point is around expertizaje comparison, through abstract or concrete knowledge. Usually, we resort to experts when it comes to make some important decisions, which requires comparisons, aggregations and meet the acceptance or rejection of such and such propositions.

From the interview and application of surveys, a list of elements was obtained, considered as determinants of exchange risk exposure. Based from the point out by the group of experts, this information was classify into two sets: one called CAUSES, and other EFFECTS, used to build direct impact matrices, forming the Cartesian plane $(C * E)$. Onward, the notation c_i for CAUSES $i = 1, 2, 3, \ldots, 21$ is used. In add the notation e_j for the EFFECTS $j = 1, 2, 3, \ldots, 21$ as shown in Table 2.

Table 2. Variables proposed by the expert group

	Causes		Effects
c_1	Economic policy	e_1	Contraction in demand
c_2	High prime material costs	e_2	Decrease in income
c_3	Importing prime material	e_3	Increase in costs
c_4	Instability of the international economy	e_4	Increase in stocks
c_5	Absence of information	e_5	Unemployment
c_6	Inflation	e_6	Decrease in production
c_7	Globalization	e_7	Budget problems
c_8	Variation in interest rates	e_8	Price instability
c_9	Transnational enterprises	e_9	Risk of bankruptcy
c_{10}	Absence of culture of protection	e_{10}	Poor debt recovery
c_{11}	Foreign exchange financing	e_{11}	Social insecurity
c_{12}	Foreign investments	e_{12}	Additional costs hedging
c_{13}	Poor financial planning	e_{13}	Decision making problems
c_{14}	Oil prices	e_{14}	Prime material scarcity
c_{15}	Entrepreneur attitude	e_{15}	Speculation
c_{16}	Informal trade	e_{16}	Decrease in investments
c_{17}	Contracts without protection clauses	e_{17}	Incapacity to pay
c_{18}	Accounts receivable in foreign currency	e_{18}	Increase in debt (pmx/dll)
c_{19}	Absence of financial capacity	e_{19}	Inflation
c_{20}	Fluctuations of other currencies	e_{20}	Labor problems
c_{21}	Taxes	e_{21}	Reengineering buying/Selling

Source: Authors (2014)

Although naturally divergence is present when the reviews it is, the information provided by the expert group, presented certain convergences, verified through a simple statistic, from the number of citations obtained for each variable, recording the frequency of occurrence, the result is divided by the number of respondents experts, and up to what is defined as a percentage of acceptance (Table 3).

Table 3. «Pre-composition»: variables and percentage of acceptance

	Variables	Expert 1	Expert 2	Expert 3	Expert 4	Expert 5	Expert 6	Expert 7	Expert 8	Expert 9	Expert 10	Vote	%
c_1	Economic policy	X	X	X	X	X	X	X	X			8	.80
c_2	High prime material costs	X		X	X	X	X					5	.50
c_3	Importing prime material	X	X		X			X	X	X		6	.60
c_4	Instability of the international economy	X	X			X	X		X			5	.50
c_5	Absence of information			X		X	X	X	X			5	.50
c_6	Inflation	X						X	X	X	X	5	.50

Source: Authors (2014)

Meet the percentage of acceptance for each variable allowed to have a preview called «pre-composition», is to say previous results to the application process composition *maxmin*. With this, are detected the variables that experts consider influential over the degree of exposure to foreign exchange risk faced by SMEs. In a respect there are any statistical technique is used. The acceptance rate is not take in account when applying the experts valuated the relationship established in the ordered pairs that constitute the incidence matrices.

4.2 Construction of Original Incidence Matrices

The valuation of ordered pairs that constitute the box matrices $[CE]$, $[CC]$, $[EE]$, using the scale called «endecadaria» whose values are in the range $[0, 1]$, was the starting point for the construction of matrices aggregated. As noted by Kaufmann and Gil Aluja [22], you must add the opinions of the expert group to concentrate on a single valuation set of propositions, giving rise to aggregate direct incidence matrices $[CE^A]$, $[CC^A]$, $[EE^A]$. The aggregation process may be do using any operator. However, we chose to use the theory of expertones [21], generating an experton for each variable. An experton is the generalization of probability when the cumulative probabilities are replace by monotonically decreasing intervals [21].

The values obtained through Experton, correspond to rows that represent each of the matrices aggregate direct incidence. For example, in the $[CE^A]$ it is observed that the relationship between the elements of C y E, in some cases gives rise to a certain, but not total reciprocity, in the sense that when c_1 is related to e_2 also e_2 is related to c_1 but the intensity of the relationship is not the same, so the antisymmetry property is presented, (Table 4).

Table 4. Aggregate matrix cause-effect (CE^A) of direct incidence

	E_1	E_2	E_3	E_4	E_5	E_6	E_7	E_8	E_9	E_{10}	E_{11}	E_{12}	E_{13}	E_{14}	E_{15}	E_{16}	E_{17}	E_{18}	E_{19}	E_{20}	E_{21}
C_1	.7	.5	.5	.4	.7	.6	.4	.5	.3	.4	.7	.4	.3	.1	.3	.5	.3	.5	.8	.4	.1
C_2	.5	.4	.9	.3	.5	.8	.3	.7	.4	.3	.2	.2	.4	.5	.4	.4	.4	.3	.8	.3	.3
C_3	.2	.2	.4	.2	.3	.2	.5	.2	.2	.1	.1	.6	.3	.4	.4	.3	.4	.5	.5	.2	.6
C_4	.5	.5	.6	.2	.4	.5	.2	.7	.2	.2	.3	.8	.2	.3	.4	.5	.2	.4	.6	.2	.6
C_5	.2	.2	.3	.2	.2	.2	.5	.1	.7	.5	.1	.4	.8	.2	.1	.5	.2	.2	.1	.4	.2
C_6	.3	.4	.3	.2	.7	.3	.1	.4	.5	.2	.4	.6	.4	.3	.2	.3	.4	.5	.4	.3	.6
C_7	.8	.8	.9	.5	.7	.7	.5	.9	.5	.5	.5	.3	.5	.2	.7	.6	.6	.6	1.0	.5	.2
C_8	.4	.4	.5	.2	.3	.5	.4	.6	.3	.3	.3	.6	.5	.2	.6	.7	.4	.5	.6	.2	.2
C_9	.4	.3	.2	.4	.6	.5	.1	.2	.6	.2	.1	.1	.3	.1	.2	.7	.2	.4	.1	.3	.4
C_{10}	.1	?	.6	.2	.2	.1	.2	.2	.6	.3	.3	.6	.4	.2	.1	.2	.4	.4	.1	.2	.2
C_{11}	.2	.2	.5	.1	.1	.1	.4	.2	.4	.3	.1	.7	.3	.3	.2	.4	.4	.8	.2	.1	.2
C_{12}	.1	.1	.3	.1	.3	.2	.5	.2	.2	.1	.1	.5	.2	.1	.3	.5	.2	.6	.1	.1	.3
C_{13}	.1	.6	.7	.3	.4	.6	.7	.1	.5	.3	.2	.5	.6	.1	.2	.6	.6	.7	.3	.4	.3
C_{14}	.3	.4	.6	.2	.4	.4	.1	.7	.2	.2	.3	.2	.4	.2	.4	.5	.3	.3	.8	.3	.2
C_{15}	.2	.4	.3	.3	.4	.2	.6	.1	.6	.4	.3	.3	.5	.1	.3	.4	.4	.6	.2	.5	.3
C_{16}	.7	.7	.2	.6	.6	.7	.2	.4	.6	.3	.8	.1	.2	.1	.2	.5	.3	.2	.1	.5	.4
C_{17}	.0	.1	.4	.1	.1	.1	.4	.1	.6	.3	.2	.7	.2	.1	.3	.2	.4	.6	.1	.2	.1
C_{18}	.1	.1	.1	.1	.1	.2	.4	.3	.4	.2	.1	.6	.2	.2	.3	.2	.2	.2	.1	.1	.3
C_{19}	.1	.5	.3	.2	.4	.6	.8	.1	.7	.2	.2	.5	.5	.2	.2	.8	.8	.8	.1	.4	.2
C_{20}	.3	.5	.5	.2	.3	.4	.4	.5	.3	.1	.1	.7	.3	.2	.3	.5	.4	.6	.5	.2	.5
C_{21}	.6	.7	.7	.1	.7	.5	.4	.3	.4	.4	.4	.5	.3	.1	.2	.6	.4	.5	.6	.3	.1

Source: Authors (2014)

In another case when there are a relationship in which the lines coincide in number and essentially the elements of the columns, are filled with one main diagonal [13], as seen in the Table 5; this means that the relationship is total, and manifests itself in the reflexive property. In addition to the above, as this matrix is subjected to the process of

Table 5. Original incidence matrix cause - cause (CC^A)

	C_1	C_2	C_3	C_4	C_5	C_6	C_7	C_8	C_9	C_{10}	C_{11}	C_{12}	C_{13}	C_{14}	C_{15}	C_{16}	C_{17}	C_{18}	C_{19}	C_{20}	C_{21}
C_1	1	.7	.5	.4	.2	.4	.8	.8	.4	.2	.5	.7	.3	.6	.4	.5	.2	0.5	.4	.5	.9
C_2	.3	1	.8	.4	.2	.2	.9	.5	.3	.2	.4	.5	.3	.3	.2	.4	.2	.3	.4	.4	.4
C_3	.4	.7	1	.6	.2	.6	.5	.5	.6	.3	.8	.4	.4	.4	.3	.3	.4	.2	.3	.7	.4
C_4	.6	.7	.6	1	0.3	.5	.6	.7	.5	.4	.5	.5	.3	.8	.2	.4	.4	.5	.5	.8	.3
C_5	.3	.3	.2	.3	1	.2	.3	.3	.2	.6	.4	.2	.8	.2	.6	.3	.5	.2	.3	.1	.1
C_6	.7	.4	.6	.6	.2	1	.4	.5	.6	.1	.5	.7	.3	.6	.4	.4	.4	.6	.3	.7	.2
C_7	.7	.9	.5	.4	.2	.3	1	.8	.5	.2	.5	.4	.3	.6	.2	.4	.2	.4	.5	.5	.5
C_8	.7	.6	.5	.7	.2	.5	.8	1	.3	.2	.7	.7	.4	.4	.4	.4	.3	.5	.4	.6	.4
C_9	.4	.4	.6	.4	.2	.5	.4	.3	1	.2	.8	.8	.2	.5	.3	.4	.4	.7	.2	.6	.5
C_{10}	.1	.2	.2	.3	.2	.1	.1	.2	.2	1	.2	.2	.7	.1	.4	.2	.5	.3	.4	.3	.1
C_{11}	.4	.4	.7	.3	.2	.5	.3	.6	.8	.2	1	.8	.3	.2	.3	.1	.3	.4	.3	.6	.3
C_{12}	.6	.3	.5	.4	.2	.7	.4	.5	.8	.4	.7	1	.2	.4	.4	.3	.3	.5	.3	.6	.5
C_{13}	.2	.2	.2	.3	.6	.2	.2	.2	.1	.7	.3	.2	1	.2	.3	.3	.4	.2	.7	.2	.2
C_{14}	.8	.7	.5	.7	.2	.7	.8	.6	.6	.2	.4	.4	.2	1	.1	.3	.3	.3	.2	.8	.4
C_{15}	.2	.2	.2	.2	.5	.3	.2	.3	.3	.7	.3	.3	.6	.1	1	.4	.5	.3	.4	.2	.2
C_{16}	.5	.3	.2	.3	.2	.4	.4	.2	.2	.5	.2	.1	.4	.3	.4	1	.3	.1	.4	.4	.6
C_{17}	.4	.2	.3	.3	.3	.3	.2	.3	.4	.7	.4	.5	.5	.3	.5	.2	1	.3	.2	.2	.1
C_{18}	.3	.3	.4	.4	.2	.3	.4	.5	.6	.3	.6	.6	.2	.3	.3	.2	.3	1	.4	.5	.2
C_{19}	.3	.2	.3	.2	.3	.3	.4	.3	.3	.4	.4	.2	.5	.1	.4	.3	.2	.3	1	.3	.3
C_{20}	.7	.5	.6	.7	.2	.7	.5	.6	.5	.4	.6	.5	.3	.6	.2	.4	.4	.6	.5	1	.2
C_{21}	.6	.6	.5	.3	.1	.2	.5	.3	.5	.2	.2	.5	.3	.3	.3	.5	.1	.2	.4	.1	1

Source: Authors (2014)

MAXMIN convolution, notes also the property of transitivity; so, at present these three properties, an order relation is established.

Constructed the matrices aggregated $[CE^A], [CC^A], [EE^A]$ *maxmin* composition process was implement, which allowed obtaining the incidence matrix of second order, is to say, the relations indirect incidence.

Table 6. Secondary effects matrix (M*)

	E_1	E_2	E_3	E_4	E_5	E_6	E_7	E_8	E_9	E_{10}	E_{11}	E_{12}	E_{13}	E_{14}	E_{15}	E_{16}	E_{17}	E_{18}	E_{19}	E_{20}	E_{21}
C_1	,8	,8	,8	.7	.7	.8	.6	.8	.7	.5	.7	.6	.6	.6	.7	.7	.8	.7	.8	.7	.9
C_2	.8	.8	.9	.7	.7	.8	.6	.9	.7	.5	.7	.6	.6	.6	.7	.7	.8	.7	.9	.7	.6
C_3	.7	.7	.7	.6	.7	.7	.6	.7	.7	.5	.7	.7	.5	.6	.6	.7	.7	.8	.7	.7	.6
C_4	.7	.7	.7	.6	.7	.7	.6	.7	.6	.5	.7	.8	.6	.6	.6	.7	.7	.7	.7	.7	.6
C_5	.6	.6	.6	.5	.6	.6	.8	.6	.7	.5	.5	.6	.8	.5	.5	.7	.7	.7	.6	.6	.5
C_6	.7	.7	.7	.7	.7	.7	.6	.7	.7	.5	.7	.7	.6	.5	.5	.7	.7	.7	.7	.7	.6
C_7	.8	.8	.9	.7	.7	.8	.6	.9	.7	.5	.7	.6	.6	.6	.7	.7	.8	.7	1,0	.7	.6
C_8	.7	.7	.7	.7	.7	.7	.6	.7	.7	.5	.7	.7	.6	.6	.7	.7	.7	.7	.7	.7	.6
C_9	.6	.7	.6	.5	.7	.6	.6	.6	.7	.5	.6	.7	.5	.5	.5	.7	.7	.8	.6	.6	.6
C_{10}	.6	.6	.6	.5	.6	.6	.7	.6	.7	.5	.5	.6	.6	.5	.5	.7	.7	.7	.6	.6	.5
C_{11}	.6	.7	.6	.6	.7	.6	.6	.6	.7	.5	.6	.7	.5	.6	.6	.7	.7	.8	.6	.6	.6
C_{12}	.6	.7	.6	.5	.7	.6	.6	.6	.7	.5	.6	.7	.5	.5	.5	.7	.7	.7	.6	.6	.6
C_{13}	.6	.7	.6	.5	.7	.7	.8	.6	.7	.5	.5	.6	.7	.5	.5	.8	.7	.7	.6	.7	.5
C_{14}	.8	.8	.8	.7	.7	.8	.6	.8	.7	.5	.7	.8	.6	.6	.7	.7	.8	7	8	.7	.6
C_{15}	.6	.6	.6	.5	.6	.6	.7	.6	.7	.5	.5	.6	.6	.5	.5	.7	.7	.7	.6	.6	.5
C_{16}	.7	.7	.5	.7	.7	.7	.5	.5	.7	.5	.7	.5	.6	.5	.5	.6	.7	.7	.5	.6	.5
C_{17}	.5	.5	.6	.5	.5	.5	.5	.5	.6	.5	.5	.7	.5	.5	.5	.5	.6	.6	.5	.6	.5
C_{18}	.6	.6	.6	.5	.6	.6	.5	.6	.7	.5	.5	.7	.5	.5	.5	.6	.7	.7	.6	.6	.5
C_{19}	.5	.7	.6	.5	.7	.7	.8	.5	.7	.5	.5	.5	.7	.5	.5	.8	.7	.7	.5	.7	.5
C_{20}	.7	.6	.7	.6	.7	.6	.6	.7	.6	.5	.7	.8	.6	.6	.6	.7	.7	.7	.7	.7	.6
C_{21}	.7	.7	.7	.7	.7	.7	.6	.7	.6	.5	.7	.5	.6	.6	.6	.7	.7	.7	.7	.7	1

Source: Authors, using FuzzyLog program (2014)

$$[CE^A] \circ [CC^A] \circ [EE^A] = [M*]^3 \tag{1}$$

The first step is convolution $[CE^A] \circ [CC^A]^3$ from which arises [M]; to achieve this, take the first row of $[CE^A]$ and the first column of $[CC^A]$, the minimum value of the ordered pairs contrasted, represented by the operator ($\wedge$) are selected; step followed, the maximum value represented by the operator ($\vee$) was selected. This step is perform for each of the pairs represented on the matrix. For example, $(c_1, e_1), (c_1 e_2), (c_1 e_3), \ldots$, $(c_1 e_n)$ to complete the new matrix [M]. The next step is the convolution $[M] \circ [EE^A]$, where arises [M*] (Eq. 1).

Table 7. Forgotten effects matrix

	E_1	E_2	E_3	E_4	E_5	E_6	E_7	E_8	E_9	E_{10}	E_{11}	E_{12}	E_{13}	E_{14}	E_{15}	E_{16}	E_{17}	E_{18}	E_{19}	E_{20}	E_{21}
C_1	.1	.3	.3	.4	.0	.2	.2	.2	.3	.3	.1	.2	.3	.3	.3	.0	.3	.1	.2	.2	.4
C_2	.1	.2	.0	.3	.1	.0	.1	.1	.2	.1	.4	.2	.3	.0	.0	.0	.1	.1	.0	.1	.1
C_3	.4	.2	.3	.1	.3	.3	.1	.3	.3	.3	.3	.0	.2	.2	.1	.2	.2	.2	.0	.3	.0
C_4	.2	.1	.1	.2	.1	.2	.2	.0	.2	.2	.1	.0	.1	.1	.0	.0	.2	.0	.1	.2	.0
C_5	.5	.4	.3	.3	.4	.4	.1	.5	.0	.1	.5	.0	.0	.2	.2	.1	.4	.3	.6	.3	.2
C_6	.3	.1	.2	.3	.0	.0	.3	.1	.0	.2	.1	.0	.0	.3	.2	.3	.2	.1	.1	.1	.0
C_7	.0	.0	.0	.2	.0	.1	.0	.0	.0	.0	.0	.0	.0	.2	.0	.0	.2	.1	.0	.0	.0
C_8	.5	.3	.2	.4	.4	.3	.3	.3	.3	.3	.5	.0	.2	.2	.4	.3	.3	.4	.1	.4	.0
C_9	.2	.1	.0	.3	.2	.1	0.1	.0	.1	.0	.2	.0	.1	.2	.0	.0	.0	.0	.0	.2	.1
C_{10}	.1	.0	.2	.2	.0	.1	.1	.3	.0	.3	.2	.4	.3	.4	.3	.0	.2	.2	.3	.1	.1
C_{11}	.5	.3	.1	.3	.4	.4	.3	.3	.0	.3	.2	.1	.2	.3	.2	.3	.0	.1	.4	.4	.2
C_{12}	.5	.3	.1	.3	.5	.3	.2	.3	.2	.3	.5	.0	.2	.1	.4	.3	.1	.0	.3	.4	.3
C_{13}	.4	.4	.4	.4	.3	.4	.2	.3	.6	.4	.5	.2	.2	.3	.3	.2	.3	.2	.2	.4	.2
C_{14}	.3	.0	.0	.0	.0	.0	.0	.5	.0	.0	.4	.0	.0	.2	.1	.1	.0	.0	.3	.0	.3
C_{15}	.3	.3	.1	.5	.3	.3	.3	.1	.4	.4	.5	.3	.4	.5	.4	.2	.3	.3	.0	.4	.4
C_{16}	.3	.1	.2	.2	.2	.2	.1	.3	.0	.1	.4	.2	.1	.4	.1	.1	.1	.1	.3	.1	.1
C_{17}	.1	.2	.4	.2	.3	.2	.1	.0	.4	.3	.3	.3	.2	.0	.3	.4	.3	.3	.1	.3	.1
C_{18}	.3	.3	.3	.3	.5	.2	.3	.2	.3	.4	.3	.3	.3	.2	.2	.2	.1	.1	.1	.3	.3
C_{19}	.4	.3	0,2	.2	.3	.1	.1	.3	.2	.4	.3	.4	.3	.3	.3	.2	.2	.2	.3	.3	.2
C_{20}	.1	.0	.3	.1	.0	.0	.3	.1	.1	.2	.0	.3	.3	.4	.1	.0	.2	.4	.3	.1	.2
C_{21}	.5	.3	.2	.3	.4	.4	.1	.1	.0	.3	.3	.0	.3	.4	.1	.2	.0	.0	.4	.3	.0

Source: Authors, using FuzzyLog program (2014)

Making convolution process manually, demands great effort and involves a number of difficulties in operation with a high risk of making mistakes. Therefore, a computer program called FuzzyLog was use. Your use represented the feasibility of information management in large quantities, the virtual decrease the margin of error and ensuring access to accurate and speedy results. Using FuzzyLog[3], we proceed to create *the matrix of secondary effects and forgotten effects,* through following general formula:

---

[3] This is a calculation program, that allow work with fuzzy models to capture the called forgotten effects at the causality relations.

$$\mu(c_n, b_k) = \vee_m(\mu(c_n, e_m) \wedge \mu(e_m, b_k)) \tag{2}$$

The cumulative values for each ordered pair that makes up the matrix of the secondary effects obtained, thereupon we proceed to calculate the forgotten effects matrix $[M**] = [M*] - [CE^A]$; to demonstrate the degree of forgetting suffered by elements not considered in the initial analysis, obtaining the results shown in Tables 6 and 7.

Table 8. Effects degree of forgetfulness

Affecting	Affected	Initial value	Accumulated value	Forgotten value
Absence of information	Inflation	$\mu i(c_5 \to e_{19}) = 0,2$	$\mu a(c_5 \to e_{19}) = 0,8$	$\mu o(c_5 \to e_{19}) = 0,6$
Poor financial planning	Risk of bankruptcy	$\mu i(c_{13} \to e_9) = 0,2$	$\mu a(c_{13} \to e_9) = 0,8$	$\mu o(c_{13} \to e_9) = 0,6$
Accounts receivable in foreign currency	Unemployment	$\mu i(c_{18} \to e_5) = 0,3$	$\mu a(c_{18} \to e_5) = 0,8$	$\mu o(c_{18} \to e_5) = 0,5$
Entrepreneur attitude	Social insecurity	$\mu i(c_{15} \to e_{11}) = 0,3$	$\mu a(c_{15} \to e_{11}) = 0,8$	$\mu o(c_{15} \to e_{11}) = 0,5$
Poor financial planning	Social insecurity	$\mu i(c_{13} \to e_{11}) = 0,2$	$\mu a(c_{13} \to e_{11}) = 0,7$	$\mu o(c_{13} \to e_{11}) = 0,5$

Source: Authors 2014

The values given in Table 7 represent the forgotten effects or incidences left aside with greater of forgetting, as presented in Table 8.

This procedure can be representing by the following equation: $\mu o(c_n \to e_m) = \mu a(c_n \to e_m) - \mu i(c_n \to e_m)$.

While $\mu o(c_n \to e_m)\mu o(c_n \to e_m)$ is closer to the upper limit of [.1], then the degree of forgetfulness is greater; therefore, it is an element of considerable importance in the formulation or restructuring of strategies.

Table 9. Variables with % of acceptance *vs.* Variables with degree of forgetfulness

	Precomposition results		Postcomposition final results
c_1	Economic policy	c_5	Absence of information
c_2	High prime material costs	c_{13}	Poor financial planning
c_3	Importing prime material	c_{18}	Accounts receivable in foreign currency
c_4	Instability of the international economy	c_{15}	Entrepreneur attitude
c_5	Absence of information	c_{13}	Poor financial planning
c_6	Inflation		

Source: Authors (2014)

In the Table 9, are contrasted the result a priori, conceived by the expert group, and those obtained by applying the forgotten effects model.

4.3   Graphs Incidence of Elements Interposed

Found forgotten effects. However, note Gil and Luis (2011), it is important to represent relationships sagittately way with two objectives: first, know the route of incidence and, second, to find intermediate influencing variables that enhancing and accumulate

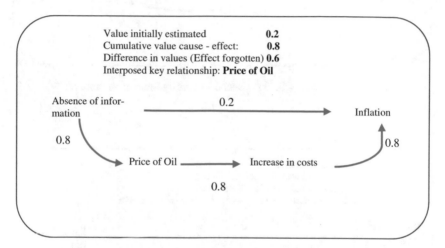

Fig. 1. Absence of information – Inflation Source: Authors (2014)

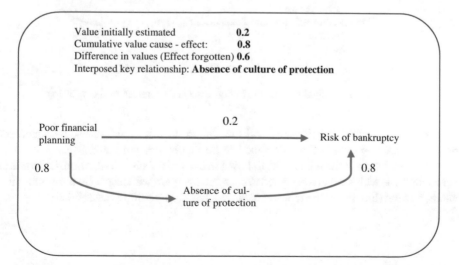

Fig. 2. Poor financial planning – Risk of bankruptcy Source: Authors (2014)

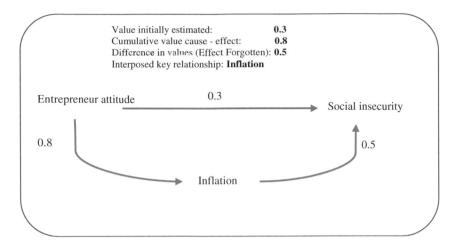

Fig. 3. Entrepreneur attitude – Social insecurity Source: Authors (2014)

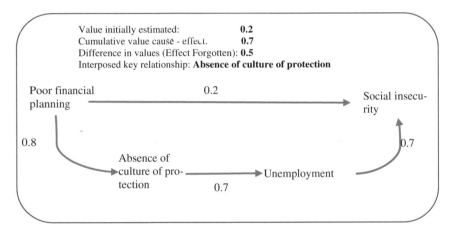

Fig. 4. Poor financial planning – Social insecurity Source: Authors (2014)

effects in the causality relationship, and that failure to meet them, cause management programs risk exposure that are designed to be ineffective and expensive.

The description of the relationship, shown in each of the Figs. 1, 2, 3 and 4: the initial value contrasts which experts assigned to each ordered pair, with respect to the cumulative value, and the difference that represents the degree of forgetting is established.

Finally, the interposed element is displayed, whose knowledge allows substantially improve the effectiveness of the strategy to be implemented in response to the problem of study.

5 Conclusions

Today, rapid mutability in economic, financial and social conditions, establishes a climate of uncertainty, which requires a change of mentality in decision makers.

The application of fuzzy model through of the Forgotten Effects theory brings significant convenience; can contribute to scientific analysis of exchange risk exposure, an alternative way to traditional analysis models, highlighting the opinion and partic-ipation of experts. Facilitates the interpretation of subjective information, by passing the objectivity of a numeric value. Faculty the identification of elements interposed, keys to designing strategies. In this sense, emerges a new optical analysis of the management of foreign exchange risk exposure, and finding no prior evidence of fuzzy models in this field of science, its use intend to be a contribution to scientific devel-opment in this area knowledge.

In general, research can be a reference for future projects, which models are located in the fields of mathematics of uncertainty. The generosity of the forgotten effects model empower its application in areas such as sustainability, marketing, quality control, economics, finance, politics, medicine, and among others in the administration of foreign exchange risk exposure.

It is important to consider, that their articulation and efficiency are, based on the immediate environment of the valuators, temporality and the characteristics of the object of study. Since they influence the way to perceive the context and approach to the problem; therefore, it is recommend defining the incidence matrices, considering time and public, clearly representing the human condition of economies.

References

1. Adler, M., Dumas, B.: Exposure to currency risk: definition and measurement. Finan. Manag. 41–50 (1984)
2. Allayannis, G., Ofek, E.: Exchange rate exposure, hedging, and the use of foreign currency derivatives. J. Int. Money Finan. **20**(2), 273–296 (2001)
3. Barcellos de Paula, L.: Modelos de gestión aplicados a la sostenibilidad empresarial (2011)
4. Bartram, S.M.: What lies beneath: foreign exchange rate exposure, hedging and cash flows. J. Bank. Finan. **32**(8), 1508–1521 (2008)
5. Chaieb, I., Mazzotta, S.: The unconditional and conditional exchange rate exposure of US firms (No. 11–15). Swiss Finance Institute (2010)
6. De Urdanivia, C.M.A.D.: ¿Por qué probar en econometría? Política y cultura **17**, 363–376 (2002)
7. Dhanani, A.: The management of exchange-rate risk: a case from manufacturing industry. Thunderbird Int. Bus. Rev. **46**(3), 317 (2004). ABI/INFORM Complete
8. Doukas, J.A., Hall, P.H., Lang, L.H.: Exchange rate exposure at the firm and industry level. Finan. Mark. Inst. Instrum. **12**(5), 291–346 (2003)

9. Flood Jr., E., Lessard, D.R.: On the measurement of operating exposure to exchange rates: a conceptual approach. Finan. Manag., 25–36 (1986)
10. Geissler, C.: Tendencias recientes en Administración de Riesgo. Corporate Finance May 126, ABI/INFORM complete 11 (1995)
11. Gento, A., Lazzari, L., Machado, E.: Reflexiones acerca de las matrices de incidencia y la recuperación de efectos olvidados. Cuadernos del CIMBAGE 4 (2012)
12. Gil, A.J.: Towards a new paradigm of investment selection in uncertainty. Fuzzy Sets Syst. 84(2), 187–197 (1996)
13. Gil, A.J.: Elementos para una teoría de la decisión en la incertidumbre (1999)
14. Gil, A.J.: La matemática borrosa en economía y gestión de empresas. Matematicalia: revista digital de divulgación matemática de la Real Sociedad Matemática Española 1(3), 5 (2005)
15. Gil, L.A.M., Barcellos De Paula, L.: Una aplicación de la metodología de los efectos olvidados: los factores que contribuyen al crecimiento sostenible de la empresa. Cuadernos del CIMBAGE 12 (2012)
16. Gil, L.A.M., Luis, B.: Identificación de los atributos contemplados por los clientes en una estrategia CRM utilizando el modelo de efectos olvidados. Cuadernos del CIMBAGE 13, 107–127 (2011)
17. Gil, L.A.M.: El análisis financiero en la incertidumbre. Ariel (1993)
18. Gil, L.A.M.: Nuevas Estrategias para el análisis financiero en la empresa. Ariel (2001)
19. Jegher, S.: Estructura flexible en la administración del riesgo financiero. Administración del Riesgo 46(1) (1999). ABI/INFOR
20. Jorion, P.: The exchange-rate exposure of US multinationals. J. Bus. 63(3), 331–345 (1990)
21. Kaufmann, A.: Theory of expertons and fuzzy logic. Fuzzy Sets Syst. 28(3), 295–304 (1988)
22. Kaufmann, A., Gil, A.J.: Técnicas especiales para la gestión de expertos (1993)
23. Kaufmann, A., Gil, A.J.: Técnicas operativas de gestión para el tratamiento de la incertidumbre. Hispano Europea (1987)
24. Kaufmann, A., Gil, A.J.: Modelos para la investigación de efectos olvidados. Milladoiro (1988)
25. Kaufmann, A., Gil, A.J.: Las matemáticas del azar y de la incertidumbre; elementos básicos para su aplicación en economía. Centro de Estudios Ramón Areces (1990)
26. Martínez, E.F.J., Hervás, M.C., Torres, J.M., Martínez, E.A.C.: Modelo no lineal basado en redes neuronales de unidades producto para clasificación. Una aplicación a la determinación del riesgo en tarjetas de crédito. Revista de Métodos Cuantitativos para la Economía y la Empresa 3, 40–62 (2007)
27. Martínez, S.P.: Metodología para la medición de la exposición económica al riesgo de cambio: una revisión. Información Comercial Española 780, 63–79 (1999)
28. Muñoz, P.M., Avilés, O.E.: Incorporation of fuzzy logic to the Black-Scholes model in exchange option pricing. In: Fourth International Workshop on Knowledge Discovery, Knowledge Management and Decision Support. Atlantis Press (2013)
29. Pascale, R.: Decisiones financieras (1995)
30. Popov, V., Stutzmann, Y.: How is foreign exchange risk managed? An empirical study applied to two Swiss companies. University of Lausanne (2003)
31. Preciado, L.B., España, C.L.F., López, C.J.A.: Gestión del riesgo cambiario en una compañía exportadora. Estudios Gerenciales 27(121), 219–238 (2011)
32. Odegaard, B.A., Priestley, R.: Linear and nonlinear exchange rate exposure and the price of exchange rate risk. In: EFA Berlin Meetings Presented Paper (2002)
33. Rejda, G.E.: Principles of Risk Management and Insurance. Pearson Education India (1997)
34. Ranhema, A.: Finanzas Internacionales. Deusto, Barcelona, España (2007)
35. Rodriguez, R.M.: Measuring and controlling multinationals' exchange risk. Finan. Anal. J. 35(6), 49–55 (1979)

36. Sánchez, B.G.: Introducción a la Econometría. Facultad de Economía (2000)
37. Shapiro, A.C.: Defining exchange risk. J. Bus. **50**(1), 37–39 (1977)
38. Spanos, A.: Statistical Foundations of Econometric Modeling. Cambrige University Press, Cambrige (1986)
39. Santaulària, D.V.: ¿Qué es la Econometría? Acta Universitaria **16**(3), 47–51 (2006)
40. Wallace, J.: Las mejores prácticas en la administración de riesgos. TMA J. ABI/INFORM **25** (1998)
41. Zadeh, L.A.: Fuzzy sets. Inf. Control **8**(3), 338–353 (1965)

On the Security of Stream Ciphers with Encryption Rate $\frac{1}{2}$

Michele Elia[(✉)]

Politecnico di Torino, Turin, Italy
michele.elia@polito.it

Abstract. Based on the geometry of finite projective planes, a secret-key encryption scheme which offers an exactly computable degree of secrecy is described. This target is achieved at the cost of an encryption rate equal to $\frac{1}{2}$, as in one-time-pad encryption, but the devised scheme avoids the burden of exchanging and destroying long keys. It is also shown that knowledge of pieces of plain text does not significantly reduce the degree of secrecy; further, the cost of possible plain-text attacks is under the designer's control, and can be made as high as desired.

Keywords: Key stream · Encryption · Secret key · Security
Projective planes

1 Introduction

In conventional cryptography there are three players, namely, sender, receiver, and attacker. Sender and receiver share a common secret key that, mainly for practical purposes, is universally assumed to be short with respect to the message length; that is, one-time-pad or similar schemes are not usually considered. The attacker does not know the secret key, but he knows the encryption mechanism (i.e. encryption rule and structure of the key-stream generator); his target is to decrypt messages and perhaps to recover the secret key. If this second task is successful, we say that the system has been broken.

In his famous paper [8], Shannon proved that Vernam's enciphering (e.g. one-time-pad [12]) achieves perfect secrecy at the cost of encryption rate $\frac{1}{2}$, with the extra burden that the key stream should be used only once and safely destroyed. In [1], with a different target, the condition for perfectly detecting any deception is shown, and in [9] a perfect deception method based on finite plane geometries is described. Based on these ideas, in this paper finite projective planes, [10] and [11], are used to show that perfect secrecy may be achieved in practice, obviously at transmission rate $\frac{1}{2}$, but without extra management costs. Here perfect secrecy should be understood in the loose sense, that the encrypted message gives no information about the plain text. The scheme is also robust enough to resist plain-text attacks, although in this case the probability of breaking the system is well defined and under the designer's control.

© Springer International Publishing AG, part of Springer Nature 2018
A. M. Gil-Lafuente et al. (Eds.): FIM 2015, AISC 730, pp. 400–406, 2018.
https://doi.org/10.1007/978-3-319-75792-6_29

2   Encryption Scheme

The mechanism may be described by referring to every finite field; however, it will be presented considering the binary field and its extensions because of their large number of applications.

Let $B = \{b_1, b_2, \ldots, b_i, \ldots\}$ be an information binary stream consisting of statistically independent and equally probable symbols, that is the plain message has been binary encoded and ideally compressed so that all redundancy is removed. Consider the sequence $M = \{m_1, m_2, \ldots, m_j, \ldots\}$ obtained by grouping every k consecutive bits of B into blocks $m_j \in F_2^k$, which are profitably also considered as elements of F_q, with $q = 2^k$ [4], [2]. Let $H = \{h_1, h_2, \ldots, h_i, \ldots\}$ be a binary key-stream [3] used to define a sequence $K = \{s_1, s_2, \ldots, s_i, \ldots\}$ of pairs $s_i = [k_{2i}, k_{2i+1}]$ of blocks of k consecutive bits each; each block is still considered to be an element of F_q. The encrypted message consists of pairs of symbols $[e_{2i}, e_{2i+1}]$ of F_q, which are obtained by means of a function f from $F_q \times F_q^2$ into F_q^2

$$[e_{2i}, e_{2i+1}] = f(m_i, [k_{2i}, k_{2i+1}]) \ ,$$

that is, f combines an information symbol m_i with a pair $[k_{2i}, k_{2i+1}]$ of symbols obtained from H. The function f is defined via a projective plane $PG(2, q)$ over F_q; this plane has $q^2 + q + 1$ points of coordinates (x, y, z), and the same number of lines of equation $ax + by + cz = 0$ and identified by triples (a, b, c), such that every line contains exactly $q + 1$ points, and through every point pass exactly $q + 1$ lines. Each point $(x, y, 1)$, excluding that at infinity (i.e. $z \neq 0$), of a fixed line $\mathcal{E}$ of equation $a_0 x + b_0 y + c_0 z = 0$, called the "equator line", represents a message symbol with the identification $x = m_i$.

Enciphering/Deciphering Mechanism. Sender and receiver share

1. The equator line $\mathcal{E} \in PG(2, q)$, with $b_0 \neq 0$, which is kept secret as part of the secret key.
2. A key-stream generator S_{gen} which is a finite state machine [5] that produces a long binary stream H. The structure of S_{gen} is assumed to be publicly known, but its initial state S_0 is assumed to be secret as part of the secret-key. The generated bit stream H is kept secret on both sender and receiver sides.

Enciphering. A message symbol $m_i \in F_q$ is enciphered as follows:

1. Draw a line $\mathcal{L}$ through the points

    A:= $(m_i, y_i, 1) \in \mathcal{E}$, where $y_i = -\frac{c_0 + a_0 m_i}{b_0}$, and
    B:= $(k_{2i}, k_{2i+1}, 1)$, obtained by considering blocks of $2k$ bits of the sequence H.

$\mathcal{L}$ has affine equation

$$y = y_i + \frac{k_{2i+1} - y_i}{k_{2i} - m_i}(x - m_i) \ . \tag{1}$$

2. Select at random a point $(e_i, g_i, 1) \in \mathcal{L}$ (the point at infinity is excluded), that is e_i is chosen as the value of a random variable ξ having probability distribution $\frac{1}{q-2}$ over the set $F_q - \{m_i, k_{2i}\}$, and (1) is used to compute

$$g_i = k_{2i+1} + \frac{k_{2i+1} - y_i}{k_{2i} - m_i}(e_i - k_{2i}) \,. \tag{2}$$

3. The block $[e_i, g_i] \in F_q^2$ is the encrypted symbol for m_i.

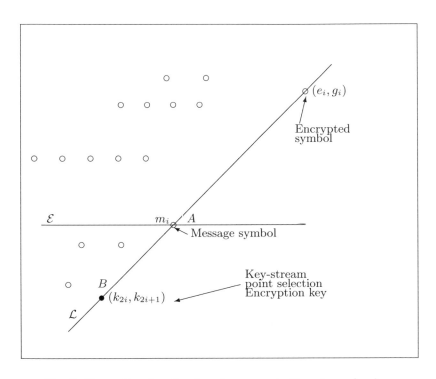

Fig. 1. Encryption function based on a projective plane of order q

Deciphering. A received encrypted message $[e_i, g_i]$ is deciphered as follows:

1. Draw the line $\mathcal{L}$ passing through the points $(e_i, g_i, 1)$ and $(k_{2i}, k_{2i+1}, 1)$.
2. Find the abscissa m_i of the intersection between the equator $\mathcal{E}$ and $\mathcal{L}$ as

$$m_i = k_{2i} + \frac{k_{2i+1} - y_i}{k_{2i+1} - g_i}(e_i - k_{2i}) \,,$$

which is the deciphered message.

Clearly, the encryption rate is $\frac{1}{2}$. At first glance, message and encrypted text sequence seem to be uncorrelated, a property that will be formally proved under reasonable realistic constraints.

3 Cryptanalysis

In this section the cryptanalysis of the proposed encryption scheme is considered, firstly treating cipher-text-only attack, and then plain-text attack.

In cipher-text-only attack, the target of the attackers is to find the message $M = \{m_1, m_2, \ldots, m_i, \ldots\}$, knowing the sequence $E = \{[e_1, g_1], [e_2, g_2], \ldots, [e_i, g_i], \ldots\}$ of encrypted symbols, and the encryption mechanism, but knowing neither the equator line, nor the key-stream.

In plain-text attack, under the most favorable circumstances the attackers know the message sequence M, the sequence E of encrypted symbols, the encryption rule, and the structure of the key-stream generator. The target is to find the secret key, which in this case consists of the equation of the equator line $\mathcal{E}$, and the initial state S_0.

Cipher-text-only Attack. The attacker has the following information, that

1. $e_0, e_1, \ldots, e_i, \ldots$ are known values of a random variable ξ defined over an unknown subset of F_q, with probability distribution $\frac{1}{q-2}$.
2. $m_0, m_1, \ldots, m_i, \ldots$ are the unknown values of a random variable η defined over F_q, with probability distribution $\frac{1}{q}$
3. $y_0, \ldots, y_i, \ldots$ are the unknown ordinates of the sequence of points $(m_i, y_i, 1)$ in $\mathcal{E}$ that are deterministically computed from m_i using the equation of $\mathcal{E}$ as

$$y_i = -\frac{c_0 + a_0 m_i}{b_0}$$

where a_0, b_0, c_0 are secret parameters.

4. $g_0, g_1, \ldots, g_i, \ldots$ are known values of a random variable θ which are computed using an unknown function of the random variables ξ and η, as shown in equation (1), that is

$$\theta = \frac{-a_0 \xi \eta + A_i \xi + B_i \eta + C_i}{b_0 \eta + D_i},$$

where A_i, B_i, C_i, and D_i are unknown parameters that change with i.

Theorem 1. *Under the above conditions, the stream $\{[e_1, g_1], \ldots, [e_i, g_i], \ldots\}$ of encrypted symbols is a random sequence of equally probable symbols which are the values of two statistically independent random variables $[\xi, \theta]$ having joint probability distribution $\frac{1}{q(q-2)}$.*

Proof. The sequence of symbols e_i consists of statistically-independent values of a random variable ξ with probability distribution $\frac{1}{q-2}$ by definition. The corresponding g_i, given by equation (2), is a function of two random variables ξ and η, which are statistically-independent variables by hypothesis. The random variable η assumes q statistically-independent values with probability distribution $\frac{1}{q}$. The probability $p\{\xi = \beta_1, \theta = \beta_2\}$ is then computed as

$$p\{\xi = \beta_1, \theta = \beta_2\} = \sum_{\beta_3 \in F_q} p\{\xi = \beta_1, \theta = \beta_2 | \xi = \beta_1, \eta = \beta_3\} p\{\xi = \beta_1\} p\{\eta = \beta_3\} .$$

However, by equation (2) there is a single value β_3 corresponding to given β_1 and β_2; thus the conditional probability is always zero, except for a single value of β_3, when it is 1; in conclusion

$$p\{\xi = \beta_1, \theta = \beta_2\} = p\{\xi = \beta_1\}p\{\eta = \beta_3\} = \frac{1}{q-2}\frac{1}{q}\ .$$

This theorem shows that the encrypted sequence consists of statistically-independent and equally probable symbols, thus does not give any information on the content of the plain message.

In terms of information measure [7], when the information symbols m_i are not known, the entropy of the encrypted sequence $E = \{[e_i, g_i], [e_i, g_i], \ldots, [e_i, g_i], \ldots\}$ is $H(E) = \log_2 q + \log_2(q-2)$, which is very close to twice the entropy $H(M) = \log_2(q)$ of the source alphabet, as in perfect encryption. In this latter case, only a single symbol (with associated entropy $\log_2 q$) is known to the attacker, since it is transmitted on a public channel, while the second symbol (with associated entropy $\log_2 q$), transmitted on a secure secret channel, is unknown to the attacker.

Plain-text Attack. The attacker, besides the information as in the cipher-text-only attack, also knows the message sequence M, and his target is to recover the secret equator line, and if possible the whole secret key-stream, or equivalently the secret initial state S_0 of the key-stream generator. Although, in these conditions, it is possible to mount an algebraic attack that reduces the complexity to a already specified lower bound, it is useful that this bound can be controlled by the system designer.

Theorem 2. *Assume all the hypotheses of Theorem 1 with the further condition that the sequence M is known. Furthermore, suppose that the sequence H is generated by a stream cipher of known structure, but unknown initial state. Therefore the complexity for obtaining the initial state S_0, and thus the equator line, is of order $O(q)$.*

Proof. If m_i is known, from equation (1) we obtain an equation connecting the interlaced sequences k_{2i+1} and k_{2i} of elements of F_q, derived from the key-stream H

$$g_i m_i + k_{2i+1}e_i - (m_i + g_i)k_{2i} = 0 \qquad \forall i \geq 0\ .$$

This is a linear relation connecting k_{2i} and k_{2i+1}, all other parameters being known. The attack is described making the conservative, though unlikely, assumption that knowing k consecutive bits of H, i.e. an element $\beta \in F_q$, we may easily obtain the initial state T_i of the key-stream generator that produces β. Therefore we may use the generator with initial state T_i to produce $2k$ consecutive bits, that is β and γ. Let α be a primitive element of F_q. Under these assumptions, the following exhaustive attack, needing only two plain text symbols $\{m_s, m_{s+1}\}$ and the correspondent encrypted symbols $\{[e_s, g_s], [e_{s+1}, g_{s+1}]\}$, may be devised:

1. Set $i = s$ and $j = 0$
2. With $j = j + 1$, set $\beta = \alpha^j$ and $k_i = \beta$.
3. Obtain the corresponding initial state $S_0(\beta)$, (easy to compute by assumption)
4. Re-obtain β and γ (consisting of bits between position $k + 1$ and $2k$) by the stream generator started with initial state $S_0(\beta)$,
5. Check whether

$$g_i m_i + \gamma e_i - (m_i + g_i)\beta \overset{?}{=} 0 ,$$

then $\qquad \begin{cases} \text{if YES, then GO TO next step} \\ \text{if NO, then GO TO step 2} \end{cases}$

6. Obtain y_i from the pair $[\beta, \gamma]$ using equation (2). Set $X_i = m_i$ and $Y_i = y_i$
7. If (i=s) set $i = s + 1$ and GO TO Step 2, OTHERWISE output $[X_s, Y_s]$ and $[X_{s+1}, Y_{s+1}]$ and STOP.

Using the two points $[X_s, Y_s]$ and $[X_{s+1}, Y_{s+1}]$, obtain the equator line $\mathcal{E}$.

As a consequence of this theorem, if k is chosen sufficiently large (e.g. > 200) no exhaustive attack is possible, since $2^k > 10^{60}$ attempts are not feasible, and the number of computations is even larger if realistic key-stream generators are used. In conclusion, the reconstruction of the key-stream is not feasible, and nor can the equator line be obtained. Clearly, the strength of the system is not comparable to that of the one-time-pad, but it is certainly competitive with that of conventional stream-ciphers, against plain-text-attacks.

4   Conclusions

It has been shown that a particular sort of perfect secrecy is achievable at the same rate as Vernan's one-time-pad. More precisely, the probability of guessing message blocks composed of k bits is $\frac{1}{2^k}$, independently of the encrypted stream which does not yield information on the message facilitating crypto-analytic attacks. Unlike schemes based on secretly exchanged whole key-streams, the system has the advantage of avoiding conservation and safe destruction of long keys; as a counterpart, it has the disadvantage of being vulnerable to plain text attacks. However, in this last circumstance, the scheme is at least as hard to break as any common stream-cipher. The complexity of enciphering and deciphering the proposed scheme is small and easy to implement; the expensive part is the key-stream generator. However this generator needs not be designed to satisfy stringent constraints, because security does not reside in its robustness against cryptanalysis. Therefore, the scheme may be of practical interest [6] because it is not expensive to implement, it may work fast, and many aspects of the robustness against attacks (plain-text attacks) are under the designer's control.

Acknowledgments. The author would like to gratefully acknowledge Dr. Marco Coppola and Dr. Guglielmo Morgari, both from Telsy Elettronica e Telecomunicazioni (Torino - Italy), for many valuable discussions and constructive suggestions, which have greatly improved the accuracy of the results and the quality of the paper.

References

1. Gilbert, E.N., MacWilliams, F.J., Sloane, N.J.A.: Codes which detect deception. Bell Labs Tech. J. **53**, 405–424 (1974)
2. Hoffstein, J., Pipher, J., Silverman, J.H.: An Introduction to Mathematical Cryptography. Springer, New York (2008)
3. Knuth, D.E.: The Art of Computer Programming, vol. 2. Addison-Wesley, Massachusetts (1981)
4. Lidl, R., Niederreiter, H.: Finite Fields. Addison-Wesley, Massachusetts (1983)
5. Rueppel, R.A.: Analysis and Design of Stream Ciphers. Springer, Heidelberg (1986)
6. Schneier, B.: Applied Cryptography. Wiley, New York (1995)
7. Shannon, C.E.: A mathematical theory of communication. BSTJ **27**, 379–423, 623–656 (1948)
8. Shannon, C.E.: Communication theory and secrecy systems. Bell Labs Tech. J. **28**, 656–715 (1949)
9. Sloane, N.J.A.: Error-correcting codes and cryptography. In: Klarner, D.A. (ed.) The Mathematical Gardner, pp. 346–382. Prindle, Weber and Schmidt (1981)
10. Vajda, S.: The Mathematics of Experimental Designs. Charles Griffin & Company, London (1967)
11. Vajda, S.: Patterns and Configurations in Finite Spaces. Charles Griffin & Company, London (1967)
12. Vernam, G.S.: Cipher printing telegraph system for secret wire and radio telegraphic communications. J. AIEE **45**, 109–115 (1926)

The Inverse Problem of Foreign Exchange Option Pricing

Baiqing Sun[1], Nataliya Yi Liu[2], and Junzo Watada[3,4](✉)

[1] Harbin Institute of Technology, School of Management, 13 Fayuan Street,
Nangang District Harbin 150001, Heilongjiang, China
[2] School of Business Administration, Liaoning Technology University, Jinzhou,
Heilongjiang, China
[3] Department of Computer and Information Sciences,
PETRONAS Institute of Technology, 32610 Seri Iskandar, Perak, Malaysia
junzow@osb.att.ne.jp
[4] Waseda University, Kitakyushu, Japan

Abstract. When investors apply foreign exchange options to avoid foreign exchange risk, the key issue is how to use a reasonable mathematical model to determine the price of foreign exchange options. At present, volatility is mostly determined by using subjective estimations and sample calculations. Therefore, different sources of volatility result in large differences between calculated price and the actual price of the exchange options, which will affect the strategy selection and the actual return of investors. Meanwhile, in the actual foreign exchange market, the normal distribution-based return rate cannot express the fat tail situation of volatility. This paper clarifies the inverse problem based on the fat tail of return rates of foreign exchange in option pricing. The inverse problem plays a pivotal role in determining the form of implied volatility and fluctuations in the value scope. First, this paper summarizes the present state of research on the inverse problems of foreign exchange option pricing. Second, this paper explains the basic theory of foreign exchange options, and deduce a positive foreign exchange option pricing problem, that is, G-K foreign exchange option pricing. At the same time, it proposes the inverse problem of foreign exchange option pricing, then puts forward a foreign exchange option pricing model based on the t-distribution and the inverse problem which is more appropriate to the actual foreign exchange market. Last the paper exemplifies the foreign exchange option pricing inverse problem based on fat tail of exchange rate return by using the numerical differential algorithm, which solves implied volatility.

Keywords: Foreign exchange option pricing · Fat tail
Inverse problem · Volatility

J. Watada—Emeritus Professor

© Springer International Publishing AG, part of Springer Nature 2018
A. M. Gil-Lafuente et al. (Eds.): FIM 2015, AISC 730, pp. 407–426, 2018.
https://doi.org/10.1007/978-3-319-75792-6_30

1   Introduction

In the 1980s, foreign exchange option has became a new financial derivative product to avoid exchange risk. After more than 30 years of its development, foreign exchange option has become a more important financial derivative tool in many countries. In the foreign exchange option market, investors apply their assumption of risk in foreign exchange transactions by means of buying or selling foreign exchange options. In applying foreign exchange options to avoid foreign exchange risk, the key issue of the investors is to know the price of foreign exchange option through reasonable mathematical models. Therefore, the pricing problem of foreign exchange option has become the core topic of foreign exchange option research. At present, the research studies on pricing methods of foreign exchange option7ujn mainly concern with closed types decomposition methods derived from.

In 1983, based on the Black-Scholes (or B-S) model [1], German and Kolhagern derived the pricing question of European-style Exchange Option through Merton formula, naming it GK model [2]. This classic one is the first pricing model of Exchange Option which has been proposed. The following models are its deviants. Then, many scientists presented foreign exchange option pricing models which are more close to the real foreign exchange market. For example, Amin, Jarrow thought interest rate is not constant, thus they built a stochastic interest rate model, where the foreign exchange option pricing model is based on the stochastic interest rate [3]. Hause, Levy constructed the foreign exchange option pricing model with transactional cost. Harikumar, Maria proposed the foreign exchange option pricing model of GARCH classification. For the exchange rate income has peak and fat-tail and not the standard normal distribution. Because of this, foreign exchange option pricing models are proposed based on the exchange rate of fat-tail return. These improved foreign exchange option pricing model.

At present, in many scientific studies, people only know the natural results, but it is hard to understand the boundary conditions or variables influenced from such results. The inverse problem is to study the unknown variable or boundary of this underlying problem under the condition of knowing some problems' result. That is, the inverse problem can solve the boundary conditions and variable of influences that conventional methods cannot solve. In this situation, the research of this problem is of important significance and of very urgency. Now, also some scholars have used the inverse problem research method to study the influence of option pricing model volatility on the pricing, but no specific research was studied on effects of the implied volatility to the foreign exchange option pricing.

2   List of Variables Used

We summarize the variables used in the paper as follows.

S: Underlying asset price
C: European bullish option price
T: Underlying asset due time

K: Option executive price
P: European bearish option price
t: Current time point
r: Risk-free interest rate
dx: Continuous Brownian motion process
μ: Expected rate of basic asset value
σ: Volatility of the basic asset price
p,p': Pay Cash Dividend
q,q': Pay Cash Dividend
G: Hedge combination
n: Number of samples
ν: Degree of freedom
i: maybe, not necessary to write here.
j: maybe, not necessary to write here.
k: maybe, not necessary to write here.

3   Foreign Exchange Option Pricing (G-K Model)

Many factors can affect the price of foreign exchange option. The classic G-K model is obtained from the B-S model. Foreign exchange option pricing is modeled as the amended Merton model which is the extension model of B-S model.

3.1   Merton Model

The Assumed Condition of the Model. Black and Shcoles placed the following assumptions before obtaining the pricing formula:

(1) Underlying asset price S defers to log-normal random process as

$$dS = \mu S dt + \sigma S dt \tag{1}$$

(2) In the option due date T, the risk-free interest rate r and the volatility of underlying asset price, σ, are constant values;

(3) Investors in the market transfer the risk operation without transaction cost S;

(4) Constant payment of cash dividend p, q of underlying stock in the option due date T are certain percentage valuation of share price;

(5) Financial market does not provide opportunities without risk arbitrage;

(6) They can have a business operation to the underlying asset;

(7) They can proceed bearish operation and further subdivided the underlying asset. In other words, it can proceed vacant transaction;

The Deduced Model. Using small dividend of C, t and x, the price of the option is $C(S,t)$, which is the function of S and t. According to Ito Lemma, the random process of option pricing is written as follows:

[**The option price:**]

$$dC = \frac{\partial C}{\partial S}\sigma S dx + (\mu S \frac{dC}{dS} + \frac{1}{2}\sigma^2 S^2 \frac{\partial^2 C}{\partial S^2} + \frac{\partial C}{\partial t})dt \tag{2}$$

[**Discretization:**]

The Eq. (2) can be discretized in the following:

$$\Delta C = \frac{\partial C}{\partial S}\sigma S \Delta x + (\mu S \frac{C}{dS} + \frac{1}{2}\sigma^2 S^2 \frac{\partial^2 C}{\partial S^2} + \frac{\partial C}{\partial t})\Delta t. \tag{3}$$

[**Construction of hedging combination G:**]

The Construction of hedging combination G is written as follows:

$$G = C - \frac{\partial C}{\partial S}S \tag{4}$$

where $\frac{\partial C}{\partial S}$ expresses the short-selling stock shares.

We can get the hedge combination G as risk-free combination in the dt time.

The model assumed to have cash dividend allocation during the validity of the option, and the quota is $q\frac{\partial C}{\partial S}S\Delta t$, where $\frac{\partial C}{\partial S}S\Delta t$ stands for the share holding value, and q is dividend yielded in every unit time. Therefore, G is written as

$$\Delta G = (\frac{\partial C}{\partial t} + \frac{1}{2}\sigma^2 S^2 \frac{\partial^2 C}{\partial S^2} - q\frac{\partial C}{\partial S}S)\Delta t. \tag{5}$$

We know that the hedge combination in the tiny time is risk-free, so the total variable should be equal to the risk-free pay $rG\Delta t$:

$$\Delta G = (\frac{\partial C}{\partial t} + \frac{1}{2}\sigma^2 S^2 \frac{\partial^2 C}{\partial S^2} - q\frac{\partial C}{\partial S}S)\Delta t = r(C - \frac{\partial C}{\partial S})\Delta t. \tag{6}$$

According to the above it solves and obtains European option pricing formula of the underlying stock that pay constantly cash dividend:

[**Discretization:**]

$$C(S,t) = Se^{-q(T-t)}N(d_1) - Ke^{-r(T-t)}N(d_2) \tag{7}$$

Among

$$d_1 = \frac{\ln(S/K) + (r - q + \frac{1}{2}\sigma^2)(T - t)}{\sigma\sqrt{T-t}} \tag{8}$$

$$d_2 = d_1 - \sigma\sqrt{T-t}$$
$$= \frac{\ln(S/K) + (r - q - \frac{1}{2}\sigma^2)(T - t)}{\sigma\sqrt{T-t}} \tag{9}$$

σ is the pricing volatility of the underlying stock, and $\sigma = \sqrt{var(\frac{dS}{S})}$

3.2 Foreign Exchange Option Pricing Formula According to the Merton Model

Suppose the exchange S follows $S = S(C_1/C_2)$, C_1 and C_2 represent the two exchange values, and the ratio of them indicates C_2 per unit C_1 value. According to this theory. so the exchange rate S follows the random process

$$dS = \mu S dt + \sigma S dx \tag{10}$$

It also satisfies to lemma. Suppose $C(S,t)$ is the call option of exchange C_1 to the exchange, C_2 and the exchange C_1 deposit constantly payment risk-free interest rate r_f, the exchange C_1 is regarded as the foreign exchange, and the corresponding exchange C_2 is the own exchange. According to the new quantity, the risk-free interest rate r_f of foreign exchange C_1 is equal to constantly pay cash dividend q, so we replace the q with r_f of Merton model, and get the call option pricing formula of European foreign exchange option in the following:
[Discretization]

$$C(S,t) = Se^{-r_f(T-t)}N(d_1) - ke^{-r(T-t)}N(d_2) \tag{11}$$

$$d_1 = \frac{\ln(S/K) + (r - r_f + \frac{1}{2}\sigma^2)(T-t)}{\sigma\sqrt{T-t}} \tag{12}$$

$$d_2 = d_1 - \sigma\sqrt{T-t}$$
$$= \frac{\ln(S/K) + (r - r_f - \frac{1}{2}\sigma^2)(T-t)}{\sigma\sqrt{T-t}} \tag{13}$$

Now, the σ in the formula is the underlying exchange rate volatility. From the above formula (10) we know that a lot of factors affect the underlying exchange option pricing, including the due date executive price K, the due date T and the current time t, the underlying exchange rate volatility σ, the due date exchange rate S, and the risk-free interest rate of two countries. Certainly, the risk-free interest rate of two countries do not change in the certain time, so we treat r and r_f as constant.

According to (10), (11), (12), we get the direct problem of the foreign exchange option pricing: Is it possible by using the known exchange rate S, the due date T, the executive price K, exchange rate volatility σ, to solve the call option $C(S,t)$ of the exchange C_1 to the exchange C_2 on the due date T?

4 The Inverse Problem of the Foreign Exchange Option Pricing

We can see that the exchange volatility σ is an important parameter in foreign exchange option pricing formula. When the changes happen, the foreign exchange option price will change largely, in other words, the exchange volatility is very sensitive to the foreign exchange option price. In the foreign exchange market, people usually would like to know the exchange rate volatility. From the current foreign exchange option market we can the option price of the same underlying exchange on the different due dates and the different executive prices. We can get the future size of the implied volatility. Normally, people treat the volatility of the underlying asset resulted from single option price as implied volatility, it is also called the exchange implied volatility of the underlying exchange rate volatility resulted from single option price.

The inverse problem of the foreign exchange option pricing is how to use the current underlying price getting from the foreign exchange option market deduce the implied volatility process of the underlying currency.

4.1   Inverse Problem of General Foreign Exchange Option Pricing

In the previous section, we explained σ is a constant obtained from the proposed formula. However, strictly speaking, this assumption cannot be accepted that the volatility is constant in the actual foreign exchange market. With different executive price, the implied volatility shows a particular change rule, that is for the same underlying asset and the due date option, assumed the other condition is invariable, then we can get the U curve. For this curve looks like smiling mouth, so we called it "volatility smile" [27]. And in fixed executive price, the volatility shows a bottom right deflective trend with different due date. It is called a "volatility deflection" curve. The two kinds of curve are all a mode to predict and estimate the future volatility. Because the assumption is not established yet that the volatility is constant. Let us assume that $\sigma = \sigma(S, t)$. using the volatility function of variables S, t. So in the risk neutral measure the random process of the exchange price can be modified as follows:

$$\frac{dS}{S} = (r - r_f)dt + \sigma(S, t)dW \tag{14}$$

dW follows normal distribution. That is $dW = \epsilon\sqrt{dt}, \epsilon \sim N(0, 1)$. So European foreign exchange bullish option price fits to the following formula:

$$\frac{\partial C}{\partial t} + \frac{1}{2}\sigma^2(S, t)S^2\frac{\partial^2 C}{\partial S^2} + (r - r_f)S\frac{\partial C}{\partial S} - rC = 0 \quad (S, t) \in R^+ \times [0, T] \tag{15}$$

$$C(S, T) = \max(S - K, 0) \equiv (S - K)^+ \tag{16}$$

Now, we can ask such a question: Knowing the option price of the foreign exchange option market, how can we obtain the future underlying exchange volatility under the existing condition? Such a question is the inverse problem of foreign exchange option pricing.

When European foreign exchange option is given as an example, then the inverse problem of foreign exchange option pricing is:

Suppose $C_t = C(S_t, t; K, T)$ is the foreign exchange option pricing, the following relations are satisfied:

$$\begin{cases} \dfrac{\partial C_t}{\partial t} + \dfrac{1}{2}\sigma^2(S_t, t)S_t^2\dfrac{\partial^2 C_t}{\partial S_t^2} + (r - r_f)S_t\dfrac{\partial C_t}{\partial S_t} - rC_t = 0 \\ C_T = (S_T - K)^+ \end{cases} \tag{17}$$

So supposing that $t = t^*(0 \le t^* \le T)$. $S - S^*$, the known current market foreign exchange option price $C(S', t^*; K_i.T_i) = C^*(K_i, T_i)$, $(i = 1, 2, \cdots, N)$, the inverse problem is to inversely solve the foreign exchange volatility $\sigma = \sigma(S, t)$.

4.2   G-K Formula Inverse Problem Based on Dupire Formula

According to the previous section we result in the foreign exchange option pricing process. Let us suppose European bullish option value $C = C(S_t, t; K, T, \sigma)$, and the following should be satisfied:

$$
\begin{cases}
\dfrac{\partial C}{\partial t} + \dfrac{1}{2}\sigma^2(S, t)S^2 \dfrac{\partial^2 C}{\partial S^2} + (r - r_f)S\dfrac{\partial C}{\partial S} - rC = 0 \\
C(S, T; K, T, \sigma) = \max(S - K, 0) \\
C(0, t; K, T, \sigma) = 0 \\
\lim\limits_{S \to \infty} \dfrac{\partial C}{\partial S}(S, t; K, T, \sigma) = e^{-r_f(T-t)}
\end{cases}
\tag{18}
$$

Now we need to solve the inverse problem given in the previous section.

Among the research studies on the option pricing inverse problem, Dupire is the first one who asked the question of solving methods. Based on the symmetry of the transfer probability density, transferred the option pricing problem to a dual equation about K, T, and obtained the famous Dupire formula. So the inverse problem of the foreign exchange option pricing we use solved the following formula:

Definition 1. *Suppose*

$$
Lp = -\frac{\partial p}{\partial t} + \frac{1}{2}\sigma^2(S, t)S^2 \frac{\partial^2 p}{\partial S^2} + (r - r_f)S\frac{\partial p}{\partial S} - rp,
\tag{19}
$$

and conjugate operator Lq is

$$
Lq = -\frac{\partial q}{\partial T} + \frac{1}{2}\frac{\partial^2}{\partial K^2}(\sigma^2(K, T)K^2 q) - (r - r_f)\frac{\partial}{\partial K}(Kq) - rq
\tag{20}
$$

Lp and Lq satisfy $\displaystyle\int_\Omega (qLp - pLq)dx = 0 \qquad \forall p, q \in \mathbb{C}_0^\infty(\Omega).$

Lemma 1. *The following relation holds:*

$$
G(S, t; K, T) = \frac{\partial^2 C}{\partial K^2}(S, t; K, T)
\tag{21}
$$

And $G(S, t; K, T)$ is the solution of Eq. (16), and the following relations are satisfied:

$$
LG = 0, \qquad G(S, T; K, T) = \delta(S - K)
\tag{22}
$$

Theorem 1. *If we take $G(S,t;K,T)$ for K, T function, then Eq. (16) is conjugate equation solution, and if $G^*(K,T;S,t) = G(S,t;K,T)$, so*

$$\begin{cases} LG^* = 0 \\ G^*(K,t;S,t) = \delta(K - S) \end{cases} \tag{23}$$

Proof. According to the definition, we know the following:

$$\begin{aligned}
0 &= \int_0^\infty dx \int_{t+\epsilon}^{T-\epsilon} [G^*(x,y;S,t)LG(x,y;K,T) - G(x,y;K,T)LG^*(x,y;S,t)]dy \\
&= \int_0^\infty dx \int_{t+\epsilon}^{T-\epsilon} \left\{ \frac{\partial}{\partial y}(G^*G) + \frac{1}{2}\frac{\partial}{\partial x}[\sigma^2 x^2 G^* \frac{\partial G}{\partial x}] \right. \\
&\quad \left. -\frac{1}{2}\frac{\partial}{\partial x}[G\frac{\partial}{\partial x}(\sigma^2 x^2 G^*)] + (r - r_f)\frac{\partial G}{\partial x}(xGG^*) \right\} dy
\end{aligned} \tag{24}$$

When $x \to 0, \infty$, $\sigma^2 x^2 G^* \frac{\partial G}{\partial x} \to 0$, $G\frac{\partial}{\partial x}(\sigma^2 x^2 G^*) \to 0$, $xGG^* \to 0$, then we have the following:

$$\int_0^\infty G^*(x,T-\epsilon;S,t)G(x,T-\epsilon;K,T)dx = \int_0^\infty G^*(x,t+\epsilon;S,t)G(x,t+\epsilon;K,T)dx \tag{25}$$

Let us take the limitation $\epsilon \to 0$, according to $C(S,T;K,t) = \delta(S - K)$ we obtain the following:

$$G^*(K,t;S,t) = \delta(K - S)\int_0^\infty G^*(x,T;S,t)\delta(x - K)dx = \int_0^\infty \delta(x - S)G(x,t;K,T)dx \tag{26}$$

That is, $G^*(K,T;S,t) = G(S,t;K,T)$ is obtained.   □

To derive C about K, T equation, let us substitute the Eq. (19) to Eq. (20), and let us suppose that K is in $[K,\infty]$ to solve twice integration. Finally we can obtain the following relation:

$$\begin{cases} -\frac{\partial C}{\partial T} + \frac{1}{2}K^2\sigma^2(K,T)\frac{\partial^2 C}{\partial K^2} - (r - r_f)K\frac{\partial C}{\partial K} - r_f C = 0 \\ C|_{T=t} = \max(S - K, 0) \end{cases} \tag{27}$$

According to (21), $\sigma(K,T)$ can be solved as follows:

$$\sigma(K,T) = \sqrt{\frac{\frac{\partial c}{\partial T} + (r - r_f)K\frac{\partial C}{\partial K} + r_f C}{\frac{1}{2}K^2\frac{\partial^2 C}{\partial K^2}}} \tag{28}$$

Formula (22) is based on Dupire formula: which results in the explicit formulation about the exchange rate volatility for G-K equation inverse problem. But in the actual foreign exchange market, the distribution of the exchange rate returns sequence with fat-tail feature has fat-tail, showing that traditional normal distribution assumption does not reflect this property, there will be errors to determine the volatility according to above explicit. So we need to put forward an assumption more closer to the actual condition and to study its inverse problem.

4.3   Foreign Exchange Option Pricing Model Derivation Based on t-distribution

The Process of Foreign Exchange Option Pricing Model Derivation Based on t-Distribution. An important assumption of the classical G-K model is the forward rate content.

$$\frac{dS}{S} = \mu dt + \sigma_t dW \qquad dW = \epsilon\sqrt{dt}, \ \ \epsilon \sim N(0,1), \tag{29}$$

As the exchange rate return sequential empirical distribution has fat-tail, the above exchange rate random process cannot solve the problem. Thus, in this section, we will describe the exchange rate pricing process satisfied t-distribution process and adjust it closer to the actual exchange rate change. That exchange rate price process is satisfied:

$$\frac{dS}{S} = \mu dt + \sigma_t dz \qquad dz = \epsilon\sqrt{dt}, \ \ \epsilon \sim t_\nu, \tag{30}$$

ν is the degree of freedom of t-distribution, t_ν density function is

$$f(\epsilon) = \frac{\Gamma(\frac{1}{2}(\nu+1))}{\sqrt{\nu\pi}\Gamma(\frac{1}{2}\nu)}(1+\frac{\epsilon^2}{\nu})^{-\frac{1}{2}(\nu+1)}, \tag{31}$$

where $\Gamma(\bullet)$ is gammar function.

Suppose forward rate price process satisfies (23), that is $dS = \mu S dt + \sigma_t S dz$, $dz = \epsilon\sqrt{dt}, \ \epsilon \sim t_\nu; f(S,t)$ shows a foreign exchange option in the moment t.

By this process, generated income is

$$\frac{dS(t)}{S(t)} = \mu dt + \sigma dz \tag{32}$$

First consider the property of income generated from random differential equation. Using extension form of ITO formula based on t-distribution and $\ln S(x)$, we get:

$$d\ln S(t) = (\mu - \frac{m\sigma^2}{2})dt + \sigma dz \tag{33}$$

two terminals integration simultaneously, and get:

$$S(t) = S(0)\exp\{(\mu - \frac{m\sigma^2}{2})dt + \sigma z\} \tag{34}$$

Consider on Δt yield property.

$$\begin{aligned}\frac{\Delta S(t)}{S(t)} &= \frac{S(t+\Delta t) - S(t)}{S(t)} = \frac{S(t+\Delta t)}{S(t)} - 1 \\ &= \exp\{(\mu - \frac{m\sigma^2}{2})\Delta t + \sigma(z(t+\Delta t) - z(t))\} - 1\end{aligned} \tag{35}$$

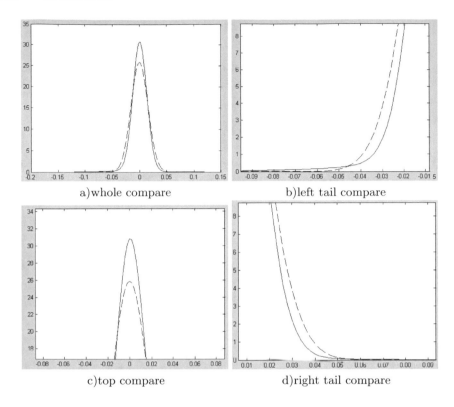

a)whole compare   b)left tail compare

c)top compare   d)right tail compare

Fig. 1. Comparison of data distribution

Because of $e^x \approx 1 + x + \dfrac{x^2}{2}$, we neglect higher order terms of Δt, and approximate the above quantity as follows:

$$
\begin{aligned}
\frac{\Delta S(t)}{S(t)} &\approx (\mu - \frac{m\sigma^2}{2})\Delta t + \sigma(z(t + \Delta t) - z(t)) \\
&\quad + \frac{1}{2}\sigma^2(z(t + \Delta t) - z(t))^2 \\
&\approx [\mu + \frac{1}{2}(\epsilon^2 - m)\sigma^2]\Delta t + \sigma\epsilon\sqrt{\Delta t}
\end{aligned}
\tag{36}
$$

ϵ is t-distribution random variable. The mean of the above equation is as follows::

$$
E(\frac{\Delta S(t)}{S(t)}) = \mu\Delta t
\tag{37}
$$

$$
Var(\frac{\Delta S(t)}{S(t)}) = m\sigma^2\Delta t
\tag{38}
$$

To expand Ito (ITO) lemma, for $f(S,t)$ is developed second Taylor expansion at (S_0, t_0), we get:

$$df = ((r - r_f)S\frac{\partial f}{\partial S} + \frac{\partial f}{\partial t} + \frac{1}{2} + \frac{\partial^2 f}{\partial S^2}\sigma_t^2 S^2\frac{\nu}{\nu - 2})dt + \sigma_t S\frac{\partial f}{\partial S}dz \qquad (39)$$

Because dz is t-distribution, so df is t-distribution.

The spot rate in the actual option has fat-tail property, so spot rate price process on content t-distribution, formula deduced in the normal distribution can be established: Now we suppose two kind spot rates g and h, and g subject to $dg = \mu g dt + \sigma_t g dz$, h subject to $dh = \omega h dt + \rho_t h dz$, where $dz = \epsilon\sqrt{dt}$, $\epsilon \sim t_\nu$, r is risk-free rate. For arbitrage opportunity, it must satisty the following relation:

$$\frac{\mu - t}{\sigma_t} = \frac{\omega - r}{\rho_t} \qquad (40)$$

Using the other European bullish option value: $f(S,t) = C(S,t)$, Eq. (25) can be changed as:

$$\begin{aligned}
dC &= (\mu S\frac{\partial C}{\partial S} + \frac{\partial C}{\partial t} + \frac{1}{2} + \frac{\partial^2 C}{\partial S^2}\sigma_t^2 S^2\frac{\nu}{\nu - 2})dt + \sigma_t S\frac{\partial C}{\partial S}dz \\
&= [\frac{1}{C}(\mu S\frac{\partial C}{\partial S} + \frac{\partial C}{\partial t} + \frac{1}{2}\frac{\partial^2 C}{\partial S^2} + \sigma_t^2 S^2\frac{\nu}{\nu - 2})]C dt + (\frac{1}{C}\sigma_t S\frac{\partial C}{\partial S})C dz
\end{aligned} \qquad (41)$$

This option drift is evaluated as:

$$\alpha = \frac{1}{C}\left(\mu S\frac{\partial C}{\partial S} + \frac{\partial C}{\partial t} + \frac{1}{2}\frac{\partial^2 C}{\partial S^2}\sigma_t^2 S^2\frac{\nu}{\nu - 2}\right) \qquad (42)$$

Volatility is $\beta = \frac{1}{C}\sigma_t S\frac{\partial C}{\partial S}$. And SB is a kind of derivative securities, B is treasury bill of risk-free to offer exchange in the host country, r_f is the risk-free rate to solve differential SB, we get $d(SB) = SdB + BdS$, and because dS satisfies the following relation:

$$dS = \mu S dt + \sigma_t S dt, \qquad dB = r_j B dt \qquad (43)$$

So we get $\frac{d(SB)}{SB} = (\mu + r_f)dt + \sigma_t dz$. Let us substitute α and β to the formula (26):

$$\frac{(\mu + r_f) - r}{\sigma_t} = \frac{\alpha - r}{\beta}, \qquad (44)$$

then we obtain the following:

$$rC = (r - r_f)S\frac{\partial C}{\partial S} + \frac{\partial C}{\partial t} + \frac{1}{2}\sigma_t^2 S^2\frac{\nu}{\nu - 2}\frac{\partial^2 C}{\partial S^2}, \qquad (45)$$

and

$$\begin{aligned}
m &= \sqrt{\frac{\nu}{\nu - 2}} \\
&= (r - r_f)S\frac{\partial C}{\partial S} + \frac{\partial C}{\partial t} + \frac{1}{2}\sigma_t^2 + S^2 m^2\frac{\partial^2 C}{\partial S^2}
\end{aligned} \qquad (46)$$

$C(S,t)$ satisfies Eq. (27) and the boundary condition, that is:

$$\left.\begin{array}{l} C(S,T) = \max(S - K, 0) \\ C(0,t) = 0 \\ C(S,t) \sim S, \qquad \qquad \text{for} \quad S \to \infty \end{array}\right\} \qquad (47)$$

Now to explain Eq. (28) three boundary conditions, $C(S,T) = \max(S-K,0)$ is the bullish option price condition paid by investors on the due date T, if $S > K$, the investors execute the option treaty, so they can get profit $S - K$. If $S < K$, so the investors does not execute the option treaty. $C(0,t) = 0$. It represents if the exchange rate is zero in anytime, the bullish option price is zero no matter how the time changes. The third condition is that when S is large enough $C \sim S$.

To solve Eqs. (27) and (28), we can solve European bullish option price on the moment t:

$$\begin{aligned} C(S,t) &= XD(x,\tau) \\ &= Se^{-r_f(T-t)}N(d_1') - Xe^{-r(T-t)}N(d_2) \end{aligned} \qquad (48)$$

$m = \sqrt{\dfrac{\nu}{\nu - 2}}$, ν is t_ν distributive freedom.

$$d_1' = \frac{\ln(\dfrac{S}{K}) + (r - r_f + \dfrac{1}{2}\sigma_t^2 m^2)(T - t)}{\sigma_t m \sqrt{T - t}} \qquad (49)$$

$$d_2' = \frac{\ln(\dfrac{S}{K}) + (r - r_f - \dfrac{1}{2}\sigma_t^2 m^2)(T - t)}{\sigma_t m \sqrt{T - t}} = d_1' - \sigma_t m \sqrt{T - t} \qquad (50)$$

And according to FX option put-call parity formula: $P = C + Xe^{-r(T-t)}N(-d_2) - S(t)e^{-r_f(T-t)}$ can result in the corresponding foreign exchange bearish option price formula:

$$P(S,t) = Xe^{-r(T-t)}N(-d_2') - S(t)e^{-r_f(T-t)}N(-d_1') \qquad (51)$$

Compared the result with the traditional G-K model, foreign exchange option pricing model based on t-distribution just turns the volatility σ to $\sigma_t m$. There is no differences in the form. But $m = \sqrt{\dfrac{\nu}{\nu - 2}}$, where ν is the degree of freedom of t-distribution. It controls the distribution tail thickness function. We can see that foreign exchange option pricing model based on return fat-tail is more severe than the normal distribution volatility. So the improved pricing model more truly reflects the actual option value in foreign exchange market than the traditional G-K model.

Estimation of t-Distributional Degree of Freedom. The degree of freedom ν controls t-distributional tail thickness, but we know the degree of freedom is not similar to different exchange rate return. We not only deduce the model, but also reckon the degree of freedom. In general, we adopt maximum likelihood estimation for the degree of freedom, suppose the standard exchange rate return is r'_t, $r'_t = \dfrac{r_t}{\sigma_t}$, and r'_t denotes exchange rate return historical date. The standard exchange rate return order satisfies:

$$f(r'_t) = \frac{\Gamma(\frac{1}{2}(\nu+1))}{\sqrt{\nu\pi}\,\Gamma(\frac{1}{2}\nu)}(1+\frac{r'^2_t}{\nu})^{\frac{1}{2}(\nu+1)} \tag{52}$$

So its maximum likelihood estimation is

$$l(\nu) = \Pi^t_{i=t-n+1}\frac{\Gamma(\frac{1}{2}(\nu+1))}{\sqrt{\nu\pi}\,\Gamma(\frac{1}{2}\nu)}(1+\frac{r'^2_t}{\nu})^{-\frac{1}{2}(\nu+1)}, \tag{53}$$

n is the number of extractive samples.

Numerical method can deal with maximum likelihood estimation optimization problem, but numerical method has better divergent and poor convergence, when the degree of freedom for higher order derivative of the model, with huge calculation. It is easier to come out algorithm error and cannot show the expression. Thus, we use moment to estimate the t-distribution with the degree of freedom.

Now we suppose m_i is the i sample moment, according to t-distribution symmetry, when i is odd number we can get $m_i = 0$, but

$$m_2 = \frac{\hat{\nu}}{\hat{\nu}-2}, \qquad m_4 = 3\frac{\hat{\nu}^2}{(\hat{\nu}-2)^2}[\frac{\hat{\nu}}{2}-2]^{-1}+1, \tag{54}$$

so we obtain: $\hat{\nu} = 6\rho^{-1}+4$

When the number of samples is small, moment estimations less effective than maximum likelihood estimation, but when the number of samples is sufficiently large, the two kinds of estimations should be consistent.

Solving the Foreign Exchange Option Pricing Inverse Problem Based on t-Distributional. In the former section, we present the general inverse problem of foreign exchange option pricing model and obtained the solution. In the section, we also use the same method to propose the inverse problem of foreign exchange option pricing based on the exchange rate return fat-tail. Certainly, volatility is depend on S, t variables of the corresponding improved model, we can get:

$$\sigma = \sigma(S,t). \tag{55}$$

Correspondingly, our exchange rate price random process becomes :

$$\frac{dS}{S} = (r-r_f)dt + \sigma_t(S,t)dz \tag{56}$$

dz follows t-distribution. We explained the forward direct problem process $dz = \epsilon\sqrt{dt}$ and $\epsilon \sim t$ in the previous section, so improved foreign exchange option pricing fits the following formula:

$$(r - r_f)S\frac{\partial C}{\partial S} + \frac{\partial C}{\partial t} + \frac{1}{2}\sigma_t^2 S^2 \frac{\nu}{\nu - 2}\frac{\partial^2 C}{\partial S^2} - rC = 0 \qquad \text{for } (S, t) \in R^+ \times [0, T) \tag{57}$$

$$C(S, T) = \max(S - K, 0) \tag{58}$$

So we present the inverse problem of improved foreign exchange model. We know option price in the foreign exchange option market. Let us discuss how to solve the unknown future underlying exchange volatility using the present market known conditions.

Mathematical formula showing the inverse problem of improved model: Suppose $C_t = C(S_t, t; K, T)$ is the improved foreign exchange option price, then it satisfies the following condition:

$$\begin{cases} (r - r_f)S_t\dfrac{\partial C_t}{\partial S_t} + \dfrac{\partial C_t}{\partial t} + \dfrac{1}{2}\sigma_t^2 S_t^2 \dfrac{\nu}{\nu - 2}\dfrac{\partial^2 C_t}{\partial S_t^2} - rC_t = 0 \\ C_t(S, T) = \max(S_T - K, 0) \end{cases} \tag{59}$$

So our solution is: when $t = t^* (0 \le t^* \le T)$, $S = S^*$, using the current market foreign exchange option price $C(S^*, t^*; K_i, T_i) = C^*(K_i, T_i)$, $(i = 1, 2, \cdots, N)$. Let us solve foreign exchange volatility $\sigma_t = \sigma_t(S, t)$.

5   Solving Foreign Exchange Option Pricing Inverse Problem Based on t-distribution

In the previous section, we presented the foreign exchange option pricing inverse problem based on t-distribution. Then let us solve the inverse problem. Suppose European bullish option price $C = C(S, t; K, T, \sigma_t)$, it satisfies the following:

$$\begin{cases} \dfrac{\partial C}{\partial t} + \dfrac{1}{2}\sigma_t^2(S, t)S^2 \dfrac{\nu}{\nu - 2}\dfrac{\partial^2 C}{\partial S^2} + (r - r_f)S\dfrac{\partial C}{\partial S} - rC = 0 \\ C(S, T; K, T, \sigma_t) = \max(S - K, 0) \\ C(0, t; K, T, \sigma_t) = 0 \\ \lim\limits_{S \to \infty} \dfrac{\partial C}{\partial S}(S, t; K, T, \sigma_t) = e^{-r_f(T-t)} \end{cases} \tag{60}$$

So the solution of the foreign exchange option pricing inverse problem based on t-distribution is presented above.

Definition 2. *suppose*

$$Lp' = \frac{\partial p'}{\partial t} + \frac{1}{2}\sigma_t^2(S, t)S^2 \frac{\nu}{\nu - 2}\frac{\partial^2 p'}{\partial S^2} + (r - r_f)S\frac{\partial p'}{\partial S} - rp' \tag{61}$$

Conjugate operator Lq' is

$$Lq' = \frac{\partial q'}{\partial T} + \frac{1}{2}\frac{\nu}{\nu - 2}\frac{\partial^2}{\partial K^2}(\sigma_t^2(K, T)K^2 q') - (r - r_f)\frac{\partial}{\partial K}(Kq') - rq' \tag{62}$$

Correspondingly Lp' and Lq' satisfy $\int_{\Omega}(q'Lp' - p'Lq')dx = 0 \ \forall p', q' \in \mathbb{C}_0^{\infty}(\Omega)$. Then, the following: is satisfied:

Lemma 2. *According to*

$$G_t(S, t; K, T) = \frac{\partial^2 C}{\partial K^2}(S, t; K, T), \tag{63}$$

where $G_t(S, t; K, T)$ is the solution of Eq. (32). The following is satisfied:

$$LG_t = 0, \qquad G_t(S, T; K, T) = \delta(S - K) \tag{64}$$

Theorem 2. *If we take $G_t(S, T; K, T)$ for K, T function, so it is Eq. (32) conjugate equation solution. If $G_t^*(K, T; S, t) = G_t(S, t; K, T)$,*

$$\begin{cases} LG_t^* = 0 \\ G_t^*(K, t; S, t) = \delta(K - S) \end{cases} \tag{65}$$

Proof. According to the Definition 2, we know the following relations:

$$
\begin{aligned}
0 &= \int_0^{\infty} dx \int_{t+\epsilon}^{T-\epsilon} [G_t^*(x, y; S, t)LG_t(x, y; K, T) - G_t(x, y; K, T)LG_t^*(x, y^*, S, t)]dy \\
&= \int_0^{\infty} dx \int_{l+\epsilon}^{T-\epsilon} \{\frac{\partial}{\partial y}(G_t^* G_t) + \frac{1}{2}\frac{\partial}{\partial x}[\sigma_t^2 x^2 G_t^* \frac{\partial G}{\partial x}] \\
&\quad - \frac{1}{2}\frac{\partial}{\partial x}[G_t \frac{\partial}{\partial x}(\sigma_t^2 x^2 G_t^*)] + (r - r_f)\frac{\partial G_t}{\partial x}(xG_t G_t^*)\}dy
\end{aligned} \tag{66}
$$

When $x \to 0, \infty$, $\sigma_t^2 x^2 G_t^* \frac{\partial G_t}{\partial x} \to 0$, $G_t \frac{\partial}{\partial x}(\sigma_t^2 x^2 G_t^*) \to 0$, $xG_t G_t^* \to 0$, furthermore we get

$$\int_0^{\infty} G_t^*(x, T - \epsilon; S, t)G_t(x, T - \epsilon; K, T)dx = \int_0^{\infty} G_t j(x, t + \epsilon; S, t)G_t(x, t + \epsilon; K, T)dx \tag{67}$$

According to $\epsilon \to 0$ and $C(S, T; K, T) = \delta(S - K)$, we obtain the following relation:

$$
\begin{aligned}
G_t^*(K, t; S, t) &= \delta(K - S)\int_0^{\infty} G_t^*(x, T; S, t)\delta(x - K)dx \\
&= \int_0^{\infty} \delta(x - S)G_t(x, t; K, T)dx
\end{aligned} \tag{68}
$$

$G_t^*(K, T; S, t) = G_t(S, t; K, T)$, certification finish. $\qquad\square$

Let us take C for K, T equation, and substitute the Eq. (35) into Eq. (36), and according to K on $[K, \infty]$, we solve two integrals as follows:

$$\begin{cases} -\frac{\partial C}{\partial T} + \frac{1}{2}K^2\sigma_t^2(K, T)\frac{\nu}{\nu - 2}\frac{\partial^2 C}{\partial K^2} - (r - r_f)K\frac{\partial C}{\partial K} - r_f C = 0 \\ C|_{T=t} = \max(S - K, 0) \end{cases} \tag{69}$$

By using Eq. (37), we can solve $\sigma_t(K,T)$ as,

$$\sigma_t(K,T) = \sqrt{\dfrac{\dfrac{\partial C}{\partial T} + (r - r_f)K\dfrac{\partial C}{\partial K} + r_f C}{\dfrac{1}{2}K^2\dfrac{\nu}{\nu - 2}\dfrac{\partial^2 C}{\partial K^2}}} \tag{70}$$

According to the above formula, we can get the volatility when K, T function $C(S^*, t^*; K, T, \sigma_t)$ is a continuous function. But the known foreign exchange option price is discrete showing in the actual foreign exchange market. Therefore, if discrete points are substituted to the above formula there exits certain error, and it is not appropriate to calculate with the error differential coefficient, and doesnot satisfy the inverse problem solving conditions. Thus, if we would like to use this formula, we must construct a well-posed numerical differential algorithm to calculate the volatility.

6   The Propose and Proof of Numerical Differential Method

Suppose the function $y = y(x)$, and $x \in [0,1]$ and n is an integer number,
And let us suppose $\Delta = \{0 = x_0 < x_1 < \cdots < x_n = 1\}$ in the interval $[0,1]$ isometric division, and $h = x_{i+1} - x_i = \dfrac{1}{n}$

Suppose that we know $y(x)$ in the point x_i approximation is $\tilde{y}_i$, and

$$|\tilde{y}_i - y(x_i)| \leq \delta, \qquad\qquad i = 1, 2, \cdots, n. \tag{71}$$

suppose that in end points x_0, x_n two points have no disturbance, that is:

$$\tilde{y}_0 = y(0), \qquad \tilde{y}_n = y(1) \tag{72}$$

We suppose that we can find a function $f_*(x)$, take $f_*^{k-1}(x)$ for a approximation of function $y^{(k-1)}$, k is integer number.

Because numerical differentiation problem is an ill-defined problem, so we adopt the regularization method to solve. We defined regularization functional as:

$$\Phi(f) = \frac{1}{n-1}\sum_{j=1}^{n-1}(\tilde{y}_j - f(x_j))^2 + \alpha\|f^{(k)}\|_{L^2(0,1)}^2, \tag{73}$$

where α is regularization parameter.

Problem 1.

$$\min_{f \in \Im}\Phi(f) = \Phi(f_*) \tag{74}$$

$$\Im = \{g|g(0) = y(0), g(1) = y(1), g \in H^k(0,1)\} \tag{75}$$

Lemma 3. *Suppose $f(x)$ is continuous function defined in the interval $(0,1)$, if for any $\phi \in C_0^\infty(0,1)$. $\int_0^1 f(x)\phi(x) = 0$, $p \le 1$, p is integer and $f(x)$ is polynomial of $p-1$ degree.*

Theorem 3. *In problem 1, minimal element f_* is divided on Δ $2k-1$ order spline.*

Proof.

$$f(x;t) = f_*(x) + t\eta(x) \tag{76}$$

$\eta(x) \in \mathfrak{F}$, consider function $F(t) = \Phi(f_*(x) + t\eta(x))$, so have

$$F(t) = \frac{1}{n-1} \sum_{i=1}^{n-1} (\tilde{y}_i - f_*(x_i)^t \eta(x_i))^2 + \alpha \int_0^1 (f_*^{(k)}(x) + t\eta^{(k)}(x))^2 dx \tag{77}$$

Therefore,

$$F'(0) = \frac{1}{n-1} \sum_{i=1}^{n-1} (f_*(x_i) - \tilde{y}_i)\eta(x_i) + 2\alpha \int_0^1 f_*^{(k)}(x)\eta^{(k)}(x)dx = 0 \tag{78}$$

Special y take $\eta_i(x)$ content

$$\eta(k) \in \begin{cases} C_0^\infty(x_{i-1}, x_i], i = 1, 2, \cdots, n-1 \\ C_0^\infty(x_{i-1}, x_i], i = n \end{cases} \tag{79}$$

According to $\eta_i(x)$, its domain is N. Let us define η as:

$$\eta = \begin{cases} \eta_i(x), x \in N, \\ 0, \quad x \in (0,1)\backslash N \end{cases} \tag{80}$$

We can obviously see $\eta \in C_0^\infty(0,1)$, so $\eta \in \mathfrak{F}$, According to η substitute formula (45), we can obtain the following::

$$\int_0^1 f_*^{(k)}(x)\eta^{(k)}(x)dx = 0 \tag{81}$$

According to Lemma 2, we can get that f_* is polynomial of $2k-1$ degree. □

Based on the above theorem, we give f_* specific form as:

(1) Because f_* is $2k-2$ differentiable function, so to $i = 1, 2, \cdots, n-1$, we have

$$f_*^{(j)}(x_i+) = f_*^{(j)}(x_i-), \quad j = 0, 1, \cdots, 2k-2 \tag{82}$$

(2) on the point x_i, $i = 1, 2, \cdots, n-1$, suppose f_* satisfies the following conditions and these conditions are continuous:

$$f_*^{2k-1}(x_i+) - f_*^{(2k-1)}(x_i-) = \frac{(-1)^k}{\alpha(n-1)}(\tilde{y}_i - f(x_i)) \tag{83}$$

(3) f_* satisfies boundary condition as follows:

$$f_*^{(j)}(0+) - f_*^{(j)}(1-) = 0, \quad j = k, k+1, \cdots, 2k-2 \tag{84}$$

Now we structure sample band as follow and prove the minimal element in the above problem 1.

Lemma 4. *Suppose g is $2k - 2$ order function square integral function, it is satisfied condition $g(0) = g(1) = 0$, Therefore, we have,*

$$\int_0^1 g_*^{(k)} f_*^{(k)} dx = \frac{1}{\alpha(n-1)} \sum_{i=1}^{n-1} \{g(x_i)(\tilde{y}_i - f(x_i))\} \tag{85}$$

Proof. let us use Theorem 3 and integration by parts get

$$\begin{aligned}
\int_0^1 g^{(k)} f_*^{(k)} dx &= (-1)^{k-1} \sum_{i=1}^{n} \int_{x_{i-1}}^{x_i} g' f_*^{2k-1} dx \\
&= (-1)^{k-1} \sum_{i=1}^{n} f_*^{2k-1} \big|_{[x_{i-1}, x_i]} g(x) \big|_{x_{i-1}}^{x_i} \\
&= \frac{1}{\alpha(n-1)} \sum_{i=1}^{n-1} (g(x_i)(\tilde{y}_i - f(x_i))) \\
&= (-1)^k \sum_{i=1}^{n-1} g(x_i)(f_*^{(2k-1)}(x_i + 1) - f_*^{(2k-1)}(x_i-)) \\
&\quad (-1)^k f_*^{(2k-1)}(x_0+)g(x_0) - (-1)^k f_*^{(2k-1)}(x_n-)g(x_n)
\end{aligned} \tag{86}$$

Because, $g(0) = g(1) = 0$, and $f_*^{(2k-1)}$ in x_i skip, so we have:

$$\int_0^1 g^{(k)} f^{(k)} dx = \frac{1}{\alpha(n-1)} \sum_{i=1}^{n-1} (g(x_i)(\tilde{y}_i - f(x_i))) \tag{87}$$

$\square$

Theorem 4. *Suppose Φ is the functional of regularization method definition, to any content $f \in \Im$ function have $\Phi(f_*) \leq \Phi(f)$ established, in other words, f_* is problem1minimal element.*

Theorem 5. *When $n \geq k+1$, f_* exists.*

Proof. Suppose f is also a minimal element in the problem 1, so $\Phi(f_*) = \Phi(f)$, and $\Phi(f_* - f) = 0$,

According to the definition, we can get:

$$||f^{(k)} - f_*^{(k)}||_{L^2_{(0,1)}}^2 = 0 \tag{88}$$

$$f(x_i) = f_*(x_i), \qquad i = 1, 2, \cdots, n-1 \tag{89}$$

Equation (49) can result in $f^{(k)} = f_*^{(k)}$, so $f_* - f$ is $k - 1$ order polynomial. According to Eq. (50), we can get that $f_* - f$ have $n-1$ zero-points, so according to algebraic basic theorem, we can get that when $n \geq k+1$. $f = f_*$,

In the actual problem, according to numerical differentiation we often take $k = 2, 3$. Say we take $k = 2$ as an example. We have the following error estimation theorem.

Theorem 6. *when* $k = 2$, *if* f_* *is problem 1 minimal element, so*

$$||f_*^t - y^t||_{L^2(0,1)} \leq (2h + 4\alpha^{1/4} + \frac{h}{\pi})||y^n||_{L^2(0,1)} + h\sqrt{\frac{\alpha}{\delta^2}} + \frac{2\delta}{\alpha^{1/4}} \tag{90}$$

Then, we still take $k = 2$ *as an example, we obtain the theorem of the algorithm to the risk recognition ability.*

Theorem 7. *when* $k = 2$, *suppose* x_0 *is discontinuous point of the function* $f_*(x)$, *so for any open interval* I *including* x_0. *When* $h \to 0$, $\delta \to 0$, *we can establish* $||f_*^{\#}||_{L^2(I)} \to \infty$.

7   Conclusion

This paper puts forward the improved model of foreign exchange option pricing, that is the foreign exchange option pricing model based on the exchange rate return fat-tail, and to deduce the direct problem of the model and analyze the yield distribution curve problem. Then, we put forward the corresponding inverse problem, and proceed detailed derivation of the inverse problem. The derivation in the theory is established, but in actual market, only with discrete points, this derivation have errors in the practical application process. To eliminate the error and use this formula reasonably to calculate the volatility, we construct a numerical differential algorithm after calculating expression and proceed a series of theory certification to this algorithm, proving the rationality of the differential value algorithm.

Acknowledgment. We should say thanks to the support of Key Project of National Natural Science Foundation in China (No.71271069), National Key Technology R&D Program of the Ministry of Science and Technology (No.2012BAH81F03, No.2012BAH66F01), Humanities and Social Sciences Foundation of Chinese Ministry of Education (No. 10YJC860040), National Soft Science Foundation in China (No.2008GXS5D113).

References

1. Mittnik, S.: VaR implied tail-correlation matrices. Econom. Lett. **122**(1), 69–73 (2014)
2. Garman, M.B., Kohlhagen, S.W.: Foreign currency options values. J. Int. Money Financ. **2**(3), 231–237 (1983)
3. Amin, K., Ng, V.: Option valuation with systematic stochastic volatility. J. Financ. **48**, 881–910 (1993)
4. Baehelier L.: Theoriedelas Peculation [D]. Sorbonne, Paris (1900)

5. Black, F., Scholes, M.: The pricing of options and corporate liabilities. J. Econ. **81**(3), 637–654 (1973)
6. Cox, J.C., Ross, S.A., Rubinstein, M.: Option pricing: a simplified approach. J. Financ. Econ. **7**, 229–263 (1979)
7. Harikumar, T., de Boyrie, M.E.: Evaluation of black-scholes and GARCH Models using currency call options data. Rev. Quant. Financ. Account. **23**(4), 299–312 (2004)
8. Merton, R.C.: Theory of rational option pricing. Bell J. Econ. Manag. Sci. **4**(1), 141–183 (1973)
9. Harrison, M., Pliska, S.: Martingales and stochastic integrals in the theory of continuous trading. Stochast. Processes Appl. **11**, 215–260 (1981)
10. Wolf, A., Hessel, C.: Pricing options on foreign exchange with a preset exchange rate. J. Math. Financ. **2**(3), 351–373 (2012)
11. Yang, S.J., Lee, M.K., Kima, J.H.: Pricing vulnerable options under a stochastic volatility model. Appl. Math. Lett. **34**, 7–12 (2014)
12. Swishchuk, A., Tertychnyi, M., Elliott, R.: Pricing exchange derivatives with Markov-modulated Lévy dynamics. Insur. Math. Econ. **57**, 67–76 (2014)
13. Hein, T., Bernd, H.: On the nature of ill-posedness of an inverse problem arising in option pricing. Inverse Prob. **19**(6), 1319–1338 (2003)
14. Dufé, D., Huang, M.: Swap rates and credit quality. J. Financ. **51**, 921–950 (1996)
15. Rong, T., Jun, C.: Applications of equivalent martingale measures model in pricing option on foreign currency. J. East China Norm. Univ. **2**, 27–32 (2003)
16. Dihong, C., Xiangyu, Y.: A foreign currency option pricing model. Theory Pract. Financ. Econ. **22**, 59–60 (2001)
17. Genxin, X.: Pricing analysis of European call foreign exchange options under Vasicek interest rate model. J. Tongji Univ. **34**(4), 2–4 (2006)
18. Kun, D., An-Xing, W.: Pricing foreign exchange option under restriction. J. Eng. Manag. **28**(1), 89–93 (2014)
19. Xing, Y., Yang, X.: Equilibrium valuation of exchange options under a jump-diffusion model with stochastic volatility. J. Comput. Appl. Math. **280**, 231–247 (2015)
20. Suna, Q., Xub, W.: Pricing foreign equity option with stochastic volatility. Phys. A **437**, 89–100 (2015)
21. Leung, K.S., YingWong, H., Ng, H.Y.: Exchange option pricing with Wishart process. J. Comput. Appl. Math. **238**, 156–170 (2013)
22. Rongda, C.: A foreign exchange option pricing model based on the exchange rate return. Oper. Manag. **15**(3), 6–11 (2006)
23. Yonggang, Y., Zhijian, H.: Foreign Exchange Options. Wuhan University Press, WuHan (2004)
24. Yang, L., Jianning, Y., Zuicha, D., Xiaoliang, D.: An inverse problem of zero-coupon bond pricing. J. Sichuan Univ. **44**(6), 1201–1205 (2007)
25. XiaoHong, C., Demian, P.: On nonlinear ill-posed inverse problems with applications to pricing of defaultable bonds and option pricing. Sci. China, Ser. A: Math. **52**(6), 1157–1168 (2009)
26. Liu, Y., JianNing, Y., Zuicha, D.: An inverse problem of determining the implied volatility in option pricing. J. Math. Anal. Appl. **4**, 16–31 (2008)
27. Chuchu, H.: The economic determinants of foreign exchange option smiles. Xiamen University (2014)

Author Index

© Springer International Publishing AG, part of Springer Nature 2018
A. M. Gil-Lafuente et al. (Eds.): FIM 2015, AISC 730, pp. 427–428, 2018.
https://doi.org/10.1007/978-3-319-75792-6

Printed in the United States
By Bookmasters